7th National Conference on Processing and Characterization of Materials (NCPCM 2017)

IOP Conference Series: Materials Science and Engineering
Volume 338

Rourkela, India
8 - 9 December 2017

Editors:

Ashok Kumar Mondal
Archana Mallik

ISBN: 978-1-5108-6159-6
ISSN: 1757-8981

Printed from e-media with permission by:

Curran Associates, Inc.
57 Morehouse Lane
Red Hook, NY 12571

proceedings.com

Some format issues inherent in the e-media version may also appear in this print version.

Printed by Curran Associates, Inc. (2018)

Institute of Physics
Dirac House, Temple Back
Bristol BS1 6BE UK

Phone: 44 1 17 929 7481
Fax: 44 1 17 920 0979

techtracking@iop.org

Additional copies of this publication are available from:

Curran Associates, Inc.
57 Morehouse Lane
Red Hook, NY 12571 USA
Phone: 845-758-0400
Fax: 845-758-2633
Email: curran@proceedings.com
Web: www.proceedings.com

Table of contents

Volume 338

7th National Conference on Processing and Characterization of Materials (NCPCM 2017)

8–9 December 2017, National Institute of Technology Rourkela, India

Accepted papers received: 15 March 2018
Published online: 10 April 2018

Preface

7th National Conference on Processing and Characterization of Materials (NCPCM 2017)

Photographs of NCPCM 2017

Editors of the Work

We Thank our Sponsors

Peer review statement

Papers

Effect of Thermal cycles and Dimensions of the Geometry on Residual stress of the Alumina-Kovar Joint

Srishti Mishra, Snehanshu Pal, Swapan Kumar Karak, Sejal Shah, M. Venakata Nagaraju and Arun Kumar Chakraborty.....1

Optimisation of Ferrochrome Addition Using Multi-Objective Evolutionary and Genetic Algorithms for Stainless Steel Making via AOD Converter

Kishore Kumar Behera and Snehanshu Pal.....10

Effect of CNT addition on cure kinetics of glass fiber/epoxy composite

A O Fulmali, R Kattaguri, K K Mahato, R K Prusty and B C Ray.....16

Multi-Response Optimization of WEDM Process Parameters Using Taguchi Based Desirability Function Analysis

Himadri Majumder and Kalipada Maity.....21

Machinability assessment of commercially pure titanium (CP-Ti) during turning operation: Application potential of GRA method

Akhtar Khan and Kalipada Maity.....27

Adsorption Studies of Gadolinium ion on Graphitic Carbon Nitride

S K Kuila and T K Kundu.....33

Pore Structure Characterization in Concrete Prepared with Carbonated Fly Ash

Sanjukta Sahoo.....39

Effect of ageing time and temperature on corrosion behaviour of aluminum alloy 2014

Vikas Gadpale, Pragya N. Banjare and Manoranjan Kumar Manoj.....45

Parametric Analysis of Pulsed Laser Melting Process

S Sukumar and S P Kar.....51

Pore Size Control in Aluminium Foam by Standardizing Bubble Rise Velocity and Melt Viscosity

G Avinash, V Harika, Ch Sandeepika and N Gupta.....57

Hardness - Yield Strength Relation of Al-Mg-Si Alloys

Aluru Praveen Sekhar, Supriya Nandy, Kalyan Kumar Ray and Debdulal Das.....63

Placer Deposits of Brahmagiri Coast, Odisha - A New Resource For Industrial Heavy Minerals

Sunita Routray, Jagadeeswari Vanamu and Ranjita Swain.....69

Effects of Welding Parameters on Mechanical Properties in Electron Beam Welded CuCrZr Alloy Plates

Sanjib Jaypuria, Nirav Doshi and Dilip Kumar Pratihar.....75

Study of distribution of Carbon nanotube in Al-CNT nanocomposite synthesized via Spark-Plasma sintering

A. Maiti and T. Laha.....81

Effect of thermally reduced graphene oxide on dynamic mechanical properties of carbon fiber/epoxy composite

Nitai Chandra Adak, Suman Chhetri, Naresh Chandra Murmu, Pranab Samanta and Tapas Kuila.....87

Estimation of Effective Directional Strength of Single Walled Wavy CNT Reinforced Nanocomposite

Krishnendu Bhowmik, Pranav Kumar, Niloy Khutia and Amit Roy Chowdhury.....93

Dynamic Response of Functionally Graded Carbon Nanotube Reinforced Sandwich Plate

Kulmani Mehar and Subrata Kumar Panda.....99

Electrodeposition of near stoichiometric $CuInSe_2$ thin films for photovoltaic applications

Ramkumar Chandran and Archana Mallik.....105

Numerical Flexural Strength Analysis of Thermally Stressed Delaminated Composite Structure under Sinusoidal Loading

C K Hirwani, S Biswash, K Mehar and S K Panda.....110

Lifetime Prediction of Nano-Silica based Glass Fibre/Epoxy composite by Time Temperature Superposition Principle

Abhijeet Anand, Poulami Banerjee, Rajesh Kumar Prusty and Bankin Chandra Ray.....116

Experimental and numerical investigations on the temperature distribution in PVD AlTiN coated and uncoated Al_2O_3/TiCN mixed ceramic cutting tools in hard turning of AISI 52100 steel

Ch Sateesh Kumar, Saroj Kumar Patel and Anshuman Das.....122

Study of effective utilization of iron ore sinter through arc plasma

Biswajit Swain, S K Samal, M K Mohanty, A Behera and S C Mishra.....128

Mechanical and Tribological Characteristics of the AMC, Prepared by P/M Route along with Thermo-Mechanical Treatment

Sambit Kumar Mohapatra, Kalipada Maity, Subrat Kumar Bhuyan and Mantra Prasad Satpathy.....134

Parametric Investigation on Microstructure and Mechanical Properties of Ultrasonic spot welded Aluminium to Copper sheets

Mantra Prasad Satpathy, Kasinath Das Mohapatra, Ananda Kumar Sahoo and Susanta Kumar Sahoo.....140

DFT Studies on Interaction between Lanthanum and Hydroxyamide

Anindita Pati, T. K. Kundu and Snehanshu Pal.....146

Flexible nano-GFO/PVDF piezoelectric-polymer nano-composite films for mechanical energy harvesting

Monali Mishra, Amritendu Roy, Sukalyan Dash and Somdutta Mukherjee.....152

Influence of thermo-mechanical treatment in ferritic phase field on microstructure and mechanical properties of reduced activation ferritic-martensitic steel

Prakash, J. Vanaja, K. Laha and G. V. S. Nageswara Rao.....158

The influence of void and porosity on deformation behaviour of nanocrystalline Ni under tensile followed by compressive loading

Md. Meraj, Shradha Nayak, Kumar Krishanjeet and Snehanshu Pal.....164

Mechanical Properties and Wear Characteristics Al-ZrO2-SiCp and Graphite Hybrid Metal Matrix Composites

S.K. Nayak, T. Mahanta, J.K. Sahoo and A. Mishra.....170

Numerical simulation of residual stress in laser based additive manufacturing process

Bibhu Kalyan Panda and Seshadev Sahoo.....176

Characterization of the Acoustic Radiation Properties of Laminated and Sandwich Composite Panels in Thermal Environment

Nitin Sharma, Trupti Ranjan Mahapatra, Subrata Kumar Panda and Pruthwiraj Sahu.....182

Development and Sliding Wear Response of Epoxy Composites Filled with Coal Mine Overburden Material

Prithika Das, Alok Satapathy and M. K. Mishra.....188

Study on effect of tool electrodes on surface finish during electrical discharge machining of Nitinol

Anshuman Kumar Sahu, Suman Chatterjee, Praveen Kumar Nayak and Siba Sankar Mahapatra.....194

Synthesis of Graphene-wrapped Silicon Nanoparticles by Ultra-sonication

P G Kale, Mihir Kumar Sahoo, Nishit Kumar Sadual and S K Behera.....200

Buckling analysis of SMA bonded sandwich structure – using FEM

Pankaj V Katariya, Arijit Das and Subrata K Panda.....205

Nano-Indentation of Aluminium Reinforced Metallic Glass Composites: A Molecular Dynamics Study

D Yadav, P Gupta and N Yedla.....211

Fabrication and characterization of nano-Y_2O_3 and Al_2O_3 dispersed W-Ni alloys by mechanical alloying and pressureless conventional sintering

V. R. Talekar, A. Patra and S. K. Karak.....217

Simulation of Texture Evolution during Uniaxial Deformation of Commercially Pure Titanium

B. Bishoyi, M.K. Debta, S.K. Yadav, R.K. Sabat and S.K. Sahoo.....223

Optimization of Dimensional accuracy in plasma arc cutting process employing parametric modelling approach

Deepak kumar Naik and K.P. Maity.....229

Effect of fiber content on the thermal conductivity and dielectric constant of hair fiber reinforced epoxy composite

Bishnu Prasad Nanda and Alok Satapathy.....235

Density Functional Theory Study of Structural and Electronic Properties of γ'-Ni_3Al and γ''-Ni_3Nb

D. K. Gorai and T. K. Kundu.....239

Mechanical and microstructural characterization of W–Cu FGM fabricated by one-step sintering method through PM route

Rajat Gupta, Rohit Kumar, A.K. Chaubey, Shailesh Kanpara and S.S. Khirwadkar.....245

Effects of Mn addition on microstructure and hardness of Al-12.6Si alloy

Prosanta Biswas, Surajit Patra and Manas Kumar Mondal.....251

Evaluation of the optical characteristics of c-axis oriented zinc oxide thin films grown by sol gel spin coating technique

K. Baisakh, S. Behera and S. Pati.....257

Investigation of consolidation kinetics and microstructure evolution of Al alloys in direct metal laser sintering using phase field simulation

Bubloom Bimal Satpathy, Jyotirmoy Nandy and Seshadev Sahoo.....263

Microstructure and Current-Voltage Characteristics of Erbium Oxide Doped Multicomponent Zinc Oxide Varistors

Samarpita Roy, Tapatee Kundu Roy and Debdulal Das.....269

Thermal shock effect on Mechanical and Physical properties of pre-moisture treated GRE composite

A.P. Chakraverty, A. B. Panda, U. K. Mohanty, S. C. Mishra and B. B. Biswal.....275

Effect of alumina on grain refinement of Al-Si hypereutectic alloys

J Majhi, S K Sahoo, S C Patnaik, B Sarangi and N K Sachan.....281

Industrial Application of Valuable Materials Generated from PLK Rock-A Bauxite Mining Waste

Ranjita Swain, Sunita Routray, Abhisek Mohapatra and Biswa Ranjan Patra.....286

Comparison of wear behaviour and mechanical properties of as-cast Al6082 and Al6082-T6 using statistical analysis

Sandhya Rani Rana, A.B. Pattnaik and S.C. Patnaik.....292

Sea water Corrosion of Nickel based Plasma Spray Coating

M Parida, S P Nanda, S K Bhuyan and S C Mishra.....298

Continuum simulation of heat transfer and solidification behavior of AlSi10Mg in Direct Metal Laser Sintering Process

Akash Ojha, Mihir Samantaray, Dhirendra Nath Thatoi and Seshadev Sahoo.....304

Effect of Bauxite addition on Adhesion Strength and Surface Roughness of Fly ash based Plasma Sprayed Coatings

S K Bhuyan, S Samal, D Pattnaik, A Sahu, B Swain, T K Thiyagarajan and S C Mishra.....310

Effect of carbonization temperatures on biochar formation of bamboo leaves

D Pattnaik, S Kumar, S K Bhuyan and S C Mishra.....316

Effect of reduced graphene oxide-carbon nanotubes hybrid nanofillers in mechanical properties of polymer nanocomposites

Kadambinee Sa, Prakash C. Mahakul, B.V.R.S. Subramanyam, Jagatpati Raiguru, Sonali Das, Injamul Alam and Pitamber Mahanandia.....322

Experimental investigation of various surface integrity aspects in hard turning of AISI 4340 alloy steel with coated and uncoated cermet

Anshuman Das, S. K. Patel, Ch. Sateesh Kumar and B. B. Biswal.....328

Effect of severely thermal shocked MWCNT enhanced glass fiber reinforced polymer composite: An emphasis on tensile and thermal responses

K. K Mahato, A.O. Fulmali, R Kattaguri, K. Dutta, R K Prusty and B C Ray.....334

Optical & current-voltage characteristics of dip coated TiO_2 thin film

S. Roy, N. Tripathy, P.K. Sahu and J.P. Kar.....340

Erosion wear response of epoxy composites filled with steel industry slag and sludge particles: A comparative study

Abhilash Purohit and Alok Satapathy.....345

Mechanical properties enhancement and microstructure study of Al-Si-TiB_2 in situ composites

S K Sahoo, J Majhi, A B Pattnaik, J K Sahoo and Swagat Das.....350

Influence of Temperature on Pyrolysis of Biomass

S Kumar, D Pattnaik, S K Bhuyan and S C Mishra.....356

Synthesis of quaternary chalcogenide CZTS nanoparticles by a hydrothermal route

S Das, K Sa, P C Mahakul, J Raiguru, I Alam, BVRS Subramanyam and P Mahanandia.....362

Preparation of few layer graphene sheets (FLGS) prepared by an electrochemical method

I Alam, K Sa, S das, J Raiguru, BVRS Subramanyam, PC Mahakul and P Mahanandia.....368

IOP Conf. Series: Materials Science and Engineering **338** (2018) 011001 doi:10.1088/1757-899X/338/1/011001

PREFACE

The 7th National Conference on Processing and Characterization of Materials (NCPCM 2017) was organized by the Department of Metallurgical and Materials Engineering, National Institute of Technology Rourkela, India during 08-09 December 2017.

The advanced processing of materials and their subsequent characterization are now-a-days gaining attention owing to the advancement of technology and need of high-quality materials for critical applications. The purpose of the conference was to bring recent developments in materials processing and characterization in the Indian context. About 120 participants and delegates from various organizations in India attended the conference. There were eight technical sessions of about 83 contributory talks along with seven keynote lectures. The conference became a platform for the interaction and exchange of ideas among the scientists, practicing engineers and young researchers from various academic institutions, research laboratories and industries from all over India engaged in frontiers of processing and characterization of conventional as well as advanced materials. This volume contains the selected full-length technical papers presented at the conference NCPCM 2017.

We would like to thank all the participants, delegates, members of the organizing committee and the sponsors for helping us to successfully conduct the conference NCPCM 2017.

February 2018 Editors

IOP Conf. Series: Materials Science and Engineering **338** (2018) 011002 doi:10.1088/1757-899X/338/1/011002

PHOTOGRAPHS OF NCPCM 2017

IOP Conf. Series: Materials Science and Engineering **338** (2018) 011002 doi:10.1088/1757-899X/338/1/011002

IOP Conf. Series: Materials Science and Engineering **338** (2018) 011002 doi:10.1088/1757-899X/338/1/011002

IOP Conf. Series: Materials Science and Engineering **338** (2018) 011003 doi:10.1088/1757-899X/338/1/011003

EDITORS OF THE WORK

Dr. Ashok Kumar Mondal
Department of Metallurgical and Materials Engineering,
National Institute of Technology Rourkela, Rourkela-769008, Odisha, India
Email: mondala@nitrkl.ac.in

Dr. Archana Mallik
Department of Metallurgical and Materials Engineering,
National Institute of Technology Rourkela, Rourkela-769008, Odisha, India
Email: archanam@nitrkl.ac.in

ORGANISING COMMITTEE

PATRON: Prof. Animesh Biswas, Director, NIT Rourkela

Chairman : Prof. Smarajit Sarkar, MME, NIT Rourkela
Convener : Prof. Ashok Ku. Mondal, MME, NIT Rourkela
Co-Convener : Prof. Archana Mallik, MME, NIT Rourkela
Treasurer : Prof. D. Nagarajan, MME, NIT Rourkela

Members:
Prof. Bipin Bihari Verma, MME, NIT Rourkela
Prof. Bankim Chandra Ray, MME, NIT Rourkela
Prof. Subash Chandra Mishra, MME, NIT Rourkela
Prof. Sudipta Sen, MME, NIT Rourkela
Prof. Mithilesh Kumar, MME, NIT Rourkela
Prof. Anindya Basu, MME, NIT Rourkela
Prof. Debasis Chaira, MME, NIT Rourkela
Prof. Santosh Kumar Sahoo, MME, NIT Rourkela
Prof. Syed Nasimul Alam, MME, NIT Rourkela
Prof. Krishna Dutta, MME, NIT Rourkela
Prof. Swapan Kumar Karak, MME, NIT Rourkela
Prof. Natraj Yedla, MME, NIT Rourkela
Prof. Snehanshu Pal, MME, NIT Rourkela
Prof. Anshuman Patra, MME, NIT Rourkela
Prof. Rajesh Kumar Prusty, MME, NIT Rourkela
Prof. Ajit Behera, MME, NIT Rourkela

IOP Conf. Series: Materials Science and Engineering **338** (2018) 011003 doi:10.1088/1757-899X/338/1/011003

NATIONAL ADVISORY COMMITTEE

Prof. Indranil Manna, Director, IIT Kanpur, India
Dr. Indranil Chattoraj, Director, NML Jamshedpur, India
Prof. P. P. Chattopadhyay, Director, NIFFT Ranchi, India
Prof. K. K. Ray, IIEST Shibpur, India
Prof. Subrata Chatterjee, IIEST Shibpur, India
Prof. Subodh Kumar, IISc Bangalore, India
Prof. Sunil Mohan, IIT BHU Varanasi, India
Prof. Uday Chakkingal, IIT Madras, India
Prof. K. S. Ghosh, NIT Durgapur, India
Prof. Chandan Srivastava, IISc Bangalore, India
Prof. T. K. Kundu, IIT Kharagpur, India
Prof. Joysurya Basu, IIT BHU Varanasi, India
Prof. Jatin Bhat, VNIT Nagpur, India
Prof. S. Kumaran, NIT Trichy, India
Prof. Uday K. Bhat, NIT Surathkal, India
Prof. N. Narsaiah, NIT Warangal, India
Dr. Chiradeep Ghosh, TATA Steel Jamshedpur, India
Dr. Apu Sarkar, BARC Mumbai, India

IOP Conf. Series: Materials Science and Engineering **338** (2018) 011004 doi:10.1088/1757-899X/338/1/011004

WE THANK OUR SPONSORS

- Science and Engineering Research Board (SERB), India
- Council of Scientific and Industrial Research (CSIR), New Delhi, India
- Board of Research in Nuclear Sciences (BRNS), India
- Avadh Rubber Prop Madras Elastomers Limited, Lucknow, India
- AIMIL Ltd., Kolkata, India
- Measure India Corporation Pvt. Ltd., Kolkata
- Spranktronics, Bangalore, India
- Anita Enterprises, Jamshedpur, India

IOP Conf. Series: Materials Science and Engineering **338** (2018) 011005 doi:10.1088/1757-899X/338/1/011005

Peer review statement

All papers published in this volume of *IOP Conference Series: Materials Science and Engineering* have been peer reviewed through processes administered by the proceedings Editors. Reviews were conducted by expert referees to the professional and scientific standards expected of a proceedings journal published by IOP Publishing.

Effect of Thermal cycles and Dimensions of the Geometry on Residual stress of the Alumina-Kovar Joint

Srishti Mishra*, Snehanshu Pal*, Swapan Kumar Karak*, Sejal Shah, M. Venakata Nagaraju**Arun Kumar Chakraborty****

* Metallurgical and Materials Engineering Department, National Institute of Technology, Rourkela-769008, India

** ITER-India, Indian Plasma Research, Block-A, Sangath Skyz, Bhat-Koteshwar Road, Koteshwar, Ahmedabad – 380 005, Gujarat, India

Corresponding author Email : srishtimishra1992@gmail.com

Abstract: Finite element method is employed to determine the effect of variation of residual stress with dimension and the stress generated under its working condition along the Kovar. 3 different dimensions of Alumina-Kovar joint with height to diameter ratio of 3/10, using TiCuSil as a filler material. Transient Structural Analysis is carried out for three different dimensions (diameter× height) (i) 60mm× 20mm (Geometry 1) (ii) 90mm × 20mm (Geometry 2) (iii) 120mm × 20mm (Geometry 3). A comparative study has been carried out between the residual stresses developed in the brazed joint that have undergone 5 thermal cycles subsequent to brazing and that between the brazed joint. The heating and cooling rates from the brazed temperature is 10°C/ up to room temperature. The brazing temperature and holding time considered for the analysis are 900°C and 10 minutes. Representative Volume Element (RVE) model is used for simulation. Sparse Matrix Direct Solver method is used to evaluate the results, using Augmented Lagrange method formulation in the contact region. All the simulations are performed in ANSYS Workbench 15.0, using solver target Mechanical APDL. From, the above simulations it is observed high concentration of residual stress is observed along the filler region i.e. in between Alumina and Kovar, as a result of difference in coefficient of thermal expansion between Alumina and Kovar. The residual stress decreases with increasing dimensions of the geometry and upon application of thermal cycles, subsequent to brazing.

Keywords: Finite element method, residual stress, brazed joint, thermal cycles, RVE model

1. Introduction

Superior mechanical properties, corrosion resistance , thermal stability, electronic insulation, relatively low manufacturing cost and chemical inertness of Al_2O_3 ceramic has grabbed the attention in electronics, aerospace, nuclear, automotive, biomedical and tool industries for years [1-3] (Table 1). However the brittleness and processing inefficiency of Al_2O_3 has restrained its widespread application [4]. The primary prerequisite for broadening the dimensions of application, is to reliably join Al_2O_3 ceramic to metals [5,6]. Such kind of dissimilar material joining is highly challenging, as the difference in Coefficient of Thermal Expansion (CTE), chemical properties and Modulus of Elasticity (MoE) between the joining components, results in generation of residual stress [7,8]. This invokes technical and scientific community to perform extensive research for mitigating thermal stress in Al_2O_3-metal joints. Different bonding/joining methods such as active brazing, ultrasonic brazing [9], friction welding [10],explosive welding [11],diffusion bonding[12], transient liquid phase bonding [13], ultrasonic nano welding [7] are generally used for joining Al_2O_3 and metal. Among these methods active brazing is considered as the most applicable method to join Al_2O_3 to metal owing to its cost-effectiveness, requirement of relatively low joining temperature, superior bonding properties [14] and good flexibility with respect to shapes of joining components [15-17]. However, difference in thermal and mechanical properties leads to generation of thermal stress develops near a ceramic-metal interface both during fabrication and application. This phenomenon can deteriorate the strength of active brazed Al_2O_3-metal joint [18].

Table1. Probable application of Metal-Alumina brazed joints

Ceramic/Metal/Filler			Applications	References
Ceramic	Filler	Metal		
Al_2O_3	Ag-Cu-Ti	TiAl	Fabrication in nuclear reactor	[19]
Al_2O_3	Ag-Pd-Ti	Kovar	Hydrogen separation from natural gas	[20]
Al_2O_3	Ag-ABA	Inconel	Electrical feed through in neutron sensors	[21]

A true understanding of deformation behavior of metal-ceramic brazed joint and its dependence on brazing parameters are highly necessary. In order to determine the strength of the metal-ceramic brazed joint it is important to determine variation in residual stress of the joint with variation in various brazing parameters such as heating and cooling rates, brazing temperature, brazing cycles (heating-cooling cycles). In this regard Finite Element Analysis (FEA) is an effective tool to determine an optimum range of brazing parameters for achieving highest mechanical strength. Various investigators have attempted the Finite Element Analysis of ceramic-metal joints to study the effect of various geometrical shapes of the joining components, thickness of interlayer, coefficient of thermal expansion of various interlayer and brazing parameters [18, 22-24]. In this paper, we have attempted to carry out the Finite Element Analysis of Alumina-Kovar brazed joint using Ticusil filler and study the effect of thermal cycles and diameter of the joining components.

2. Model description

Continuum models are considered for calculation of residual stress that is developed in a joined Al_2O_3-Kovar specimen, when brazed and then cooled down to room temperature. Since non-uniform cooling is considered in this simulation study therefore equations of temperature and stress equilibrium both are solved. Dissimilar materials are assumed to be perfectly bonded at the interfaces. Numerical solutions are obtained employing Workbench 15.0. Workbench 15.0 employs finite element method to obtain the solutions to the partial differential equations in Langragian form. Quadrilateral elements are considered for all the simulations with quadratic interpolation and reduced integration. All the materials are assumed to be elastic.

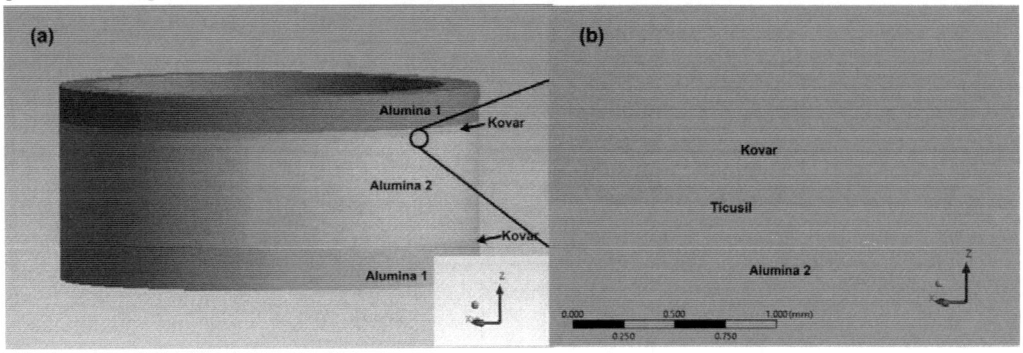

Figure 1. Geometry construction (a) Alumina-Kovar annulus ring (b) Magnified portion showing Ticusil filler

to be perfectly elastic in this current simulation study. Transient thermal analysis is opted for determining the temperature distribution across the body as a result of brazing, at the end of brazing process. Static structural analysis is carried out for determining the residual stress across the body. 3D geometries (refer to Figure 1) are created for the simulation, in ANSYS Workbench as per the dimensions given in Table 2.

Table 2. Dimensions of the various geometries

Dimensions of various geometry	Inner Diameter (mm)	Outer Diameter (mm)	Kovar Thickness (mm)	Filler Thickness (mm)	Alumina 1Thickness (mm)	Alumina 2 Thickness (mm)
Geometry No.1	50	60	1	0.03	5	10
Geometry No.2	75	90	1	0.03	5	10
Geometry No.3	100	120	1	0.03	5	10

Material properties considered for the simulation are given in Table 3.

Table 3. Properties of Alumina Kovar and TiCuSil

Properties of materials	Alumina	Kovar	TiCuSil
Young's Modulus(in GPa)	350	138	85
Poisson's Ratio	0.22	0.317	0.36
Density(kg/m^3)	3900	8350	9400
Coefficient of Thermal Expansion(/^{o}C)	8.4e-6	11.5e-6	18.5e-6
Thermal conductivity(W/m^{o}C)	3	17.3	219
Specific Heat(J/Kg^{o}C)	880	481.6	280

IOP Conf. Series: Materials Science and Engineering **338** (2018) 012001 doi:10.1088/1757-899X/338/1/012001

Following geometry creation comes the meshing, which involves dividing the whole geometry into small elements, as shown in Figure 2.

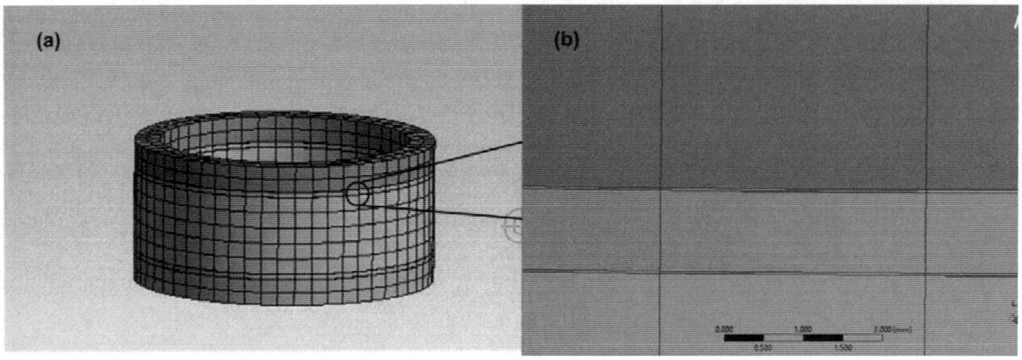

Figure 2. Typical finite element mesh for an interlayer calculation (a) entire mesh (b) near the interlayer Alumina-Kovar- interface showing interface

The size of the mesh determines the simulation time of the program. After meshing requisite boundary condition are applied for carrying out the analysis. Sparse Matrix Direct Solver method is used to evaluate the results and Augmented Lagrangian formulation is applied here for contact formulation. For carrying out thermal analysis a heating rate of 10°C/min is applied to the outer surface of all the components 30°C-900°C, with a holding time of 10minutes at 900°C subsequently cooling the whole assembly back to 30°C. The optimized operating parameters have been chosen on the basis of various literature reviews which will be applied to the experimental conditions carried out in the near future. For carrying out the structural analysis the displacement of top and bottom faces of the assembly is constrained along the Z-axis. Gravitational force is applied in the negative Z direction the whole body, as would be constrained by the sample holder in the brazing furnace. The application of load and boundary condition is presented in Figure 3.

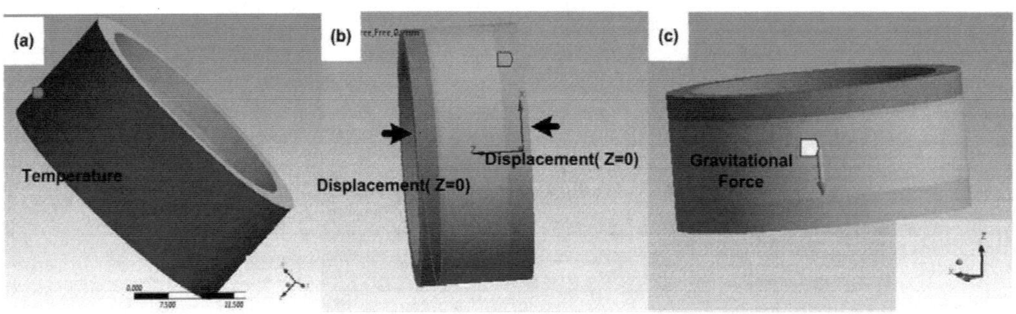

Figure 3. Application of load to the geometry (a) Thermal load during Transient Thermal Analysis (b) Application of zero displacement on both end faces during Static Structural Analysis (c) Application of Gravitational force during Static Structural analysis

To study the effect of thermal cycles on residual stresses three sets of simulations are carried out. One at the end of brazing process followed by cooling, one after conducting 5 thermal cycles between 600°C and

200°C subsequent to brazing followed by cooling to room temperature and the last one after conducting 10 thermal cycles between 600°C and 200°C subsequent to brazing followed by cooling to room temperature. The heating and cooling rates are given in Table 4.

Table 4. Brazing parameters for the analysis

Brazing Temperature(in °C)	Holding Time(in °C)	Heating/ Cooling Rate(°C/min)
900	10	10

The von-mises stress or the residual stress is determined according to the given below equation. The detail study of this criterion for von Mises yield criterion have been reported in literature [25-29].

$$F = \sqrt{J_{2D}} - \sigma_y = 0 \tag{1}$$

where, J_{2D} is expressed as the second invariant of deviatoric stress tensor, S_{ij} and σ_y are the yield stress in the simple tension

3. Results and Discussion

3.1. Variation of thermal stress with geometry

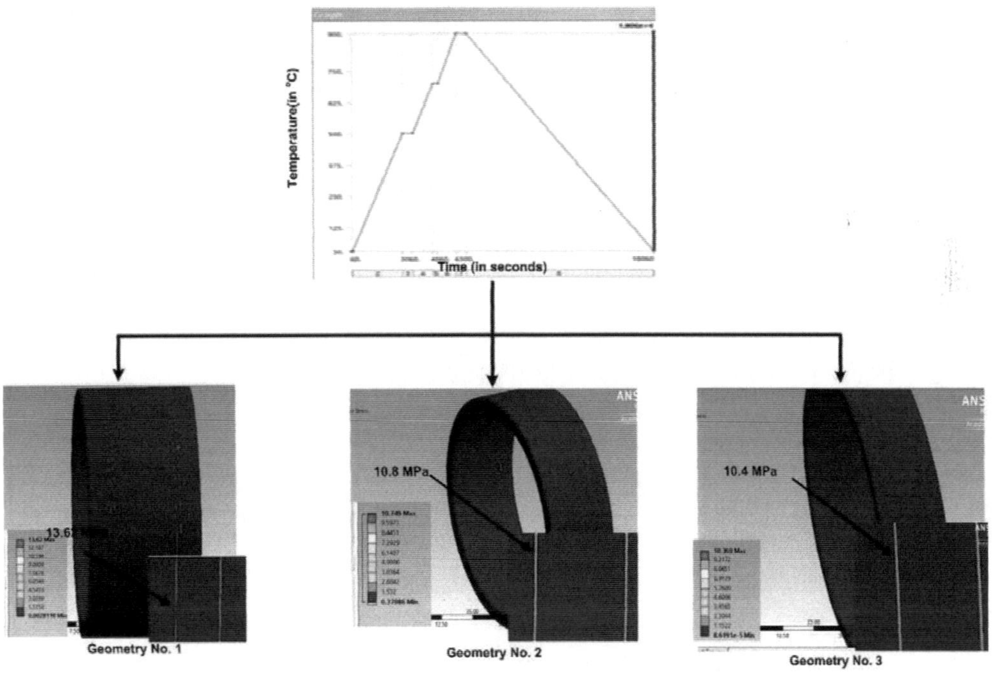

Figure 4. Variation of residual stress with varying geometry size

From the simulations (Figure 4) it is evident that the thermal stresses are high along the Alumina-Kovar joint owing to mismatch in CTE. It is also observed that the stresses are higher along the curvature. One keen observation however, the residual stress decreases with increasing diameter size. The decrement is residual stress might be due to decreasing curvature with increasing diameter.

3.2. Variation of thermal stress with thermal cycles

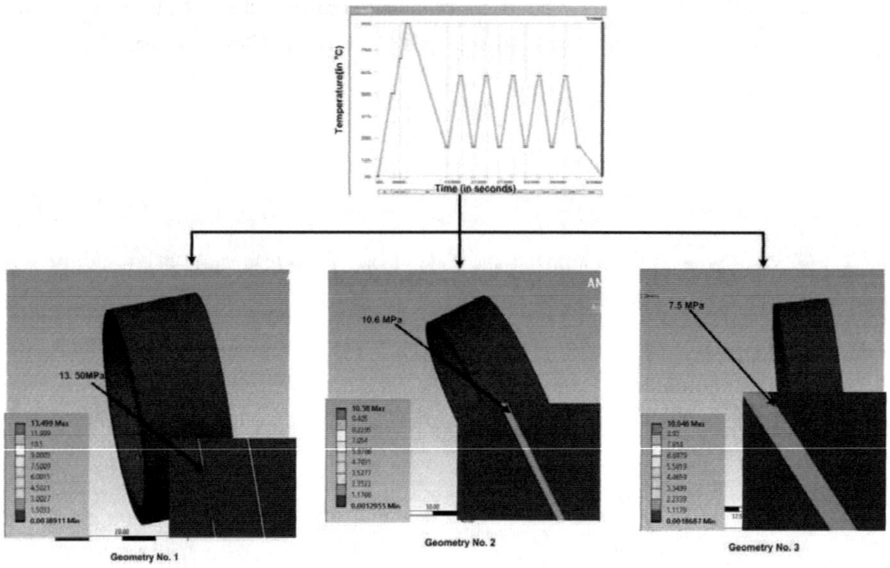

Figure 5. Distribution of residual stress after 5 thermal cycles

The residual stress decreases at the end of 5 thermal cycles (Figure 5) by an amount of 0.2 MPa for Geometry No. 1 and Geometry No. 2. However, for Geometry No. 3 decreases from 10.4 MPa to 7.5 MPa. Figure 6 shows the distribution of residual stress after 10 thermal cycles. It might be observed that the concentration of highest stress is not uniform across the filler. This might be due to the larger mesh size chosen for the study. A finer mesh may result is uniformity in stress concentration.

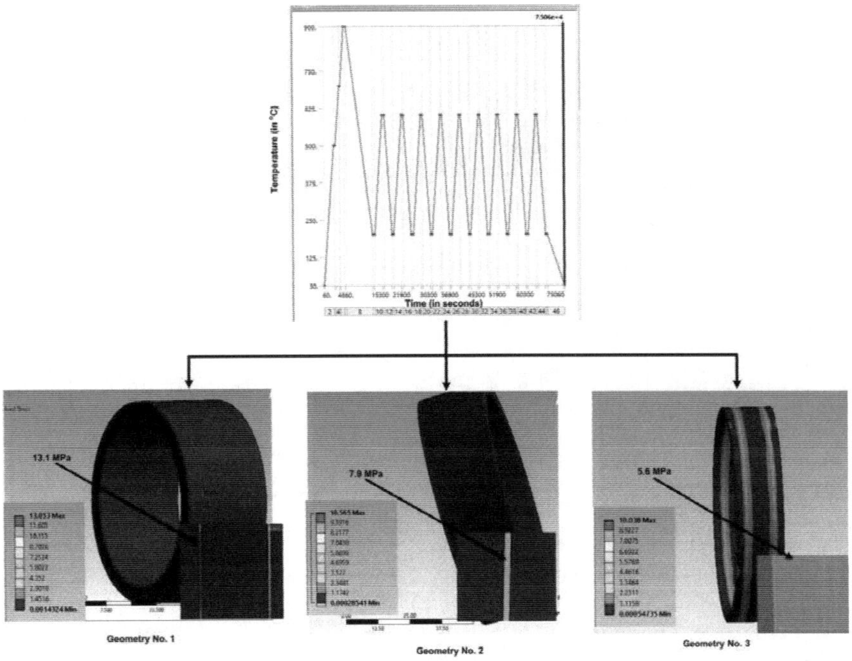

Figure 6. Distribution of residual stress after 10 thermal cycles

The application of cyclic heating cooling cycles leads to alleviation of residual stresses caused due to difference in coefficient of thermal expansion during cooling. A comparative study of effect of thermal cycles on the 3 geometry is given in Figure 7 for all the three geometries. It is evident from the graph that the effect of thermal cycles enhances with increasing diameter of the brazing components.

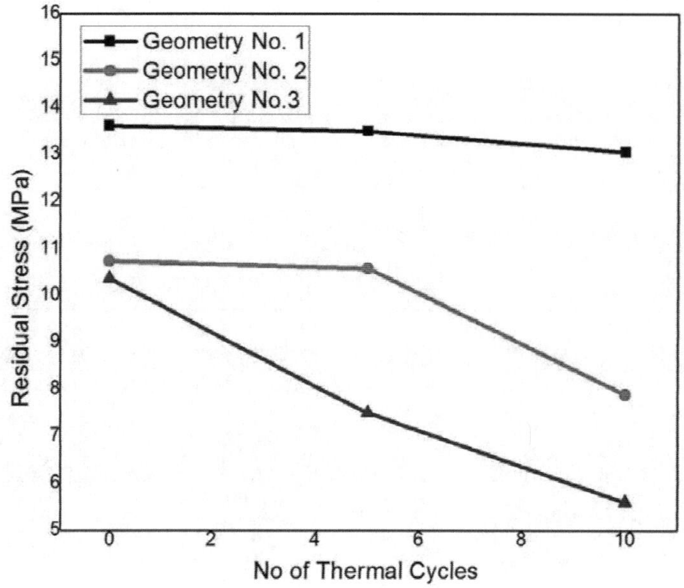

Figure 7. A comparative study of distribution of residual stress with varying geometry

4. Conclusions

Residual stress of the Alumina-Kovar brazed joint is carried out using Finite element analysis (Workbench 15.0) with varying geometry sizes and thermal cycles. The effect of geometry and thermal cycles on residual stress is determined. It is inferred that with increasing radius of the geometry and with increasing thermal cycles the residual stress decreases. With increasing radius the curvature of the geometry decreases as a result the residual stresses decreases. Similarly the stress generated due to CTE is released upon application of cyclic thermal cycles. The number of stress cycles may be increased to determine an optimum number of thermal cycles where the residual stress decreases to a minimum and thereby apply those number of thermal cycles in actual experimental conditions for the above mentioned geometry.

References

[1] Li J, Pan W, Yuan Z, Chen Y 2008 Appl. Surf. Sci **254** 4584.
[2] Kar A, Mandal S, Venkateswarlu K., Ray A K 2007 Mater. Charact **58** 555.
[3]Komarov, Sergey V, Romankov, Sergey E 2014 J. Eur. Ceram. Soc **34** 2
[4] Rohde, Südmeyer M I, Urbanek A M Ceram. Int 2009 **35** 333.
[5] Cao Y, Yan J, Ning Li, Y. Zheng, C. Xin, Effects of brazing temperature on microstructure and mechanical performance of Al2O3/AgCuTi/Fe–Ni–Co brazed joints, J. Alloy. Compd. 650 (2015) 30–36.
[6]A. Kar, M. Ghosh, A.K. Ray, Effect of interfacial thickness and residual stress on the mechanical property of the alumina-stainless steel braze joint interface, Mater. Sci. Eng. A 498 (2008) 283–288
[7]Zhao B, Jiang G, Qi H. Joining aluminum sheets with conductive ceramic films by ultrasonic nanowelding.Ceram.Int(2016);42:8098-101.

[8]Xin C, Liu W, Li N, Yan J, Shi S. Metallization of Al2O3 ceramic by magnetron sputtering Ti/Mo bilayer thin films for robust brazing to Kovar alloy. Ceram.Int92016);42:9599-604.
[9]Chen X, Xie R, Lai Z, Liu L, Yan J, Zou G. Interfacial Structure and Formation Mechanism of Ultrasonic-assisted Brazed Joint of SiC Ceramics with Al12Si Filler Metals in Air.J.Mater. Sci. Technol.(2016).
[10]Zimmerman J, Wlosinski W, LindemannZR.Thermo-mechanical and diffusion modelling in the process of ceramic–metal friction welding.J. Mater.Process. Technol(2009);209 (4): 1644–53.
[11]Hokamoto K, Fujita M, Shimokawa H, Okugawa H. A new method for explosive welding of Al/ZrO2 joint using regulated underwater shock wave.J.Mater.Process.Technol(1999); 85 (1-3): 175–9.
[12]Jung Y-I, Park J-H, Kim H-G, Park D-J, Park J-Y, Kim W-J. Effect of Ti and Si Interlayer Materials on the Joining of SiCCeramics.Nucl. Eng. Technol(2016);48(4):1009-14.
[13]Pan R, Wei H, Lin T, He P, Sekulic DP, Wang Q, Duan X. Homogenization of the zirconium carbide–titanium interface domain. Scr. Mater (2016);112:42-5.
[14]Asthana R, Singh M. Joining of partially sintered alumina to alumina, titanium, Hastealloy and C–SiC composite using Ag–Cu brazes. J. Eur. Ceram. Soc(2008);28: 617–31.
[15]Lin KL, Singh M, Asthana R. Interfacial characterization of alumina-to-alumina joints fabricated using silver-copper-titanium interlayers. MaterCharact(2014);90:40-51.
[16]Ding WF, Xu JH, Chen ZZ, Miao Q, Yang CY. Interface characteristics and fracture behavior of brazed polycrystalline CBN grains using Cu-Sn-Ti alloy. Mater SciEng A (201;559:629-34.
[17]Shi J M, Feng J C, Tian X Y, Liu H. , Zhang L X.. Interfacial microstructure and mechanical property of ZrC-SiC ceramic and Ti6Al4V joint brazed with AgCuTi alloy. J. Eur. Ceram. Soc(2017); 37(8): 69-2778
[18] Williamson, R. L., Rabin, B. H., & Byerly, G. E. (1995). FEM study of the effects of interlayers and creep in reducing residual stresses and strains in ceramic-metal joints. Composites Engineering, 5(7), 851-863.
[19]Laik A, Mishra P, Bhanumurthy K, Kale G B, Kashyap B P.Microstructural evolution during reactive brazing of alumina.to Inconel 600 using Ag-based alloy. Acta Mater.(2013);61(1): 126-38
[20]Zhu W, Chen J, Jiang C, Hao C, Zhang J. Effects of Ti thickness on microstructure and mechanical properties of alumina–Kovar joints brazed with Ag–Pd/Ti filler.Ceram.Int(2014); 40(4): 5699-705.
[21]Niu G B, Wang D P, Yang Z W, Wang Y. Microstructure and mechanical properties of Al2O3 ceramic and TiAl alloy joints brazed with Ag–Cu–Ti filler metal. Ceram.Int2016;42(6) :6924-34.
[22] Kirchner, H. P., Conway, J. C., & Segall, A. E. (1987). Effect of Joint Thickness and Residual Stresses on the Properties of Ceramic Adhesive Joints: I, Finite Element Analysis of Stresses in Joints. Journal of the American Ceramic Society, 70(2), 104-109.
[23] Williamson, R. L., Rabin, B. H., & Drake, J. T. (1993). Finite element analysis of thermal residual stresses at graded ceramic- metal interfaces. Part I. Model description and geometrical effects. Journal of Applied Physics, 74(2), 1310-1320.
[24] Lee, S. B., & Kim, J. H. (1997). Finite-element analysis and X-ray measurement of the residual stresses of ceramic/metal joints. Journal of Materials Processing Technology, 67(1-3), 167-172.
[25] Desai, C. S. & Siriwardane H. J. (1984) Constitutive Laws for Engineering Materials Prentice-Hall, Englewood Cliffs, NJ.
[26] Desai, C.S. (1984). "Hierarchical Single Surface and the Disturbed State Constitutive Models with Emphasis on Geotechnical Applications," Chap. 5 in Geotechnical Engineering: Emerging Trends in Design and Practice, K.P. Saxena, (ed), Oxford & IBH Publ. Co., New Delhi, India
[27] Desai C.S., Somasundaram S., & Frantziskonis G. (1986) A hierarchical approach for constitutive modelling of geologic materials, International Journal for Numerical and Analytical Methods in Geomechanics 10(3) 225-257.
[28] Desai C. S. (1995). "Constitutive Modelling Using the Disturbed State as Microstructure Self-Adjustment Concept," Chapter in Continuum Models for Materials with Microstructure, H.B. Mühlhaus (ed), John Wiley and Sons, U.K
[29] Desai C.S. (2001) Mechanics of Materials and Interfaces: The Disturbed State Concept, CRC Press, Boca Raton, Florida, USA.

Optimisation of Ferrochrome Addition Using Multi-Objective Evolutionary and Genetic Algorithms for Stainless Steel Making via AOD Converter

Kishore Kumar Behera, Snehanshu Pal *

Metallurgical and Materials Engineering Department, National Institute of Technology Rourkela, Rourkela 769008, India

Corresponding author Email : snehanshu.pal@gmail.com

Abstract: This paper describes a new approach towards optimum utilisation of ferrochrome added during stainless steel making in AOD converter. The objective of optimisation is to enhance end blow chromium content of steel and reduce the ferrochrome addition during refining. By developing a thermodynamic based mathematical model, a study has been conducted to compute the optimum trade-off between ferrochrome addition and end blow chromium content of stainless steel using a predator prey genetic algorithm through training of 100 dataset considering different input and output variables such as oxygen ,argon ,nitrogen blowing rate, duration of blowing, initial bath temperature ,chromium and carbon content, weight of ferrochrome added during refining. Optimisation is performed within constrained imposed on the input parameters whose values fall within certain ranges. The analysis of pareto fronts is observed to generate a set of feasible optimal solution between the two conflicting objectives that provides an effective guideline for better ferrochrome utilisation. It is found out that after a certain critical range, further addition of ferrochrome does not affect the chromium percentage of steel. Single variable response analysis is performed to study the variation and interaction of all individual input parameters on output variables.

Keywords: Stainless steel making, Optimisation, Genetic algorithm, Ferrochrome

1. Introduction

AOD steel making is a complex metallurgical process which involves intricate interplay of multiphase reactions, turbulent reacting flow, mass, momentum & heat transfer as well as numbers of heterogeneous chemical reactions among slag, metal, gas and solid phases. So under extreme operating conditions, it is hard to accurately predict the operating parameters and complex interaction between them in real time .Random fluctuations combined with statistical variations in the composition and properties of various input material and various controlling parameters even add to further complexity. In order to overcome this problem, data-driven approaches on the noisy and clumsy data are increasingly being leveraged to analyse and optimise the steel making process [1-2]. Removal of impurity element from molten steel to maximum extent is desirable in steel making process. Addition of chromium in stainless steel above 12% produces a thin layer of stable and passive surface film of chromium oxide. This oxide protective film passivates the metal and prevents corrosion in extreme corrosive media. With increase in percentage of chromium, stability of that oxide film increases resulting in higher corrosion resistance and wear resistance [3-4]. Kim, Jeong Kil, et al reported that for Ti-stabilized ferritic stainless steels, Increase in Chromium content improved inter granular resistance even with temperature and time for the sensitization becomes higher [5]. In addition to chromium, other alloying elements are also added in order to obtain required chemical and mechanical properties of the final products. Since all types stainless steel grade contain some amount of chromium, it is economical to use chromium containing raw materials such as stainless steel scrap or ferrochrome for production of stainless steel .Because of the availability and suitability, Ferroalloys are primary choice of addition for Stainless steel making. As occurrence of raw materials of ferrochrome is quiet limited on a global scale, production cost of ferroalloy and price of low carbon ferrochrome is high [6-7]. Chromium percentage of stainless steel is augmented by the amount of ferrochrome added during initial stages of oxygen blowing. As ferrochrome addition increases, cost per ton of steel production also increases which makes selling of the product unviable in competitive commercial market. Also with decrease in ferrochrome addition, chromium percentage also decreases which is undesirable since it affects the mechanical properties of stainless steel. So obtaining high chromium content in stainless steel with minimum ferrochrome addition is an industrial problem related to optimising the system and such problems are difficult to solve in conventional way [8-10]. Simultaneous optimisation of the above two competing and incommensurable objective is required to find a pareto optimal solution as uniformly as possible for which evolutionary computing techniques are being leveraged efficiently these days [11-12]. Normally there is error in the neural network model as the under fitted network does not capture the trends in the dataset due to large bias and those over fitted network does not generalise to the training data due to large variation [13]. To overcome this problem, Concept of multi objective genetic algorithm based on Evolutionary Neural Network (EvoNN) was developed by Pettersen et al [14]. Since AOD converter steel making data are very noisy and nonlinear, EvoNN generates several neural network meta models that tends to eliminate the arbitrary noisy data. Here the optimised neural network models are created by multi objective predator prey algorithm through arbitrarily generated population of neural nets. Multi objective algorithm selects the best model corresponds to an optimal trade-off between training error and neural network complexity [15-16]. Scope of data driven modelling and optimisation processes based on evolutionary technique has numerous application in the field of steel making as these processes are involved in large number of input as well as output parameters making it extremely difficult to evaluate pareto optimality in an effective manner [17-18].

In the present investigation, an attempt has been made to achieve the optimisation of two conflicting objectives simultaneously in presence of multiple constraints imposed on the system. The two conflicting objectives are

1. Minimisation of ferrochrome addition in AOD converter stainless steel making
2. Maximisation of end blow chromium content of stainless steel in AOD converter

The input variables used to optimise above two conflicting objective functions are 1st stage blowing rate,2nd stage blowing rate ,3rd stage blowing rate, reduction stage nitrogen blowing rate, reduction stage argon blowing rate, initial bath temperature, time of ferrochrome addition, initial carbon and chromium percentage and different time period of blowing.

2. Methodology

In this study, by training the dataset obtained from the mathematical model through Evolutionary Neural Network (EvoNN), bi-objective optimisation task has been carried out. Evolutionary Neural Network is a process of neural network construction that tends to create an optimum trade-off between training error and complexity of the models developed from a population of dataset without the risk of under-fitting or over-fitting. The trade-off curve between these two criterion deals with a flexible architecture from which a suitable one is chosen by using the Corrected Akaike Information Criterion (AICC) [19].

Multi-layer Neural Networks are commonly applied machine learning algorithm that converts predictive models by learning the patterns in the obtained dataset. The architecture of neural networks are composed of input layer, output layer and hidden layer. The input layer is partitioned into number of input cells which receive the input signal and multiply them as per their own weight. A set of hidden nodes called as hidden layer receive their input from input layers and feedforword it to the output layers. Each nodes of the hidden layer respond multiple nodes of previous layer to different extent and is multiplied by their connection strength to a unique weight and added directly with a bias value. The total is processed by an activation function and leaves the nodes as output. This process continues until these responses combine to generate a final response at the output layer and leaves as prediction for input variable. The network then compares against the actual dependent variable during the iterative learning process. If any standard error exists between these values, the weights are recalibrated between the hidden nodes and input nodes to better learn the process. Iteration is repeated until neural network is able to estimate accurate prediction for most of the observations [20-21].

Using meta models generated from training of dataset by EvoNN, pareto optimality between the conflicting objectives is calculated by predator prey algorithm. This algorithm consists of a small numbers of preys chosen from the dataset. They are arranged in an interconnected two dimensional network. The algorithm works as the predator population swallows the weakest prey in its neighbourhood. The prey is allowed to breed indiscriminately but the predator is restricted for breeding. The annihilation mechanism is governed by some sets of hunting rules present in a Moore's neighbourhood. The population grows exponentially in absence of predator and predator starves in absence of prey population. As the predator continues to consume the weakest prey by moving around the lattice, prey population surviving after several generation of predator attack evolves with stronger attributes which leads to generation of Pareto frontier [22-23]. The pareto frontier generated by EvoNN algorithm does not help to predict the effect of each individual variables on the objective function .So Single variable analysis is required for investigating relationship amongst the clinical input variable. To examine the conclusiveness of a response variable towards the input variable, a simple and intuitive procedure is followed. In this approach, systematic variation of the output variable has been examined with the concerned input variable keeping other independent variables constant. If response of output variable is synchronised as a qualitatively similar pattern of variation as the nature of the input variable, then they are considered to be correlated by a direct relationship. If the input variable has negative response towards the output variable, then they are assumed to inversely correlated. If one part of the response is towards the positive side and remaining part is in the negative side, then the response is assumed to be of mixed type in nature [24]. Various parameters used in predator prey algorithm and EvoNN are listed in **Table 1**.

Table 1. Parameters defined in learning module of EvoNN and PPGA during optimisation

Training	**Values**
Probability of node exchange	0.8
Probability of mutation	0.1
Maximum no of nodes	5
Initial prey population	500
Preferred prey population	500
Predator population size	30
Number of generations	100
Probability of prey movement	0.5
Lattice dimension	50 x 50
Number of hidden nodes	3
Optimization	
Initial prey population	500
Preferred prey population	500
Predator population size	30
Number of generations	100
Probability of prey movement	0.8
Probability of mutation	0.1
Lattice dimension	50 x 50

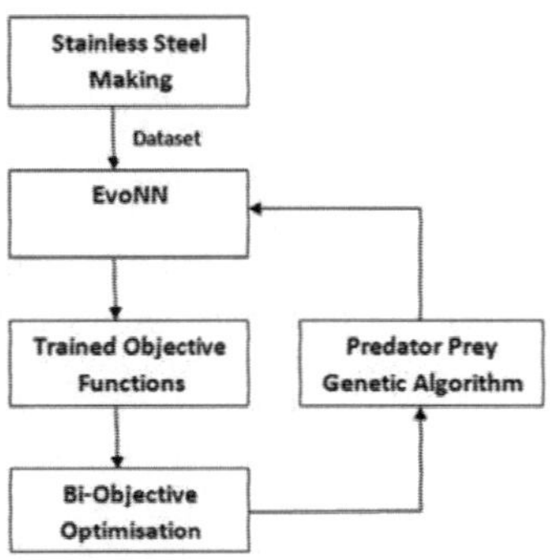

Figure 1: Flow chart of overall process.

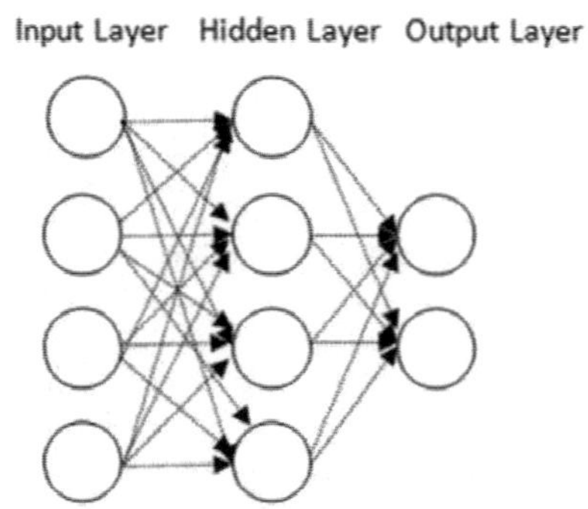

Figure 2: Simplified model of neural Network

Referring the mathematical model developed in the research paper of Patra et al., 2016 [25] for 204Cu grade stainless steel for predicting Final composition of stainless steel in reference to blowing time in an AOD converter. Using collection of input variables constrained to values within certain ranges given in **Table 2**, multiple simulations are run programmatically.100 dataset is generated using those input data ranges. Using different combination of input values, final carbon content is varied in between 0.0003 wt% to 0.19751 wt%. The chromium percentage is varied between 14.47 wt % to 26 wt%.

Table 2. Range of variables for optimising Ferrochrome content and chromium percentage of stainless steel

Sl. No	Input Parameter	Unit	Lower Boundary	Upper Boundary
1	1st stage Oxygen Blowing rate	Nm3/minute	90	130
2	Initial bath Temperature	°C	1510	1570
3	2nd stage blowing rate	Nm3/minute	15	25
4	3rd Stage blowing rate	Nm3/minute	5	10
5	Silicomanganese addition	MT	2.51	4.98
6	Nitrogen blowing rate	Nm3/minute	0	25
7	Argon blowing rate	Nm3/minute	0	35
8	Initial Carbon content	wt %	1.49	2
9	Initial Chromium content	wt %	15	16.1
10	Time of ferrochrome addition	Minute	2	8
11	1st stage blowing time	Minute	10	13
12	2nd stage blowing time	Minute	10	14
13	3rd stage blowing time	Minute	12	14
14	Nitrogen blowing time	Minute	0	4
15	Argon blowing time	Minute	0	5
16	Ferrochrome addition	MT	2.5	7.98
17	End blow Carbon content	wt%	0.0003	0.19751
18	End Blow Chromium content	wt%	14.47	26

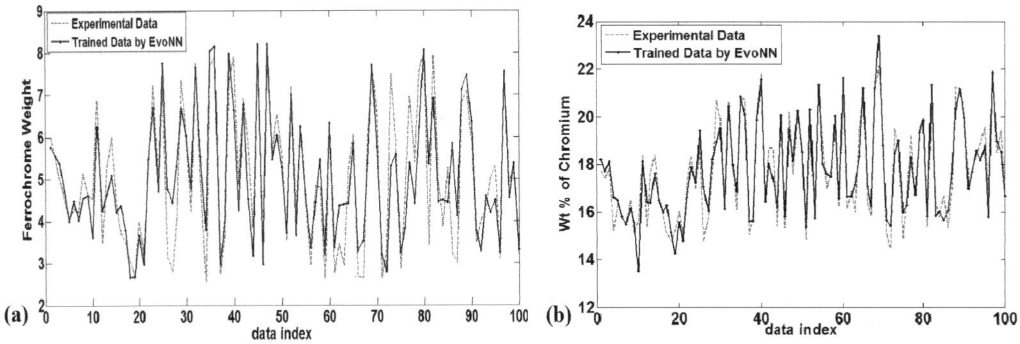

Figure 3. Setlog files of two objectives trained through data training module using predator prey algorithm for (a) amount of ferrochrome added during steel refining (b) end blow chromium content of stainless steel

The flowchart depicted in **Figure 1** illustrates the step by step procedure used in the optimisation process named as minimisation of Ferrochrome addition and maximisation of end blow chromium content of stainless steel by EvoNN. The generated dataset are trained and meta-models are created using EvoNN algorithm. With proper parameter setting, the same input parameter is run twice individually for each objective function to create two meta models through data training modules. **Figure3** represents the optimised meta models with lowest error after training the original experimental dataset. Optimisation study is carried out between the two conflicting objective functions using predator prey genetic algorithm.

3. Result and Discussion

In the current work, end blow chromium content of stainless steel has been maximised resulting minimisation of ferrochrome addition during refining of molten metal. First part of the dataset was used for neural network training purposes. The data is optimised using EvoNN and pareto front is generated by resolving tradeoffs between the two selected objectives. The pareto front in **Figure 4** translates a complete set of unique optimal solution into intuitive visual representation and each point of the curve precisely approaches a single pareto optimal solution between two concerned objectives in a constrained design search space. All the pareto points evaluated by Multi EvonNN are optimised to have their individual set of sixteen input variables and majority of the points are positioned in the user's area of interest on the pareto frontier. Each point of the pareto front leads to an optimum potential solution between ferrochrome addition and end-blow chromium percentage for which any improvement in one objective function leads to impairment of the other objective function. It is observed from the plot of pareto set that increase in weight of ferrochrome addition up to 7 MT results increase in end blow chromium percentage of steel till 25% and after that chromium content becomes gradually constant even if more ferrochrome is added to the molten metal.

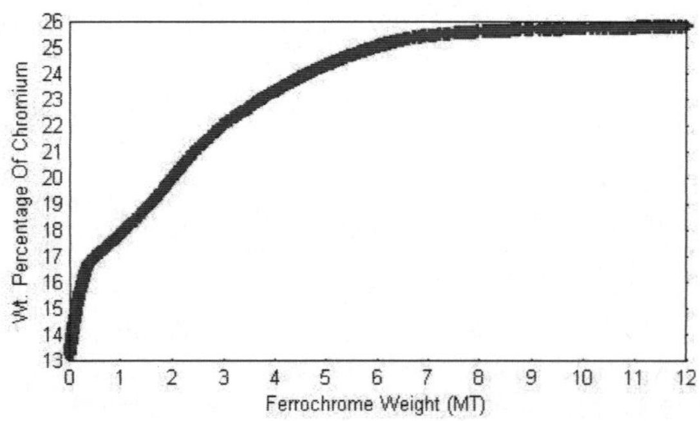

Figure 4. Pareto Fronts between weight of ferrochrome addition and end blow chromium content computed using Evolutionary Neural Network

Table 3. Single variable response of individual input variables on objective functions

Sl. No	Variable/Output	Ferrochrome Weight	Chromium Content
1	1st stage Oxygen Blowing rate	Direct	Direct
2	Initial bath Temperature	Nil	Nil
3	2nd stage blowing rate	Direct	Inverse
4	3rd Stage blowing rate	inverse	Nil
5	Silicomanganese addition	Nil	Inverse
6	Nitrogen blowing rate	Direct	Direct
7	Argon blowing rate	Nil	Nil
8	Initial Carbon content	Inverse	Inverse
9	Initial Chromium content	Inverse	Direct
10	Time of ferrochrome addition	Inverse	Nil
11	1st stage blowing time	Direct	Nil
12	2nd stage blowing time	Nil	Nil
13	3rd stage blowing time	Nil	Nil
14	Nitrogen blowing time	Direct	Direct
15	Argon blowing time	Direct	Direct
16	End blow Carbon content	Direct	Direct

The single variable response analysis in **Table 3** suggests that increase in first stage oxygen blowing rate increases consumption of ferrochrome during refining and subsequently increases chromium content of stainless steel. Initial bath temperature has no effect on both the objective functions. Second stage blowing rate has direct relation with ferrochrome weight but has inverse relation on the final chromium content. With increase in blowing rate during third stage of refining, ferrochrome consumption decreases .But it has no effect on final chromium content of steel. Initial carbon and chromium content has inverse relation with the ferrochrome addition .But higher initial chromium content leads to high end blow chromium content. Increase in ferrochrome addition time requires less amount of ferrochrome during refining but has no effect on final chromium content. First stage blowing time proportionally affects the utilisation of ferrochrome but has no effect on chromium content of steel. Similarly Second stage and third stage blowing time has no effect on both the objective functions. Nitrogen blowing time and argon blowing time has a positive response on both the concerning objective function as increase in blowing time of requires more ferrochrome addition and subsequently increase the chromium content of stainless steel.

4. Conclusion

Effective utilisation of ferrochrome during stainless steel making in an AOD converter has been investigated using the principle of multi objective evolutionary and genetic algorithm. Constructing data-driven models, optimisation work has been conducted to minimize the ferrochrome addition and simultaneously maximize the end blow chromium content of stainless steel, satisfying some conflicting requirements. A set of input operating variables considered in this investigation are 1st stage blowing rate ,initial bath temp, 2nd stage blowing rate, 3rd stage blowing rate, silicomanganese weight, nitrogen blowing rate ,Argon blowing rate, Initial carbon %, initial chromium %, time of ferrochrome addition, 1st stage blowing time, 2nd stage blowing time, 3rd stage blowing time, Nitrogen blowing time, Argon blowing time and end blow % C. Using the data collected from the mathematical model, two meta models has been developed for two objective functions. Using predator prey algorithm, the meta models has been used to construct a pareto frontier where each point on the pareto curve leads to a best fitness solution. The optimisation work confirms that first and second stage blowing rate, nitrogen blowing rate, nitrogen blowing time & argon blowing time are most important influential parameters among all parameters inspected here. The pareto frontier reveals that with increase in ferrochrome addition, chromium content of stainless steel increases up to 25 wt% .After that, further addition of ferrochrome does not affect the chromium content of stainless steel. This study provides an insight towards the impact of important process parameter for determining the effective utilisation of Ferrochrome during stainless steel making.

5.References

[1] Miletic I, Boudreau F, Dudzic M, Kotuza G, Ronholm L, Vaculik V & Zhang Y. 2008 Can. J. Chem. Eng. 86 5
[2] Li J N, Li Y, Liu F, Bie S P, Tang C C & Zhang X L 2013 Adv. Mater. Res Trans Tech Publications.
[3] Crane F A A, Charles J A, & Furness J 1997 Selection and use of engineering materials. Butterworth-Heinemann.
[4] Trepanier C, Tabrizian M, Yahia L H, Bilodeau L & Piron D L 1998. J. Biomed. Mater. Res, 43 4
[5] Kim J K, Kim Y H, Lee J S & Kim K Y 2010 Corros. Sci 52 5
[6] Riekkola-Vanhanen, M. 1999
[7] Shih C C, Shih C M, Su Y Y, Su L H J, Chang M S, & Lin S J 2004 Corros. Sci. 46 2
[8] Fontana M G 2005 Corrosion engineering Tata McGraw-Hill Education.
[9] Ou T Y, Cheng C Y, Chen P J & Perng C 2016 Computers & Industrial Engineering 101
[10]Oliveira L S D , Saramago S F 2010 J. Braz. Soc. Mech. Sci. Eng, 32 1
[11]Chiandussi G,Codegone M,Ferrero S & Varesio F. E. 2012 Computers & Mathematics with Applications 63 5
[12] Curry B & Morgan P H 2006 European Journal of Operational Research 170 2
[13] Pettersson F, Chakraborti N & Saxén H 2007 Applied Soft Computing 7 1
[14] Berenyi P, Valyon J, & Horváth G. 2001 Neural Lecture notes in computer science 269
[15] Coello Coello C A, & Becerra R. L. 2009 Mater. Manuf. Processes 24 2
[16] Chugh T, Chakraborti N, Sindhya K, & Jin Y 2017 Mater. Manuf. Processes 1-7
[17] Chakraborti N 2014 Critical assessment: Mater. Sci. Technol, 30 11
[18]T Tohge,Y Fujita and T Watanabe: Proc. 4th Process Technology Conf., Iron &Steel Society, 1984
[19] Chakraborti N 2016 Computational Approaches to Materials Design: Theoretical and Practical Aspects 346
[20]Miller G F, Todd P M, & Hegde S U 1989 In ICGA 89 379
[21] Whitley D, Starkweather T & Bogart C 1990 Parallel computing 14 3
[22]Li X 2003,In International Conference on Evolutionary Multi-Criterion Optimization(Springer)
[23] Deb K 2001 Multi-objective optimization using evolutionary algorithms (John Wiley & Sons)
[24] Helle M, Pettersson F, Chakraborti N & Saxén H 2006 Steel Res. Int, 77 2
[25] Patra S, Nayak J, Singhal L K & Pal S 2017 Steel Res. Int. 88 5

IOP Conf. Series: Materials Science and Engineering **338** (2018) 012003 doi:10.1088/1757-899X/338/1/012003

Effect of CNT addition on cure kinetics of glass fiber/epoxy composite

A O Fulmali*, R Kattaguri, K K Mahato, R K Prusty, B C Ray

Composites Materials Group, Department of Metallurgical and Materials Engineering, National Institute of Technology, Rourkela, India-769008

***Corresponding author Email: abhinavfulmali@gmail.com**

Abstract: In present time, developments in reinforced polymer composites have acquired preferential attention for high performance and high precision applications like aerospace, marine and transportation. Fibre reinforced polymer (FRP) composites are being substituted because of their low density, higher strength, stiffness, impact resistance, and improved corrosion resistance. Further laminated composites exhibit superior in-plane mechanical properties that are mostly governed by the fibers. However, laminated FRP composites suffer from poor out of plane properties in some applications. These properties can further be improved by the addition of Nano fillers like carbon nanotube (CNT), graphene and so on. Curing cycle plays a very important role in drawing out the optimum property of glass fiber/epoxy (GE) composite. It is expected that the cure kinetics can further be altered by addition of CNT due to its higher aspect ratio. The main objective of this work is to study the effect of CNT addition on cure kinetics of GE composite as multi-segment adsorption of polymer takes place on the CNT surface. In this study effects of curing parameters on mechanical properties and glass transition temperature of CNT embedded glass fiber/epoxy composite (CNT-GE) has been evaluated. For this study control GE and CNT-GE (with 0.1 wt. %) laminates were fabricated using hand lay-up technique followed by hot compression. The curing parameters that were considered in the present investigation were temperature (80°C, 110°C, and 140°C) and time (0.5 hr, 3 hr and 6 hr). For different combination of above mentioned temperature and time, samples of GE and CNT-GE composites were post cured. Mechanical properties were determined by flexural testing using 3 point bending fixture on INSTRON-5967 and thermal properties i.e. glass transition temperature (Tg) determined by Differential Scanning Calorimeter (DSC) to evaluate the effects of curing parameters. For CNT-GE samples, No much variation observed in flexural modulus with increase in post curing temperature and time, but swift increment was observed in flexural strength at 140°C with increase in post cure time. Elevation in Tg observed with increase in temperature and time duration of post curing; highest Tg noted at 140°C-6hr. Optimum post curing parameters for CNT-GE composite observed to be 140°C-6hr.

1. Introduction

When two or more materials are combined together the resultant material have improved properties as compared to the properties of original ingredient material, this resultant material is called as composite material. In recent time, due to increase in need for a material with combination of different properties, material having high strength with less weight or high corrosion resistance with less maintenance cost, fiber reinforced polymer(FRP) composite could be the answer. FRP composite is comprised of matrix polymer called epoxy and reinforcement which is glass fibers. Fibers have less diameter and long length resulting in higher aspect ratio which provides the composite, high strength and stiffness. Whereas epoxy provides shape and prevent these fibers from damage. Due to its high flexural strength, high flexural modulus and high corrosion resistance, it is the desired material for applications like space vehicles, automobile, marine and so on[1]. Fabrication of glass fiber/epoxy(GE) composite laminate goes through many different steps which decides the end properties. One of these steps is curing cycle, during which polymerization means cross linking of polymer takes place[2]. Once epoxy is polymerized it cannot be brought back to the initial condition, it changes from liquid to solid state permanently. Hardener is used to accelerate this cross linking. With correct selection of curing parameters; one can obtain the optimum properties of end product[3]. Two of the most important curing parameters are curing temperature and time duration which are considered for the present study. Still during its life span these properties degrade with time. To improve these properties further, one method is to integrate second phase particulate in epoxy. Nano fillers like carbon nanotube(CNT), graphene and many more can be added for this purpose[4]. CNT have high aspect ratio which results in high specific surface area. Fiber/polymer interface is the weakest site in FRP composite, from where most failure initiate. Better interaction can be expected between CNT, polymer and fiber, resulting in improved interfacial bonding. It is possible that CNT addition may hinder the polymer-polymer bonding and promote polymer-CNT-polymer bonding and increasing its mechanical and thermal properties[5]. A relationship is observed between glass transition temperature(Tg) and degree of curing, hence differential scanning calorimetry(DSC) is done to evaluate Tg[6]. To evaluate changes in mechanical properties, flexural testing is done. It could be possible that CNT addition may affect the cure kinetics of the GE composite; to study this effect the current research is made.

2. Materials required and Experimental procedure

2.1 *Materials required*

Epoxy used for this study was diglycidyl ether of BisphenolA (DGEBA) and the hardener was Triethylene tetra amine (TETA), both were supplied by Atul Industries Ltd, Gujarat, India. The CNT used has diameter of ~10nm and length of ~5μm, purchased from sigma Aldrich. Glass fiber used is an E-type, 3 K plain weave and having 15μm fiber diameter, manufactured by Owens Corning, India.

2.2 *Fabrication of CNT embedded GE composite*

For the optimization of post curing temperature, laminates of glass fiber/epoxy (GE) composite and 0.1 wt. % of CNT filled glass fiber/ epoxy(CNT-GE) composite were fabricated. Initially CNT (0.1 wt. % of epoxy) was dispersed in acetone of 160mL and stirred for 0.5hr at a speed of 1000rpm. Then the epoxy was preheated for 0.5hr at a temperature of 70°C and stirred at a speed of 400rpm for better fluidity which facilitates the homogeneous dispersion of added CNT in the epoxy. Preheating of the epoxy was done; Parallely sonication of acetone + CNT suspension was done for 0.5hr to reduce agglomeration. After this acetone + CNT suspension was transferred to the flask which contains the preheated epoxy. Now the complete solution was stirred for 2.5hr at a temperature of 70°C and speed of 1000 rpm, to evaporate the acetone completely from the solution. Again sonication is done for 1hr followed by degassing for 18hr to remove the gases or air bubbles that are formed during the process of stirring and sonication. Laminate was fabricated by using hand layup method followed by hot pressing. Laminate were cut for the flexural testing as per the ASTM D7264standards using a diamond cutter.

2.3 *Experimental method*

2.3.1 *Post curing schedule*

Control GE and CNT-GE composite samples were post cured at temperatures of 80°C, 110°C and 140°C in a hot air oven for 0.5hr, 3hr and 6hr and then allowed to be annealed.

2.3.2 *Flexural testing*

This testing was performed on INSTRON 5967 UTM using 3 point bending fixture to evaluate the flexural strength. 1mm/min loading rate was used and test was performed at room temperature(25°C). Span length was kept equal to 16 x thickness of sample and overhang of 8 mm was provided.

2.3.3 *Differential scanning calorimetry (DSC)*

Differential scanning calorimetry is the technique used to study the effect of addition of CNT on the glass transition temperature(T_g) of the fiber reinforced composite. Change in the slope of the heat flow vs. temperature curve gives the glass transition temperature. This test was performed from 30°C to 180°C at heating rate of 5°C/min. Tg provides the range of temperature during which transition of epoxy from hard glassy state to soft rubbery state occurs.

3. Results and discussion

3.1 *Effect of temperature and time on flexural strength and flexural modulus*

Figure 1 shows the Variation in flexural strength with post curing time at different curing temperature for GE and CNT-GE samples. In figure 1 (a), at post curing temperature of 80°C no considerable changes observed in flexural strength with increasing curing time. Thermal energy required for better cross linking may not be available at this temperature for different time duration. Which is not the same case at post curing temperature of 110°C and 140°C, increment in flexural strength observed till 3hr of curing time, but for 6hr no notable change occurred at both temperatures. In figure 1(b), For CNT-GE samples increment in flexural strength can be observed as compared to control GE samples. Due to CNT addition, it promotes interfacial bonding which results in effective stress transfer from matrix to reinforcement. Energy supplied for the same reason goes on increasing with curing temperature and time. For CNT-GE composite maximum flexural strength noted at post curing parameters of 140°C-6hr.

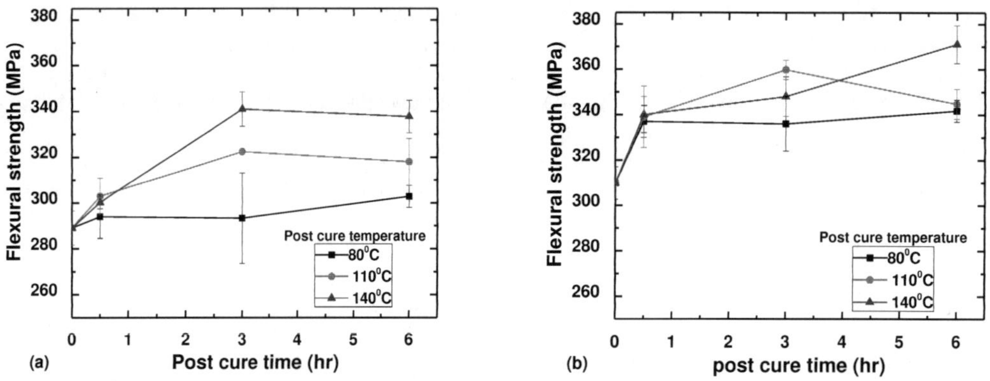

Figure 1.Variation in flexural strength with post curing time at different curing temperature a) for GE samples b) for CNT-GE samples

Figure 2 shows the Variation in flexural modulus with post curing time at different curing temperature for GE and CNT-GE composite. In figure 2(a) and 2(b), No notable change can be observed in flexural modulus for GE and CNT-GE samples, with increasing curing temperature and time. As flexural modulus is mostly governed by the glass fibers, further cross linking of the polymers does not have influence on overall flexural modulus of the composite.

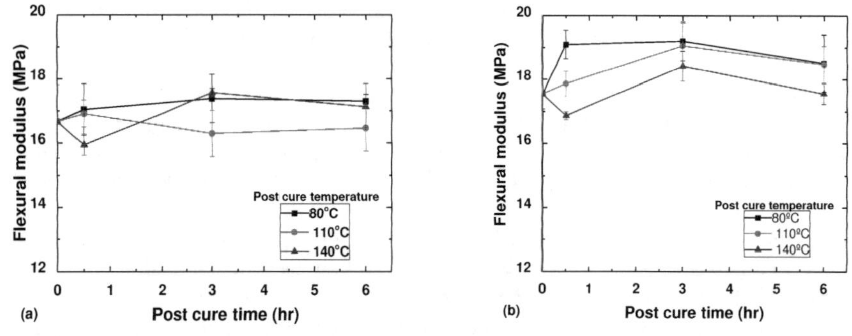

Figure 2. Variation in flexural modulus with post curing time at different curing temperature a) for GE samples b) for CNT-GE samples

3.2 *Effect of temperature and time on Tg*

Deviation in Tg with post curing time(hr) at various post curing temperatures can be observed in figure 3. Tg is acquired from the initiation of alter in slope of the heat flow against temperature. It was noticed that the Tg for GE samples is slightly higher than that of CNT-GE samples. CNT presence hinders the polymer-polymer chain interaction due to multi-segment adsorption of polymer on CNT surface, which hampers development of cross linking resulting in decreased cross link density of polymer-polymer chain, which leads to reduced Tg. In figure 3 (a) for GE samples, it can be observed that for all duration of post curing time, Tg is higher at post curing temperature of 140°C as compared to post curing temperature of 80°C and 110°C. It increases rapidly for all the temperatures till 3hr and after that the rate of increment decreases. Same goes for CNT-GE samples; there is an increase in Tg for all the temperatures till 3hr of post curing, but after 3hr to 6hr there is no considerable increment in Tg for post curing temperature of 80°C and 110°C. For post curing temperature of 140°C there is a continuous increase in Tg with increase in the post cure time showing no sign of saturation. Highest Tg noted for GE samples and CNT-GE samples was at post curing parameters of 140°C-6hr. The flexural strength for the CNT-GE samples are the highest for post curing at 140°C-6hr, confining better post curing and improved interfacial bonding.

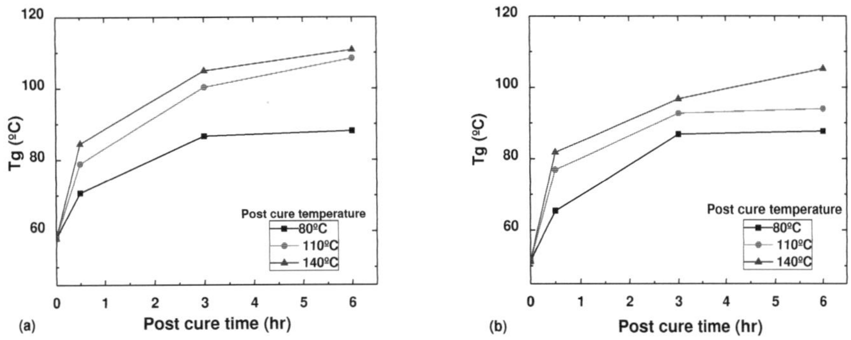

Figure 3. Deviation in Tg with post curing time at different curing temperature a) for GE samples b) for CNT-GE samples

4. Conclusion

The following conclusions can be drawn from the current study

1. Post curing at temperature 110°C and 140°C provides sufficient amount of activation energy for further cross linking of polymer which is not available at lower temperature like 80°C, that would be the reason for no considerable increment in the flexural strength and modulus, for the samples which are post cured at 80°C.
2. Optimum post curing parameters i.e. post curing temperature and curing time is observed to be 140°C and 6hr respectively for CNT-GE composite. Presence of CNT improved interfacial bonding resulting in increased flexural strength due to better stress transfer from epoxy to glass fiber. No considerable change in flexural modulus was observed as it mostly depends on glass fiber, not as much on cross linking of polymers.
3. Tg for CNT-GE samples observed to be lower than GE samples because of multi-segment adsorption of polymers on CNT surface, resulting in reduced polymer cross linking density which hinders the heat flow. Highest Tg value noted for GE and CNT-GE samples post cured at temperature and time duration of 140°C and 6hr respectively.

Acknowledgments

The authors convey their sincere thanks to CSIR India (project code – 22(0735)/17/EMR-II) and NIT Rourkela for providing necessary financial and infrastructural support to carry out this work successfully, in addition the technical association of Mr. Rajesh Patnaik in conducting various experimental work is deeply appreciated.

References

[1] B. C. Ray and D. Rathore, "Durability and integrity studies of environmentally conditioned interfaces in fibrous polymeric composites: Critical concepts and comments," *Adv. Colloid Interface Sci.*, vol. 209, pp. 68–83, Jul. 2014.
[2] P. K. Kumar, N. V. Raghavendra, and B. K. Sridhara, "Optimization of infrared radiation cure process parameters for glass fiber reinforced polymer composites," *Mater. Des.*, vol. 32, no. 3, pp. 1129–1137, Mar. 2011.
[3] W. Stark, "Investigation of the curing behaviour of carbon fibre epoxy prepreg by Dynamic Mechanical Analysis DMA," *Polym. Test.*, vol. 32, no. 2, pp. 231–239, Apr. 2013.
[4] M. J. Shukla, D. S. Kumar, D. K. Rathore, R. K. Prusty, and B. C. Ray, "An assessment of flexural performance of liquid nitrogen conditioned glass/epoxy composites with multiwalled carbon nanotube," *J. Compos. Mater.*, vol. 50, no. 22, pp. 3077–3088, Sep. 2016.
[5] R. K. Prusty, D. K. Rathore, and B. C. Ray, "CNT/polymer interface in polymeric composites and its sensitivity study at different environments," *Adv. Colloid Interface Sci.*, vol. 240, pp. 77–106, Feb. 2017.
[6] G. Furtos, L. Silaghi-Dumitrescu, M. Moldovan, B. Baldea, R. Trusca, and C. Prejmerean, "Influence of filler/reinforcing agent and post-curing on the flexural properties of woven and unidirectional glass fiber-reinforced composites," *J. Mater. Sci.*, vol. 47, no. 7, pp. 3305–3314, Apr. 2012.

MULTI-RESPONSE OPTIMIZATION OF WEDM PROCESS PARAMETERS USING TAGUCHI BASED DESIRABILITY FUNCTION ANALYSIS

Himadri Majumder[1]; Kalipada Maity[2]*
[1,2] Department of Mechanical Engineering
National Institute of Technology, Rourkela-769008, Odisha, India

[1]himu.nita@gmail.com
[2]*Corresponding author: kpmaity@gmail.com

Abstract
Shape memory alloy has a unique capability to return to its original shape after physical deformation by applying heat or thermo-mechanical or magnetic load. In this experimental investigation, desirability function analysis (DFA), a multi-attribute decision making was utilized to find out the optimum input parameter setting during wire electrical discharge machining (WEDM) of Ni-Ti shape memory alloy. Four critical machining parameters, namely pulse on time (T_{ON}), pulse off time (T_{OFF}), wire feed (WF) and wire tension (WT) were taken as machining inputs for the experiments to optimize three interconnected responses like cutting speed, kerf width, and surface roughness. Input parameter combination T_{ON}= 120 µs., T_{OFF}= 55 µs., WF= 3 m/min. and WT= 8 kg-F were found to produce the optimum results. The optimum process parameters for each desired response were also attained using Taguchi's signal-to-noise ratio. Confirmation test has been done to validate the optimum machining parameter combination which affirmed DFA was a competent approach to select optimum input parameters for the ideal response quality for WEDM of Ni-Ti shape memory alloy.

Keywords: Desirability function analysis; shape memory alloy; Taguchi; WEDM.

1. Introduction

WEDM becomes very popular non-traditional process (NTM) because of its capability to machine complex shapes and geometry and difficult-to-cut conductive material (1). In his thermo-electrical process, WEDM material is removed by a continuous highly focused spark energy developed in between electrode and workpiece (2, 3). Any electrically conductive material irrespective of its hardness can precisely machine using WEDM process. Due to its unique ability, WEDM application dramatically increased in tool and die-making industry, automobile sector, aerospace and nuclear industry where high precise end product is needed.

In this competitive production field, selection of best input parameter setting for desired responses is very difficult due to the presence of variety of choices, contradictory nature of output, etc. To offer this selection process, several distinctive multi-attribute decision making (MADM) procedures are now available (4-11). Among different MADM approach, desirability function analysis (DFA) is very easy to execute as because DFA does not contain complicated mathematical theory or computation (12).

Kanlayasiri et al. (13) used Box–Behnken design based DFA technique to decide an optimum cutting combination for surface roughness (Ra) and dimensional accuracy for WEDM of K460 tool steel. Bhaumik et al (14) used MADM approach GRA coupled with DFA to obtain the optimum level of the input parameters for higher material removal rate and lower tool wear rate (TWR). Gopalakannan et al.

(15) optimized EDM input parameters using DFA approach to maximize MRR and minimize electrode wear rate (EWR) and Ra.

Due to its properties like good biocompatibility, superior fatigue performance, Ni-Ti shape memory alloy is widely used in automotive, biomedical, robotics, aerospace and civil structures etc (16, 17). In this research, MADM approach desirability function analysis has been practiced to elect ideal machining parameter combination for WEDM responses namely average cutting speed, average kerf width and average Ra for Ni-Ti shape memory alloy. Using Taguchi's signal to noise (S/N) ratio, attempts were further made to decide the optimum parameter combination for each input parameter viz. pulse-on time (T_{ON}), pulse-off time (T_{OFF}), wire feed (WF) and wire tension (WT).

2. Experimental setup and Data Collection

A number of experiments was practiced on four axis CNC WEDM (EIPULS 15) following L27 orthogonal array. Different significant machining variables like pulse-on time (T_{ON}), pulse-off time (T_{OFF}), wire feed (WF) and wire tension (WT) were considered (see Table 1) to find out key machinability aspects namely cutting speed (CS), kerf width (KW) and surface roughness (Ra). Brass wire (diameter 0.25 mm) and de-ionized water was used as wire electrode and dielectric medium respectively. From 1.2 mm thickness plate of Ni-Ti shape memory alloy, 5 mm length was cut by each input setting. Using Taylor Hobson 3D profilometer surface roughness or arithmetic mean roughness (Ra) were find out and average value from three measurement were taken as final Ra value. From machine monitor the cutting speed were noted down during each experiment and average value from three measurement were taken for final cutting speed value. The kerf width for each setting were measured using Toolmaker microscope and average value from three measurement were taken as final value.

Table 1. Input variables and their levels

Machining Parameter	Denotation	Unit	Level		
			Low	Medium	High
Pulse ON Time (T_{ON})	A	μs	110	115	120
Pulse OFF Time (T_{OFF})	B	μs	50	55	60
Wire Feed (WF)	C	m/min	3	5	7
Wire Tension (WT)	D	kg-F	4	6	8

3. Methodology

In this experimental work, a MADM model desirability function analysis has been applied to optimize different correlated responses of WEDM of Ni-Ti shape memory alloy.

3.1 Desirability function analysis

Following steps need to follow for the proposed methodology:

1st Step: Identification of the purposes and its characteristics and decision matrix preparation to represent the performance characteristics with respect to different variables.

2nd Step: Calculation of the desirability index (d_i) value for individual response. d_i value depends on the character of the respective output.

- When output character is "Larger-is-better"

$$d_i = \begin{cases} 0, & x \le x_{min} \\ \left(\dfrac{x - x_{min}}{x_{max} - x_{min}}\right), & x_{min} \le x \le x_{max}, r \ge 0 \\ 1, & x \ge x_{min} \end{cases} \quad (1)$$

here, d_i=Single desirability index;

X_{min} and X_{max} are the lowest and highest values of x;

R= Weight.

- When output character is "smaller-is-better"

$$d_i = \begin{cases} 1, & x \le x_{min} \\ \left(\dfrac{x - x_{max}}{x_{min} - x_{max}}\right), & x_{min} \le x \le x_{max}, r \ge 0 \\ 0, & x \ge x_{max} \end{cases} \quad (2)$$

3rd Step: using the following equation the overall desirability grade (d_G) is evaluated:

$$d_G = \left(d_1^{w_1} * d_2^{w_2} * d_3^{w_3} * .... * d_i^{w_i}\right)^{\frac{1}{w}} \quad (3)$$

here, d_i = Single desirability index;
w_i = weight assigned to individual response respectively;
w = sum of all individual weights.
Highest d_G value is reflected as ideal setting of corresponding input variables.

4. Results and discussions

With varied combination of input parameters Total 27 number of experiments were executed following Taguchi's L_{27} orthogonal array. Different experimental settings and their respective out comes are shown in Table 2.

Table 2. Responses from respective input parameter combination

Run no.	T_{ON} (µs)	T_{OFF} (µs)	Wire Feed (m/min)	Wire Tension (kg-F)	CS (mm/min)	KW (mm)	Ra (µm)
1	110	50	3	4	0.792	0.493	2.870
2	110	55	5	6	0.585	0.491	3.030
3	110	60	7	8	0.425	0.521	3.653
4	115	50	5	8	0.965	0.511	3.925
5	115	55	7	4	0.878	0.498	3.641
6	115	60	3	6	0.876	0.472	3.506
7	120	50	7	6	1.815	0.502	3.014
8	120	55	3	8	1.681	0.492	2.667
9	120	60	5	4	1.215	0.593	3.814
10	110	50	3	4	0.760	0.472	2.730
11	110	55	5	6	0.560	0.502	3.840
12	110	60	7	8	0.496	0.517	3.752
13	115	50	5	8	1.019	0.507	3.879
14	115	55	7	4	0.867	0.472	3.876
15	115	60	3	6	0.870	0.466	3.418
16	120	50	7	6	1.904	0.527	3.017
17	120	55	3	8	1.547	0.498	2.517
18	120	60	5	4	1.187	0.601	3.793
19	110	50	3	4	0.770	0.481	2.806
20	110	55	5	6	0.577	0.52	3.661
21	110	60	7	8	0.408	0.523	3.850
22	115	50	5	8	0.950	0.520	4.015
23	115	55	7	4	0.860	0.458	3.271
24	115	60	3	6	0.896	0.497	3.507
25	120	50	7	6	1.785	0.533	2.867
26	120	55	3	8	1.480	0.481	2.473
27	120	60	5	4	1.079	0.574	3.698

Using desirability function analysis, the optimum combination of the input parameters was determined. From the different correlated responses, "larger-is-better" criteria was executed for average cutting speed whereas "smaller-is-better" criteria were executed for average kerf width (KW) and average surface roughness (Ra). The desirability index (d_i) for average cutting speed was calculated using Eq. 1 and on the contrary the same for average kerf width and Ra were calculated using Eq. 2. The weightage for each responses were taken as 33%. After calculating the d_i value for each respective responses, the overall desirability grade (d_G) was calculated using Eq. 3. The distinct desirability index and overall desirability grade for individual responses are shown in Table 3. In DFA, highest value of d_G signifies the optimum combination. From the Table 3, it is clearly understandable that run no. 26 has the highest d_G value with 0.84548. DFA gives the optimum input parameter setting which is T_{ON}: 120 µs., T_{OFF}: 55 µs., WF: 3 m/min. and WT: 8 kg-F.

Table 3. Optimization using DFA

Run no.	Individual Desirability index (d_i)			d_G	Rank
	CS	KW	Ra		
1	0.25668	0.75525	0.74254	0.52748	10
2	0.11832	0.76923	0.63878	0.39108	14
3	0.01136	0.55944	0.23476	0.11678	24
4	0.37233	0.62937	0.05837	0.24259	20
5	0.31417	0.72028	0.24254	0.38372	15
6	0.31283	0.90210	0.33009	0.45692	12
7	0.94051	0.69231	0.64916	0.75263	4
8	0.85094	0.76224	0.87419	0.82925	2
9	0.53944	0.05594	0.13035	0.16080	23
10	0.23529	0.90210	0.83333	0.56459	7
11	0.10160	0.69231	0.11349	0.20310	21
12	0.05882	0.58741	0.17056	0.18375	22
13	0.40842	0.65734	0.08820	0.29076	17
14	0.30682	0.90210	0.09014	0.29582	16
15	0.30882	0.94406	0.38716	0.48681	11
16	1	0.51748	0.647211	0.69700	5
17	0.76136	0.72028	0.97147	0.81237	3
18	0.52072	0	0.14397	0	25
19	0.24198	0.83916	0.78405	0.54532	8
20	0.11297	0.56643	0.22957	0.24838	19
21	0	0.54546	0.10700	0	25
22	0.36230	0.56643	0	0	25
23	0.30214	1	0.48249	0.52969	9
24	0.32620	0.72727	0.32944	0.43120	13
25	0.92046	0.47552	0.74449	0.69072	6
26	*0.71658*	*0.83916*	*1*	*0.84548*	*1*
27	0.44853	0.18881	0.20558	0.26270	18

After finding the overall optimum input parameter setting, attempts were further made to get the optimum parameter setting for single quality characteristic using Taguchi's signal to noise (S/N) ratio (18). Taguchi's main effect plot of S/N ratio for higher cutting speed, smaller kerf width and Ra with respect to WEDM input parameters viz. T_{ON}, T_{OFF}, WF and WT are shown in Figures. 1-3. For higher cutting speed, parameter combination T_{ON}= 120 µs., T_{OFF}= 50 µs., WF= 3 m/min. and WT= 6 kg-F was found to yield the preferred outcomes (Figure 1). On the other hand, the parameter setting T_{ON}= 115 µs., T_{OFF}= 55 µs., WF= 3 m/min. and WT= 6 kg-F was found optimum for smaller kerf width (Figure 2). In the same way, for the smaller Ra, by using Taguchi's S/N ratio, the optimum input parameter combination was found as T_{ON}= 120 µs., T_{OFF}= 55 µs., WF= 3 m/min. and WT= 6 kg-F (Figure 3). It is obvious from the figures that specific responses has its individual specific parameter setting.

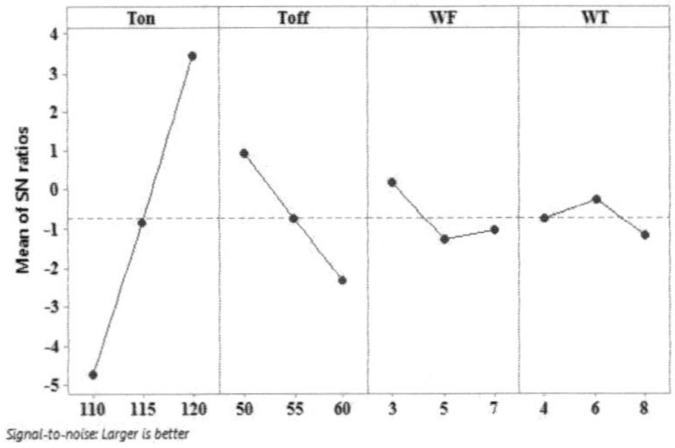

Figure 1. Main effects plot for S/N ratio for cutting speed.

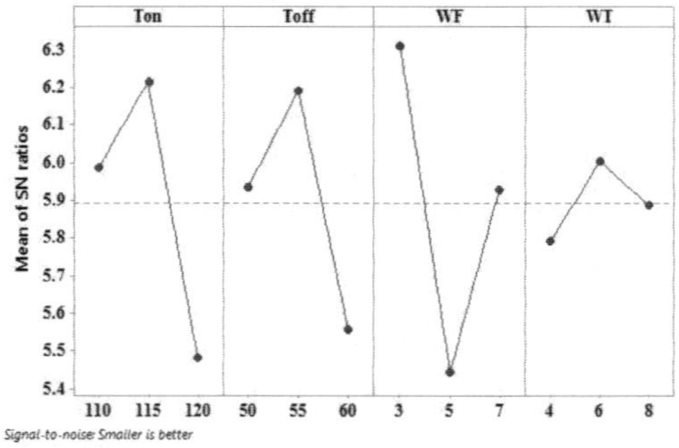

Figure 2. Main effects plot for S/N ratio for kerf width.

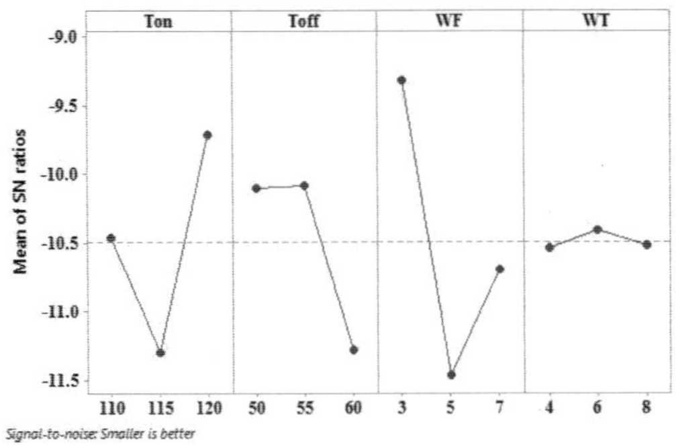

Figure 3. Main effects plot for S/N ratio for Ra.

4.1 Confirmation test

Table 7 indicates that, among 27 experiments, experiment no. 26 has the highest overall desirability grade (d_G) which specifies that the input parameter combination of A3-B2-C1-D3 has the optimum multiple performance characteristic. According to the optimum parameter combination, confirmation test was conducted to figure out the quality characteristics for WEDM of Ni-Ti shape memory alloy (see Table 4). Confirmation test shows improvement in the quality of WEDM responses.

Table 4. Confirmation test.

	Taguchi-DFA	**Confirmation test**
Optimum combination	A3-B2-C1-D3	A3-B2-C1-D3
Average cutting speed	1.48	1.52
Average kerf width	0.481	0.397
Average surface roughness	2.473	2.464

5. Conclusion

According to Taguchi's L_{27} orthogonal array, Ni-Ti shape memory alloy was machined by WEDM process. Experimental results were optimized using a desirability function analysis technique. From the investigation results, the following conclusions might be drawn:

a) Desirability function analysis comprises a lesser amount of mathematical calculations and very easy to execute compare to other optimization strategies. Consequently, DFA will be useful for the engineers who does not have very strong statistical background.
b) Experiment no. 26 has the highest overall desirability grade (d_G) which specifies that the input parameter setting of T_{ON}= 120 µs., T_{OFF}= 55 µs., WF= 3 m/min. and WT= 8 kg-F has the optimum multiple performance characteristic.
c) Optimum process parameters for each preferred responses were determined by the Taguchi S/N ratio.
d) Parameter combination of T_{ON}= 120 µs., T_{OFF}= 50 µs., WF= 3 m/min. and WT= 6 kg-F was optimum for higher cutting speed. Whereas process parameter setting T_{ON}= 115 µs., T_{OFF}= 55 µs., WF= 3 m/min. and WT= 6 kg-F was optimum for smaller kerf width. T_{ON}= 120 µs., T_{OFF}= 55 µs., WF= 3 m/min. and WT= 6 kg-F was found optimum for smaller Ra.

The accuracy of the proposed optimization technique is limited to the studied environment and parameter range. It can be improved by conducting more experiments with varying range of machining parameter and machining environment. DFA can also be employed as a future work for other WEDM responses like corner deviation, MRR, etc. with high accuracy.

IOP Conf. Series: Materials Science and Engineering **338** (2018) 012004 doi:10.1088/1757-899X/338/1/012004

References

1. Majumder H, Maity K. *Journal of Advanced Manufacturing Systems.* 2017;16(02):81-99.
2. Majumder H, Paul T, Dey V, Dutta P, Saha A. *Measurement.* 2017;107:19-30.
3. Kumar A, Majumder H, Vivekananda K, Maity K. *Materials Today: Proceedings.* 2017;4(2):2194-202.
4. Majumder H, Saha A. *Decision Science Letters.* 2018;7(2):143-56.
5. Saha A, Majumder H, *International Journal of Engineering Research in Africa;* 2016: Trans Tech Publ.
6. Saha A, Mondal SC. *Measurement.* 2016;94:46-59.
7. Khan A, Maity K. *International Journal of Engineering Research in Africa.* 2016;22:94.
8. Khan A, Maity K. *International Journal of Engineering Research in Africa.* 2016;23.
9. Saha A, Mondal SC. *Measurement.* 2017;102:80-9.
10. Saha A, Mondal SC. *Transactions of the Indian Institute of Metals.* 2017;70(6):1491-502.
11. Buddala R, Mahapatra SS. *Journal of Industrial Engineering International.2017;*1-16.
12. Sait AN, Aravindan S, Haq AN. *The international journal of advanced manufacturing technology.* 2009;43(5-6):581.
13. Kanlayasiri K, Jattakul P. *Precision Engineering.* 2013;37(3):556-61.
14. Bhaumik M, Maity K. *Decision Science Letters.* 2017;6(4):387-94.
15. Gopalakannan S, Senthilvelan T. *Measurement.* 2013;46(8):2705-15.
16. Theisen W, Schuermann A. *Materials Science and Engineering: A.* 2004;378(1):200-4.
17. Manjaiah M, Narendranath S, Basavarajappa S. *Transactions of Nonferrous Metals Society of China.* 2014;24(1):12-21.
18. Saha A, Majumder H. *Journal of King Saud University-Engineering Sciences.* 2016. https://doi.org/10.1016/j.jksues.2016.11.002

IOP Conf. Series: Materials Science and Engineering **338** (2018) 012005 doi:10.1088/1757-899X/338/1/012005

Machinability assessment of commercially pure titanium (CP-Ti) during turning operation: Application potential of GRA method

Akhtar Khan[1], Kalipada Maity[2]*

[1,2] Department of Mechanical Engineering, National Institute of Technology Rourkela, Odisha, India

Email address: akhtarkhan00786@gmail.com, kpmaity@gmail.com*
Corresponding author: kpmaity@gmail.com

Abstract. This paper explores some of the vital machinability characteristics of commercially pure titanium (CP-Ti) grade 2. Experiments were conducted based on Taguchi's L9 orthogonal array. The selected material was machined on a heavy duty lathe (Model: HMT NH26) using uncoated carbide inserts in dry cutting environment. The selected inserts were designated by ISO as SNMG 120408 (Model: K313) and manufactured by Kennametal. These inserts were rigidly mounted on a right handed tool holder PSBNR 2020K12. Cutting speed, feed rate and depth of cut were selected as three input variables whereas tool wear (VBc) and surface roughness (Ra) were the major attentions. In order to confirm an appreciable machinability of the work part, an optimal parametric combination was attained with the help of grey relational analysis (GRA) approach. Finally, a mathematical model was developed to exhibit the accuracy and acceptability of the proposed methodology using multiple regression equations. The results indicated that, the suggested model is capable of predicting overall grey relational grade within acceptable range.

Keywords: Tool wear; surface roughness; commercially pure titanium; Taguchi method; GRA; Multiple regression analysis

1. Introduction

Recently, titanium and its alloys are gaining enormous attention from several industries due to their attractive inherent properties such as highest strength-to-weight ratio, low density, superior corrosion resistance and excellent bio-compatibility [1-3]. Because of the aforesaid qualities these alloys are widely used in space craft, aerospace, marine, medical and chemical processing industries [4]. Therefore, titanium alloys are receiving an appreciable attention around the globe and the researchers are focusing on exploration of various machinability aspects of these alloys. In spite of the afore-mentioned activities, the investigations on titanium alloys are sturdily limited because of the high cost and difficulty associated with their extraction. In addition to that, low thermal conductivity and high chemical reactivity also act as a barrier during the study of the key machining characteristics of titanium alloys. Poor thermal conductivity of these alloys restricts high speed machining and hence resulted in low production rate. According to the available literature on titanium machining, the suggested values of the spindle speed during titanium machining ranges from 30 to 60 m/min when using uncoated carbide inserts. On the other hand, high speed machining of these alloys causes high cutting temperature owing to rapid tool wear. Tool failure at its pre-mature stage also contributes in diminishing the quality of the machined surface as well as the dimensional accuracy of the end product. Furthermore, high chemical reactivity of titanium alloys introduces several defects such as built-up edge formation, chipping, development of shear cracks etc. which in turn curtails the life cycle of the cutting tool material. In such situation, an appropriate selection of machining parameter becomes necessary in order to attain an efficient machining performance without compromising the quality. An optimization technique plays key role in attaining the aforesaid objectives.

In the past few decades, several techniques were employed to identify the influence of cutting variables on the key machining performance in order to confirm the quality of the end product in association with high productivity. In addition to that, various statistical prediction tool were also suggested to approximate the key machining responses such as response surface methodology (RSM), artificial neural network (ANN), genetic algorithm (GA), multiple regression analysis (MRA) etc. Sahoo [5] proposed Taguchi based regression model to predict surface roughness (Ra) while turning AISI D2 hardened steel. The results indicated that the suggested model was effectively capable of estimating Ra with a confidence level of 95%. In a different study experimental study Asilturk and Cunkas [6] also used multiple regression model and artificial neural network for prediction surface roughness when turning AISI 1040 steel. They concluded that the ANN model was more efficient in comparison to MRA approach. However, the results might be limited to the studied range of machining parameters. Kant & Sangwan [7] in their study, recommended ANN coupled with GA to predict surface while turning AISI 1060 steel in dry cutting environment. The results revealed that the proposed methodologies were capable enough in estimating and minimizing the surface roughness of the machined part. Khan & Maity [8] studied the impact of machining variables on cutting force, surface roughness, material removal rate and machining temperature. They utilized MRA approach for prediction and desirability function analysis (DFA) method for optimizing the multiple responses. In addition to that, they also proposed some multi-criteria decision making (MCDM) based techniques to identify the optimal parametric combination of input variables [9, 10]. Moreover, a large number of literature is also available dealing with multi-objective problems while machining a wide range of work materials[11-13]

Although several statistical prediction and optimization tools were reported in the past decades to predict the surface quality of various materials such as steel alloy, nickel base alloy, titanium alloy (particularly Ti-6Al-4V, grade 5), machining characteristics of commercially pure titanium (CP-Ti) is not adequately addressed so far. Therefore, the present investigation aimed at estimating tool wear and surface roughness while machining CP-Ti grade 2 using uncoated carbide inserts. A series of experiments were conducted based on Taguchi's L9 orthogonal array. The aforesaid performance measures i.e. VBc and Ra were minimized with the help of grey relational analysis (GRA) method. At the end, a mathematical model was developed to verify the acceptability of the proposed methodology using multiple regression equations.

2. Material and methods

2.1 Experimental details

A cylindrical bar, made of commercially pure titanium (CP-Ti grade 2) having diameter 50 mm and length 500 mm was selected as work material. Table 1 illustrates the chemical composition of workpiece. This work part was turned on a heavy duty lathe (Manufactured by Hindustan Machine Tools, India; Model: NH26) using a square shaped uncoated carbide insert in dry cutting environment. A series of experiments were carried out according to Taguchi's L_9 orthogonal array. Machining length was kept fixed as 250 mm. for each experimental run. The selected inserts were designated by ISO: SNMG 120408 (Manufactured by Kennametal, Grade: K313). During machining operation, these inserts were tightly mounted on a right handed tool holder (ISO: PSBNR 2020K12, Manufactured by WIDIA). A stereo zoom optical microscope was utilized to measure the wear on the flank surfaces of the cutting inserts. In order to confirm the statistical accuracy, the measurement was recorded at three different locations on the flank face of each insert and the average value was calculated for further analysis. Similarly, the roughness of the turned part was measured with the help of a roughness testing device namely Talysurf (Model: Taylor Hobson, Surtronic *3+*). The values of arithmetic surface roughness were also noted at three different locations around the circumference of the machined surface for better accuracy and the average value was considered as final R_a during the investigation. Table 2 illustrates the experimental layout and the outcomes of the present study.

Table 1. Chemical composition of CP-Ti grade II

Element	Carbon	Nitrogen	Oxygen	Iron	Hydrogen	Titanium
Wt. (%)	0.08-0.1	0.03	0.25	0.30	0.015	99.3

Table 2. Experimental results

Run	*v* *(m/min)*	*f* *(mm/rev)*	*d* *(mm)*	VBc (μm)	Ra (μm)
1	35	0.08	0.1	73.26	1.68
2	35	0.12	0.3	66.52	1.82
3	35	0.16	0.5	80.04	1.92
4	70	0.08	0.5	112.32	1.72
5	70	0.12	0.1	84.68	1.88
6	70	0.16	0.3	92.22	1.92
7	105	0.08	0.3	132.52	2.04
8	105	0.12	0.5	224.46	2.24
9	105	0.16	0.1	192.48	2.12

2.2 Grey relational analysis (GRA)

In the present investigation, Taguchi based grey relational analysis (GRA) approach was used to determined optimal parametric combination of the process parameters in order to minimize the tool wear and surface roughness. For the purpose, the outcomes of the experiment i.e. measured values of VB_c and R_a were initially normalized between 0 and 1 by converting them into a non-dimensional quantity. Data normalization was performed using Equation (1)

$$X_i^*(k) = \frac{\max yi(k) - yi(k)}{\max yi(k) - \min yi(k)} \quad (1)$$

where, *i = 1,2…m*; *k= 1,2..n*; *m* is the total number of experimental runs and *n* is the number of output responses; *minyi(k)* and *maxyi(k)* are the minimum and maximum values of the output response under consideration. $X_i^*(k)$ is the obtained normalized value.

In the next step, the grey relational coefficient was calculated for each machining parameter related to their normalized values. This was done using Equation (2)

$$\xi_i(k) = \frac{\Delta_{\min} + \tau\Delta_{\max}}{\Delta_{0i}(p) + \tau\Delta_{\max}} \quad (2)$$

where, $\xi_i(k)$ represents the grey relational coefficient, τ is the identification coefficient (or distinguishing factor) which varies from 0 to 1 but it is generally taken as 0.5. For the sensitivity checking one can examine all the values of τ between zero to one. *Δmin* and *Δmax* are the minimum and maximum quality loss values of the normalized sequence.

Finally, the value of grey relational grade (GRG) was computed by averaging the values of grey relational coefficient corresponding to each experimental run by employing Equation (3) as shown below.

$$\gamma_i = \frac{1}{n}\sum_{k=1}^{n}\zeta_i(k) \quad (3)$$

here, γ_i represents the grey relational grade for the corresponding sequence.

Thus, a preference order were assigned to each experimental run by arranging the GRG values in descending order. The experimental run, corresponding to the highest d_G value is referred as the best parametric setting to acquire the desired output.

3. Results and discussion

Initially, all the data sets corresponding to each experimental run were normalized using Equation (1) in order to obtain a dimensionless quantity between 0 and 1. Table 3 depicts the normalized data matrix along with quality loss function for both the selected responses.

Table 3. Normalized data and quality loss function

Run	Normalized Value		Absolute Value (Delta)	
	VB_c	R_a	VB_c	R_a
Ideal Sequence	1	1	1	1
1	0.957	1.000	0.043	0.000
2	1.000	0.750	0.000	0.250
3	0.914	0.571	0.086	0.429
4	0.710	0.929	0.290	0.071
5	0.885	0.643	0.115	0.357
6	0.837	0.571	0.163	0.429
7	0.582	0.357	0.418	0.643
8	0.000	0.000	1.000	1.000
9	0.202	0.214	0.798	0.786

Both the performance characteristics i.e. tool wear and surface roughness were characterized as "smaller-is-better" which necessitates minimum values of the same. Thus, in the next step, grey relational coefficients were calculated by considering equal weights for both the responses with the help of Equation (2). Finally, the grey relational grades (GRG) were determined using equation (3) and listed in Table 4 which indicates all the designated cut qualities of the work material. In this way, multi-objective optimization problem has been converted into a single objective. The parametric combination related to the highest grey relational grade is referred as the best combination within the studied range of process parameters. From the table, it is revealed that experiment number 1 has highest grey relational grade and hence the corresponding parametric setting i.e. cutting speed = 35 m/min, feed rate = 0.08 mm/rev and depth of cut = 0.1 mm can be said as the best combination to attain minimum tool wear and surface roughness. However, this might be limited to the studied range of process variables and the selected machining conditions. At the end, a mathematical model was developed with the help multiple regression analysis (MRA) to exhibit the suitability of the proposed methodology. The following equation shows the relationship between input parameters and GRG (Equation 4):

$$GRG = 1.3743 - 0.0061\times v - 1.9835\times f - 0.2188\times d \quad (4)$$

Table 4. Grey relational grade

Run	Grey relational coefficients		Grey relational grade(GRG)	Rank
	VB_c	R_a		
1	0.921	1.000	0.961	*1*
2	1.000	0.667	0.833	2
3	0.854	0.538	0.696	5
4	0.633	0.875	0.754	3
5	0.813	0.583	0.698	4
6	0.754	0.538	0.646	6
7	0.545	0.438	0.491	7
8	0.333	0.333	0.333	9
9	0.385	0.389	0.387	8

Figure 1 shows the comparison between calculated and predicted GRG values for each experimental run. It is evidently visualized from the figure that, the predicted values are in good agreement with calculated one. Moreover, predicted values of GRG were observed to be higher for most of the experimental runs. Thus, it can be recommended that, MRA coupled with GRA can be successfully employed for solving multi-objective optimization problems in various academic as well as research institutions within the specified range of process variables.

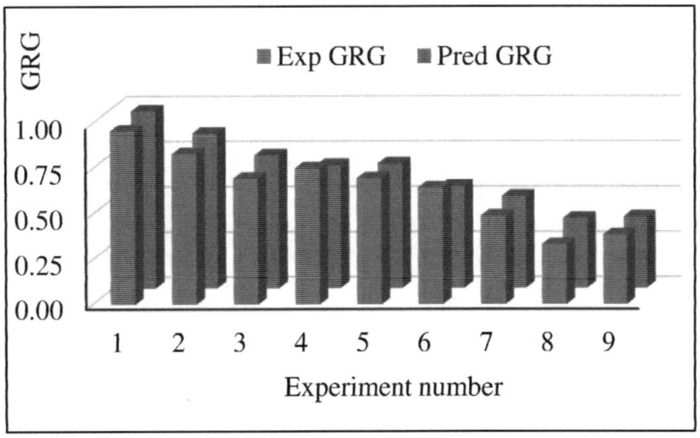

Figure 1. Comparison graph between predicted and calculated GRG values

Finally, the adequacy of the regression model was checked by performing analysis of variance (ANOVA) test. The test was conducted for a significance level of 95% which implies that, if the values of probability of acceptance (*P*-value) is ≤ 0.05, then the model is said to be statistically significant. In addition to that, influence of each process parameter was assessed by calculating their percentage contribution on developing the model for predicting GRG. Table 5 illustrates the results of ANOVA test of the current investigation. From the table, it is clearly visualized that the cutting speed is the most influencing machining variable followed by feed rate on GRG which explains 80.16% and 11.12% contribution respectively. On the contrary, depth of cut was found to be insignificant parameter as the P-value related to it, is equal to 0.1344 which is more than 0.05.

Table 5. Results of ANOVA test

Source	DOF	SS	MS	*F*-value	*P*-value	% Contribution
Regression	3	0.3217	0.1072	29.7256	**0.0012**	
v	1	0.2724	0.2725	75.5235	**0.0003**	80.16
f	1	0.0378	0.0378	10.4688	**0.0230**	11.12
d	1	0.0115	0.0115	3.1845	0.1344	3.38
Error	5	0.0180	0.0036			5.29
Total	8	0.3398	R-Sq = 94.69%			100.00

4. Conclusions

The present paper proposes GRA method to acquire an optimal parametric combination when machining CP-Ti grade II in dry cutting environment. On the basis of the experimental results attained during the investigation, the following conclusions may be drawn.

1. The highest grey relational grade of 0.961 was noticed for experimental run 1 accordance of which the optimal combination of machining variables are cutting speed = 35 m/min, feed rate = 0.08 mm/rev and depth of cut = 0.1 mm.
2. A close relationship was seen between the predicted and calculated GRG values which describes the adequacy of the MRA approach.
3. Cutting speed was observed as the most influencing machining parameter on GRG with a percentage contribution of 80.16%, followed by feed rate which described as percentage contribution of 11.12%.
4. The suggested methodology can be effectively implemented for solving multi-objective optimization problems associated with various real time manufacturing systems. However, it might be limited to the specified range of process variables and cutting conditions.

References:

1. Bhaumik, S., C. Divakar, and A. Singh, (1995) *Materials & Design*. **16**(4): p. 221-226.
2. Machado, A. and J. Wallbank (1990), *Proceedings of the Institution of Mechanical Engineers, Part B: Journal of Engineering Manufacture*. **204**(1): p. 53-60.
3. Yang, X. and C. Richard Liu (1999), *Machining Science and Technology*. **3**(1): p. 107-139.
4. Adams, K. (1994) *Titanium 1994 Products and Applications Proceedings of the Technical Program from the 1994 International Conference*.
5. Sahoo, A. (2014), *International Journal of Industrial Engineering Computations*. **5**(2): p. 295-304.
6. Asiltürk, I. and M. Çunkaş (2011), *Expert Systems with Applications*. **38**(5): p. 5826-5832.
7. Kant, G. and K.S. Sangwan (2015), *Procedia CIRP*. **31**: p. 453-458.
8. Khan, A. and K. Maity (2017), *International Journal of Materials and Product Technology*, **54**(4): p. 306-331.
9. Khan, A. and K. Maity (2016),*International Journal of Engineering Research in Africa*, 2016. **23**.
10. Khan, A. and K. Maity (2016), *International Journal of Engineering Research in Africa*, 2016. **22**: p. 94.
11. Majumder, H. and K. Maity (2017), *Journal of Advanced Manufacturing Systems*. **16**(02): p. 81-99.
12. Majumder, H., et al. (2017), *Measurement*. **107**: p. 19-30.
13. Saha, A. and H. Majumder (2016). *International Journal of Engineering Research in Africa*. Trans Tech Publ.

Adsorption Studies of Gadolinium ion on Graphitic Carbon Nitride

S K Kuila*[1] and T K Kundu[1]
[1]Department of Metallurgical and Materials Engineering, Indian institute of Technology Kharagpur, West Bengal, India-721302
*E-mail: saikatkuila@gmail.com

Abstract:
Bulk graphitic carbon nitride (g-C_3N_4) is synthesized by thermal decomposition of urea and used as an adsorbent for gadolinium ion (Gd^{3+}) from aqueous solution. Adsorption capacity of g-C_3N_4 is found to be influenced by initial Gd^{3+} concentration, solution pH and contact time. Adsorbed $Gd^{3+}$is separated from g-C_3N_4 by ultracentrifuge. Initial and Gd ion accumulated g-C_3N_4 adsorbent are characterized by X-ray diffraction technique (XRD) for phase identification, UV-visible and Fourier transform infrared (FTIR) spectroscopy for adsorption characteristics and optical property, scanning electron microscopy (SEM) for morphological behaviour along with energy dispersive X-ray spectroscopy (EDS) for elemental study. HNO_3 (0.1M), NaOH (0.1M) and de-ionized water are used for desorption and around 97% quantitative recovery of Gd ion is observed.

1. Introduction

Rare earth elements have an emerging application in magnet, catalyst, electronic, functional glass, battery and other advanced technological applications [1-2]. G-C_3N_4 is a metal free tris-triazine based polymer having good thermal stability, good electronic property, basic surface functionalities and H-bonding motifs [3]. G-C_3N_4, as polymer based adsorbent is used for separation and recovery of different heavy metal ions such as Pb^{2+} ,Cu^{2+}, also graphene oxide, a carbon based compound is used for different rare earth ions adsorption [4-5]. Effect of controlling parameters on adsorption capacity, such as, solution pH, ion concentration, temperatures and contact times have also been studied [6-7]. Many separations and purification techniques, such as, ion exchange, solid phase extraction, liquid liquid extraction, nano filtration etc. are have been used to separate or extract the rare earth ions from aqueous solution [8-10]. The route of bulk g-C_3N_4 synthesis is polymeric condensation of different precursors such as melamine, dicynamide, urea and thiourea etc. at different temperatures [11-13]. Adsorption capacity and photocatalytic effects with different physical, optical, structural, thermal and surface properties of g-C_3N_4 and modified (hybridized) g-C_3N_4 are characterized by different techniques; XRD, UV-vis spectroscopy, FTIR, TG-DSC, XPS and BET etc. [11-17]. Electronic properties of the CO_2 adsorbed g-C_3N_4 are presented computationally including band gap and density of states [18]. In our present study g-C_3N_4 has been synthesized by thermal decomposition of urea. Adsorption study of rare earth ion (Gd^{3+}) on g-C_3N_4 is presented in our paper which is not reported any earlier literature. The initial and Gd^{3+}accumulated g-C_3N_4 adsorbent are characterized by X-ray diffraction technique (XRD) for phase identification, UV-visible and Fourier transform infrared (FTIR) spectroscopy for adsorption characteristics and optical property and scanning electron microscopy (SEM) for morphological behaviour along with energy dispersive X-ray spectroscopy (EDS) for elemental study. Effects of controlling parameters (initial concentration, solution pH and contact times) on adsorption capacity are discussed in batch wise experiment.

2. Materials and methods

G-C_3N_4 is synthesized by heating of synthetic urea (Merck, 99.9% purity). 100 mg urea is heated to 560°C in programmable muffle furnace (Pt-Rh thermocouple) in three different segments (heated to 250°C and kept it hold for 1 hr, and then heated to 350°C and kept it hold for 2 hrs and finally heated to 560°C and kept it hold for 2 hrs) where the intermediate heating rate is maintained at 2°C/min. Bulk g-C_3N_4 powder is collected after furnace cooling. 0.01gm of high purity analytical grade

$Gd(NO_3)_3.6H_2O$ (Alfa Aesar, 99.9% purity) is added with 1 litre of deionised water and the solution pH is obtained as 6. $Gd(NO_3)_3$ solution (at pH = 6) along with adsorbent g-C_3N_4 (at constant concentration of g-C_3N_4 (0.02 g/L)) is stirred for 5 to 180 min contact times at room temperature. The effect of different doses of bulk g-C_3N_4 powder (0.02 g/L, 0.1 g/L, 0.2 g/L and 0.3 g/L) in $Gd(NO_3)_3$ solution at constant pH value 6 is also studied for fixed contact time of 40 min. The effect of different solution pH values (from 4 to 10) on adsorption capacity of constant concentration of g-C_3N_4 (0.02 g/L) for $Gd^{3+}$is also studied for 40 min contact time. The solution pH values are adjusted by adding nitric acid (HNO_3) and sodium hydroxide (NaOH) into the solution. After the stirring each set of solution along with adsorbent is sonicated for 2 hours. After completion of adsorption process, Gd^{3+} along with g-C_3N_4 is separated by centrifuge (8000 rpm, 20 min). Adsorption capacity (q) is measured by using the following formula [4].

$$q = \frac{(Initial\ Gd\ concentration\ \text{-}Final\ Gd\ concentration\ ) \times Total\ aqueous\ volume}{Mass\ of\ the\ g-C3N4} \quad (1)$$

Recovery of Gd^{3+} ion is calculated by desorption experiment. Desorbing solution is prepared by using HNO_3 (0.1M), NaOH (0.1M) and de-ionized water. Gd^{3+} accumulated g-C_3N_4 is mixed with 30 mL of the desorbing solution. Whole sample is stirred for 60 min at room temperature. The sample is separated by ultracentrifuge (8000 rpm, 20 min) and final adsorbed concentration is measured by atomic absorption spectrometer (AAnalyst 200, Perkin Elmer). Recovery percentage is calculated according to the following equation [5].

$$\text{Recovery}\ (\%) = \frac{(concentration\ \times volume\ of\ the\ desorbing\ solution\ )\ at\ desorption}{(concentration\ \times volume\ of\ the\ initial\ solution\ )\ at\ adsorption} \times 100 \quad (2)$$

Initial and Gd ion accumulated g-C_3N_4 are identified by a Bruker D8 Advance X-ray diffraction (XRD) unit using Cu-Kα radiation. A continuous scan mode from 10° to 60° is used to collect data from a 0.02 mm sample width and 1° min^{-1} scan rate. Ultraviolet (UV)-visible (vis) absorption spectra of the g-C_3N_4, adsorbent along with $Gd^{3+}$is recorded by 'Orbit' single beam visible spectrophotometer. A Fourier-transform infrared spectroscopic (FTIR) analysis is carried out by using the KBr pellet technique in range of 4000 to 400 cm^{-1} in Thermofisher scientific, Nicolet 6700 FTIR spectroscope. The morphology of g-C_3N_4 adsorbent, adsorbent along with $Gd^{3+}$is studied by Zeiss-Merlin EVO 60 scanning electron microscope with Oxford EDS detector using secondary electron mode. The microscope works with tungsten filament and maximum acceleration voltage of 30kV. For making the powder sample conductive, sputter coater of gold-palladium target (make: POLARON, model: SC7620) was used. Area scanning and microanalysis of different EDS spots with elemental study are analyzed by energy dispersive X ray spectroscopy (EDS).

3. Result and Discussion

3.1. Structural, physical and morphological characterization of adsorbent along with adsorbed Gd^{3+} ion

X ray diffraction pattern of the initial and Gd ion accumulated g-C_3N_4 is presented in figure 1. The peaks identified for g-C_3N_4 is at 2θ=27.1° and 13° with the corresponding planes (002) and (100) respectively, determined by comparing with JCPDS standard. In (g-C_3N_4+ Gd^{3+}) sample, g-C_3N_4 peak is convoluted with $Gd(NO_3)_3$ peak at 2θ=29° and shifted towards right due to adsorption of Gd^{3+} ion onto the g-C_3N_4 surface. Ultraviolet-visible (UV-Vis) spectra of both adsorbent and Gd^{3+} accumulated adsorbent are presented in figure 2. Due to the interaction of Gd^{3+} with g-C_3N_4 absorbance peak at 440 nm of pristine g-C_3N_4 is shifted to 435 nm for Gd^{3+} absorbed g-C_3N_4. FTIR spectra of g-C_3N_4 are showing major chemical bands within 800 cm^{-1} to 1700 cm^{-1} wavenumber as shown in figure 3. Among three major spectrums, 1280 cm^{-1} and 1632 cm^{-1} are due to aromatic heptazine derive C-N stretching bonds (sp^3) [12,13,15] and 3396 cm^{-1} due to N-H stretching sp^2 bond [12,17]. 810 cm^{-1} and 883 cm^{-1}found due to tris-triazine characteristic breathing mode and N-H deformed mode respectively [13, 16, 17]. The broad peak 2900 cm^{-1} to 3600 cm^{-1} is for residual free N-H stretching and physical absorption of moisture on g-C_3N_4 surface in O-H form [12, 13, 17]. Gd ion accumulated g-C_3N_4 has not shown any drastic change in their characteristic peaks. It indicates that the adsorption occurred due to physical bonding. Scanning electron microscopy (SEM) images of g-C_3N_4 and Gd^{3+} adsorbed g-C_3N_4 are shown in figure 4.(a) and figure 5.(a) respectively. Morphological behaviour and surface is not changed much before and after the adsorption. Only the Gd ions are adsorbed physically on g-C_3N_4 surface which is clearly shown in (Gd accumulated adsorbent) figure 5.(a). Energy dispersive X-ray spectra (EDS) of corresponding SEM are shown in figure 4.(b) and figure 5.(b). Elemental contributions from EDS study are presented in weight % and atomic % in table no.1.

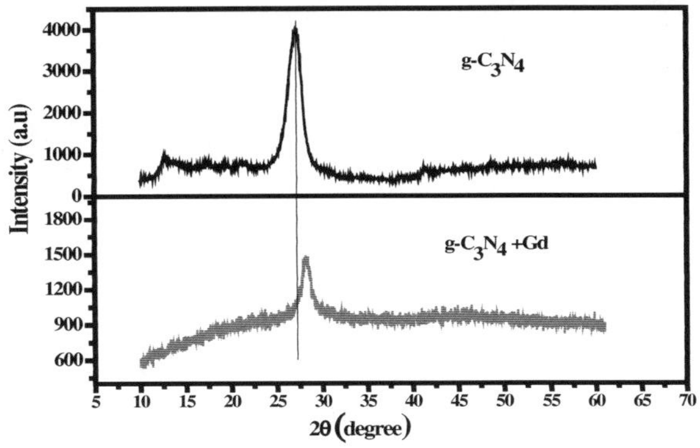

Figure 1. X-ray diffraction pattern for adsorbent g-C_3N_4 along with Gd adsorbed g-C_3N_4 [solution pH = 6, g-C_3N_4 concentration = 0.02 g/L at room temperature, stirring time = 40 min]

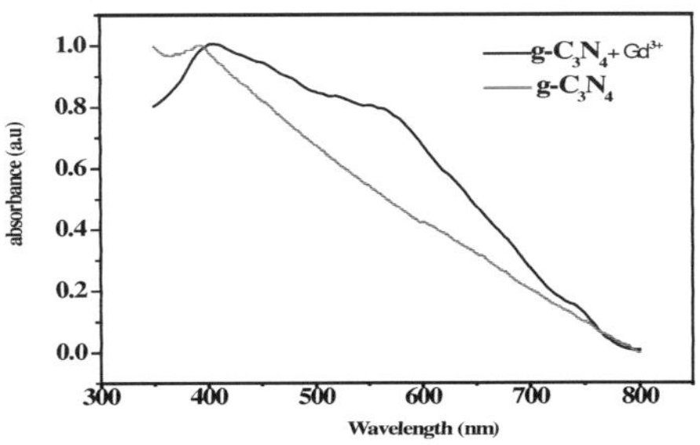

Figure 2. UV-Vis absorption spectrum for adsorbent g-C_3N_4 along with Gd adsorbed g-C_3N_4 [solution pH=6, g-C_3N_4 concentration = 0.02 g/L at room temperature, stirring time = 40 min]

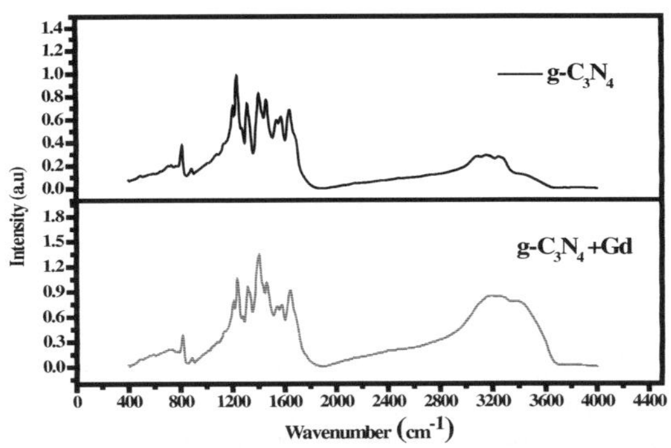

Figure 3. FTIR image for adsorbent g-C_3N_4 along with Gd accumulated g-C_3N_4 [solution pH = 6, g-C_3N_4 concentration = 0.02 g/L at room temperature, stirring time = 40 min]

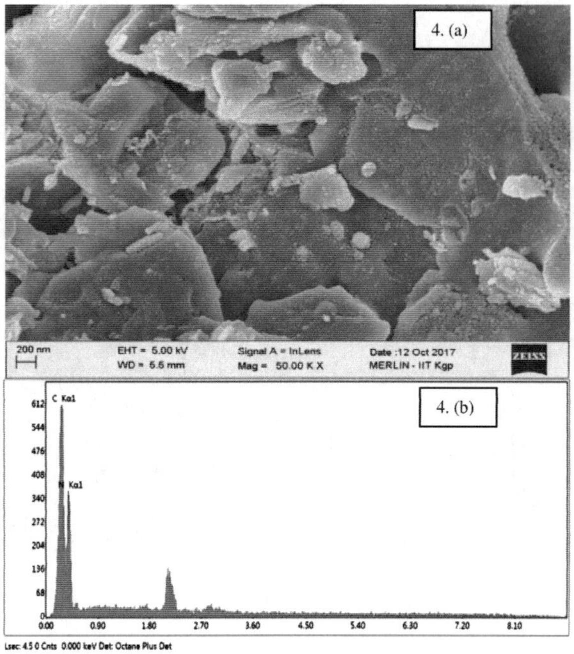

Figure 4.(a) Scanning Electron microscopy image for adsorbent g-C_3N_4. **4.(b)**Corresponding energy dispersive X-ray spectra (EDS). [Solution pH = 6, g-C_3N_4 concentration = 0.02 g/L at room temperature, stirring time = 40 min]

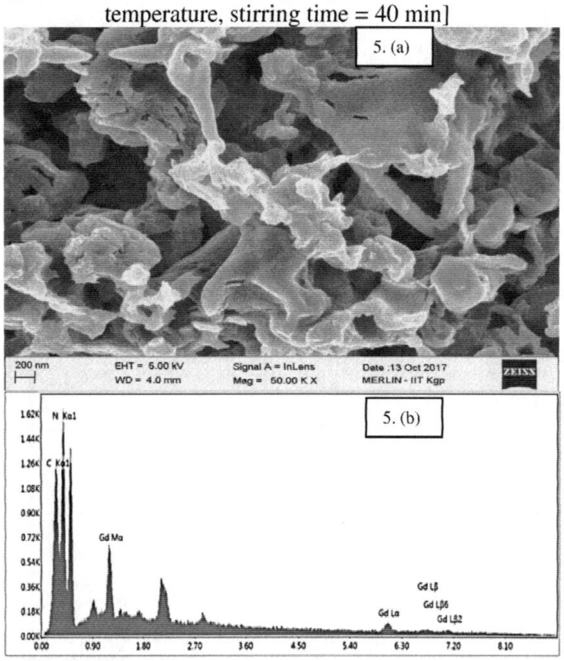

Figure 5.(a) Scanning Electron microscopy image for adsorbent g-C_3N_4 along with (Gd^{3+}) **5.(b)** Corresponding energy dispersive X-ray spectra (EDS). [Solution pH = 6, g-C_3N_4 concentration = 0.02 g/L at room temperature, stirring time = 40 min]

Table1. Energy dispersive X-ray spectroscopy results after and before the adsorption

$g\text{-}C_3N_4$+ Gd^{3+} (after adsorption)			$g\text{-}C_3N_4$ (before adsorption)		
Element	Weight %	Atomic %	Element	Weight %	Atomic %
C	30.29	36.72	C	48.56	52.40
N	60.02	62.38	N	51.44	47.60
Gd	9.69	0.90			

3.2. Effect of initial contact time and concentration at fixed solution pH

Adsorption capacities are calculated by using eq.1 at different contact times at constant pH value 6 and fixed concentration of $g\text{-}C_3N_4$ dose (0.02 g/L) in $Gd(NO_3)_3$ solution and the same are shown in figure 6. (a). Adsorption reached at equilibrium after 40 min of stirring. The effect of adsorbent doses (0.02 g/L, 0.1 g/L, 0.2 g/L and 0.3 g/L) in $Gd(NO_3)_3$ solution at constant pH value 6 for 40 min contact time is shown in figure 6.(b). The adsorption capacities are decreasing by the increasing of initial doses. Recovery percentages are increasing from 94.85 to 97.2% by increasing the initial dose of $g\text{-}C_3N_4$ as surface area are increasing with increasing $g\text{-}C_3N_4$ is shown in figure 7.(a). When the available binding sites are more with respect to less Gd^{3+} ion in solution then the adsorption capacities are decreased with the increasing dose.

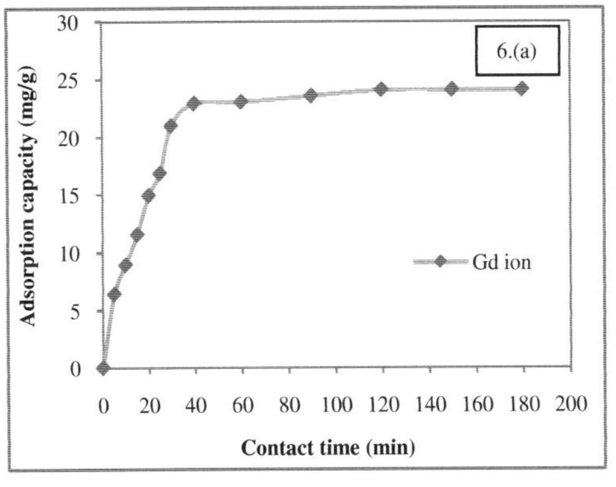

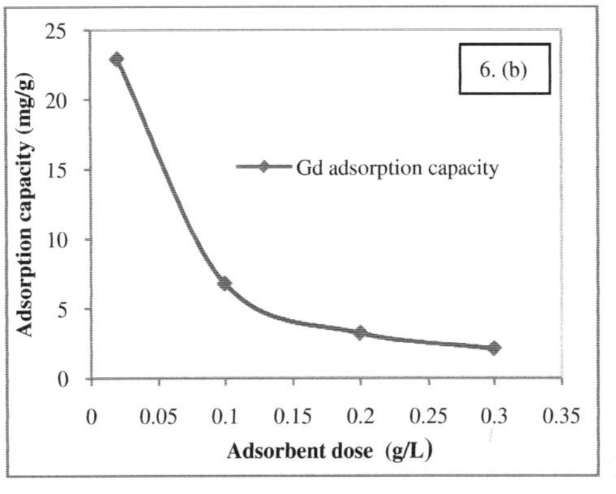

Figure 6.(a) Effect of contact times [solution pH = 6, concentration of $g\text{-}C_3N_4$ = 0.02 g/L at room temperature] **6. (b)** Effect of different concentrations [solution pH = 6, contact time = 40 min] of $g\text{-}C_3N_4$ on the adsorption capacities of Gd^{3+}

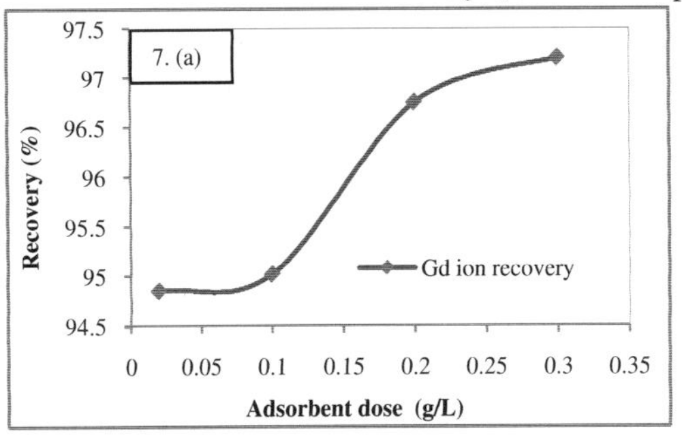

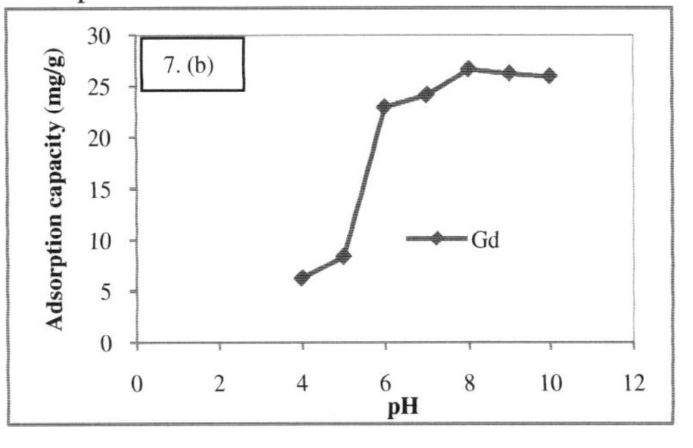

Figure 7.(a) Effect of different concentrations [solution pH = 6, contact time = 40 min] of $g\text{-}C_3N_4$ on the Gd^{3+} recovery percentages. **7. (b)** Effect of solution pH on the adsorption capacity of Gd^{3+} on $g\text{-}C_3N_4$ [concentration of $g\text{-}C_3N_4$ =0.02 g/L at room temperature, contact time = 40 min]

3.3. Effect of solution pH at fixed concentration of g-C_3N_4 at fixed contact time

The solution pH values are important parameters for adsorption capacity of Gd^{3+} onto the g-C_3N_4 adsorbent surface. Adsorption capacities at different pH values at fixed concentration of g-C_3N_4 (0.02 g/L) at room temperature for 40 min stirring are shown in figure 7. (b). Adsorption capacities are increasing by increasing of pH values from 4 to 8. At low pH values there are excessive additions of H^+ protons on nitrogen surface. By increasing the initial pH values the deprotonation rate is also increases up to 8. So the adsorbent surface becomes more negative up to that certain pH value of 8. The electrostatic interaction between Gd^{3+} and g-C_3N_4 plays an effective role to increase the adsorption capacity. So up to a certain limit higher pH values are always favourable in adsorption study.

4. Conclusion

Different experimental findings indicate that g-C_3N_4 has good adsorption capacity and very effective in adsorption of Gd^{3+}. Structure and morphology before and after the adsorption are characterized by XRD, UV-vis, FTIR and SEM-EDS. Morphological behaviour and surface structures are not changed much before and after the adsorption. Maximum adsorption capacity is obtained at room temperature after 40 min of stirring. FTIR study confirms that the Gd^{3+} ions are physically adsorbed which is also compared with other literatures. In FTIR there is not much change in characteristic peaks after the adsorption. These results show that the sample is stable after the adsorption and can be recovered up to 97% after the desorption experiment.

References

[1] Habashi F 1977 *Willey* **10** 1695-1743.
[2] Krishnamurty N K and Gupta C K 2015 *Taylor and Fransis CRC press* 33-70.
[3] Wena J, Xiea J, Chenb X and Lia X 2017 *Applied Surface Science* **391** 72–123,
[4] Anbia M and Haqshenas M 2015 *Int. J. Environ. Sci. Technol.* **12** 2649–64.
[5] Ashoura R, Abdelhamidc N H, Abdel-magied F A, Ahmed A, Abdel-khalekd M, Alib M, Muhammeda U M, Zouc X and Dutta J 2017 *Solvent extraction and ion exchange* **35** 91–103.
[6] Shen C , Chen C , Wenb T, Zhao Z, Wangb X and Xu A 2015 *Journal of Colloid and Interface Science* **456** 7–14.
[7] Zhu L, You L, Wang Y and Shi Z 2017 *Mater. Res. Express* **4** 75606-16
[8] Jain V.K, Pandya R. A, Pillai S. G, Agrawal Y. K. and Kanaiya P. H 2007 *Journal of Analytical Chemistry* **62** 104–12.
[9] Ashoura R M, Ahmed F A M, Ahmed A. Abdel-k, O.S. Helalya and Alia M.M. 2016 *Journal of Environmental Chemical Engineering* **4** 3114–21.
[10] Romanchuk A. Y, Kuzenkova S, Slesarev S, Tour M and Kalmyk S. N 2016 *Solvent extraction and Ion exchange* **34** 594–602.
[11] Xu H Y, Wu L C, Zhao H, Jin L G and Qi S Y 2015 *PLOS ONE* **10(11)** 1-20.
[12] Yan S C, Li Z S and Zou Z G 2009 *Langmuir* **25(17)** 10397–401.
[13] Zhang J, Zhang M, Zhang G, and Wang X 2012 *ACS Catal* **29** 40–8.
[14] Zhu B, Xia P, Ho W and Yua J 2015 **344** 188–95.
[15] Chen Y, Li J, Hong Z, Shen B, Lin B and Gao B 2014 *Phys. Chem. Chem. Phys* **16** 8106-13.
[16] Wang P, Wang Z, Jiaa L and Xiaoa Z 2009 *Phys. Chem. Chem. Phys* **11**, 2730–40.
[17] Dong F, Wanga Z, Sun Y, Ho W K and Zhang H 2013 *Journal of Colloid and Interface Science* **401** 70–9.
[18] Zhua B, Zhanga L, Xua D, Chenga B and Jiaguo Y 2017 *Journal of CO_2 utilization* **21** 327-35.

IOP Conf. Series: Materials Science and Engineering **338** (2018) 012007 doi:10.1088/1757-899X/338/1/012007

Pore Structure Characterization in Concrete Prepared with Carbonated Fly Ash

Sanjukta Sahoo*

Head, Department of Civil Engineering, KIIT Polytechnic, KIIT (Deemed to be University), Bhubaneswar: 751024, Odisha, India
*E-mail: sanjukta.sahoo7@gmail.com

Abstract. Carbon dioxide capture and storage (CCS) is a technique to address the global concern of continuously rising CO_2 level in the atmosphere. Fly ash is considered as a suitable medium for CCS due to presence of metal oxides. The fly ash which has already sequestered carbon dioxide is referred to as carbonated fly ash. Recent research reveals better durability of concretes using carbonated fly ash as part replacement of cement. In the present research pore structure characterization of the carbonated fly ash concrete has been carried out. Mercury Intrusion porosimetry test has been conducted on control concrete and concrete specimens using fly ash and carbonated fly ash at replacement levels of 25% and 40%. The specimens have been water cured for 28 days and 90 days. It is observed that porosity reduction rate is more pronounced in carbonated fly ash concrete compared to control concrete at higher water curing age. Correlation analysis is also carried out which indicates moderately linear relationship between porosity % and pore distribution with particle size and water curing.

1. Introduction

Recently the CO_2 concentration in atmosphere already crossed the value of 403.38 PPM [1] and is projected to touch the value of 800 PPM by 21st century. The predicted consequences of rising CO_2 level in the atmosphere include dramatic change in climate, leading to predictions of devastating consequences for both human and biota [2]. This motivated the research world to think about a safe and suitable strategy to mitigate the challenge. Strategies to reduce the CO_2 content of atmosphere include limiting the CO_2 production, adopting fuels having less carbon, enhancing the use of alternate energy sources [3, 4], utilising the produced CO_2 in different fields and capturing and storing the CO_2 safely through various technology. CCS (carbon capture and carbon storage) is one of the most promising approaches to reduce the carbon dioxide (CO_2) content in the atmosphere and thus the greenhouse effect [5-7]. Mineral carbonation comes under the chemical storage of CO_2, where both carbon capture and carbon storage is being done by solid carbonation. It is one of the most promising options for the permanent CO_2 fixation by forming stable carbonate mineral such as calcite, magnesite, siderite and dolomite etc. Fly ash (FA), the residue of coal power plant, has abundant availability and toxic nature. Being a hazardous pollutant, it is a great threat to the society. This needs safe disposal or productive reutilisation. FA after sequestering carbon dioxide is referred to as carbonated FA. Recently researches undertaken by the author to utilize carbonated FA as part replacement of cement in concrete have established the relevance of the cementitious nature exhibited by the material. Research indicates better alkali, acid and sulphate resistance by concrete which uses FA [8] and carbonated FA as part replacement of cement [9]. Porosity is one of the major parameters of the concrete influencing its behaviour related to both strength and durability. Pore structure is one of primary deciding factors for permeability [10] of the concrete, which influences the aggressive ingress of deteriorating ions and thus may be used as a predictor of structural durability. Concrete has a complex internal pore system. The capillary pores arising from cement particle spacing in the dispersion, smaller gel pores in inter layers of the C-S-H or other hydrate products, air voids and porous zone at aggregate cement interface [11] constitute the pore system.

The objective of the present research is to characterize the pore structure of carbonated FA concrete. The characterization will include determination of porosity % and pores distribution. The objective also includes finding the existence of correlation of these parameters with water curing age, proportion of the FA and carbonated FA used in the concrete.

2. Experimental Program

In this research work three cementitious materials such as commercially available ordinary Portland cement (43 grade as per IS: 8114-1978), low Calcium FA (ASTM Class F) from national thermal power corporation (NTPC), Kaniha, Odisha and carbonated fly ash (CFA), prepared through mineral carbonation of FA through atmospheric carbonation ([9], [12]) are used. Crushed granite (20mm maximum size) and river sand conforming to zone III of nominal maximum size of 4.75mm were used as coarse aggregate and fine aggregate respectively in the concrete mixture. Commercially available poly-carboxylic based super plasticizer and normal tap water was used in the experiment. The specific gravities of cement, FA and CFA were found as 3.1, 2.33 and 2.31, respectively. Their particle-size was analysed using Mastercizer-2000 (Malvern Instruments, Malvern, U.K.). Their respective granulometric data are also reported in Table 1.

Table 1: Granulometric data for cement, fly ash, carbonated fly ash and their size fractions

Constituent	D10 (μm)	D50 (μm)	D90 (μm)	Mean size (μm)
Cement	1.860	8.939	31.946	16.79805
Fly Ash	0.719	5.914	24.714	13.09198
Carbonated fly ash	0.060	8.329	32.416	15.89338

The mix design calculations for M30 grade concrete were performed as per IS: 10262 (BIS 2009). Total five types of concrete were prepared in the laboratory for investigation in accordance with IS: 516-1959. Control concrete (CC) was made using 100% cement as binding material and two set (Low and High) of fly ash concrete (LFC and HFC) as well as carbonated fly ash concrete (LCFC and HCFC) were cast using FA and CFA at cement replacement level of 25% and 40%, respectively. A constant water/ binder ratio of 0.38 with slump value of around 100 mm was used in the experiment. The total 24 numbers of concrete cubes (150 mm) from each type were cast to perform the compressive strength, durability and pore structure characterisation test. After 24 h, they were removed from the mould and kept in the water curing chamber for curing period of 28 and 90 days. To conduct the compressive strength test of the concrete six numbers of cubes from each type were taken out of the curing chamber and dried in shade at the end of 28 and 90 days of water curing. The test was carried out in a compressive testing machine of 2000KN capacity (CTM Digital) in the concrete laboratory of SCE, KIIT University at the loading rate of 0.2-0.4 $N/mm^2/s$. To determine the durability characteristics of the concrete against chemical attack, each type of concrete cubes were allowed to immerse in three chemicals such as 5% Sodium chloride, 5% Sodium sulphate, and 1% Sulphuric acid at the end of 28 and 90 days of water curing. After 120 days of chemical curing the compressive strength of the concrete were measured following the similar procedure. The compressive strength value was considered as the average of strength value of three samples. For the pore structure characterisation, slices of 8 mmϕ and 2mm thickness concrete specimens have been obtained from the core of the 28 and 90 days water cured concrete cube. The specimens were immersed in ethyl alcohol for eight days and dried in oven at 60^0C for mercury intrusion porosimeter (MIP) test by Quantachrome Poremaster.

3. Experimental Results

3.1 Particle size distribution:

From the granulometric data along with their size fractions as presented in Table 1 it is observed that the D_{90} and D_{50} values of all the samples are ranging between 24.741μm - 32.416μm and 5.914μm -

8.939μm, respectively, hence all the samples have the capacity of adequate hydration as all have the mean particle size less than 45μm [13, 14]. Both the FA and carbonated FA have finer particles than the cement [15] and the smaller particles add to fineness and strength on hydration as the pozzolanic reaction can be highly activated when the particle size is small.

3.2 Compressive strength of concrete under normal water and chemicals:

As presented in the Table 2, the compressive strength value for CFC though less for 28 days water curing, but with increasing water curing period it behaved in a better manner showing significant higher strength value. Further, it displayed best resistance to chemicals for both the water curing ages. Though CC performed better in initial ages, but both the FA and CFC have shown much higher resistance to chemicals compared to that of CC. This is due to the filling of smaller size pores, voids and air spaces of cement particles by the FA and carbonated FA. This results in packing of the concrete material tightly, restricting the permeation of harmful ions.

3.3 Mercury intrusion porosimetry (MIP) test:

From MIP test information on several parameters related to porosity are obtained. These parameters include total intrusion volume, total pore area, median pore radius, average pore radius, bulk density, apparent skeletal density and porosity %. The porosity % and pore size distribution are two critically studied factors which influence utility aspects of concrete specimens in structural usages. Information retrieved relating to these important parameters are tabulated in Table 2. Figure 1 and Figure 2 represent the histogram of total intrusion volume and porosity percentage in concrete specimens for 28 and 90 days of water curing period, respectively. The intrusion volume vs pore diameter plots from the machine generated data for concrete specimen water cured for 28 and 90 days have been furnished in Figure 3 and Figure 4, respectively. The pore percentage distribution in three major influencing zones for concrete specimen water cured for 28 and 90 days have been tabulated in Table 3. Figure 6 illustrates the pore percentage distribution of concrete specimens over volume, water cured for period of 28 and 90 days.

Table 2. Test results for concrete specimen water cured for 28 and 90 days

Water curing period in days	Type of concrete	Compres strength (MPa)	Compressive strength under chemical attack for 120 days (MPa)			MIP Test Results			
			Salt	Sulphate	Acid	Total intrusion volume (ml/g)	Total pore area (sqm/g)	Bulk density (gm/cc)	Porosity %
28	CC	43.62	32.35	30.89	25.99	0.0673	7.627	1.981	13.332
	LFC	39.9	34.18	34.01	28.03	0.052	6.568	1.892	9.838
	HFC	38.76	25.4	24.85	22.8	0.05	4.509	1.796	8.98
	LCFC	31.96	23.98	33.82	30.51	0.052	4.505	1.924	10.158
	HCFC	25.45	20.61	26.77	26.56	0.068	7.162	1.822	12.480
90	CC	49.36	35.94	31.41	35.64	0.0646	6.427	1.972	12.739
	LFC	51.62	47.79	40.4	43.38	0.0405	3.254	1.861	7.537
	HFC	52.01	50.07	47.23	47.39	0.0430	3.868	1.794	7.714
	LCFC	42.88	39.04	51.01	44.12	0.0420	4.642	1.931	8.110
	HCFC	38.92	35.88	44.4	43.02	0.0431	3.230	1.826	7.870

4. Discussion on MIP test results

4.1 Variation of total intruded volume and porosity %:

It can be observed from Figure 2, that FC show the best performance followed by CFC at 28 days water curing. HCFC show the highest cumulative intruded pore volume percentage as almost approaching 70%. A decrease in total intrusion volume has been observed for all the concrete specimens with increasing water curing age, the decrease of intrusion volume in CC is much low, thus indicating almost a slow release of hydrate products during 28 to 90 days water curing. However, in FC and CFC the difference in the intrusion volume is appreciably larger compared to CC. The CFC display the best performance with a higher degree release of hydrate products compared to all other concrete. This indicates accelerated chemical transition, formation and deposition of the extra hydrated product within this curing period those are responsible for densification of the CFC and FC specimens in a faster rate.

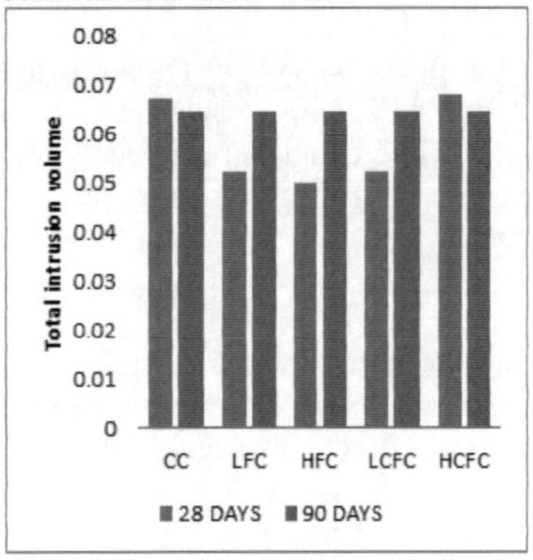

Figure 1. Histogram of total intrusion volume in concrete specimens

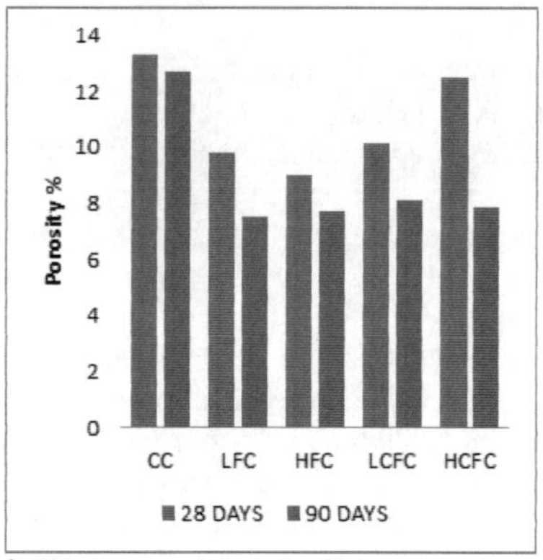

Figure 2. Histogram of porosity % in concrete specimens

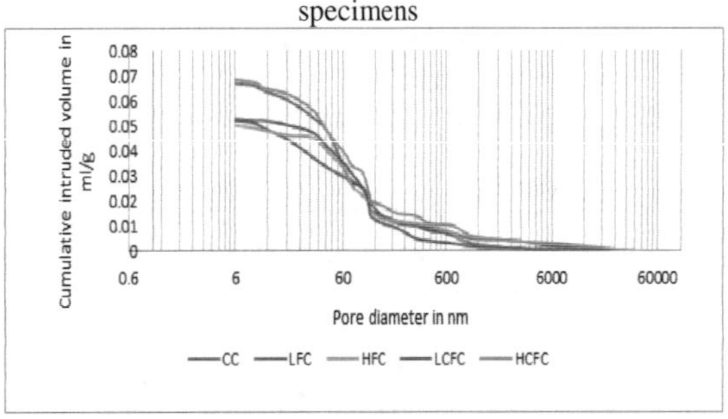

Figure 3. Cumulative intruded volume vs pore diameter for concrete specimen water cured for 28 days

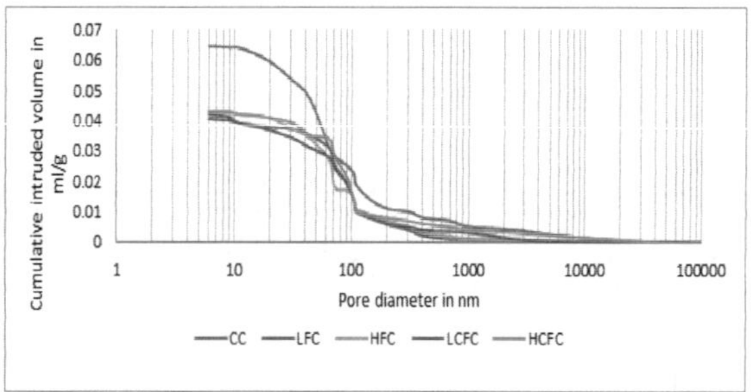

Figure 4. Cumulative intruded volume vs pore diameter for concrete specimen water cured for 90 day

From Figure 3, it can be inferred that for specimens water cured for 28 days, CC displayed the highest porosity % followed by the HCFC, LCFC and LFC, the HFC being the lowest. Low porosity associated with FC may be explained in light of denser pore structure arising from low particle size and secondary hydration of FA. With the FA addition to the cement, physically it serves as the nucleation sites for the cement hydration products. Thus, it helps the cement grain to achieve a higher degree of hydration. In addition, chemically FA particles also undergo the hydration process (pozzolanic reaction) in presence of water and $Ca(OH)_2$ thus giving the extra C-S-H gel. Due to these dual effects of FA, the extra gel formed helps in increasing the solid volume of the cement paste and adds extra solids to the concrete material. These extra solid deposited in the pore space filling the capillary pores and thus the making the microstructure packed and dense [16]. An improvement in porosity through decrease in porosity % is also noticed with increase in curing period. It is seen that the CFC performs in a better way to reduce its porosity % with water curing age in comparison to that of FA concrete and the CC. This is due to the continuous and gradual increase of hydration rate with time thus developing the microstructure more and more with the global refinement of continuous pores filling. In case of CFC along with the extra hydration products, the calcite ($CaCO_3$) present in carbonated FA, additionally fill the micro pores of the material helping dense packing of the microstructures in a higher degree.

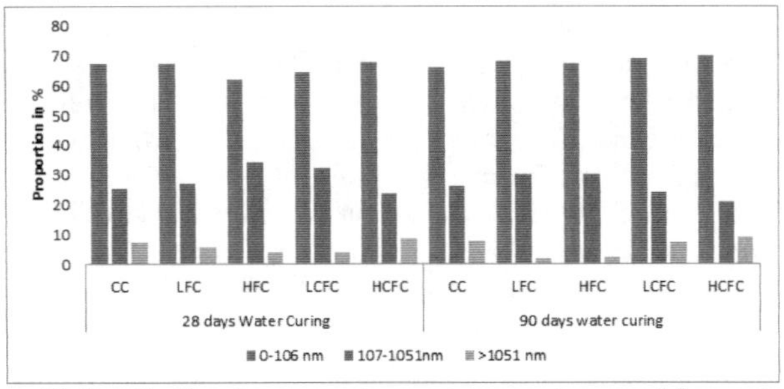

Figure 5. Pore % distribution for concrete specimen water cured for period of 28 days

4.2 Pore size distribution:

As observed from Figure 3 to Figure 5 the 28 days water cured samples, with respect to the pores in the size range from 0 to 106 nm, HCFC has the highest proportion followed by LFC, CC, LCFC and HFC. Regarding the pores in the size range of 107 to 1051 nm, HFC leads followed by LCFC, LFC, CC and HCFC. For the pore diameter > 1051 nm, HCFC has the highest proportion and LCFC, the lowest. With increase in water curing period, it is observed that pore % in larger size is also higher in CC and CFC, when compared to the FC specimens. For CC and LFC, the percentage of macro pores (107 to 1051 nm) only, increased whereas for all other concrete it decreased whereas for the mega pores of pore diameter > 1051nm, in FC only it has decreased. It is also noted that the micro pores % increases with water curing days with a faster rate in case of HFC and LCFC, and its rate is lowest for the CC. The percent of mega pores for CC also increases but for FC it is greatly reduced with curing period. It indicates the fine pores filling in case of CC is the slowest and for HFC it is the highest. It can be inferred that inclusion of FA lead to the densification of pore structure of the concrete.

Table 3. % distribution of pores in concrete specimen water cured for 28 and 90 days

Range of	28 days Water Curing					90 days water curing				
Pore sizes	CC	LFC	HFC	LCFC	HCFC	CC	LFC	HFC	LCFC	HCFC
0-106 nm	67.16	67.31	62	64.02	67.88	65.94	67.9	67.44	69.05	69.84
107-1051nm	25.37	26.92	34	32.2	23.36	26.32	29.88	30.23	23.81	20.88
>1051 nm	7.47	5.77	4	3.78	8.76	7.74	2.22	2.33	7.14	9.28

5. Correlation Analysis

To determine the existence and significance of a linear relation between dependent and independent variable, the correlation analysis is performed and a relationship between volume fraction, water curing age and porosity parameters related to the various concrete is obtained. The volume fraction denotes the ratio of volume of cementitious materials to the volume of cement in concrete specimens without any replacement. The concrete specimens prepared constitute of cement, FA and carbonated FA as cementitious materials. Assuming spherical shape of the particle, the volume of the particles can be computed using the Eqs. 1-3. V_i is the volume of any cementitious particle, whose radius is r_i, x_i is the proportion of the cementitious particle in terms of percentage and r_c is the mean particle size of the OPC 43. $VF = \sum x_i V_i / Vc;\ where,\ V_i = 4/3\pi r_i^3 and\ V_c = 100 \times 4/3\pi r_c^3$ (1)

The Pearson product moment correlation coefficient (PCC) between two parameters is calculated as ratio of the covariance to the product of standard deviations of the two variables. The PCC of porosity % with respect to the water curing age and volume fraction is provided in Table. 4

Table 4. Pearson Correlation coefficient for porosity % with Volume fraction and Water Curing Age

	Volume fraction	Water curing age	Porosity %
Volume fraction	1		
Water curing age	-9.59215E-18	1	
Porosity %	0.616162478	-0.509437784	1

The above table indicates a moderately positive correlation between volume fraction and porosity % and thus, an increase in volume fraction will result in higher porosity %. This clearly establishes that, the replacement of cement with smaller particles of FA or carbonated FA will reduce the porosity % in concrete specimen thus will reduce the permeability. Reduction in permeability results in improved resistance against the chemical attack. The experimental results of reduced strength loss in FC and CFC against acid, sulphate and chloride attack are set of the evidences. Table 5 offers the PCC relating the volume fraction and water curing age with the distribution of pore sizes. The data offer a moderate positive correlation between volume fraction and the pore % at higher size. The distribution is more towards higher size pores at higher volume fraction *i.e.* in CC. Similarly, water curing age has a moderate positive correlation with the pore size distribution favouring to smaller pores.

Table 5. Pearson correlation coefficients for pores distribution to water curing age and volume fraction

	Volume fraction	Water curing age
Volume fraction	1	
Water curing age	-9.59215E-18	1
Pore distribution 0-106 nm	0.290566975	0.533294782
Pore distribution 107-1051 nm	-0.586587493	-0.270035599
Pore distribution >1051 nm	0.681325048	-0.043190589

6. Conclusion

The conclusions drawn from the porosity test performed on the concrete specimens are outlined in the following lines. The porosity percentage at 28 days of water curing is the least for the high FC and the highest for the control concrete. However, at 90 days of water curing the porosity % remained high for control concrete whereas both the FC and CFC show almost nearly equal value with very low porosity %. With water curing age, the porosity of control concrete slightly decreased whereas the rate of decrease in case of FC was found medium and for CFC was the fastest. This may be predicted due to the high rate of hydration in case of CFC and FC in the later water curing age. Further, with water curing age the total intrusion pore volume decreases at high rate in case of CFC than that of FC and CC. The Pore distribution in CFC and FC tends to be more in the finer zone ($<10^6$nm) compared to those in control concrete. From the correlation analysis, it was concluded that a moderately positive linearity exists between volume fraction and porosity % and a moderately negative linearity exists between water curing period and porosity %.

Reference

[1] Earth's CO_2 Home Page 2017 "Atmospheric CO_2." *Mauna Loa Observatory, Hawaii* (NOAA-ESRL)

[2] Patz J A, Epstein P R, Burke T A, Balbus J M 1996 *J. Amer. Med. Assoc.* **275** 217-223

[3] Yang H, Xu Z, Fan M, Gupta R, Slimane R B, Bland A E, Wright I 2008 *J. Environ. Sci.* **20** 14-7

[4] Mikkelsen M, Jorgensen M, Krebs F C 2010 *Energy Environ. Sci.* **3** 43–81

[5] Yu K M K, Curcic I, Gabriel J, Tsang S C E 2008 *Chem. Sus. Chem.* **1** 893–899

[6] Fu C, Gundersen T 2012 *Energy Proc.* C **16** 1806–1812

[7] Leung D Y C, Maroto-Valer M M 2014 *Renewable Sustainable Energy Rev.* **39** 26–443

[8] Sahoo S, Das B B, Rath A K, Kar B B 2015 Indian J of Science and Technology **8** 72266 1-12

[9] Sahoo S, Das B B, Mustakim S 2016 *J. of Mater. Civ.Eng.*,2017, **29,** 04016242 1-12

[10] Marsh B K, Day R L, Boner D G 1985 *Cement. Concrete. Res.* **15(6)** 1027–1038

[11] Liu J, Tang K, Qiu Q, Pan D, Lei Z, Xing F 2014 *Materials* **7** 6646-6659.

[12] Muduli S D, Nayak B D, Dhal N K, Mishra B K 2014 *Greener J. Physical Sciences* **4(1)** 1-6

[13] Mehta P K 1986 *Concrete structure, properties of materials* (Prentince-Hall, New Jersy, USA)

[14] Binici H, Cagatiy I H, Shah T 2007 *Built environment* **43** 1318 – 1325

[15] Ergogdu K, Turker P 1994 *Cement concrete research* **24 (4)** 791-796

[16] Ma H 2014 *Journal of Porous Materials* **21** 207–215.

IOP Conf. Series: Materials Science and Engineering **338** (2018) 012008 doi:10.1088/1757-899X/338/1/012008

Effect of ageing time and temperature on corrosion behaviour of aluminum alloy 2014

Vikas Gadpale[1], **Pragya N. Banjare**[2] and **Manoranjan Kumar Manoj**[3]
[1, 2 and 3] Department of Metallurgical Engineering, National Institute of Technology Raipur, Raipur, India.
E-mail: vgadpale.phd2014.meta@nitrr.ac.in

Abstract. In this paper, the effect of corrosion behaviour of aluminium alloy 2014 were studied by potentiodynamic polarization in 1 mole of NaCl solution of aged sample. The experimental testing results concluded that, corrosion resistance of Aluminum alloy 2014 degraded with the increasing the temperature (150°C & 200°C) and time of ageing. Corroded surface of the aged specimens was tested under optical microscopes for microstructures for phase analysis. Optical micrographs of corroded surfaces showed general corrosion and pitting corrosion. The corrosion resistance of lower ageing temperature and lower ageing time is higher because of its fine distribution of precipitates in matrix phase.

***Index Terms*—Al Alloy 2014, Ageing, Corrosion test**

1. Introduction

It is generally accepted that the ageing treatment and precipitation hardening mechanism can be enhancing the corrosion behaviour of the Al-alloy because of the uniform distribution of fine precipitates, Phases and homogenization of microstructures [1-2]. The main affecting factors on microstructures of aluminum alloy are alloying element /chemical composition, ageing time (AT) and temperature. These factor can be also affect the corrosion behaviour of aluminum alloy [3-5]. Corrosion is the one of the important factor for failure of material. This reason researchers are studying corrosion properties of Aluminum (Al) alloy. Liu et al. reported that copper dissolves in the α-Al matrix that increases corrosion potential. After the T6 heat treatment of 2014 Al alloy, Al_2Cu phase forms which has cathodic nature relative to the α-Al solution. As a result, it increases driving force for pitting corrosion [6]. All results shown in different researchers into the field of corrosion behaviour of aluminium alloy [7-10].

2. Material and Experimental

2.1. Material

The Alloy 2014 was purchased from Perfect Metal Work Bangalore. Alloy were extruded rod of 12 mm diameter and its chemical composition is listed in Table 1.

Table 1. As Received Aluminum alloy 2014 Chemical composition

Elements	Al	Cu	Mn	Si	Mg	Fe
Weight %	Balance	5.17	1.04	1.76	0.63	0.34

IOP Conf. Series: Materials Science and Engineering **338** (2018) 012008 doi:10.1088/1757-899X/338/1/012008

2.2. Ageing treatment

This alloy rod was machined into 12 mm diameter and 10 mm length for corrosion samples. These samples are used for Heat treatment process. First, we did solution treatment (ST) of all samples at 545°C for three hours and subsequent quenched in water medium in room temperature and then ageing was done at two temperature (150°C & 200°C) for 4, 8, 12, 16 and 24 hrs. The corrosion behavior of alloys was explored foe as received and after ageing treatment samples. For microstructural analysis, specimens were first belt grinded and polished with different grade of emery papers (microgrits 220 to 1200) and finally with cloth polishing in alumina paste. After cloth polishing samples were etched with Keller's reagent for 30 seconds and washed. The metallographic study was carried out on Zeiss microscope.

2.3. Corrosion test

Corrosion tests were carried out on Versa stat 3 corrosion testing machine shown in Fig. 1. All the sample were cleaned in ethanol before the corrosion test. These cleaned sample were attached in the flat cell and the was filled with 250ml of electrolyte solution. Concentration of NaCl electrolyte was 1 mole/litre. After filling electrolyte solution in flat cell, working electrode, counter electrode and refence electrode were connected respectively with the corrosion testing sample, platinum mesh electrode and reference electrode tube. The potentiodynamic polarization test were performed with 1 mV/s scan rate. Open circuit potentials of the Al Alloy 2014 were tested in ranging from -900 mV to -300 mV with a 1 mV/s of scan rate. After corrosion test corrode surface structure study was carried out on Zeiss microscope.

2.4. Hardness Test

Hardness test of as received sample, and (4, 8, 12, 16, 20 & 24 hrs.) aged samples are tested in Shimadzu Micro Vickers hardness test machine in 1Kgf load. Each sample have 5 hardness reading and the average of these are presented in graphical from in Fig. 2.

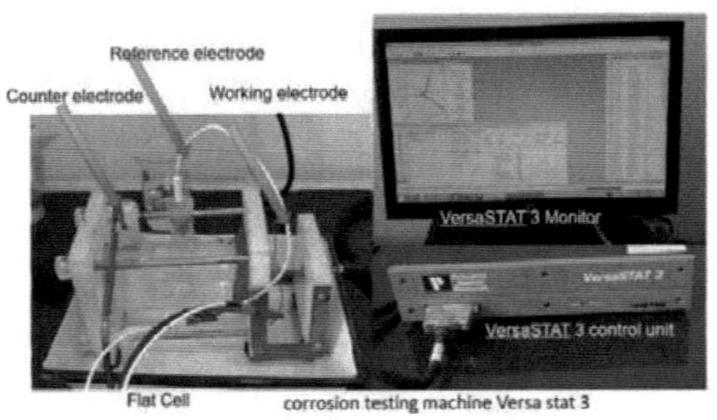

Figure 1. Versa stat 3 corrosion testing machine.

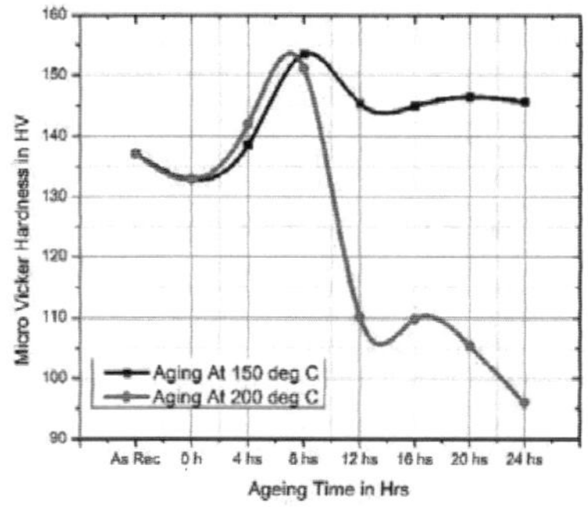

Figure 2. Vickers micro hardness of aluminum alloy as received, ST and aged samples

3. Results and Discussion

3.1. Metallographic Investigation

Microstructures of Al Alloy 2014 in as received, ST and aged conditions were taken in 500X magnification have been shown in Fig. [3-4]. In Fig. 3 we can see the elongated grain structures in of as received sample, due to the extruded condition. After ST the elongated grains were converted into uniform grains that are shown in Fig. 3-ST. The phases present in microstructures are of two colours, one is brighter and another phase slightly darker. These are because of alloying elements existing in matrix phase. The brighter phases mainly contain aluminium rich phase and the slight dark phases contain intermetallic phases which reacts easily to etchant. During the ageing treatment process the effect on the microstructures of the sample observed that the ageing temperature and time increases the size of precipitates.

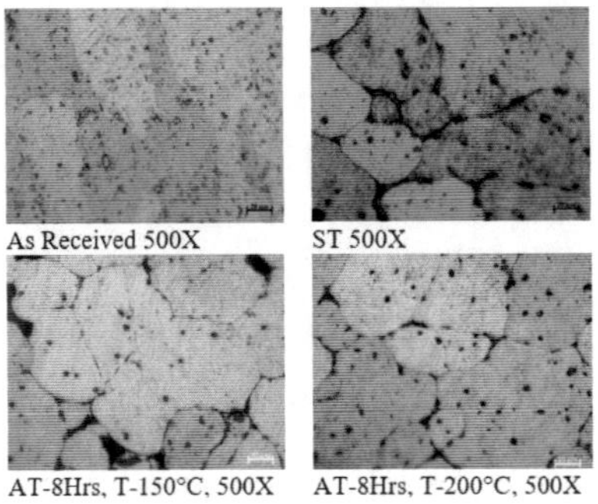

Figure 3. Microstructure of the aged sample at (150°C and 200°C) temperatures and time periods (As received, ST and 4 hrs.).

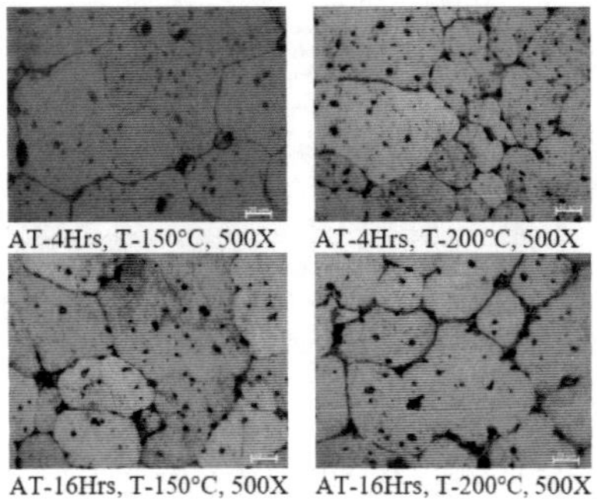

Figure 4. Microstructure of the aged sample at (150°C and 200°C) temperatures and time periods (8 and 16 hrs.).

3.2. Hardness

In Fig. 2 illustrated that result of ageing treatment process on hardness of Al-alloy. This figure, we could easily interpret that, initially hardness value (Hv) increases as the time increases (upto 8 hrs) but beyond this time, Hardness (Hv) decrease. This may be due to the uniformed distribution of fine precipitate in the matrix phase (upto 8 hrs). Increasing the time of ageing these fine precipitates are increase in size and degrade the properties of hardness. Increasing the temperature of ageing process increases the precipitate's kinematics due to this samples aged at higher temperature has lower hardness in compare to the samples aged at lower temperature. From hardness testing results we know that the maximum hardness was 154 Hv at 150°C and 8 hrs. of ageing time.

3.3. Potentiodynamic polarization

The Fig. [5-6] shows the Potentiodynamic polarization curves of Al Alloy 2014 of two different ageing temperature in 1 mole/litre of NaCl electrolyte solution. Potentiodynamic polarization test curves shows similar trend for all samples. The cathodic branches in the polarization curves showing the Tafel behavior and by applying Tafal exploration method we get the values of I_{corr} and V_{corr}, which are shown in table 2. Comparison of potentiodynamic polarization curve in 1mole of NaCl electrolyte solution on same ageing time with respect to 150°C and 200°C ageing temperature are shown in Fig. [7-10]. In all curves as the temperature and time increases the current density increases and potential voltage decreases. The corrosion rate is directly proportional to current density.

Table 2 Effect on the current and potential voltage on Ageing time/ temperature

S No	Ageing time/Temp	Current (A)	Potential (V)
1	4/150	6.805	-628.519
2	8/150	11.703	-637.844
3	12/150	12.119	-660.784
4	24/150	22.617	-612.119
5	4/200	9.189	-617.745
6	8/200	18.235	-652.528
7	12/200	19.068	-612.527
8	24/200	26.174	-642.299

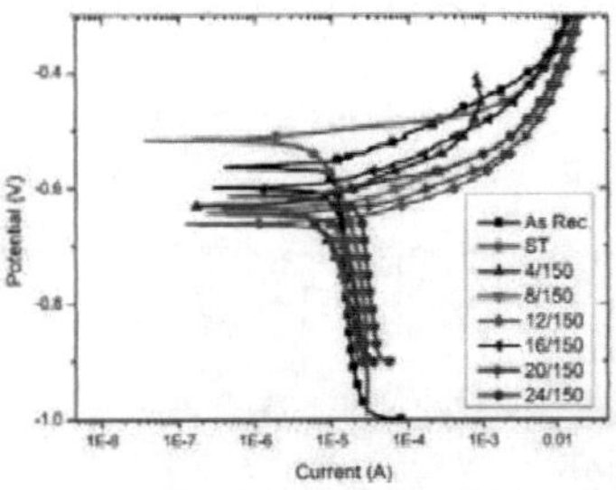

Figure 5. Polarization curves of Al Alloy 2014 as received, ST and aged sample in 150°C ageing temp.

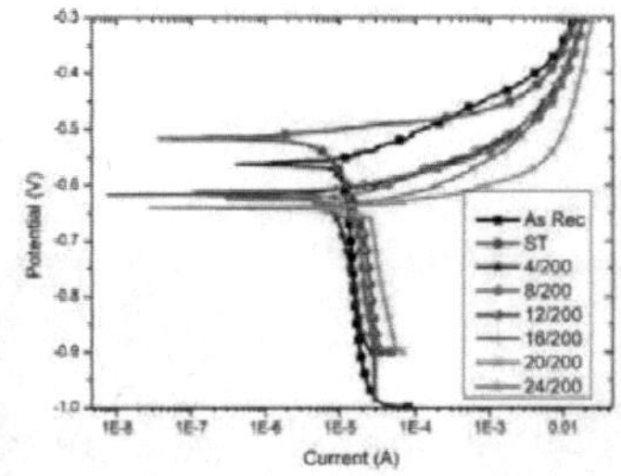

Figure 6. Polarization curves of Al Alloy 2014 as received, ST and aged sample in 200°C ageing temp

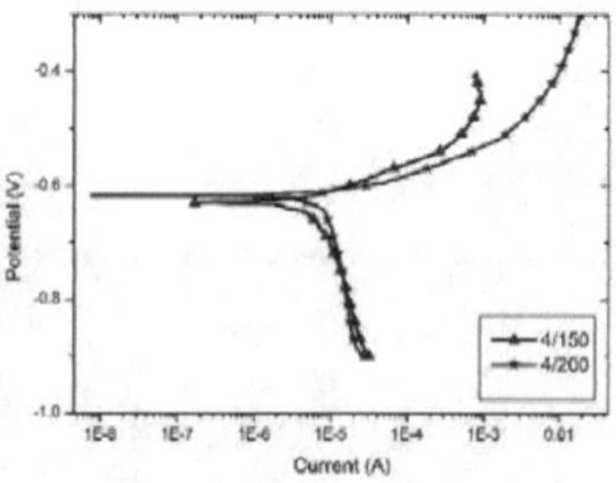

Figure 7. Polarization curves of 4hrs. of aged Al Alloy 2014 in different ageing temperature.

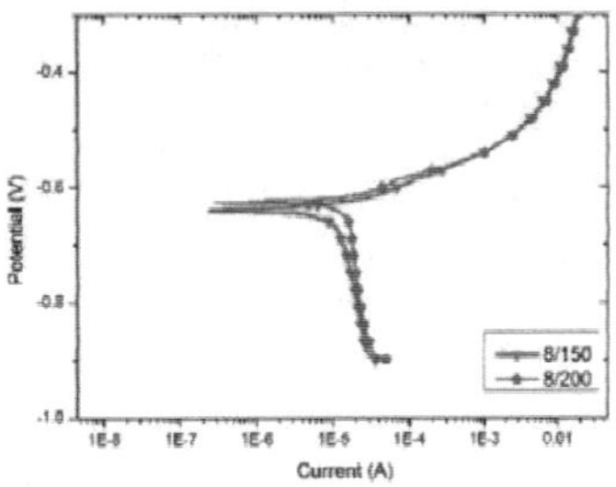

Figure 8. Polarization curves of 8 hrs. of aged Al Alloy 2014 in different ageing temperature

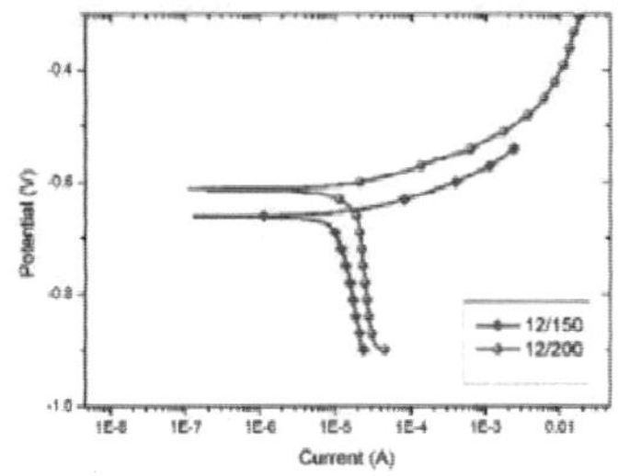

Figure 9. Polarization curves of 12 hrs. of aged Al Alloy 2014 in different ageing temperature

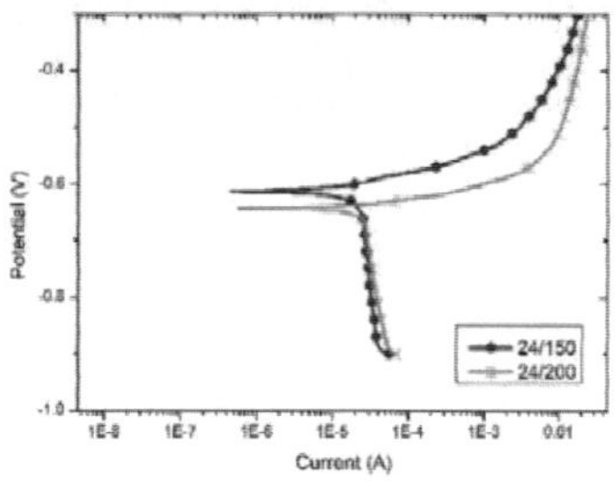

Figure 10. Polarization curves of 24 hrs. of aged Al Alloy 2014 in different ageing temperature

3.4. *Corrosion result*

All potentiodynamic polarization test samples were evidently shown in damaged on the surface all sample. This clearly shown to a certain extent of corrosion occurs.

In the variation of corrosion behavior of Al Alloy 2014 with different ageing temperature and time have been explained by the graph in Fig. 11. By comparing the corrosion rate with respect to time and temperature, we can see that the lower temperature of ageing as well as lower aging time shows lower corrosion rates. Boag et al also reported that the current density (Icorr) increases with ageing time and it increases the corrosion rate. Corrosion rate of the aged sample were high due to the presence of intermetallic phases and increases in precipitate size [10].

After corrosion test micrographs of the worn-out surface have been taken of As received and samples were aged at 150°C (8 hrs., 16 hrs. and 24 hrs., respectively) and they are shown in Fig. 12. These worn-out surfaces clearly show the pattern of the corrosion. Initially corrosion was started in the form of pits and when the ageing time was increased pitting corrosion was converted into the intergranular corrosion. Weilong et al. also reported that corrosion morphology changeover into pitting to intergranular corrosion was found in Al Alloy 2014 [11-13].

3.5. Tensile Test

Figure 13 shows the comparison between ultimate tensile strength of as received, ST and aged samples. This graph shows that the maximum Ultimate tensile strength first increase and then decreases due to over ageing. Tensile testing result shows that the maximum Ultimate tensile stress reaches to 573 MPa and Yield strength 528 MPa values at 150°C and 8 hrs of aged sample. Samples aged at higher ageing temperature (200°C), attains higher strength then as received samples but they get slight lower UTS than the samples aged at lower temperature (150°C).

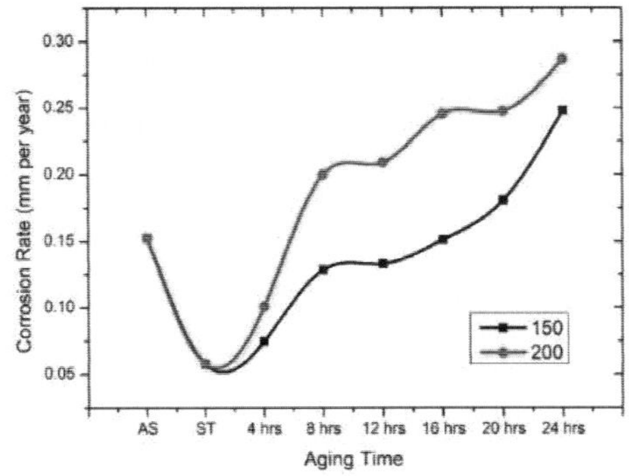

Figure 11. Effect of ageing time on the corrosion rate.

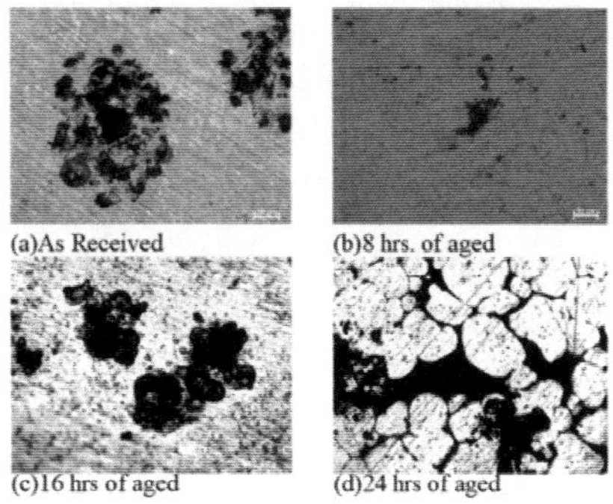

Figure 12. Macrographs of worn out surface after corrosion tests.

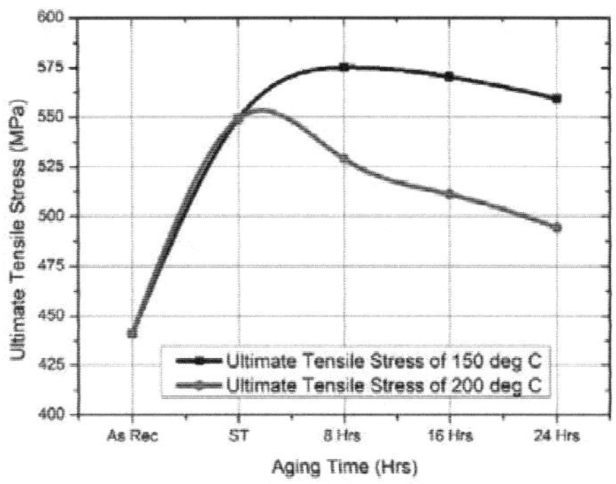

Figure 13. Tensile stress of aluminium alloy aged samples

4. Summery

1. Corrosion damage start with pitting corrosion and proceeds to convert into intergranular corrosion as ageing time increases.
2. Lower ageing temperature 150°C resulted lower corrosion rate as compare to higher ageing temperature.
3. In hardness testing result, the maximum hardness value of 154Hv was found at 150°C and 8 hrs. of ageing time.
4. In Tensile testing result the maximum Ultimate tensile strength of 575.13 MPa, Yield strength 530.77 MPa and elongation (%) of 10.26 values were found at 150°C and 8 hrs. of ageing time.

5. References

[1] D. A. P. Reis; A. A. Couto; N. I. Domingues Jr.; A. C. Hirschmann; S. Zepka; and C. de Moura Neto; 2012; "Effect of Artificial Aging on the Mechanical Properties of an Aerospace Aluminum Alloy 2024;" Defect Diffus. Forum; pp. 193–198.

[2] J. H. Jang; D. G. Nam; Y. H. Park; and I. M. Park; 2013 ;"Effect of solution treatment and artificial aging on microstructure and mechanical properties of Al-Cu alloy"; vol. 23;Trans. Nonferrous Met. Soc. China; English; pp. 631–635.

[3] *K. S. Ghosh; M. Hilal; and S. Bose; 2013; "Corrosion behavior of 2024 Al-Cu-Mg alloy of various tempers"; vol. 23; Trans. Nonferrous Met; China (English Ed.); pp. 3215–3227.*

[4] *S. K. Panigrahi and R. Jayaganthan; 2011; "Effect of ageing on microstructure and mechanical properties of bulk; cryorolled; and room temperature rolled Al 7075 alloy"; vol. 509; J. Alloys Compd.; pp. 9609–9616.*

[5] *Z. Liu; P. H. Chong; A. N. Butt; P. Skeldon; and G. E. Thompson; "Corrosion mechanism of laser-melted AA 2014 and AA 2024 alloys;" Appl. Surf. Sci.; vol. 247; no. 1–4; pp. 294–299; 2005.*

[6] *P. H. Chong; Z. Liu; P. Skeldon; and G. E. Thompson; 2014; "No Title"; vol. 3311.*

[7] *M. S. Kaiser and S. Dutta; 2014; "Corrosion Behaviour of Aluminium Engine Block in 3 . 5 % NaCl Solution"; pp. 52–58.*

[8] *L. Lacroix; C. Blanc; N. Pébère; G. E. Thompson; B. Tribollet; and V. Vivier; 2012; "Simulating the galvanic coupling between S-Al_2CuMg phase particles and the matrix of 2024 aerospace aluminium alloy"; vol. 64; Corros. Sci.; pp. 213–221.*

[9] *C. Meng; D. Zhang; H. Cui; L. Zhuang; and J. Zhang; 2014; "Mechanical properties; intergranular corrosion behavior and microstructure of Zn modified Al-Mg alloys"; vol. 617; J. Alloys Compd.; pp. 925–932;.*

[10] *A. Boag; A. E. Hughes; A. M. Glenn; T. H. Muster; and D. McCulloch; 2011; "Corrosion of AA2024-T3 Part I: Localised corrosion of isolated IM particles"; vol. 53; Corros. Sci.; pp. 17–26.*

[11] *W. Zhang and G. S. Frankel; 2003; "Transitions between pitting and intergranular corrosion in AA2024"; vol. 48; Electrochim. Acta; pp. 1193–1210.*

[12] *X. Zhang; X. Zhou; T. Hashimoto; and B. Liu; 2017; "Localized corrosion in AA2024-T351 aluminium alloy: Transition from intergranular corrosion to crystallographic pitting"; vol. 130; Mater. Charact.; pp. 230–236.*

[13] *A. E. Hughes; A. Boag; A. M. Glenn; D. McCulloch; T. H. Muster; C. Ryan; C. Luo; X. Zhou; and G. E. Thompson; 2011; "Corrosion of AA2024-T3 Part II: Co-operative corrosion";vol. 53; Corros. Sci.; pp. 27–39.*

Parametric Analysis of Pulsed Laser Melting Process

S Sukumar[1] and S P Kar[2]

[1] Ph. D Scholar, KIIT Deemed to be University, Bhubaneswar, Odisha
[2] Associate Professor, KIIT Deemed to be University, Bhubaneswar, Odisha

E-mail: skarfme@kiit.ac.in

Abstract. An axisymmetric numerical model has been developed to study the laser drilling of a cylindrical work piece under repetitive Gaussian laser pulse. Finite volume method has been used to discretize the energy equation. Convective and radiative heat losses have been considered from the laser irradiated surface. The resulting algebraic equation has been solved with the help of Tri-Diagonal Matrix Algorithm (TDMA) to know the temperature distribution in the work piece. The enthalpy porosity method has been employed to capture the moving solid-liquid interface evolving due to melting of the work piece subjected to laser energy. The laser source irradiated on the work piece has been considered as volumetric laser source instead of surface laser source which is rarely found in the existing literature. The present model is validated first with the existing literature. It has been found that, the results agree well. A parametric analysis has been done to know the effects of different laser parameters such as duty cycle, energy, pulse width and frequency etc. on the temperature field and the size of the melt pool when the work piece is subjected to repetitive laser pulse. The numerical results can provide some guidance for practical laser drilling process.

Nomenclature

a_0	beam radius
C_p	specific heat
g_L	liquid mass fraction
h	convective heat transfer coefficient
h_{sen}	sensible heat
H	total enthalpy
k	thermal conductivity
I_0	peak intensity of laser beam
L_f	latent heat of fusion
ms	millisecond
n	number of pulses
S_{LS}	volumetric laser source term
S_h	enthalpy source term
T	temperature
t	time
t_{on}	ON-time of laser
t_{off}	OFF-time of laser
r,z	coordinate directions
R	radius of the workpiece
L	length of the workpiece

Greek Symbols

α	thermal diffusivity
λ	under-relaxation factor
σ	Stefan-Boltzmann constant
Δt	size of time step
ΔV	volume of control volume
ΔH	nodal latent heat
τ	pulse width
κ	absorption coefficient

Subscripts

m	melting point
P	control volume
S	solid phase
L	liquid phase

Superscripts

n	number of iteration
0	previous time step

IOP Conf. Series: Materials Science and Engineering **338** (2018) 012009 doi:10.1088/1757-899X/338/1/012009

1. Introduction

Materials with better mechanical properties have been developed in the recent years with gradual progresses in the fields of metallurgy and material science. The traditional machining processes are not adequate enough to machine these materials properly with better accuracy and precision. Therefore, non-traditional machining processes have been developed and among these processes laser machining process has received better response from the manufacturers because of its accuracy, precision and better handling ability of the machined materials. In order to improve the performance of laser machining process, it is important to understand the laser-material interaction.

Many researchers have conducted research studies to find out the influence of various laser parameters such as peak intensity, laser energy, pulse width, duty cycle, frequency, beam radius etc. and the physical properties of the materials being machined on the laser machining process, material removal rate, type of surface being generated and grain structure following the machining process. Brent et al. [1] developed the enthalpy-porosity technique to track the solid-liquid interface during the melting of a pure metal. Basu and Srinivasan [2] investigated the effects of convection on laser melting and studied how convection changes the melt pool shape. Chen et al. [3] conducted laser heating under repetitive pulse and studied variation in temperature with radial and axial length of the workpiece model and with variations in duty cycle and frequency of the laser source. Kar and Rath [4] investigated the effect of natural convection on the shape of melt metal pool developing a two dimensional axisymmetric numerical model based on one-phase continuum mixture theory and fixed grid technique for both single and repetitive laser pulse. Qin et al. [5] developed a semi-infinite axisymmetric model to study the melting process in a metal slab subjected to Gaussian laser pulse.

In the current work, an axisymmetric numerical model has been developed. First, the model has been validated with the existing literature and then the study has been extended to investigate the effects of the laser parameters like peak intensity, energy, pulse width, duty cycle and frequency on the performance of laser melting.

2. Physical Model

A two-dimensional axisymmetric geometry has been developed for the parametric study of laser melting of a cylindrical workpiece. The laser source, being irradiated on the work surface is Gaussian in spatial variation and square in temporal variation as shown in the Figure. 1a.The schematic and computational domains have been shown in the Figure.1b. The workpiece is assumed to be at a uniform initial temperature of T_0. The laser irradiated surface is subjected to convective and radiative boundary condition while the rest of the two surfaces are being subjected to insulated boundary conditions. The radius of the workpiece has been taken as 5mm while the length has been taken as 2mm. Aluminium has been taken as the material of the workpiece. For simplifying the study, certain assumptions have been made. The thermo-physical properties of aluminium have been considered as constants, but different values have been taken for solid and liquid phases. Boussinesq's approximation holds good for density change with temperature. The laser source has been considered as volumetric in nature. Both the workpiece and laser source are assumed to be stationary. Metal removal occurs only through melting and vaporization has been avoided during the machining process. After the consideration of the above assumptions, the governing differential equations and boundary conditions have been defined.

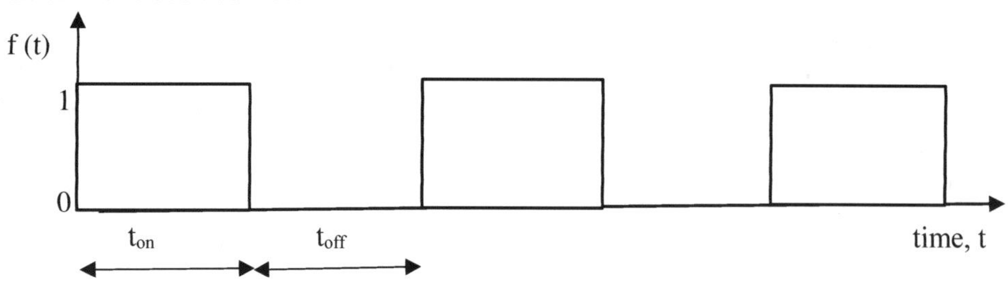

Figure 1(a). Temporal variation of laser intensity for repetitive pulse.

IOP Conf. Series: Materials Science and Engineering **338** (2018) 012009 doi:10.1088/1757-899X/338/1/012009

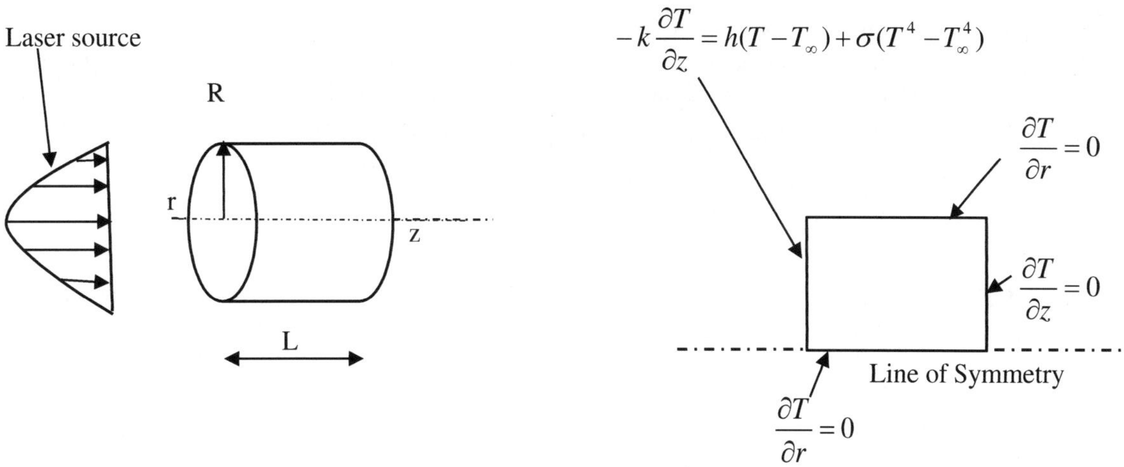

Figure 1(b). Schematic and computational model.

Governing Equations:

Energy Equation:

$$\frac{\partial}{\partial t}(\rho H) = \frac{\partial}{\partial z}(\rho\alpha\frac{\partial h_{sen}}{\partial z}) + \frac{1}{r}\frac{\partial}{\partial r}(\rho\alpha\frac{\partial h_{sen}}{\partial r}) + S_{LS} \tag{1}$$

$$\frac{\partial}{\partial t}(\rho C_p T) = \frac{1}{r}\frac{\partial}{\partial r}(kr\frac{\partial T}{\partial r}) + \frac{\partial}{\partial z}(k\frac{\partial T}{\partial z}) + S_h + S_{LS} \tag{2. 1}$$

$$S_h = -\rho_L L_f \frac{\partial g_L}{\partial t} \tag{2. 2}$$

Initial Condition:

$$T(r, z, 0) = T_0 \tag{3. 1}$$

Boundary Conditions:

$$\text{Right wall: } -k\frac{\partial T(r,L,t)}{\partial z} = 0 \tag{3. 2}$$

$$\text{Top surface: } -k\frac{\partial T(R,z,t)}{\partial r} = 0 \tag{3. 3}$$

$$\text{Bottom surface (Symmetry): } \frac{\partial T(0,z,t)}{\partial r} = 0 \tag{3. 4}$$

$$\text{Left wall: } -k\frac{\partial T(r,0,t)}{\partial z} = h\left(T(r,0,t) - T_\infty\right) + \sigma\left(T^4(r,0,t) - T_\infty^4\right) \tag{3. 5}$$

Equation (1.1) is applicable both to the solid and liquid phases and also the interface that defines the movement of solid-liquid interface. When the laser is ON, the value of f (t) is 1 while it is zero when the laser is OFF.

3. Numerical Method

The finite volume method (FVM) described by Patankar [6] has been used to discretize the equation 2.1. The resulting algebraic equations are solved using a line-by-line Tri-Diagonal Matrix Algorithm. The latent heat content has been used to capture the solid-liquid interface during melting. The iterative liquid volume fraction update equation of g_L for a control volume can be written as follows.

$$g_L^{n+1} = g_L^n + \lambda \frac{a_p \Delta t}{\rho_L L_f \Delta V_P}\left(T_P^n - T_m\right) \tag{4}$$

A control volume is considered to be completely melted when $g_L = 1$. The convergence criteria of the iterative solution has been set to 10^{-6} in the present simulation.

4. Results and discussion

The present numerical model has been validated with the existing result from the literature [3]. The temperature at the centre of the work piece at different heating time has been plotted as shown in the Figure 2. The properties of the material of the workpiece and different laser parameters have been taken as per the literature. The result of the present model agrees well with the analytical solution found in the existing literature. Therefore, the study with the help of the existing model has been extended further.

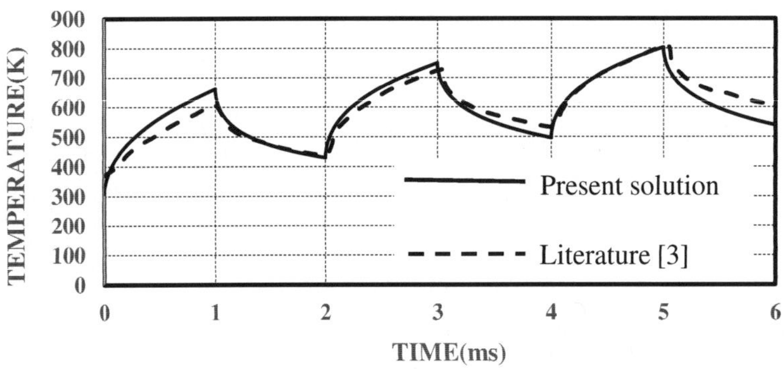

Figure 2. Comparison of temperature distribution at the centre of the workpiece with literature [3].

For the current analysis, repetitive laser pulse has been taken into consideration here. The total time of application of laser is 10ms and the number of pulses is 5. Total laser energy has been taken as 4J and frequency is 500Hz. By varying the duty cycle of the laser source as 0.25, 0.50 and 0.75, the ON and OFF- time and the peak intensity of the laser have been varied. The properties of aluminium have been taken from literature [5]. The laser beam radius has been taken as 700μm and initial temperature of the workpiece has been taken as 300K.

When the laser is ON, temperature at the point of laser application increases and decreases when laser is OFF. For the smallest duty cycle, the rise in temperature is the greatest because of the maximum peak intensity at constant energy. When duty cycle decreases, the OFF-time of laser increases that causes lowest final temperature for the lowest duty cycle. The final temperature after total heating is the highest for the maximum duty cycle which has been shown in the Figure 3. With increase in laser energy, the maximum temperature attained during heating increases.

With the laser irradiation at the centre of the workpiece, aluminium starts melting and with progress in heating period, the amount of metal removed increases which is evident from the outward movement of the liquid-solid interface with time as shown in the Figure 4.

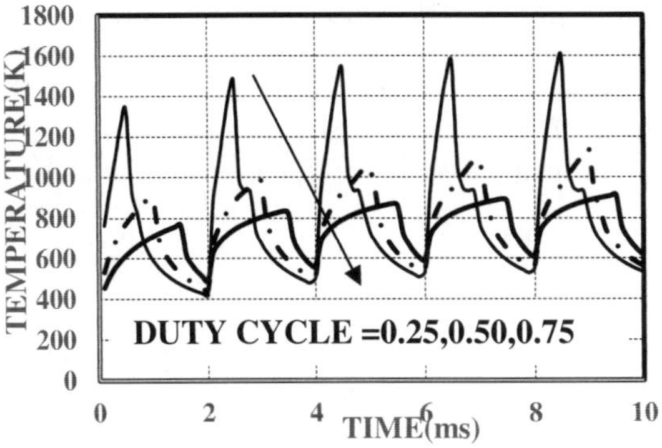

Figure 3. Variation of centre temperature with duty cycle.

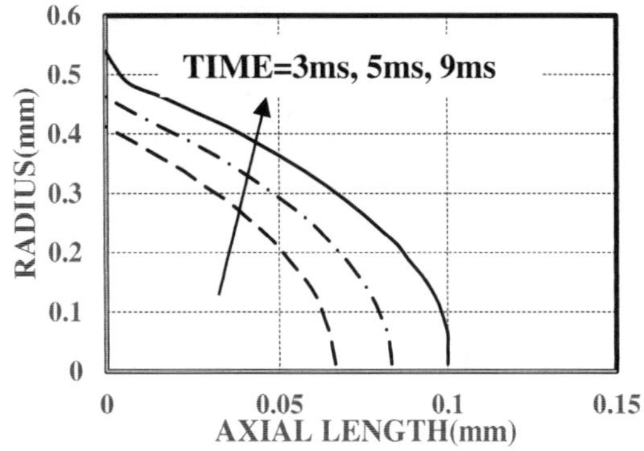

Figure 4. Movement of solid-liquid interface with time.

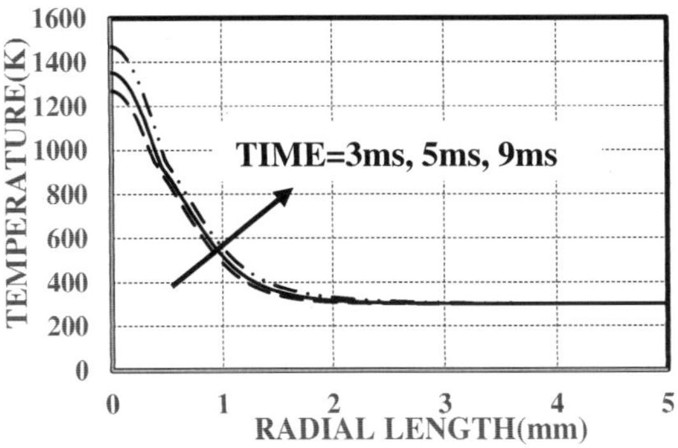

Figure 5. Variation of radial temperature with time.

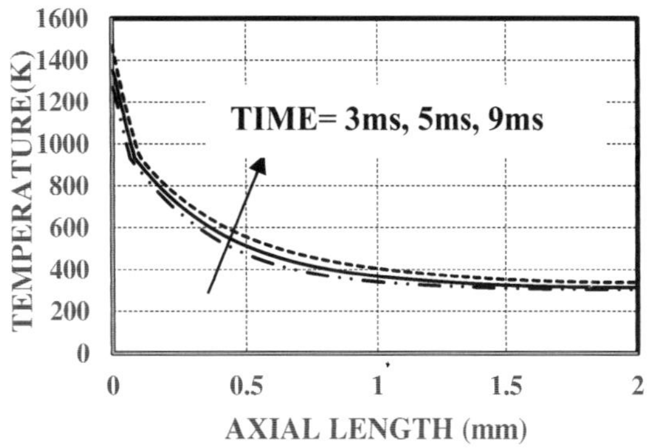

Figure 6. Variation of temperature along the axis of symmetry with time.

The workpiece was initially at 300K temperature. With heating, temperature at the point of application increases highly and conduction causes rise in temperature in the neighbouring points both in the radial and axial directions as shown in the Figure 5 and Figure 6 respectively, but beyond a certain distance from the point of application, the initial temperature is not disturbed.

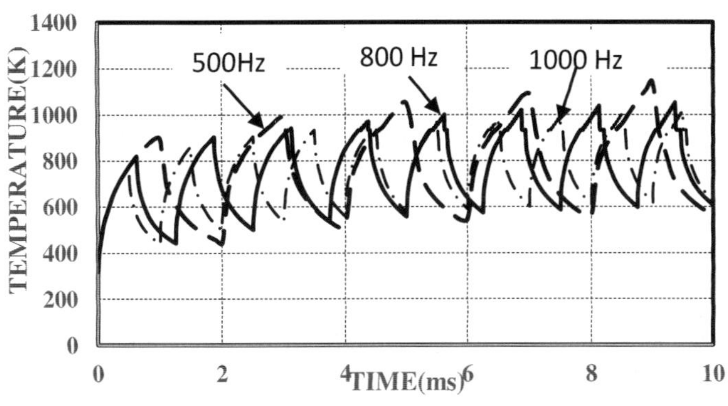

Figure 7. Variation of centre temperature with laser frequency.

With the decrease in laser frequency, the total number of pulses decreases for a particular heating time and heating time per pulse increases that results in attainment of maximum temperature during heating. Simultaneously because of increase in cooling period per pulse, the final temperature attained after total heating period decreases as shown in the Figure 7. With the increase in frequency, number of pulses increases and frequent heating causes increase in final temperature. Therefore, the maximum final temperature is attained for the highest duty cycle.

5. Conclusion

An axisymmetric model was developed and the effects of laser parameters like duty cycle, energy, pulse width and frequency on the performance of melting process were studied for a volumetric Gaussian laser pulse. It was found that with the decrease in duty cycle, the maximum temperature achieved at the centre during heating increases, but the final temperature at the end of total heating period decreases. With increase in laser energy, the maximum temperature achieved during heating increases keeping other parameters constant. Decrease in frequency causes an increase in maximum centre temperature achieved during heating, but the final temperature at the end of total heating period decreases.

In conclusion, the laser machining process can be improved by proper control of different laser parameters and the above derived results from the current numerical analysis can be utilised in practical laser melting process to achieve better results.

References

[1] Brent A D, Voller V R and Reid K J 1988 Enthalpy-Porosity Technique for modelling Convection-Diffusion Phase Change: Application to the Melting of a Pure Metal *Numerical Heat Transfer* **13** pp 297-318

[2] Basu Biswajit and Srinivasan J 1988 Numerical study of steady-state laser melting problem, *International Journal of Heat and Mass Transfer* **31** pp 2331-2338

[3] Chen Guibo, Wang Yaodong, Zhang Jiajia and Bi Juan 2017 An analytical solution for Two-dimensional modelling of repetitive long pulse laser heating material *International Journal of Heat and Mass Transfer* **104** pp 503-509

[4] Kar S P and Rath P 2014 A Fixed-grid based mixture model for pulsed laser phase change Process *Computational Thermal Sciences* **6** pp 13-26

[5] Qin Y, Wang B, Ni X W, Shen Z H and Lu J 2012 Axisymmetric Analytical solution for pulsed Laser melting of metals *Computational Materials Science* **54** pp 1-6

[6] Patankar S V 1980 *Numerical Heat Transfer and Fluid Flow* Hemisphere, London

IOP Conf. Series: Materials Science and Engineering **338** (2018) 012010 doi:10.1088/1757-899X/338/1/012010

Pore Size Control in Aluminium Foam by Standardizing Bubble Rise Velocity and Melt Viscosity

G Avinash , V Harika, Ch Sandeepika and N Gupta

Department of Metallurgical Engineering, National Institute of Technology, Raipur

Email: nehagupta.nitrr@gmail.com

Abstract. In recent years, aluminium foams have found use in a wide range of applications. The properties of these foams, as good structural strength with light weight have made them as a promising structural material for aerospace industry. Foaming techniques (direct and indirect) are used to produce these foams. Direct foaming involves blowing of gas to create gas bubbles in the melt whereas indirect foaming technique uses blowing agents as metallic hydrides, which create hydrogen bubbles. Porosity and its distribution in foams directly affect its properties. This demands for more theoretical studies, to control such cellular structure and hence properties. In present work, we have studied the effect of gas bubble rise velocity and melt viscosity, on pore size and its distribution in aluminium foam. A 15 PPI aluminium foam, prepared using indirect foaming technique having porosity ~86 % was used for study. In order to obtain metal foam, the bubble must not escape from the melt and should get entrapped during solidification. Our calculations suggest that bubble rise velocity and melt viscosity are responsible for vertical displacement of bubble in the melt. It is observed that melt viscosity opposes bubble rise velocity and help the bubbles to stay in the melt, resulting in porous structure.

1. Introduction

Metallic foams are obtained by solidifying melt containing well dispersed gas bubbles[1] and can be used as structural material since they exhibit property combinations as of low density, high surface area, high stiffness-to-mass ratio, good energy absorption, great specific strength and stiffness, excellent heat transfer capability etc.[2] Casting, powder metallurgy, metal deposition and additive manufacturing are commonly used methods to obtain these structures.[3] Open-celled metallic foams and closed-cell metallic foams are two basic foam types. Generally, Low-Pressure Infiltration (LPI) and additive manufacturing techniques are employed for open-celled foams and Gas Foaming techniques are used for closed-celled foams.[4] These porous metals find frequent use in automotive, railway, aerospace and several other applications. High quality metal foams can be obtained by developing a well controlled porous structure so as to meet the requirements of these applications. Therefore, it is much required to have good pore control during fabrication as most of the properties, such as strength, sound insulation, heat transfer, impact energy absorption etc, of these foams are extremely sensitive and are directly affected by its pore size achieved and distribution. However, it is very difficult to control pore size using gas foaming techniques due to non-uniform distribution of the bubbles. Pore size and its distribution depends melt viscosity, bubble rise velocity, surface tension, fluid buoyancy and drag velocity.[5] Therefore, it is important to control the melt viscosity during foaming process as it is seen that low melt viscosity often lead to rapid floating and escape of gas bubbles and high viscosities can cause suppression to formation of bubbles.[6]

Property combination of low specific weight with good mechanical physical properties offer significant performance gains in aluminum foams. This has lead to growing interest in the development of light and stiff foam structures for various engineering applications.[7] In particular, closed cell Aluminium foams have remarkable ability to absorb impact energy, which makes them appropriate for applications where effective utilization of impact energy is required as in car bumpers. Akiyama et.al prepared close-celled aluminium foams through gas foaming technique using Titanium Hydride (TiH_2) as foaming agent, where to control melt viscosity Calcium was added along with continuous stirring.[8] They observed that the particle size of TiH_2 and melt viscosity directly affected the porous structure and concluded that a properly controlled melt viscosity leads to controlled porosity, which is required to obtain uniform cellular structure.

In this work, we have used closed cell aluminium foam prepared using same technique by adding TiH_2 as foaming agent to create hydrogen bubbles.[9] These hydrogen bubbles are responsible for creating pores in the melt which upon solidification resulted in a porous aluminium structure. To control the pore size and its distribution inside these foams, theoretical calculations are done. For this study, the bubble size is assumed to be the same as the final pore size due to smaller bubble sizes formed. The final pore size distribution for solidified foam is obtained by calculating bubble rise velocity inside the aluminium melt as would be discussed in the following sections. It is also studied, the effect of additives as calcium on viscosity of this aluminium melt and its significance on the final pore distribution.

2. Materials and Methods

Calcium (1.8 wt.%), used as thickening agent to adjust the melt viscosity is added to Aluminum, melted at 720°C in a Heat Treat Electric furnace for 1.5 hr. Titanium hydride (1.5 wt.%) is thereafter added as foaming agent to the melt in order to form hydrogen bubbles. During this addition, the melt is stirred continuously to obtain a homogeneous melt and hence a uniform property foam material having porosity of ~ 86%. It was expected that this controlled method would result in cellular structure consisting of cells with uniform size and shape. However, we were unable to achieve the uniform cellular structure (size and shape) even after having cautions in the processing method, which finally resulted in reduced foam properties as compared to expected theoretical values, as was also observed by Suresh et al.[10] To solve this problem, we tried to study and hence control the pore distribution inside foam, using set of calculations are discussed in next section.

3. Results and Discussions

As stated earlier, pore distribution in foam strongly depends on the bubble rise velocity inside the aluminium melt. To calculate the bubble rise velocity we have considered final pore size of 1mm radius, in 15PPI foam. A mold with dimensions 15*15*15 cm^3 with 20 cm sprue length and 2 cm sprue diameter is used. Porosity is determined using relationship:

$$\%\text{Porosity} = \left[\frac{\text{Pore volume}}{\text{Bulk volume}}\right] * 100 = 86.2\% \quad (1)$$

Pore size can be controlled by controlling the diameter of hydrogen bubble generated inside the aluminium melt. Pore size distribution of foam can be controlled by controlling the rising velocity of the hydrogen bubbles. Therefore, relationship between bubble size with final pore size, and bubble rising velocity with final pore distribution inside the aluminium foam is developed here. For simplifying the calculations, since the pore size is small (1mm) we have assumed the final pore size obtained to be equal to the size of the gas bubble.

Also as stated earlier, it is known that terminal rising velocity of hydrogen gas bubbles is strongly dependant on the viscosity, surface tension, buoyancy and apparent density of melt. Rising velocity is an indicative of how fast the gas bubbles will rise inside melt and finally lead to the pore distribution inside aluminum foam upon solidification. For calculations on bubble rise velocity, we have considered isolated spherical bubbles, as per relationship provided in Table 1, which is due to the dominant surface tension effect.

Table 1. Relation between bubble size and shape

Radius (cm)	Bubble Shape
< 0.01	Solid Spheres
0.01 – 0.1	Deviations from spherical
> 0.1	Ellipsoidal

For an infinite stagnant fluid, the study of bubble rise velocity in two-phase (gas-liquid) fluid flow can be obtained using Stokes solution as:[11]

$$U_\infty = \frac{1}{18} g d_e^2 \left(\rho_l - \rho_g\right) \frac{1}{\mu_l} = 3.747\text{m/s} \quad (2)$$

g = Acceleration due to gravity (9.8 m/s^2)

d_e = Diameter of bubble (2 mm)

ρ_l = Density of Aluminium melt (2375 Kg/m^3)

ρ_g = Density of hydrogen gas (0.0893 Kg/m^3)

μ_l = Dynamic viscosity of Aluminium melt (1.38 mPa)

Considering bubble rising velocities as 3.747m/s, the pore distribution in the foam can hence be determined. For this, we have considered five similar bubbles formed at different depths inside aluminium melt as shown in Fig 1. Using rising velocities for these bubbles as 3.747m/s, the rising distance covered by them at solidification start (~ 615°C, 30 sec)[12] is 99 m. Here, it is assumed that rising velocity between two bubbles is constant over the time interval considering constant melt viscosity. The pore distribution inside the aluminium foam for these five bubbles can hence be determined precisely in this manner. Likewise we can consider large amount of bubbles forming inside the melt & hence can determine their exact distribution using the concept of rising velocity.

However, this vertical displacement of the bubbles is much higher such that, all gas bubbles formed get easily escaped into the atmosphere from the mold. So, we need to make changes in the melt such that the gas bubbles are entrapped inside the melt. As discussed in work by Akiyama et.al, calcium was used as thickening agent to increase the melt viscosity so as to hold gas bubbles inside the melt for longer time.

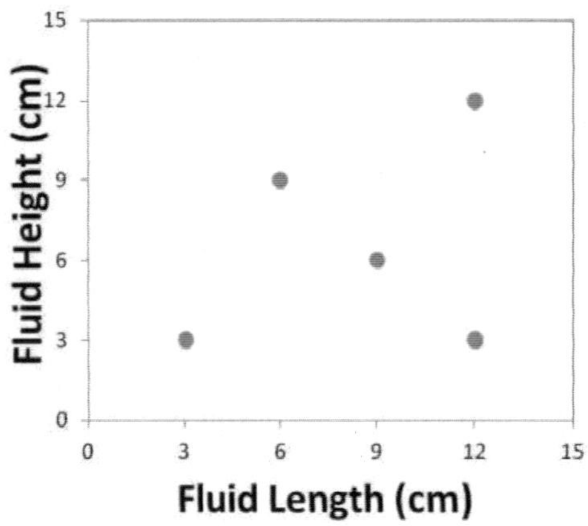

Figure1. Distribution of bubbles inside mold

The relative viscosity (μ_{rel}) representing factor by which viscosity of the liquid is increased, through the addition of the calcium (1.8 wt.%) is given as:[13]

$$\mu_{rel} = \frac{\mu_{sus}}{\mu_{liq}} \tag{3}$$

μ_{sus} = Dynamic viscosity of the suspension

μ_{liq} = Dynamic viscosity of the aluminium melt (1.38 mPa)

The relative viscosity of the melt is dependent on the solid fraction (φ). Einstein[14] derived an analytical relation for the relative viscosity for very dilute suspensions (low solid fraction $\varphi \ll 1$), as:

$$\mu_{rel} = 1 + \frac{5}{2}\varphi \tag{4}$$

Kunitz[15] proposed relation that is valid for larger values of solid fraction, $\phi >> 1$, as:

$$\mu_{rel} = \frac{1 + \frac{1}{2}\varphi}{(1-\varphi)^4} \tag{5}$$

For 1.8wt.% of calcium added to aluminium melt, the solid fraction can be found out as :

$$\varphi = \frac{wt. of\ calcium}{wt. of\ calcium + wt. of\ aluminium}$$

$$= 1.77$$

Substituting φ=1.77 in equation (5), we get μ_{rel}= 5.36

Substituting this in equation (3), we get μ_{sus}= 7.4mPa

Hence, after addition of 1.8 wt.% calcium, the aluminium melt viscosity has been increased to 7.4mPa from 1.38mPa. This increase in viscosity is high enough to hold bubbles inside the melt for longer times and hence help in achieving the desired pore structure.

4. Conclusion

This work is performed to have a better understanding & control over pore size and its distribution in aluminium foam to have better property control. The pore distribution in the melt is analyzed by calculating the vertical displacement of the bubbles which are formed using TiH_2 as foaming agent. Assuming gas bubble size as the final pore size (1 mm), since further expansion of the bubble cannot takes place once the solidification starts (~ 615°C), measurements to determine pore distribution inside aluminium foam are made. The small isolated gas bubbles rise in melt at 750°C, with rising velocity of U_∞=3.747m/s. The final vertical displacement travelled by individual gas bubble which is the product of rising velocity (U_∞) and solidification start time (30 sec) is then calculated as 99 m. This displacement of bubbles is much higher than the height of our mold used, which actually resulted in obtaining a structure with very less entrapment of pores and hence observation of undesired porous structure. Therefore, the bubble rising velocity has to be opposed by some other parameter (like melt viscosity), which must allow the gas bubble to stay in the melt for longer time, thus resulting in a porous structure. The same is confirmed by Akiyama et al., who used Calcium as thickening agent to increase the melt viscosity. The final viscosity (μ_{sus}= 7.4mPa) of melt after calcium addition is then calculated theoretically which was same as of results observed by Liqun Ma et al.(5-8mPa). This increase in melt viscosity on calcium addition will help in entrapping the gas bubbles in the melt thus resulting in desired porous structure. Hence, it is extremely important to increase melt viscosity and study the effect of viscosity for entrapping these bubbles inside the melt. In our future works, we are trying to theoretically relate the melt viscosity with the bubble rising velocity so as to entrap these bubbles by reducing their rising velocities & hence control the final pore structure observed.

References

[1] Banhart J 1999 *Europhys News* **30** 17-20.

[2] Gibson L J and Ashby M F 1997 *Cellular solids: Structure and Properties* (Cambridge, UK: Cambridge Universityn Press).

[3] Banhart J 2001 *Manufacture, characterization and application of cellular metals and metal foams* Prog. Mater. Sci. **46** 559–632.

[4] Ma L and Song Z 1998 *Cellular structure control of Aluminium foams during Foaming process of Aluminium melt* Scripta Met **39** 1523.

[5] Mário A R Talaia 2007 *Terminal Velocity of a Bubble Rise in a Liquid Column* World Academy of Science Engineering and Technology **28** 264-68.

[6] Banhart J 2006 *Metal foam:Production and stability* Adv. Eng. Mater **8(9)** 781-794.

[7] Byakova A, Gnyloskurenko S and Nakamura T 2012 *The Role of Foaming Agent and Processing Route in the Mechanical Performance of Fabricated Aluminum Foam* Metals **2** 95-112.

[8] Markaki A E and Clyne T V 2000 *Characterization of impact response of metallic foam/ceramic laminates* Mater. Sci. Technol. **16** 785–91.

[9] Akiyama S, Imagawa K, Kitahara A, Nagata S, Morimoto K, Nishikawa T and Itoh M 1987 *Foamed metal and method of producing same* US Patent 4713277 A.

[10] Suresh C and Vidyashankar S 2016 *Investigation of the Effect of Pore Size on Mechanical Properties of Open Cell Aluminium Foam Structure* International Journal of Advanced Engineering, Management and Science **2(9)** 1641-44.

[11] Stokes G G 1880 *Mathematical and Physical Papers* (Sommerfeld, London:Cambridge University Press).

[12] Bala K C and Khan R H 2014 *Rate of solidification of Aluminium casting in varying wall thickness of cylindrical metallic moulds*, Leonardo Journal of Sciences, Issue **25** 19-30.

[13] Konijn B J, Sanderink O B J, Kruyt N P 2014 *Experimental study of the viscosity of suspensions: Effect of solid fraction, particle size and suspending liquid,* Powder Technology **266** 61–69.

[14] Einstein A 1906 *Eine neue Bestimmung der Molekïldimensionen*, Ann. Phys. **19** 230–247.

[15] Kunitz M 1926 *An empirical formula for the relation between viscosity of solution and volume of solute*, J. Gen. Physiol. **9(6)** 715–725.

Hardness - Yield Strength Relation of Al-Mg-Si Alloys

Aluru Praveen Sekhar*, Supriya Nandy, Kalyan Kumar Ray and Debdulal Das

Department of Metallurgy and Materials Engineering, Indian Institute of Engineering Science and Technology, Shibpur, Howrah - 711103, West Bengal, India

*Corresponding author E-mail address: praveensekhar429@gmail.com

Abstract: Assessing the mechanical properties of materials through indentation hardness test is an attractive method, rather than obtaining the properties through destructive approach like tensile testing. The present work emphasizes on the relation between hardness and yield strength of Al-Mg-Si alloys considering Tabor type equations. Al-0.5Mg-0.4Si alloy has been artificially aged at various temperatures (100 to 250 °C) for different time durations (0.083 to 1000 h) and the ageing response has been assessed by measuring the Vickers hardness and yield strength. Correlations of the existing data from the open literature have also been reviewed. Lastly, it has been explained that the deviation in obtained relation from Tabor's equation is owing to the dislocation accumulation during indentation.

1. Introduction

Over the years, indentation hardness measurements are being used to determine several mechanical properties like yield strength, Young's modulus, fracture toughness *etc.* for different bulk materials, coatings and engineering surfaces. The hardness test is carried out in a conventional way by forcing the indenter onto the surface of the specimen. The additional stress generated by the indenter should be more than the yield strength of the material to assist the plastic flow for the permanent indentation. The strength of a material during service is often the primary concern to materials engineers. The yield strength is regularly considered essential for the design and selection of materials in various applications. Alloys for high-end structural applications need these properties as material specifications and have to frequently undergo quality assurance tests to monitor their strength [1]. However, the yield strength obtained from tensile tests is expensive, time consuming and requires a large volume of material with specified shape and dimension, and is also destructive in nature. Thus, the engineers have attempted an alternate route to estimate the mechanical properties by measuring the hardness, which is simple and semi-destructive in nature. Prior to this, an empirical relationship needs to be established between the hardness and tensile properties [1,2]. Hill et al. [3] have first correlated hardness test results with deformation properties obtained by tensile testing. Subsequently, Tabor has suggested an empirical relation between the Vickers hardness and the yield strength which has recently been theoretically explained by Tiryakioğlu et al. [4] considering the strain distribution under the indentation using the concept of representative strain. According to the knowledge of the present authors, the correlation between hardness and yield strength of age-hardenable Al-alloys, especially that of Al-Mg-Si (6xxx) system has not been systematically investigated. Considering their technological importance, it is essential to establish the hardness-yield strength relationship.

This study makes an attempt to establish the relation between the Vickers hardness (*HV*) and the yield strength (σ_{ys}) of Al-Mg-Si alloys based on experimentally generated results. The suitability of the developed relation is evaluated by using similar data from the open literature.

2. Material and Methods

The commercial Al-Mg-Si alloy was procured in the form of extruded rods with 16.5 mm in diameter. The chemical composition of this alloy (0.50 Mg, 0.43 Si, 0.18 Fe, 0.08 Mn, minor amounts of Cu,

Cr, Zn, Ti, Al in balance; all in weight %) was obtained by atomic emission spectroscopic analysis. Solution treatment of hardness and tensile (25 mm gauge length and 6.25 mm diameter) specimens were done at 525 °C for 2 h followed by ice-water quenching and then artificial aged for 0.083 to 1000 h at seven different temperatures from 100 to 250 °C at 25 °C intervals. The Vickers hardness measurements were made by using pyramidal diamond indent tester (VMHT: Leica, Germany) under an applied load of 2 kgf and for indentation duration of 15 s. An average of fifteen readings with the associated standard errors was used for subsequent analyses. Uniaxial tensile tests were carried out using a close-loop servo-hydraulic universal testing machine (Model: 8801, Instron, USA). These tests were performed at a nominal strain rate of 1 x 10^{-3} s^{-1} at room temperature (298 K). Three specimens were tested at each state of ageing and the mean values of the tensile properties are reported.

3. Background

The mean pressure under the indenter (P_m) also known as Mayer's hardness is measured by considering the applied load (L) and projected area of indentations (A_i) which is given as:

$$P_m = L/A_i \quad (1)$$

Tabor [5] has suggested that for ductile metals, the mean contact pressure (P_m) is related with the flow stress (σ_f) of the material in simple compression test as:

$$\sigma_f = P_m/C \quad (2)$$

where, C is the constraint factor. According to Hill et al. [3], the P_m acting on the surface of the wedge *i.e.,* the principal compression stress which is uniform and is given as:

$$P_m = 2\tau_c(1+\theta) \quad (3)$$

where, τ_c is the critical maximum shear stress and θ is half angle of the tip of indenter. Tabor has relied on the Von-Mises yield criterion ($2\tau_c = 1.15\sigma_{ys}$) and suggested that for the Vickers indentation the flat punch, $\theta = \pi/2$ would be a fair approximation. By substituting these in Eq. (3) one gets:

$$P_m = 2.956 \times \sigma_{ys} \quad (4)$$

From Eqs. (2) and (4), the constraint factor C is found to be 2.956 for fully work hardened materials ($\sigma_f = \sigma_{ys}$). Tabor [2] has quantified that the C value can be approximately 3 since it is found to be 2.9, 2.8 and 2.9 for Te-Pb, annealed Cu and mild steel, respectively. Therefore, the relation between HV and P_m can be expressed as:

$$HV = 0.927P_m \quad (5)$$

Combining Eqs. (4) and (5) one gets:

$$\sigma_{ys} = HV / 0.927C \quad (6)$$

The relation between HV and σ_{ys} is likely to have a slope of 0.365 and should pass through the origin when one considers C= 2.956. Fully work-hardened materials validate this relation satisfactorily [2]. However, the HV and σ_{ys} relation for strain-hardening materials like steels, Mg, Al and Cu alloys is better expressed as [1]:

$$\sigma_{ys} = \beta_0 + \beta_1 HV \quad (7)$$

In most of the cases, β_0 is negative which is due to the characteristic strain during the Vickers indentation as suggested by earlier researchers [4]. It means that the true strain corresponding to the yield strength values is greater than the characteristic strain.This investigation uses Eq. 7 as suggested by earlier studies for age-hardenable alloys [1, 4].

4. Results and Discussion

4.1 Ageing response of the selected Al-0.5Mg-0.4Si alloy

Age hardening response has been studied by measuring Vickers hardness (HV) and yield strength (σ_{ys}) at different temperatures (T_A) from 100 to 250 °C for time durations (t_A) ranging from 0.083 to 1000 h as shown in Fig. 1. The variations in HV (Fig. 1(a)) and σ_{ys} (Fig. 1(b)) indicate the typical ageing characteristics of Al-alloys. With increasing T_A, the peak HV and peak σ_{ys} values decrease, and the time to achieve the peak values are also reduced. At lower T_A and for lesser t_A, the rate of increase in

HV and σ_{ys} is found to be very low. This state of ageing is identified as highly under-aged. On further ageing, there seems to be a sharp rise in the *HV* and σ_{ys} due to the formation and dissolution of the GP Zones and β'' precipitates up to the peak ageing condition. Thereafter, the *HV* and σ_{ys} values decrease due to the coarsening of precipitates allowing the dislocations to bow off during the deformation. It is observed that at T_A=250 °C for t_A> 8 h, the *HV* and σ_{ys} values are less than the as-quenched values. This state of ageing indicates the highly over-aged condition. The obtained results are in good agreement with the reports for the similar alloy by Jiang et al. [6] and Munitz et al. [7]. However, these results deviate from the data reported by Siddiqui et al. [8] presumably due to the selection of different solutionizing condition.

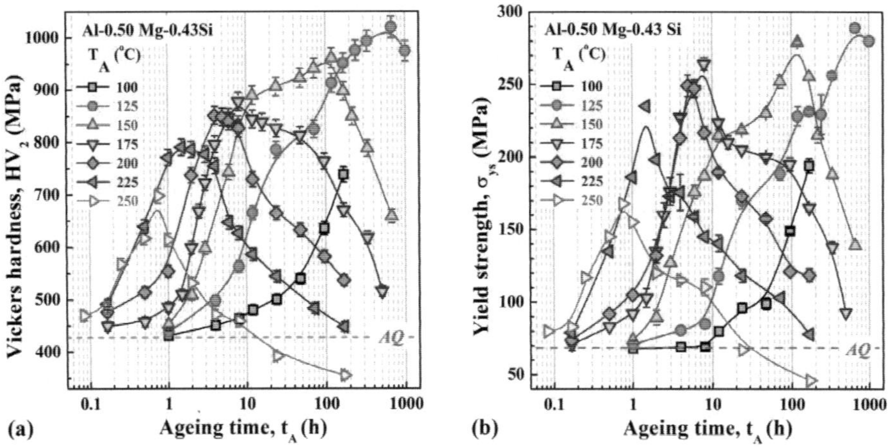

Figure 1. Variations of measured (a) hardness and (b) yield strength with ageing time for artificially aged Al-Mg-Si alloy specimens at different temperatures. Property of as-quenched (AQ) state is shown as horizontal line.

4.2 Development of correlation between hardness and yield strength

The empirical relationships developed on a theoretical basis might assist as a valuable guide. However, the correlations which are developed based on experimental results are essential for the practical application of hardness indentation as a potential alternative to the yield strength values [9]. Experimental data sets of the selected Al-0.5Mg-0.4Si alloy in addition to the data collected from twelve different studies consisting of four different Al-alloys have been studied separately for the correlations of their hardness and yield strength. Linear regression analyses have been carried out on

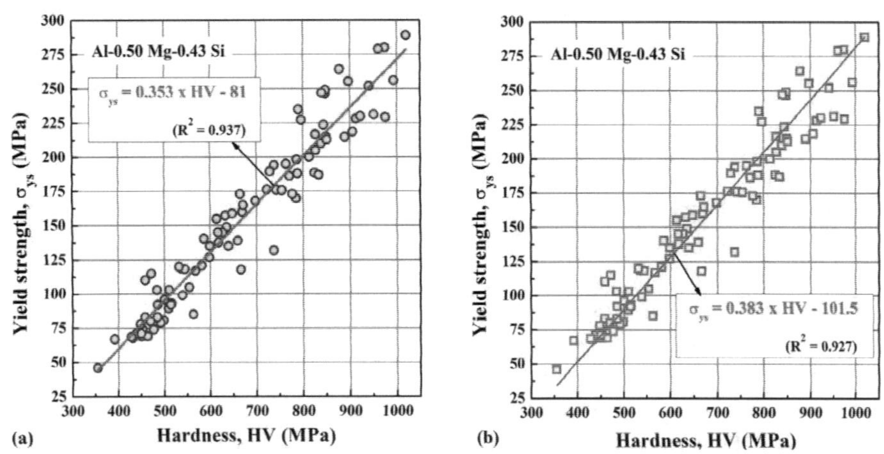

Figure 2. Correlation between the hardness and yield strength of Al-Mg-Si alloy from the present experimental data following (a) Eq. 7 and (b) the same with a constant slope of 0.383.

experimental data (Fig. 2(a)) and the best fit line is achieved with correlation coefficient (R^2) of 0.937. The estimated coefficient values of β_0 and β_1 (as in Eq. 7) and the 95% confidence interval of these values are presented in Table 1. Shaw and DeSalvo [10] have suggested that constraint factor C is 2.82 for a blunt axisymmetric indenter. Shield [11] has also proposed almost similar C value of 2.84 for a circular flat-ended punch indenter against a plane surface. However, for the Vickers indentation, Larsson [12] has reported that the C value to be 2.8 by numerical analysis using finite element method. Now, substituting the value of C=2.82, the slope β_1 is found to be 0.383 and the one suggested by Tabor is 0.365. Both the values of β_1 are found to be within the 95% confidence interval; however, the intercept β_0= 0 suggested by Tabor is not within the 95% confidence interval which contradicts to the general assumption of $P_m= 3\sigma_{ys}$ for the selected Al-alloy. As the line of slope, 0.383, is almost similar to the regression line in Fig. 2(a), it is suggested that the regression is carried out once again with a constant slope of 0.383 as shown in Fig. 2(b) which can be expressed as:

$$\sigma_{ys} = 0.383HV - 101.5 \tag{8}$$

Abdulwahab et al. [13] have reported that the hardness and yield strength values of Al-0.3Mg-7Si (A356) alloy whose relation is found to be sharply linear as demonstrated in Fig. 3(a) by a simple linear regression. Earlier, Rometsch and Schaffer [14] has attempted the direct correlations between Brinell hardness and yield strength of both A356 and A357 cast alloys and found that there is a significant amount of scatter. However, these data exhibit good linear relationships when yield strength values are correlated with Vickers hardness as shown in Figs. 3(a) and (b). The data collected

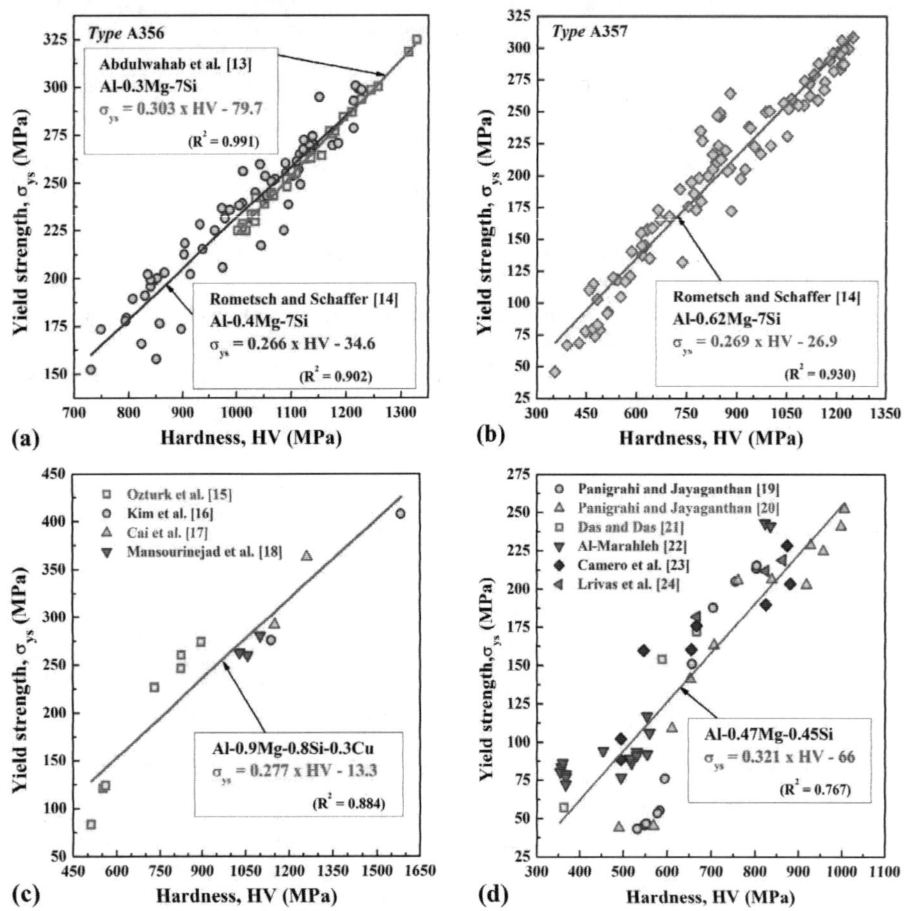

Figure 3. Correlations between the hardness and yield strength data collected from existing literature for (a) A356 (b) A357 (c) AA6061 and (d) AA6063 Al-alloys.

from several sources [15-24] of Al-0.9Mg-0.8Si-0.3Cu and Al-0.47Mg-0.45Si alloys also depict a linear relation (Figs. 3(c) and (d)). The slope (β_1), intercept (β_0) with correlation coefficient (R^2) and their corresponding confidence interval values are summarized in Table 1.

Table 1. Summary of regression analyses results between hardness and yield strength as per Eq. 7 for different Al-Mg-Si alloys.

Material	β_1	95%LCL	95%UCL	β_0	95%LCL	95%UCL	R^2
Al-0.5Mg-4Si [Present study]	0.383	-	-	-101.5	-105.12	-97.9	0.927
Al-0.3Mg-7Si [13]	0.303	0.292	0.313	-79.7	-91.28	-68.01	0.991
Al-0.4Mg-7Si [14]	0.266	0.244	0.288	-34.6	-57.43	-11.72	0.902
Al-0.62Mg-7Si [14]	0.269	0.254	0.285	-26.9	-40.57	-13.31	0.930
Al-0.9Mg-0.8Si-0.3Cu [15-18]	0.277	0.216	0.337	-13.3	-73.01	46.42	0.884
Al-0.47Mg-0.45Si [19-24]	0.321	0.273	0.367	66	-97.08	-35.16	0.767

4.3 Assessment of the model

Finally, to validate the modified Tabor's equation established in this investigation (*i.e.,* Eq. 8), the yield strength values (MPa) of the experimental data and the ones obtained from the open literature have been estimated by using the hardness values in MPa. Fig. 4 displays the plot of measured and the predicted yield strength values where most of the data are lying well within the 15% and 30% error bands. A few data points which are outside of the deviation bands are the alloys that are tested in differently solutionized and quenched conditions or highly over-aged state or involve data of differently homogenized samples. The deviations in the data are perhaps due to the higher rate of strain hardening occurring during the tests.

Tabor's equation (Eq. 6) can be replaced as shown below for fully work hardened materials ($\sigma_f = \sigma_{ys}$):

$$\sigma_f = \frac{HV}{0.927C} \tag{9}$$

Whereas, for materials which are strain hardened, the Vickers hardness indenter has some representative strain [4] under it for which the σ_f is greater than the σ_{ys} and can be expressed as:

$$\sigma_f = \sigma_{ys} + \Delta\sigma \tag{10}$$

$\Delta\sigma$ is the increase in strain during the Vickers indentation and reaches a maximum of 8% representative strain.

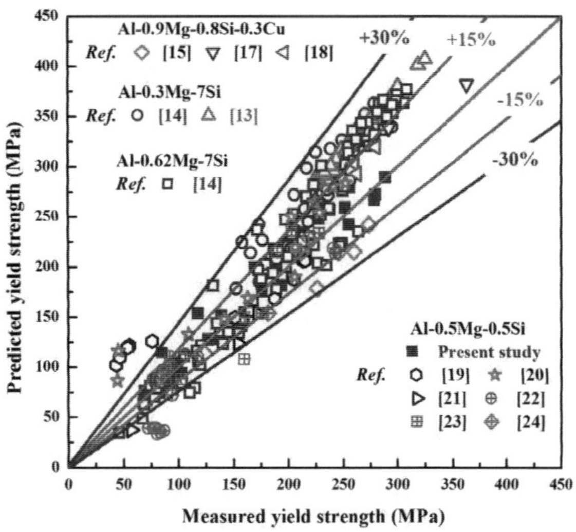

Figure 4. Yield strength values (Eq. 8) of different Al-Mg-Si alloys from the present study as well as from the literature; these are combined to compare their corresponding yield strength values.

Now, substituting Eq. 9 in Eq. 10 one gets:

$$\sigma_{ys} = \frac{HV}{0.927C} - \Delta\sigma \tag{11}$$

The above equation is the modified version of Tabor's equation (Eq. 6) with addition of the negative intercept. Tiryiakioglu et al. [4] have carried out this analysis and suggested that Eq. 11 would explain the reason behind the intercept being negative for all the strain hardened materials. The present study on correlation between hardness and yield strength demonstrates a slope of 0.383 and the negative intercept of 101.5; these values indicate the strain on the flow stress during indentation is less than the representative strain.

5. Conclusions

The correlation between hardness and yield strength of the selected Al-0.5Mg-0.4Si alloy has been established using the experimentally generated values considering the Tabor's equations. The artificial ageing response of the alloy has been studied for seven different temperatures from 100 to 250 °C for the durations of 0.083 to 1440 h. Simple linear regression analysis has been carried out between hardness and yield strength data of various age hardenable alloys collected from the existing literature and the corresponding slope and intercept values have been noted. Finally, the adequacy of the developed relationship has been validated considering the experimental hardness and yield strength values as well as using data from the literature. The deviations from the obtained correlation for some reported data have been ascribed to increase in strain hardening due to accumulation of dislocations during indentation.

Acknowledgement

The authors are grateful for the financial assistance received from the COE for *Microstructurally Designed Advanced Materials Development*, TEQIP II, IIEST Shibpur to carry out this work.

References

[1] Tiryakioğlu M., Robinson J.S., Salazar-Guapuriche M.A., et al., 2015, *Mater. Sci. Eng.A* **631** 196.
[2] Tabor D. 1951, *The Hardness of Metals*, Clarendon, Oxford, London
[3] Hill R., Lee E.H., Tupper S.J. 1947, *Proc. R. Soc. London, Ser. A* **188** 273.
[4] Tiryakioğlu M., Robinson J.S, 2015, *Mater. Sci. Eng. A* **641** 231.
[5] Tabor D.1948, *Proc. R. Soc. London, Ser. A* **192**, 247.
[6] Jiang D.M., Hong B.D., Lei T.C, et al., 1991, *Mater. Sci. Technol.* **7** 1010
[7] Munitz A., Cotler C., Talianker M. 2000, *J. Mater. Sci.* **35** 2529.
[8] Siddiqui R.A., Abdullah H.A., Al-Belushi K.R. 2000, *J. Mater. Process. Technol.* **102** 234.
[9] Busby J.T., Hash M.C., Was G.S., 2005, *J. Nucl. Mater.* **336** 267.
[10] Shaw M.C., Desalvo G.J. 1970, *Trans. ASME: J. Eng. Ind.* **92** 480.
[11] Shield R.T. 1955, *Proc.R. Soc. Lond. A* **233** 267.
[12] Larsson P.L. 2001, *Intl. J. Mech. Sci.* **43** 895.
[13] Abdulwahab M., Madugu I.A., Yaro S.A., et al., 2011, *Mater. Des.* **32** 1159.
[14] Rometsch P.A., Schaffer G.B. 2002, *Mater. Sci. Eng. A* **325** 424.
[15] Ozturk F., Sisman A., Toros S., Kilic S., Picu R.C. 2010, *Mater. Des.* **31** 972.
[16] Kim J.K., Jeong H.G., Hong S.I., Kim Y.S., Kim W.J. 2001, *Scripta Mater.* **45** 901.
[17] Cai M., Field D.P., Lorimer G.W. 2004, *Mater. Sci. Eng. A* **373** 65.
[18] Mansourinejad M., Mirzakhani B. 2012, *Trans. Nonferrous Met. Soc. China* **22** 2072.
[19] Panigrahi S.K., Jayaganthan R. 2011, *Mater. Des.* **32** 2172.
[20] Panigrahi S.K., Jayaganthan R. 2011, *Mater. Sci. Eng. A* **528** 3147.
[21] Das M., Das G. 2011, *T. Indian. I. Metals* **64** 75.
[22] Al-Marahleh G. 2006, *Am J Appl. Sci.* **3(5)** 1819.
[23] Camero S., Puchi E.S., Gonzalez G. 2006, *J. Mater. Sci.* **41** 7361.
[24] Rivas A.L., Muñoz P., Camero S., Quintero-Sayago O. 1999, *Adv. Mat. Sci. Tech* .**2** 15.

IOP Conf. Series: Materials Science and Engineering **338** (2018) 012012 doi:10.1088/1757-899X/338/1/012012

PLACER DEPOSITS OF BRAHMAGIRI COAST, ODISHA - A NEW RESOURCE FOR INDUSTRIAL HEAVY MINERALS

Sunita Routray[1], Jagadeeswari Vanamu[2] and Ranjita Swain[3]

[1, 3]Assistant Professor, C.V. Raman College of Engineering, Bhubaneswar 752054, India
[2]JRF, C.V. Raman College of Engineering, Bhubaneswar 752054, India

E-mail: sunitaroutray77@gmail.com

ABSTRACT

Characterization and beneficiation studies have been carried out on Brahmagiri placer deposits of Puri, Dist., Odisha. The studies show that this deposit contains potential amount of heavy minerals. The dune sands contain higher heavy minerals than that of beach sand. One of the dune sand samples collected from this deposit shows heavy mineral content of about 29 % also. A composite sand sample is prepared and heavy mineral recovery is studied from this by using gravity table. During recovery of heavy minerals both fresh and sea water are used to see their effect on grade and recovery of concentrate. Sea water rougher gravity table data shows better results than fresh water rougher gravity table. The gravity table study of composite sample shows a yield of 14.8% with a grade of 91.2 % and 93.2 % recovery.

***Key words**: Placer Deposits, Beach Sand, Dune Sand, Characterization, Total Heavy Minerals, Gravity Units*

1. INTRODUCTION

The beach and dune placer sands in India contain heavy minerals like ilmenite, rutile, garnet, zircon, monazite and sillimanite, which are derived from parent rocks by weathering process and transported by wind and rivers. Ilmenite-rich major beach and dune sand deposits occur in the coastal stretches of Kerala (Chavara), Tamil Nadu (Manavalakurichi, Midalam, Vayakallur), Andhra Pradesh, Maharashtra and Odisha. The coastal sand of Odisha also possesses more or less same minerals with variable mineral composition all along the coastline of 480 kms and has a beach placer heavy mineral resource of about 128 Mt (5% to 30% grade) and shares about 12% of country's heavy mineral resource. In spite of such large deposits in this State, only one mineral sand plant operates at IREL, Chhatrapur. Department of Atomic Energy, India, gives an formation about potential resource of heavy mineral deposits all along Brahmagiri to Puri coast covering with an average width of 1.91 km accounting for ilmenite deposit alone 37.98 million tons and the heavy minerals deposit ranging from 1.73% to 24.78% with an average grade of 5.76% up to the depth of 18m, but work related to R&D on beneficiation to recover heavy minerals has not been attempted so far [1-5]. Brahmagiri deposit is bounded by latitude 19° 47' N and longitude 85° 38' E and is shown in Fig 1.

An attempt is made in the present investigation is specially to study the heavy mineral distribution all along the Brahmagiri to Puri coast. In addition to this, it is also aimed to recovery total heavy minerals by using laboratory gravity tables with natural fresh water and sea water media.

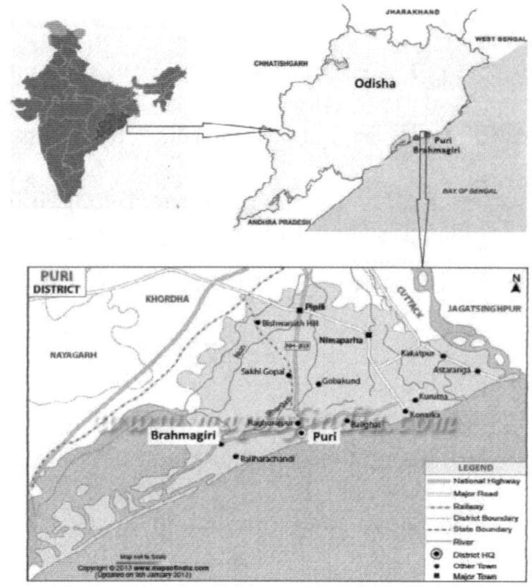

Figure1. Geographical location of Brahmagiri deposit, Odisha

2. MATERIALS AND METHODS

At first instance, samples were collected in Bramhagiri area from beach and dunes covering above 2 kms distance by a grid pattern with an equal intervals of 700 meters. At each grid around 100 kgs samples were collected and from which representative samples of each 10Kg were collected by coning and quartering. These beach placer sand samples were named from 7C to 20 C and dune sand samples were named from 7D to 20 D.

All collected samples were washed with fresh water before characterization studies to remove the salt content. The samples contain around 0.1 % salt. Physical characterization such as size analysis, true density, bulk density and angle of repose of all the samples were carried out using standard sieves. The d_{80}percent passing size of each sample has been calculated from the size analysis data. Heavy liquid separation studies were carried out using bromoform (specific gravity 2.89 g /cm^3) to determin the total heavy minerals present in the feed samples.

After complete physical characterization study mineral separation studies were carried out on a composite sample using a laboratory model gravity table (Type Concenco, The Diester Concentrator Company Ltd, Model No. Denver 13A). The composite sample was prepared by mixing three beach and three dune sand samples (7C, 14C, 19C, 7D, 10D and 12D). Physical characterization was done for the composite sample. Sink-float study was also carried out for each size fraction of the composite sample. Gravity table was used for beneficiation studies by using fresh and also sea water. The operating conditions used in the present study were feed rate 20 kg/hr, wash water 3 liter/min and deck angle 1.4°. Heavy liquid separation studies were carried out using bromoform (specific gravity 2.89 g/cm^3) for determining the total heavy minerals grade of concentrate and tailing samples.

3. RESULTS AND DISCUSSION

The placer sands of beach and dune are subjected to physical characterization including size analysis, density as well angle of repose and the data are given in Table 1. It is observed that the d_{80}

passing size is varying from 210 μm to 510 μm for beach sand samples and from 350 μm to 450 μm for dune sand samples.

Table 1: Typical physical properties of some of the Beach and Dune sand samples of Brahmagiri coast

Sl.No	Sample name	Angle of repose (°)	Density, [g/cm^3]		THM, %	d_{80} passing size, μm
			Bulk	True		
1.	7C	46.02	1.548	2.72	15.21	340
2.	8C	29.31	1.474	2.71	8.95	370
3.	10C	33.4	1.603	2.66	4.16	500
4.	12C	34.41	1.625	2.7	9.75	415
5.	15C	43.49	1.493	2.7	10.44	335
6.	19C	31.7	1.511	2.73	13.95	310
7.	20C	39.8	1.463	2.7	7.27	325
8.	7D	29.85	1.694	2.78	17.78	370
9.	8D	29.73	1.791	2.75	17.54	450
10.	10D	27.74	1.75	2.82	19.14	400
11.	12D	29.05	1.775	2.83	20.33	410
12.	15D	29.51	1.69	2.79	16.98	460
13.	17D	30.95	1.854	2.93	29.4	350
14.	20D	30.62	1.725	2.73	13.88	390

The d_{80} passing size values vs. true density, THM content are shown in Figures 2 to 5. Figure 2 and 4 represent the data for the true density of beach and dune sand minerals with d_{80} passing sizes. It can clearly be seen from Figure 2 and 4 that with increase of particle size, the true density of minerals is falling down. These experimental data are justified, as with increase in particle size the concentration of total heavy mineral content decreases as well as the percentage of gangue minerals increases, which is having specific gravity less than 2.6 gm/cc. The Figure 3 and 5 shows the variation of d_{80} passing size with total heavy mineral content for beach and dune sand samples. Both the graphs show decreasing trend i.e. with increase in d_{80} passing size, the THM % decreases. This is because; the THM is concentrated in finer fraction.

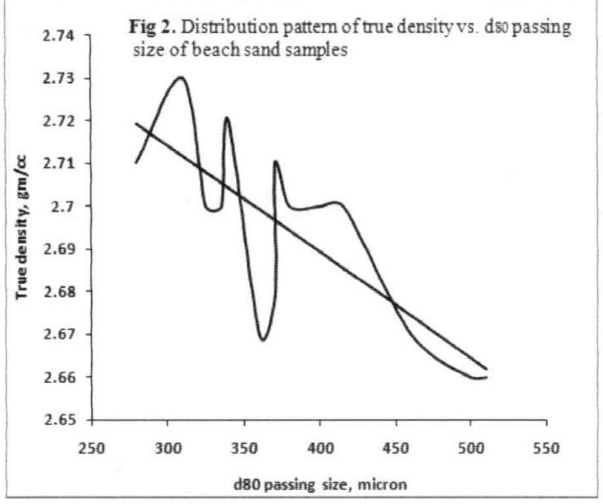

Fig 2. Distribution pattern of true density vs. d_{80} passing size of beach sand samples

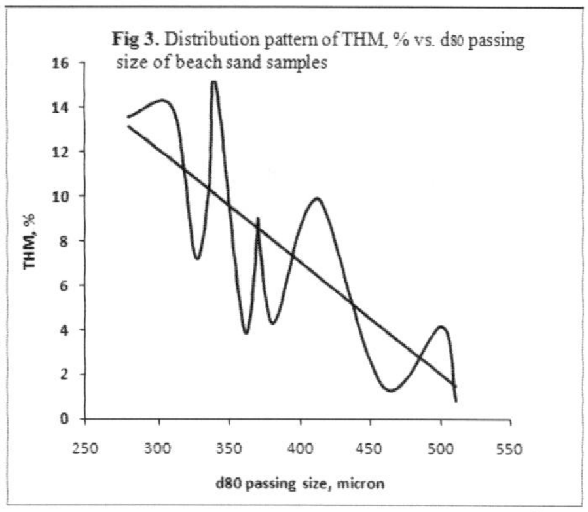

Fig 3. Distribution pattern of THM, % vs. d_{80} passing size of beach sand samples

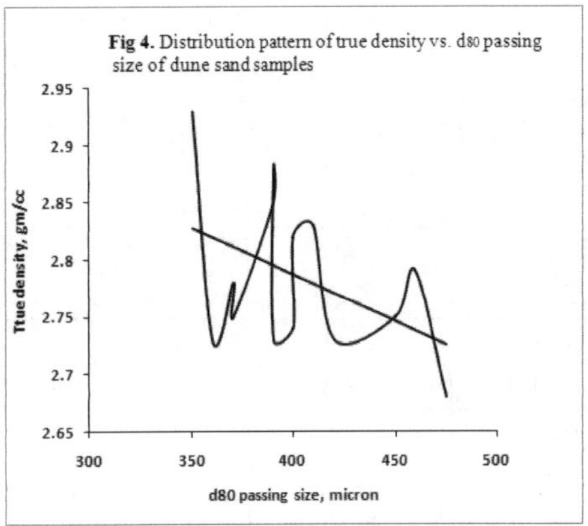

Fig 4. Distribution pattern of true density vs. d_{80} passing size of dune sand samples

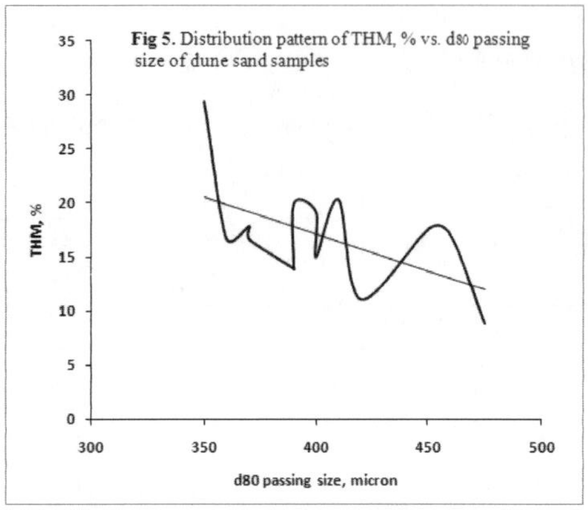

Fig 5. Distribution pattern of THM, % vs. d_{80} passing size of dune sand samples

Figure 6 shows the histogram presenting the THM content of beach and dune sand samples. It is observed in all cases that the dune sand samples contain higher THM than beach sand samples.

Figure 7 and 8 present the size analysis data for beach and dune sand sample. In Figure 7, it is observed that weight % of -425μm + 250 μm fractions are more in almost all samples. The Figure 8 also shows the similar type of behavior for dune sand samples. The next highest weight % lies in the fraction -250 μm+212 μm for both beach and dune sand samples.

The physical characterization of the composite sample is presented in Table 2. The THM content is 14.5 % and the d_{80} passing size is 370 micron. The close size sink-float study reveals that maximum percentage of THM i.e. 4.9 % lies in -150+75 micron fraction (Table 3).

The rougher gravity table operation was done using sea water at different operating conditions obtained for fresh water to assess the effect of medium on separation. The results of sea water gravity table are given in Table 4. Under the conditions Wash water 2.5 lit/ min, pulp density 15%, deck angle 1.3°, the rougher gravity table operation with sea water shows grade of 61.3% and recovery of 93.1 %. Then under the conditions of Wash water 3 lit/ min, pulp density: 20%, deck angle 1.4°, the rougher gravity table shows grade of 63.8 % and recovery of 96.4 %. Rougher gravity table operation under the conditions Wash water 3.5 lit/ min, pulp density: 25%, deck angle 1.5°shows grade of 63.0% and recovery of 95.3 %.

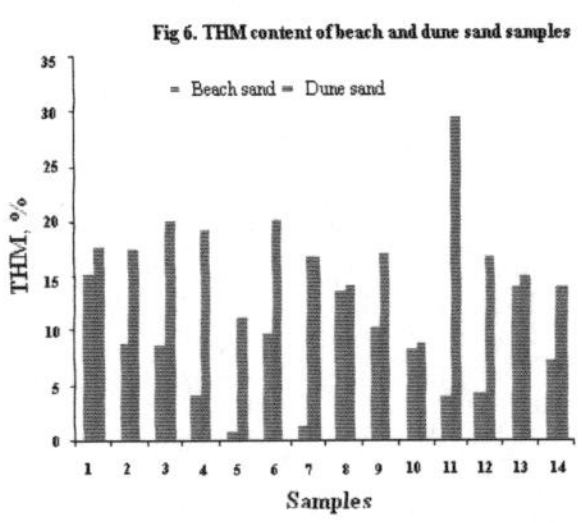

Fig 6. THM content of beach and dune sand samples

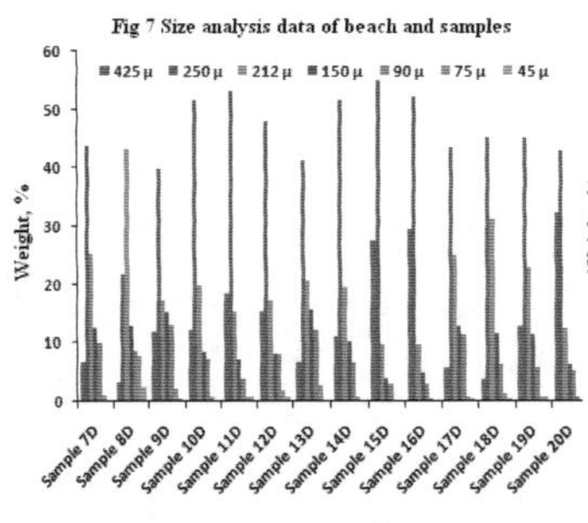

Fig 7 Size analysis data of beach and samples

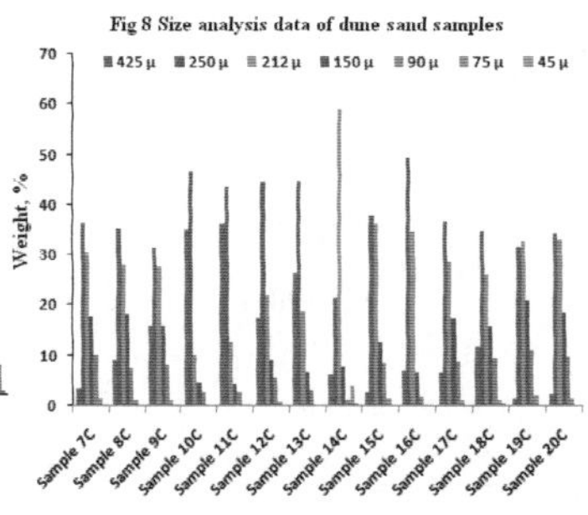

Fig 8 Size analysis data of dune sand samples

Table 2: Physical properties of composite sample	
Angle of repose (°)	28.07
Bulk density, g/cm^3	1.74
True density, g/cm^3	2.76
THM, %	14.5
d_{80} passing size, μm	370

Table 3: Sink-float result of close size fraction of the composite sample	
Size, micron	THM distribution, %
+425	0.4
-425+250	1.4
-250+212	2.1
-212+150	4.1
-150+90	4.9
-90+75	1.0
-75	0.6
Total	14.5

Table 4 Results of rougher spiral concentrator using sea water and fresh water as medium

Conditions	Sea water		Fresh water	
	Grade, %	Recovery, %	Grade, %	Recovery, %
Wash water 2.5 lit/min, pulp density 15%, deck angle 1.3°	61.3	93.1	59.8	92.3
Wash water 3 lit/min, pulp density: 20%, deck angle 1.4°	63.8	96.4	61.0	95.1
Wash water 3.5 lit/min, pulp density: 25%, deck angle 1.5°	63.0	95.3	60.2	94.1

In all cases sea water rougher gravity table produces better results than fresh water rougher gravity table. It is expected that with sea water as the separation medium, the grade and recovery of heavy minerals will be better than that of fresh water. It is because concentration criterion 'C' achieved for separation of heavy minerals is better in sea water than that for fresh water. The concentration criterion 'C' is defined as an indication whether a given ore is amenable for processing by gravity methods or not. It is expressed as $C = (d_h - d_m)/(d_l - d_m)$ -------------- Eq 1 [6] Where, d = specific gravity and Subscripts h, l and m = heavy mineral, light mineral and medium respectively. If this value is less than 1.25 practical separations is substantially impossible.

Then the composite sand sample with 14.5 % THM is subjected to number of stages of gravity table to recover THM under the optimized conditions (Table 4) of wash water 3 lit/ min, pulp density 20%, deck angle 1.4°. The rougher gravity table is again subjected to cleaner and recleaner gravity table to achieve a THM of more than 91.2 % grade. The cleaner and recleaner tailings are subjected combinely to three stages of scavenging gravity table operations to recover values from it. The tailings from all scavenging circuit are rejected. The flowsheet with mass balance is presented in Figure 9.

4. CONCLUSIONS

From the study on the assessment on characterization of physical properties of beach placer deposits of Puri district, Odisha and its application on gravity units to recover heavy minerals for metallurgical applications, the following conclusions are drawn:

- The true density of all beach placer and dune placer samples decreases with increase in d_{80} passing size
- The composite sample prepared from beach and dune sand samples contain 14.5 % THM.
- Sea water rougher gravity table shows better results than fresh water rougher gravity table.
- The gravity table study shows a yield of 14.8%, grade of 91.2 % and recovery of 93.1 %.

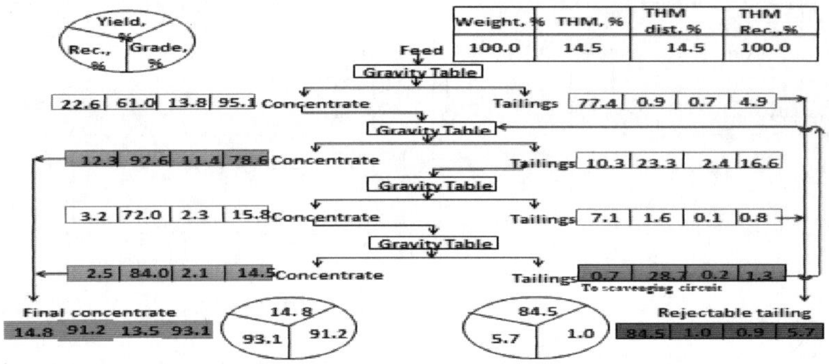

Fig 9. Flowsheet with material mass balance for recovery of total heavy minerals from placer beach sand

ACKNOWLEDGMENTS

Authors are grateful to the Chairman of C. V. Raman group of Institutions, Bhubaneswar, Odisha, India for providing support to carry out the research work. One of the authors, Ms. V. Jagadeeswari is thankful to SERB for granting JRF.

REFERENCES

[1] Azam, Ali Mir, Krishnan, S and Banerjee, D C 2001 Beach and inland heavy mineral sand investigations and deposits in India – An overview, Special issue on Beach and Inland Heavy Mineral Sand Deposits of India. Exploration & Research for Atomic Minerals **13** 1-21.

[2] AMD 2011 Atomic Minerals Directorate for Exploration and Research, www.dae.gov.in, retrieved in Dt. 07.07.2014.

[3] Jagannadha Rao K, Subramanyam A V, Abhinav Kumar, Sunil T C and Chaturvedi A K 2008 Discovery of heavy mineral-richsand dunes along the Orissa – Bengal coast of India using remote sensing techniques *J. Curr. Sci* **94** 983-85

[4] IMYB 2012, Indian Minerals Year Book, Indian Bureau of Mines, Chapter 27 (Ilmenite & Rutile).

[5] IMYB 2013, Indian Minerals Year Book, Indian Bureau of Mines, Chapter 27 (Ilmenite & Rutile).

[6] Barry A Wills and Tim Napier-Munn 2006 Gravity Concentration *Wills' Mineral Processing Technology (Seventh Edition): An Introduction to the Practical Aspects of Ore Treatment and Mineral Recovery* 231.

Effects of Welding Parameters on Mechanical Properties in Electron Beam Welded CuCrZr Alloy Plates

Sanjib Jaypuria[1*], Nirav Doshi[1] and Dilip Kumar Pratihar[1]

[1]Mechanical Engineering Department, Indian Institute of Technology Kharagpur, Kharagpur-721302, India

Corresponding author E-mail: sanjibjaypuria@gmail.com

Abstract: CuCrZr alloys are attractive structural materials for plasma-facing components (PFC) and heat sink element in the International Thermonuclear Experimental Reactor (ITER) fusion reactors. This material has gained so much attention because of its high thermal conductivity and fracture toughness, high resistance to radiation damage and stability at elevated temperatures. The objective of this work is to study the effects of electron beam welding parameters on the mechanical strength of the butt welded CuCrZr joint. Taguchi method is used as the design of experiments to optimize the input parameters, such as accelerating voltage, beam current, welding speed, oscillation amplitude and frequency. The joint strength and ductility are the desired responses, which are measured through ultimate tensile strength and percent elongation, respectively. Accelerating voltage and welding speed are found to have significant influence on the strength. A combination of low amplitude and high-frequency oscillation is suggested for the higher joint strength and ductility. There is a close agreement between Taguchi predicted results and experimental ones. Fractographic analysis of joint and weld zone analysis are carried out to study the failure behaviour and microstructural variation in the weld zone, respectively.

Keywords: CuCrZr alloy, Electron Beam Welding, Beam Parameters, Taguchi Method, Tensile Strength, Microstructure

1. Introduction

In most of the fusion reactors, such as an international thermonuclear experimental reactor (ITER), plasma facing components (PFCs) are the most crucial ones, which act as the barrier between various internal components and vacuum chamber from the direct heat of plasma irradiation. These PFCs contain one armor material with one heat sink component, which helps to transfer the heat from the armor to the water coolant. So, this heat sink material should have resistance to the high heat load, high flux energy ion and neutral particle irradiation [1]. ITER recommended Copper-Chromium-Zirconium (CuCrZr) alloy has the potential to be used as heat sink material due to the properties including high thermal conductivity, excellent welding properties, and relatively high strength. The relatively new fusion welding technology like electron beam welding (EBW) has the inherent advantage of high aspect ratio, minimal heat affected zone (HAZ) and high joint efficiency. The precipitation hardened CuCrZr alloy plates are welded using electron beam welding for industrial assembling within the ITER context [2].

Kanigalpula et al. [3] studied the effect of EBW input parameters on bead geometry, and a few mechanical properties of the bead-on-plate welded CuCrZr. Statistical regression was also carried out to find the input-output relationships and then, the same were used to construct the optimization problem in order to minimize weld-bead geometry. Gong et al. [4] conducted the laser-arc hybrid welding of pure copper and characterized the weld by microstructural variation, conductivity, and tensile strength. They found out the coarse and columnar grains in the fusion zone, which lead to lower conductivity of the weld. With the increase of heat input, the weld and HAZ area were increasing, but they found a little effect on the tensile behaviour. Durocher et al. [5] characterized the various copper alloys with electron beam welding and selected the optimized grade for electron beam welding. They found a repeated occurrence of hot cracks in the EB welded CuCrZr joint, which leads to a water leak in the heat sink element. It was also reported that EB welded CuCrZr joint has less ductility and strength than that of base metal. High zirconium (> 0.14 wt. %) content and low chromium (< 0.6 wt. %) content were recommended to increase the ductility and strength, respectively. Feng et al. [6] pointed out the problem faced with the traditional welding of CuCrZr and found a significant reduction in mechanical and electrical properties along with significant difference of microstructure in terms of grain size and precipitates between the weld and base metal. They had suggested continuous extrusion forming after flash butt welding, which leads to sub-micron scale grain through dynamic recrystallization. Siddaiah et al. [7] conducted bead-on-plate electron beam welding of AISI 304 stainless steel according to the full-factorial design of experiments with three inputs having three levels each and four weld responses. A multivariate regression analysis had been conducted to establish an input-output relationship.

From the literature review, it is found that although electron beam welding of CuCrZr is one of the preferred technologies for fabrication, it has some disadvantages like grain coarsening and strength reduction due to high power density. Thus, there is a need for the studies regarding the strength and ductility of EB welded CuCrZr joint. Apart from these, there is a need to study the effect of secondary welding input parameters like oscillation amplitude and frequency along with the primary parameters like voltage, current and welding speed. In the present study, Taguchi's experimental design based on L8 orthogonal array has been used in investigating the effects of electron beam welding input parameters on strength and dutility of the EB welded CuCrZr joint. The optimal parameters for maximization of strength and ductility have been evaluated along with a fractographic study to understand the failure behavior.

2. Experiments and Methodology

2.1. *Materials and Design of Experiments*

Precipitation hardened CuCrZr alloy plates with the dimensions of 170 mm × 50 mm × 5 mm are used as the base material for this work. The chemical compositions of the material are as follows: 0.83% Cr, 0.06% Zr and remaining copper. At first, the as-received metal is cleaned and milled to get defect-free joint. In this study, beam oscillation parameters like oscillation amplitude and frequency are used as input parameters along with accelerating voltage, beam current and welding speed. The facility used in this study is an indigenously devolved EB welding setup (80kV-12kW) located at electron beam welding center of IIT Kharagpur. Taguchi L8 orthogonal array is used here for planning the experimental runs because of ease and minimum use of resources with the minimum numbers of experimental runs. In this study, five input parameters of electron beam welding with two levels each are used to establish the relationship between the response (mechanical strength and ductility) and input parameters. Table 1 represents the input parameters of EB welding and their levels. These experimental runs designed on the basis of Taguchi L8 OA along with responses are given in Table 2.

2.2. *Post Weld Characterization Techniques*

The EBW samples are cut, metallographically polished and then etched with the solution of $FeCl_3$ and HNO_3 to reveal the microstructure. Microstructures are captured using an optical microscope (Leica, Model: S8AP0) and scanning electron microscope (Zeiss, Model: Evo18 Research). The longitudinal tensile specimens are cut from the joint with CNC wire-EDM according to the ASTM E8-M standard [8]. The tensile tests are conducted with a strain rate of 1 mm/min in Instron universal testing machine to evaluate ultimate tensile strength (UTS), yield strength (YS), % elongation (%El) and failure location. The fractured surface of the tensile specimen is studied under scanning electron microscope to investigate the behaviour of failure.

Table 1. Experimental input parameters and their level

Parameters	Symbol	Unit	Level 1	Level 2
Accelerating Voltage	V	kV	50	70
Beam Current	I	μA	80	90
Welding Speed	S	mm/min	600	1000
Oscillation Amplitude	A	mm	0.1	0.5
Oscillation Frequency	F	Hz	400	900

Table 2. Experimental runs and responses

Run	V	I	S	A	F	YS	UTS	% El	Failure location	Remark on Penetration
1	50	80	600	0.1	400	107.7	146.3	9.308	Weld metal	Partial
2	50	80	600	0.5	900	95.31	114.7	8.36	Weld metal	Partial
3	50	90	1000	0.1	400	89.25	100.2	6.144	Weld metal	Partial
4	50	90	1000	0.5	900	87.58	99.8	6.236	Weld metal	Partial
5	70	80	1000	0.1	900	118.6	193.7	12.6	Weld metal	Almost Full
6	70	80	1000	0.5	400	107	186.3	11.71	Weld metal	Almost Full
7	70	90	600	0.1	900	137.9	260	19.43	Weld metal	Full
8	70	90	600	0.5	400	121.2	248.2	18.16	Weld metal	Full

3. Results and Discussion

Three response parameters (UTS, YS, % elongation) are considered for the experimentation. However, UTS and % elongation are only considered for further statistical analysis due to the very same nature of behavior observed by the yield stress and UTS. In this study, Taguchi method is used to determine the optimal input parameters of electron beam welding to maximize the ultimate tensile strength and ductility of the joint.

3.1. *Analysis of variance (ANOVA) for UTS and % Elongation*

Analysis of variance (ANOVA) is used to investigate which EBW parameters have significant effects on the response (UTS and % elongation) of the welded CuCrZr joint. Here, the variability of the S/N ratio is evaluted by the sum of squared deviation. The total sum of squred deviation is composed of sum of squared deviation contributed by each input parameter and the error in prediction [9]. An ANOVA result for UTS is shown in Table 3. It is seen from the UTS ANOVA table that welding speed and accelerating voltage are significant parameters for deciding the joint strength. Accelerating voltage and welding current have significant contribution of 82 % and 14%, respectively, in the calculation of UTS. Although the remaining three parameters have no significant effect, welder should give priority to the amplitude of beam oscillation, which directly influences the weld width and reinforcement height of the joint. Although the contributions of the input parameters are different, a similar trend is observed from the % elongation ANOVA table. It is also observed that accelerating voltage and speed have 76% and 23% contributions, respectively, towards the ductility of the joint.

Table 3. Result of ANOVA for UTS

Source	DF	Seq SS	Adj SS	Adj MS	F	P	Percentage of Contribution
Accelerating Voltage (V)	1	65.2621	65.2621	65.2621	120.61	0.008	82.54
Beam Current (I)	1	0.0371	0.0371	0.0371	0.07	0.818	0.05
Welding Speed (S)	1	11.4028	11.4028	11.4028	21.07	0.044	14.42
Oscillation amplitude (A)	1	1.0357	1.0357	1.0357	1.91	0.301	1.31
Oscillation Frequency (F)	1	0.2458	0.2458	0.2458	0.45	0.57	0.31
Residual Error	2	1.0822	1.0822	0.5411			
Total	7	79.0658					

3.2. *Main effect plot and parametric study*

The main effects plots show the influences of different input parameters on the responses, namely UTS and % elongation. Here, MINITAB 15 is used to draw the main effects plot. Main effects plots for UTS are given in Fig 1. Accelerating voltage is found to have the maximum significant effect on UTS. It is seen from the Fig 1 that a greater strength can be achieved with high accelerating voltage, high beam current, low welding speed and low beam amplitude and frequency. As the voltage is directly related to kinetic energy of electron beam, the latter increases with the increase of former, this helps the penetrating beam to go to the higher depth of weld profile. The remark on penetration status for each weld has been shown in Table 2. Samples-7 and 8 have high values of YS and UTS because of full penetration of the weld profile. A high beam current is suggested by the plot, as beam current assists to enhance the power density of beam for a constant beam radius. A low welding speed ensures high heat input per unit length for a constant power and high heat input ensures the better fusion of the joint. Beam oscillation amplitude and frequency decide the scan profile of the beam. It is seen from the main effect plot of UTS that a low oscillation amplitude and frequency ensure the better strength of the joint. However, the differences between high and low levels of amplitude and frequency are found to be low in Fig 1. It is difficult to conclude the optimum combination of beam oscillation parameters for a sound weld profile. It is noted that almost a similar trend has been observed in the main effect plot of % elongation. The optimal parametric conditions for maximizing % elongation are high accelerating voltage, high current, low welding speed, low amplitude and high frequencies.

3.3. *Regression analysis*

Minitab 15 [10] is used to conduct linear regression analysis in order to establish the input-output relationships of the electron beam welding process. The equations for UTS and % elongation are shown in equations (1) and (2), respectively.

$$\text{UTS} = 43.98 - 2.85 \times V - 0.0681 \times I + 1.1939 \times S + 0.3598 \times A + 0.1753 \times F \quad (1)$$

$$\%\text{Elongation} = 20.48 - 3.1059 \times V - 0.1768 \times I + 1.7153 \times S + 0.2539 \times A - 0.0529 \times F \quad (2)$$

3.4. *Confirmatory test result*

Based on the main effect plot and ANOVA, the optimum set of parameters for maximizing the strength and ductility of the joint is found to be V2I2S1A1F2. Table 4 displays the confirmatory test result. The experimental result has a good agreement with the predicted one, as the percent error in prediction is found to be equal to 0.0327%.

Table 4. Confirmatory test ı results

	Optimal Parameters	
	Prediction	Experimental
Factors level	V2I2S1A1F2	V2I2S1A1F2
S/N ratio	48.2825	48.2983
Error between S/N ratio	0.0327%	

3.5. *Metallurgical studies*

UTS and % elongation of base metal are found to be equal to 397.962 MPa and 16.486%, respectively. There are significant reductions in the UTS and YS of the welded sample, as compared to that of the base metal. This statement is supported by the fact that all the welded samples are broken in the weld zone of longitudinal sub-size tensile specimens. Due to the high thermal conductivity and soft nature of this copper alloy, it is difficult to achieve the parent material's strength with any kind of fusion welding process. There is a significant amount of grain coarsening in the weld

IOP Conf. Series: Materials Science and Engineering **338** (2018) 012013 doi:10.1088/1757-899X/338/1/012013

and heat affected zones of the joint, which is displayed in Fig 2. The fractured surface of the broken welding surface is studied under scanning electron microscope to understand the failure behaviour. The fractographic image of one of the samples is shown in Fig 3. The macrograph consists of fine and uniform dimples and micro-voids, which indicates that the specimen fails in a ductile mode under the action of tensile loading.

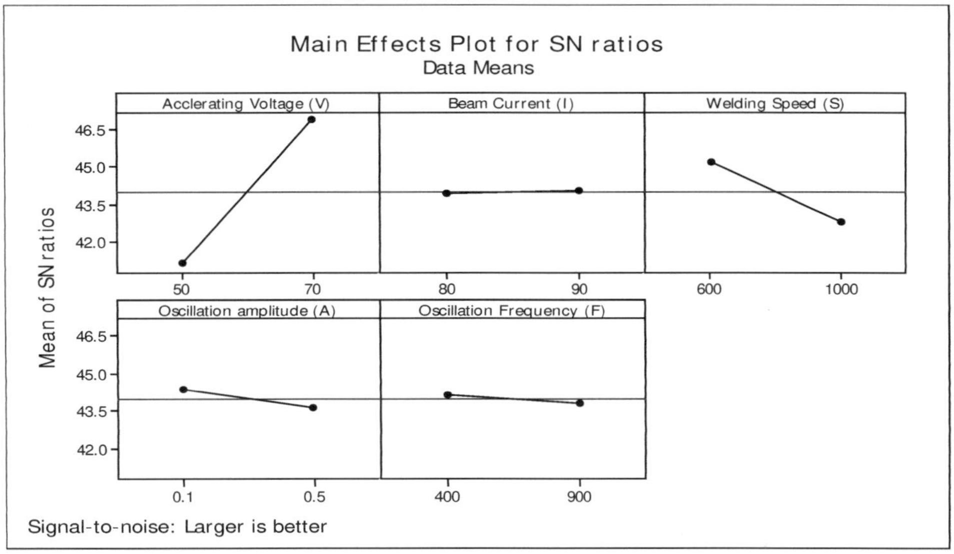

Figure 1. Main effects plot for ultimate tensile strength (UTS)

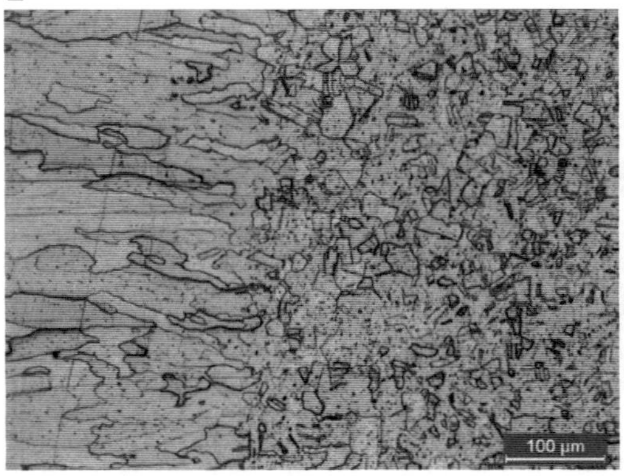

Figure 2. Microstructure of weld interface

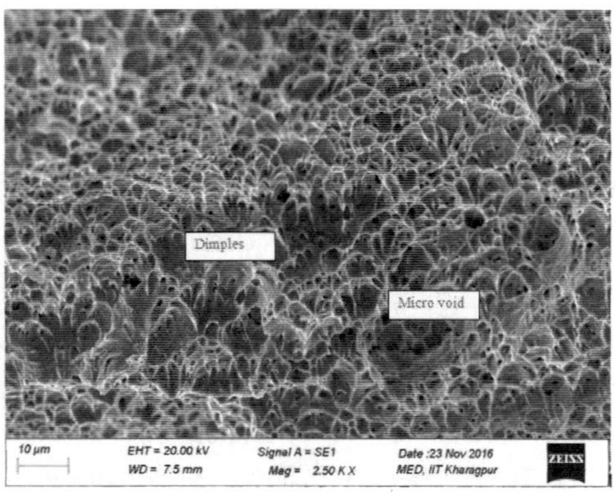

Figure 3. Fractography of tensile specimen

4. Conclusion

The aim of this work is to study the effects of EBW process parameters on the mechanical strength of butt welded CuCrZr and to optimize these parameters for achieving the maximum strength and ductility. Both accelerating voltage and welding speed have the significant contributions for achieving the maximum strength and ductility. It is concluded from the main effects plot that high levels of voltage and current, low level of welding speed, and low and high levels of oscillation amplitude and frequency, respectively, are required for the maximization of strength and ductility, for a fixed thickness of CuCrZr plate in EBW. All welded tensile samples undergo ductile failure under tensile loading, which is indicated by the presence of dimples. The strength of welded CuCrZr reduces significantly, as compared to that of base metal due to grain coarsening and generation of intense heat during EBW. There is a good agreement between the predicted and experimental results.

Acknowledgement

The authors gratefully acknowledge the funding of Board of Research Nuclear Science (BRNS), Department of Atomic Energy, Government of India to carry out this study (Sanction Number: 34/14/66/2014-BRNS /10511 Dated 19/06/2015)

References

[1] Sun, C., Wang, S., Guo, W., Shen, W. and Ge, C., 2014. Bonding Interface of W–CuCrZr Explosively Welded Composite Plates for Plasma Facing Components. *Journal of Materials Science & Technology*, **30**(12), pp.1230-1234.
[2] Durocher, A., Ayrault, D., Chagnot, C., Lipa, M. and Saikaly, W., 2007. CuCrZr alloy hot cracking during electron beam welding. *Journal of Nuclear Materials*, **367**, pp.1208-1212.
[3] Kanigalpula, P.K.C., Pratihar, D.K., Jha, M.N., Derose, J., Bapat, A.V. and Pal, A.R., 2016. Experimental investigations, input-output modeling, and optimization for electron beam welding of Cu-Cr-Zr alloy plates. *The International Journal of Advanced Manufacturing Technology*, **85**(1-4), pp.711-726.
[4] Gong, M., Kawahito, Y., Gao, M. and Zeng, X., 2017. Weld characterization of laser-arc hybrid welding of pure copper. *The International Journal of Advanced Manufacturing Technology*, **92**(1-4), pp.561-567.
[5] Durocher, A., Lipa, M., Chappuis, P., Schlosser, J., Huber, T. and Schedler, B., 2002. TORE SUPRA experience of copper chromium zirconium electron beam welding. *Journal of Nuclear Materials*, **307**, pp.1554-1557.
[6] Feng, H., Jiang, H., Yan, D. and Rong, L., 2015. Microstructure and mechanical properties of a CuCrZr welding joint after continuous extrusion. *Journal of Materials Science & Technology*, **31**(2), pp.210-216.
[7] Siddaiah, A., Singh, B.K. and Mastanaiah, P., 2017. Prediction and optimization of weld bead geometry for electron beam welding of AISI 304 stainless steel. *The International Journal of Advanced Manufacturing Technology*, **89**(1-4), pp.27-43.
[8] ASTM E8/E8M-16a Standard Test Methods for Tension Testing of Metallic Materials, ASTM International, West Conshohocken, PA, 2016, https://doi.org/10.1520/E0008_E0008M-16A
[9] Javadi, Y., Sadeghi, S. and Najafabadi, M.A., 2014. Taguchi optimization and ultrasonic measurement of residual stresses in the friction stir welding. *Materials & Design*, **55**, pp.27-34.
[10] Minitab Inc. User Manual of MINITAB Statistical Software. Release 15 State College PA 16801 USA, 2007 (http: // www.minitab.com)

Study of distribution of Carbon nanotube in Al-CNT nanocomposite synthesized via Spark-Plasma sintering

A.Maiti[1], and T. Laha[2]

[1]Department of Mechanical Engineering, C V Raman Collage of Engineering, Bhubaneswar-752054, Orissa, India

[2]Department of Metallurgical & Materials Engineering, Indian Institute of Technology Kharagpur, Kharagpur- 721302, West Bengal, India

Corresponding author Email : amaiti@cvrce.edu.in

Abstract: In the present study, first ever attempt has been made to develop physically functionalized multiwalled carbon nanotube (MWCNT) reinforced Al-11.5Si alloy nanocomposites synthesized via novel consolidation technique viz spark plasma sintering (SPS). There is a recent trend in employing carbon nanotubes (CNTs), an allotrope of carbon, as reinforcement for high strength structural metallic composite materials, as these cylindrical nano-fibers poses extremely unique mechanical properties such as very high elastic modulus (~ 300 GPa to 1.5 TPa) as well as tensile strength (~150 GPa). However, it has remained as an ever-existing problem to achieve a porosity-free nanocrystalline matrix with homogenously dispersed CNTs, owing to the very high coagulation tendency of CNTs. The gas-atomized, spherical Al-11.5Si alloy powders (1-8 μm) were subjected to high energy ball milling for the purpose of achieving nanocrystallinity in the powders. The improvement in MWCNT dispersion was effort by treating the MWCNTs with a physical surfactant, sodium dodecyl sulfate (SDS). The nano-grained ball-milled Al-Si powders with varying MWCNT content (0.5 and 1 wt%) were consolidated via spark plasma sintering in order to retain the nano-sized grains in the Al-Si matrix, attributed to the faster and highly effective sintering kinetics of the sintering techniques. FESEM study shows problem of MWCNT agglomeration persists by addition of non-SDS treated as pristine MWCNT in the composite. After treated with SDS, MWCNTs are well separated out from each other and as a result of that good morphological and mechanical property such as high hardness value obtained after analysis. Detailed TEM study of the 0.5wt% MWCNT reinforced SPS nanocomposite revealed that the distribution of CNTs in the matrix. Mechanical analysis study of the nanocomposite attributes higher hardness in case of SDS treated CNT reinforced nanocomposite owing to less agglomeration problem of the CNT in the matrix. Nano-tribological data attributed variation of surface roughness after consolidated by SPS.

1. Introduction

In automobile and aerospace industries different components are made of various grades of Aluminum based alloys owing to its lighter in weight and easy deformability. However, some of the application areas where strength of the material is prime concern, there Al alloys are failed to serve satisfactory result due to its very low load bearing capacity. To overcome this issue, various types of hard reinforcing elements such as SiC, alumina etc. is added with different proportion to the Al matrix in order to increase the strength of the composite. However, further study reveals that not only proportion but final size as well as distribution of the reinforcing elements play a vital role in the final strength of the composites. By reducing the size of the reinforcing elements, dislocation movements can block in nanometer level which ultimately increases the strength of the composite. Interestingly, density of the consolidated composite does not alter even after incorporating reinforcement elements in different proportion, sometimes it lower the density of the composite but obtained superior strength after addition. Nanometer size hard particles such as Nano-SiC, Nano-Alumina are being used as reinforcement for synthesizing nanocomposite. But, among all the hard nanoparticles, Carbon nanotubes (CNT) are chosen for the present research work as reinforcement elements due its tremendous high tensile strength and elastic module [1-5]. However, major problem with CNTs is that it has very high tendency of agglomeration owing to their very high surface energy [6]. Agglomerated CNTs are undesirable as they are hindered the proper consolidation during sintering by forming a barrier between two individual particle. As a result of that catastrophic failure of the material would be occurred from that agglomerated region of CNTs. Therefore, separation of CNTs is vital for obtaining a good consolidate microstructure as well as high strength. To meet the objective, in the present research work, Multi-walled CNTs are added in two different proportions (0.5 & 1 wt %) after treating with surfactant SDS. For comparison purpose, another two different samples with same proportion of non SDS treated CNTs (i.e. .0.5 and 1 wt%) are added with the Al alloys powder and finally consolidated via spark plasma sintering (SPS) process. Technique of SPS process is novel as here simultaneously thermal as well as mechanical energy are applied for consolidation. Due to this very good consolidated samples could be obtained after sintering. Sintered nanocomposites were analyzed through optical, SEM and finally TEM in order to study the distribution of MWCNT before and after SDS treatment. Mechanical property evaluations of the sintered samples are carried out to obtain the effect of distributed MWCNT on the composite.

2. Materials and methods

In the Present research work following are the materials obtained externally in as received condition for synthesizing nano-compostite :

(i) Gas atomized, spherical Al-Si (11.5 wt %) alloy powder with 2-10 μm diameter which was used as matrix material in the nanocomposite.

(ii) 0.5 wt% & 1 wt% multi-walled carbon nanotubes (MWCNTs) of purity >95%, 40–70 nm in diameter and 0.2–0.5 μm in length were used to reinforce the Al-Si alloy matrix.

At first, Al-Si alloy powders of desired quantity was measured separately and then poured inside a planetary ball mill (Retsch GmbH, Germany, Model: PM 100) in order to convert micron size powder to nano-meter size Al powder with continuous milling operation. Ball milling was continued for 8 h using stainless steel balls of 10 mm diameter with ball to powder ratio of 20:1 at a milling speed of 180 rpm. In order to separate out the individual MWCNT from the agglomerated as received MWCNT bundle in powder form; they were dissolved in acetone with SDS by maintaining 2:3 weight ratios. Solutions of SDS and MWCNTs mixtures were then ultra -sonicated for 20 min. Sonicated mixture was poured drop-wise via a burette in a beaker which already contained ball milled Al-Si powders dissolved in acetone, which was being continuously ultra-sonicated. Solutions were allowed for drying in air & eventually dry powder mixture (ball milled Al-Si powder and MWCNT) was consolidated via SPS (Fuji Electronic Industrial Co. Ltd, Japan, Model: SPS 515S) in a 20 m diameter graphite die. Sintering was performed at 500°C for 3 min with heating rate of 163°C/min at an applied uniaxial loading of 50 MPa.

3. Result and Discussion

3.1 MWCNT distribution in the nanocomposites: FEG-SEM result

Figure 1 shows the FEG-SEM micrographs of SPS nanocomposite with 0.5wt% and 1wt% *SDS treated and non SDS treated* MWCNT. Dispersed CNTs are observed in the Figure1a and c where it is cleared that after SDS treatment CNTs are separated out. Surfactant has ability to disperse the CNT by covering the surface as a form of micelle and subsequently lowering the surface energy of CNT [7]. Interestingly, In Figure 1b and d, it has been observed that MWCNTs were agglomerated because

CNTs are added without treated with surfactant. Besides, it is clearly observed from Figure 1b and d is that agglomerated CNTs restrict the bonding between the particles by forming a barrier.

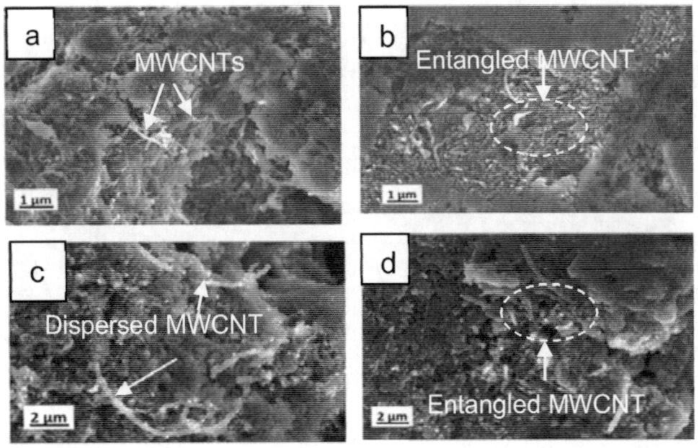

Figure 1: Distribution of (a) SDS treated 0.5 wt% MWCNT (b) non-SDS treated 0.5 wt% (c) SDS treated1 wt% (d) non-SDS treated 1wt% MWCNT reinforced SPS nanocomposite

3.2 The distribution of MWCNT in the Al-CNT Nanocomposite: TEM study

0.5wt%MWCNT reinforced SPS NC are shown in the HRTEM micrographs (Figure 2). MWCNTs agglomerated in bundle form and squeezed in between the grains. These tightly compacted MWCNTs would restrict the grain growth (Figure 2a). High magnification HRTEM micrograph of MWCNTs bundle show that MWCNTs are twisted and overlapped due to surrounding pressure of the grains.

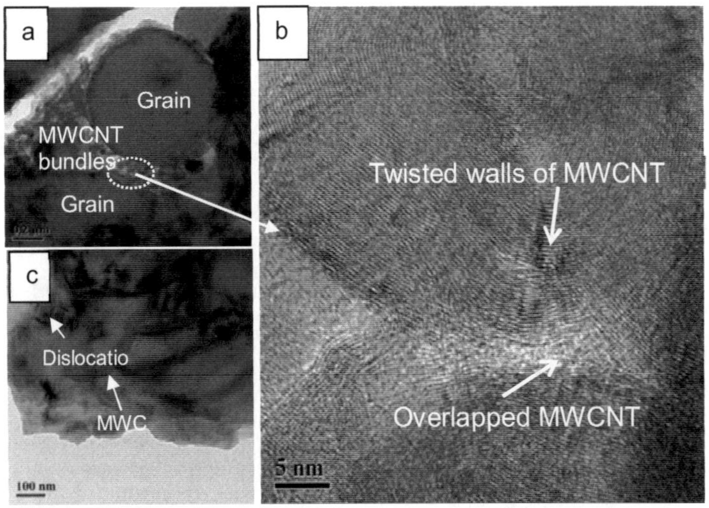

Figure 2: HRTEM micrographs showing (a) MWCNT bundles, (b) twisted and overlapped MWCNTs in between Al-Si grains, and (c) MWCNT – dislocation interaction in SPS nanocomposites with 0.5 wt% MWCNT

MWCNT-dislocation interaction can be observed in Figure 3, which indicates strengthening of the nanocomposite by virtue of dislocation movement restriction by MWCNT nano-fibers. The effect of physical functionalization on the MWCNT dispersion by treating the MWCNTs with Sodium dodecyl sulfate (SDS) is shown in Figure 3a. Although TEM images revealed only a very small portion of a sample, still it can clearly be observed that the MWCNTs were separated out very well from each other. It is worthy to mentioned here that SPS nanocomposite synthesized with pristine (not SDS treated) MWCNT exhibited agglomerated MWCNTs as shown clearly in Figure 3b.

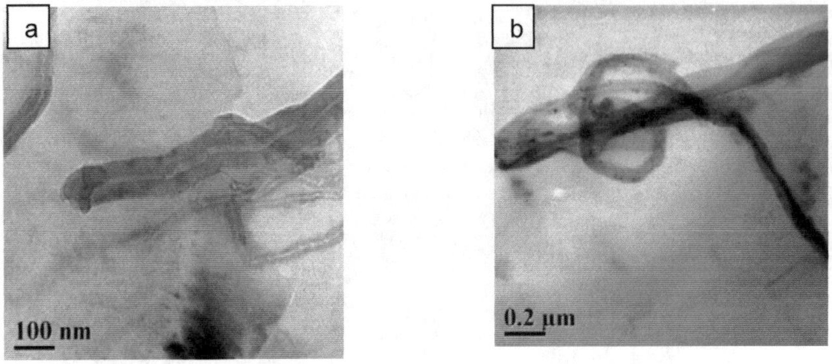

Figure 3: (a) Dispersion of SDS treated MWCNT (b) Agglomerated MWCNT

3.3 Mechanical property evaluation: hardness

Figure 4 shows the microhardness values of the SPS nanocomposite reinforced with SDS treated & non-SDS treated MWCNT. The micro-hardness increased after incorporating SDS treated MWCNT in the nanocomposite. Better distribution of the CNT in the matrix as shown in Figure 1 reason for higher consolidation and porosity attributed higher micro-hardness.

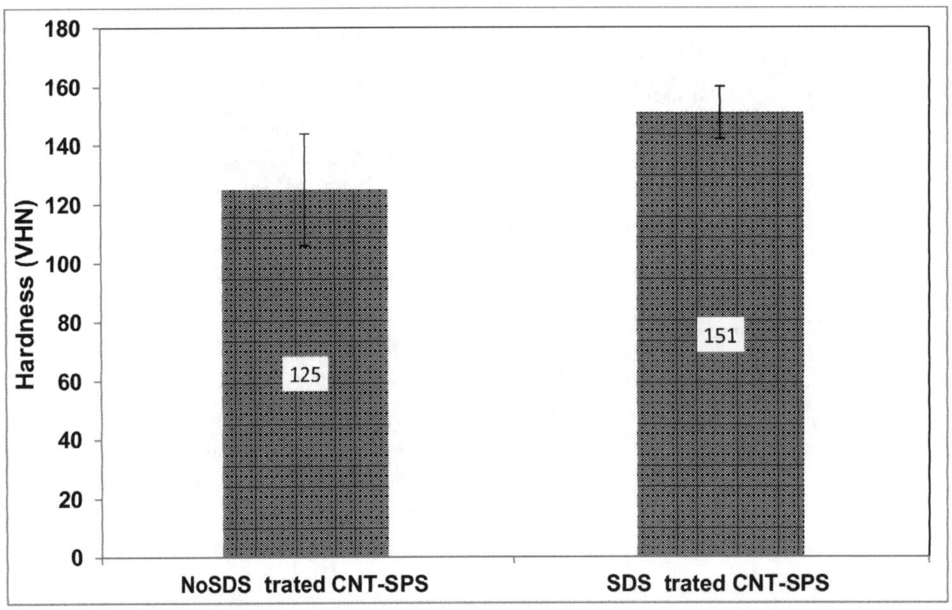

Figure 4: Micro-hardness values of the 0.5 wt% SDS & non-SDS treated MWCNT reinforced nanocomposites sintered SPS

IOP Conf. Series: Materials Science and Engineering **338** (2018) 012014 doi:10.1088/1757-899X/338/1/012014

3.4 Mechanical property evaluation: Tribological behaviour analysis by nano-scartch

Nano-scratch technique was carried out to understand the tribological behaviour of the two nanocomposites having SDS treated and non-SDS treated MWCNT as reinforcement. Figure 5 shows the variation of coefficient of friction (COF) values with a scratch distance of 4 μm at a constant load of 2000 μN. COF was measured from the ratio of the instantaneous lateral force and normal force applied during scratching. The variation of COF values along the 4 μm scratch distance in case of 0.5 wt% SDS & non SDS treated MWCNT reinforced composites are shown in Figure 5. Interestingly, variation of COF values were increased in case of non-SDS treated MWCNT reinforced nanocomposite. For example, COF of non SDS treated MWCNT reinforced composite fluctuated in between ~0.38 to 0.45 (Figure 5). But, variation is less after adding SDS treated MWCNT in the matrix. Large variation of COF data indicated the non-uniform consolidation in the nanocomposite & it is supported by the FESEM micrographs in Figure 1 where agglomerated CNTs were entangled to form a barrier for consolidation. Surface roughness or irregularities minimized after adding SDS treated MWCNT in the matrix as better consolidation obtained due to MWCNTs were separated out shown in Figure 1a.

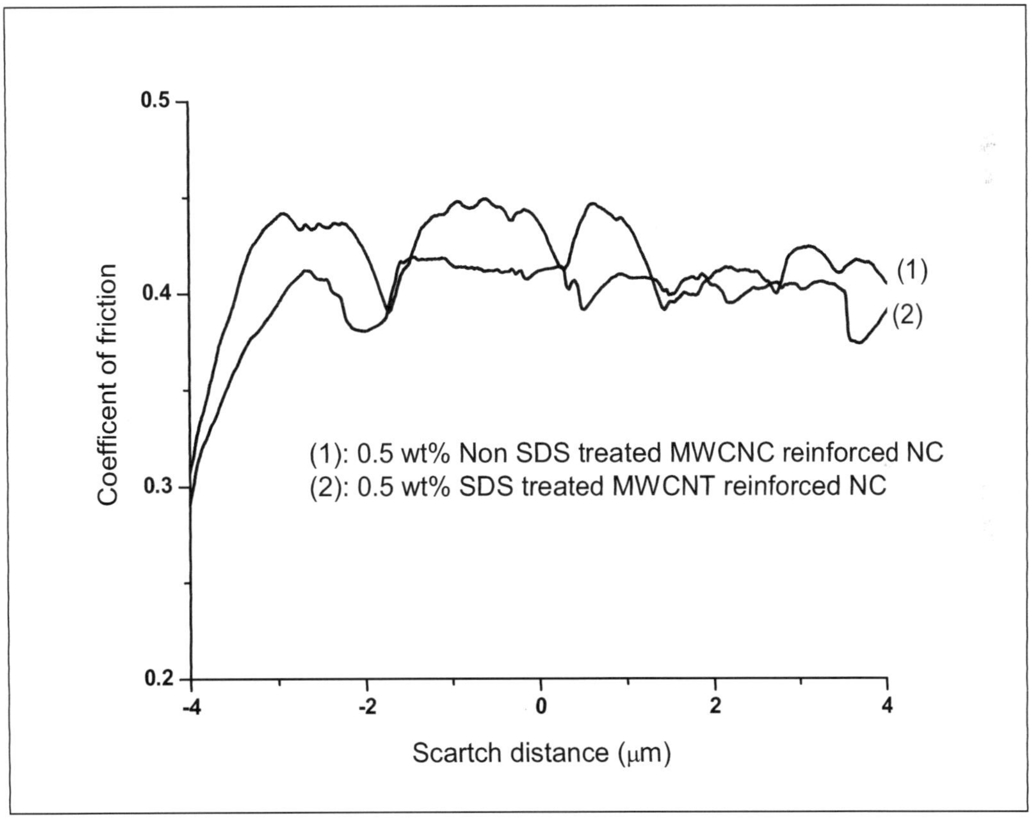

Figure 5: Variation of the coefficient of friction (COF) with scratch distance of SPS nanocomposites having 0.5wt% SDS treated & Non-SDS treated MWCNT as reinforcement

4. Conclusion

Physically functionalization of the MWCNTs improved the dispersion of these nano-fibers in the Al-Si matrix and better dispersion was achieved in the 0.5 wt% MWCNT reinforced nanocomposites, in comparison with that in 1 wt% MWCNT reinforced ones. TEM study showed that the MWCNTs retained their structural integrity without introducing major damages on the structure of the MWCNT. MWCNTs were tightly compacted along the grain boundaries and restricted the grain growth. Good adhesion between matrix and the MWCNT was observed from HRTEM study. There is further scope of densification enhancement of the nanocomposites by further improving the dispersion of MWCNTs in the matrix. Effect of other physical surfactants such as CTAB, Brij76 etc. on the dispersion of MWCNT should be studied in this context. Interestingly, SDS treated MWCNT reinforced nanocomposite elucidated better mechanical properties such as elastic modulus and micro-hardness as compared to non-treated MWCNT reinforced nanocomposite owing to the better dispersion of surfactant treated MWCNTs. Agglomerated CNTs hindered the particle to particle interaction by forming a barrier in between which eventually imparted porosity in the matrix. Increasing porosity contain in the matrix is the prime reason for degrading the mechanical properties of the nanocomposite.

5. References

[1] Pipes, R.B and Hubert, P. (2002), Helical carbon nanotube arrays: mechanical properties, Composites Science and Technology, Vol. 62, No. 3, pp. 419-428.

[2] Zhang, X.F., Zhang, X.B., Van Tendeloo, G., Amelinckx, S., Op de Beeck, M. and Van Landuyt, J. (1993), Carbon nano-tubes; their formation process and observation by electron microscopy, Journal of Crystal Growth, Vol. 130, No. 3–4, pp. 368-382.

[3] Salvetat-Delmotte, J. P and Rubio, A. (2002), Mechanical properties of carbon nanotubes: a fiber digest for beginners, Carbon, Vol. 40, No. 10, pp. 1729-34

[4] Saito, Y., Yoshikawa, T., Bandow, S., Tomita, M.and Hayashi, T. (1993), Interlayer spacings in carbon nanotubes, Physical Review B, Vol. 48, No. 3, pp. 1907-9.

[5] Baxendale, M. (2003), The physics and applications of carbon nanotubes, Journal of Materials Science: Materials in Electronics, Vol. 14, No. 10-12, pp. 657-659

[6] Lau, A.K.T and Hui, D. (2002), The revolutionary creation of new advanced materials—carbon nanotube composites, Composites Part B: Engineering, Vol. 33, No. 4, pp. 263-277.

[7] Vaisman, L., Wagner, H. D and Marom, G. (2006), The role of surfactants in dispersion of carbon nanotubes, Advances in Colloid and Interface Science, Vol. 128–130, pp. 37-46

IOP Conf. Series: Materials Science and Engineering **338** (2018) 012015 doi:10.1088/1757-899X/338/1/012015

Effect of thermally reduced graphene oxide on dynamic mechanical properties of carbon fiber/epoxy composite

Nitai Chandra Adak,[1,2] Suman Chhetri,[1,2] Naresh Chandra Murmu,[1,2] Pranab Samanta,*[1,2] Tapas Kuila *[1,2]

[1]Surface Engineering and Tribology Division, Council of Scientific and Industrial Research-Central Mechanical Engineering Research Institute, Durgapur -713209, India
[2]Academy of Scientific and Innovative Research (AcSIR), CSIR-CMERI, Campus, Durgapur- 713209, India

E-mail address: ps.iitb@gmail.com and tkuila@gmail.com

Abstract: The Carbon fiber (CF)/epoxy composites are being used in the automotive and aerospace industries owing to their high specific mechanical strength to weight ratio compared to the other conventional metal and alloys. However, the low interfacial adhesion between fiber and polymer matrix results the inter-laminar fracture of the composites. Effects of different carbonaceous nanomaterials i.e., carbon nanotubes (CNT), graphene nanosheets (GNPs), graphene oxide (GO) etc. on the static mechanical properties of the composites were investigated in detail. Only a few works focused on the improvement of the dynamic mechanical of the CF/epoxy composites. Herein, the effect of thermally reduced grapheme oxide (TRGO) on the dynamic mechanical properties of the CF/epoxy composites was investigated. At first, GO was synthesized using modified Hummers method and then reduced the synthesized GO inside a vacuum oven at 800 °C for 5 min. The prepared TRGO was dispersed in the epoxy resin to modify the epoxy matrix. Then, a number of TRGO/CF/epoxy laminates were manufactured incorporating different wt% of TRGO by vacuum assisted resin transfer molding (VARTM) technique. The developed laminates were cured at room temperature for 24 h and then post cured at 120 °C for 2 h. The dynamic mechanical analyzer (DMA 8000 Perkin Elmer) was used to examine the dynamic mechanical properties of the TRGO/CF/epoxy composites according to ASTM D7028. The dimension of the specimen was 44×10×2.4 mm^3 for the DMA test. This test was carried out under flexural loading mode (duel cantilever) at a frequency of 1 Hz and amplitude of 50 µm. The temperature was ramped from 30 to 200 °C with a heating rate of 5 °C min^{-1}. The dynamic mechanical analysis of the 0.2 wt% TRGO incorporated CF/epoxy composites showed ~ 96% enhancement in storage modulus and ~ 12 °C increments in glass transition temperature (T_g) compared to the base CF/epoxy composites. The fiber-matrix interaction was studied by Cole-Cole plot analysis. It proved the homogeneous dispersion of the epoxy resin and TRGO. The homogeneous dispersion of the TRGO in the epoxy matrix increased the overall enhancement of the dynamic mechanical properties of the hybrid composites.

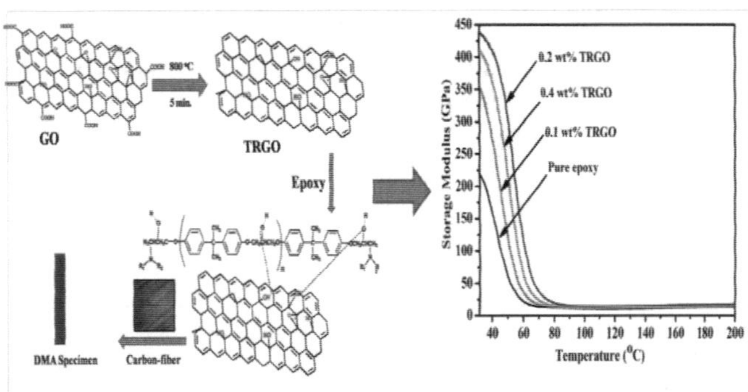

1. Introduction

Carbon fiber (CF) reinforced epoxy composites have enormous applications in light-weight composite part constructions [1]. However, the low interfacial adhesion between CF and epoxy resin creates inter-laminar fracture of the composites. Therefore, lots of works were carried out to find out the improvement in static mechanical properties of the composites by using different carbonaceous nano-structured materials i.e., carbon nanotubes (CNT), graphene nanosheets (GNPs), graphene oxide (GO) etc [2-4]. However, the incorporation of these carbonaceous nanofillers complicated the curing process of the CF/epoxy composites in presence of the dispersing solvents. In addition, the processing temperature, vacuum pressure and particle size of the nanofillers affects the cross-linking density of the composite materials. Trappe et al. cured epoxy resin at 50 and 60 °C over 15 h and noticed that the crack propagation speed of the epoxy resin cured at 60 °C is 10 times lower than that of the epoxy cured at 50 °C [5]. However, the fracture toughness of both the samples remained same. The degree of curing is usually determined by using dynamic mechanical analysis (DMA) [6]. The DMA study not only expresses the cure kinetics of the composite materials but also the damping capacity of the composites. Most of the research works are based on DMA study of the fiber reinforced epoxy composites [7, 8]. There are also a few reports on the in depth investigation of the symmetric laminated composites. Therefore, an attempt has been paid on the detail DMA study of TRGO incorporated symmetric CF/epoxy laminated composites. The GO was synthesized by modified Hummers method and then reduced through heat treatment at high temperature. The prepared TRGO was dispersed in epoxy resin for manufacturing the TRGO/CF/epoxy laminate. The VARTM process was used to manufacture the TRGO reinforced CF/epoxy laminate. The main objective of using TRGO was to restore the graphene like properties of GO. The partially reduced GO was expected to increase the interfacial interaction with the epoxy resin as well as increase the glass transition temperature (T_g) of the composites.

2. Materials and methods

Graphite flakes purchased from Sigma Aldrich, India was used to prepare the GO. Merck India supplied the chemicals i.e. H_2SO_4 (98% vv), HCl (35% v/v) and H_2O_2 (30% v/v) etc. Plain weave carbon cloth with an areal density of 200g/m^2 was purchased from Flips India Engineering (Mumbai) for reinforcement. A low viscous Bisphenol-A epoxy resin (LAPOX*C-51) and modified cycloaliphatic amine hardener (Lapox AH-428) was purchased from Atul Limited (Gujarat, India) to prepare the matrix system in this work. A heavy-duty silicon spray mold release agent and an adhesive spray were purchased from Aerol Formulation Pvt. Ltd. (New Delhi, India).

At first GO was prepared by using modified Hummers method as stated earlier work [9]. Then, about 1 g of GO was taken in a crucible and then the crucible was placed inside a muffle furnace at 800 °C for 5 min. The amount of TRGO was ~ 500 mg. In order to modify the epoxy resin, the required amount of TRGO was first dispersed in THF solvent. Similar experimental procedure was described in our previously reported work [10]. The dispersion was then poured into epoxy resin to get the 0.05, 0.1, 0.2 and 0.4 wt% TRGO incorporated epoxy matrix as compared to the total weight of resin and hardener. Then the THF was removed under reduced pressure from the TRGO/epoxy dispersion by evaporation at 85 °C for an hour. The TRGO/epoxy mixture was allowed to cool down to room temperature and stirring was continued for 4 h with a magnetic stirrer. The desired amount of curing agent was added with TRGO/epoxy mixture and stirred at 500 rpm for 2 min to obtain homogeneous dispersion. Finally, the modified TRGO/epoxy matrix was used for the manufacturing of CF/TRGO/epoxy laminate after degassing the TRGO/epoxy suspension in a vacuum chamber. The CF were cut with a dimension of 150 × 150 mm and the CF/TRGO/epoxy laminates were prepared by VARTM process. The lamina's thickness was 0.3 mm. The stacking sequence of the lamina was $[(0/90)_5]_S$. The CF/TRGO/epoxy laminates were cured at room temperature for 24 h and then post cured at 120 °C for 2 h. Finally, the rectangle shape samples were cut from the laminates for DMA study. DMA of the TRGO/CF/epoxy composite materials were carried out by using a dynamic mechanical analyzer (DMA 8000 Perkin Elmer) according to ASTM D7028. The specimen dimension was 45×10×2.5 mm^3 for the DMA test. This test was carried out under flexural loading mode (duel cantilever) at a frequency of 1 Hz and amplitude of 50 μm. The temperature was ramped from 30 to 200°C with a heating rate of 5°C min^{-1}. Field emission scanning electron microscopy (FE-SEM) images were recorded with Σigma HD, Carl Zeiss, Germany to find out the dispersion of TRGO in the epoxy resin.

3. Results and Discussion

The dispersion state of TRGO in the epoxy matrix was analyzed with FE-SEM image analysis as shown in Figure 1 (a-d). Figure 1(a) represents the surface micrograph of pure epoxy resin, which was very smooth indicating the absence of guest materials in the epoxy resin. In contrast, the surface of the TRGO/epoxy composite appeared with rough morphology due to presence of TRGO. The presence of TRGO in the epoxy resin has been shown by indicating an arrow in the Figure 1(c). The composites with 0.4 wt% of TRGO loading in the epoxy resin is shown in Figure 1 (d) and it is clearly shown that the TRGO particles were agglomerated in the epoxy resin.

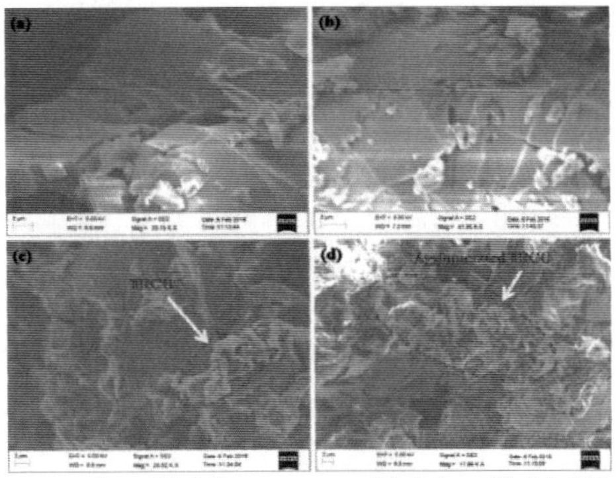

Figure 1. Surface micrographs of (a) Pure epoxy (b) 0.1 wt% (c) 0.2 wt% and (d) 0.4 wt% TRGO/epoxy dispersions

DMA investigates the curing kinetics and damping capacity of the prepared CF/epoxy and TRGO incorporated CF/epoxy composite. It measures the changes in stiffness of materials in terms of storage modulus (E'), loss modulus (E") and damping property by damping factor (tan δ). The E' signifies the stiffness or rigidity of a material and E" signifies the damping capacity of the material. The changes of E' of pure CF/epoxy and TRGO reinforced CF/epoxy composites by varying the temperature as shown in Figure 2(a). The E' increased with the addition of TRGO in the prepared composites. It indicated that the TRGO increased the stiffness of the CF/epoxy composites. The molecular mobility of the epoxy resin increased with the increase of temperature and decreased the value of E' corresponding to the temperature. The molecular mobility of the TRGO/CF/epoxy composites was lower compared to the pure CF/epoxy composite. The 0.1 wt% of TRGO incorporation in the epoxy resin did not influenced the modulus of the matrix material due to the less interactions of small number of functional groups of TRGO. The increment of TRGO loading in the CF/epoxy composites showed a significant enhancement of the composite modulus. The functional groups present at the TRGO surfaces are capable to interact with the epoxy resin and curing agent and enhanced the filler-matrix interaction. Thus, the E' increased by ~ 96% with 0.2 wt% of TRGO loading in the CF/epoxy composites as compared to pure CF/epoxy composite. The TRGO/CF/epoxy composite showed high E' at higher temperature.

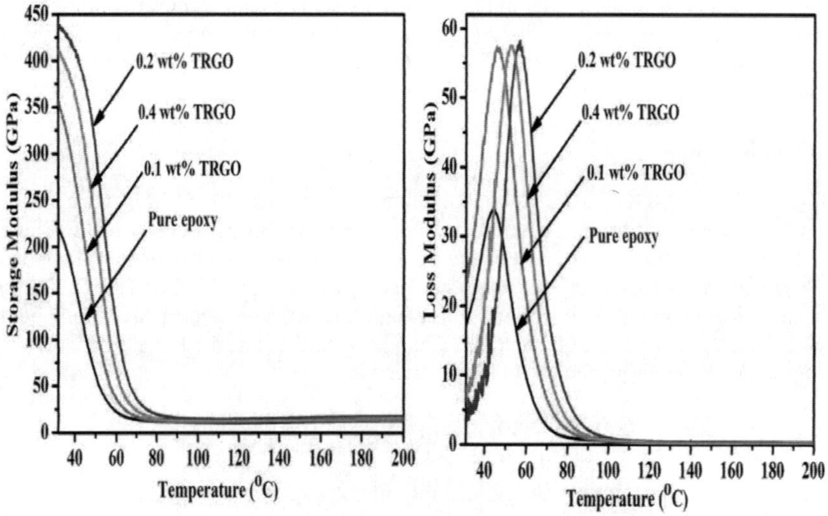

Figure 2. Variation of (a) Storage Modulus and (b) Loss Modulus of the TRGO/CF/epoxy composites as a function of temperature

The viscous response of the TRGO/CF/epoxy composite is expressed by E" and it is the ratio of stress to the strain. E" evaluates the energy dissipation of the CF/epoxy composites as heat per unit cycle under deformation. The conversion of mechanical energy into thermal energy per unit cycle can be expressed according to the equation [11],

$$\Delta E\ (\omega) = \int_0^{2\pi/\omega} \sigma(\omega)\varepsilon(\omega)dt \tag{1}$$

$$= \omega\sigma_0\varepsilon_0 \int_0^{2\pi/\omega} (\cos\omega t \cos\delta \sin\omega t + \sin\omega t \sin\omega t \sin\delta)\ dt \tag{2}$$

$$= \pi\ \varepsilon_0^2\ (\omega)\ E'' \tag{3}$$

Where, $\sigma(\omega)$ = the stress function at frequency ω with amplitude σ_0, $\varepsilon(\omega)$ = the strain function at frequency ω with amplitude ε_0, δ = the phase angle between stress and strain. Equation 3 signifies that the E" is directly proportional to energy dissipation of the material under cyclic excitation. The changes of E" with the increase of temperature of the TRGO/CF/epoxy composites is shown in Figure 2(b) and it showed narrow peak of E" for the TRGO incorporated CF/epoxy composites. The narrow peak of E" suggested less inhibition in relaxation process of the TRGO/CF/epoxy composites. The increment of E" also improved the damping capacity of the composites. The tan δ (ratio of E"/E') curve of the prepared composites is shown in Figure 3(a). The tan δ is associated with the T_g of the composites. The T_g was the maximum of the tangent of the delta peak. It was found that below T_g the damping was low as the polymer chains were in frozen state. It is renowned that the Tg increased with the increment of the cross-link density or chain stiffness in polymer films. The large amount of energy is required for the segmental movement of the closely dense polymer network i.e. the Tg value of the closely dense polymer films is greater than the loosely dense polymer. Therefore, the segmental movement and cross-linking density of the matrix material is dependent on T_g. The TRGO based CF/epoxy composites displayed higher T_g values than the pure CF/epoxy composites. The overall dynamic mechanical properties of the prepared composites were summarized in Table 1. From Table 1 and Figure 3(a), it was observed that the prepared CF/epoxy composites containing 0.1 % of TRGO were higher T_g value. Increasing the content of TRGO resulted in increase on T_g of the CF/epoxy composites up to 0.2 wt%.

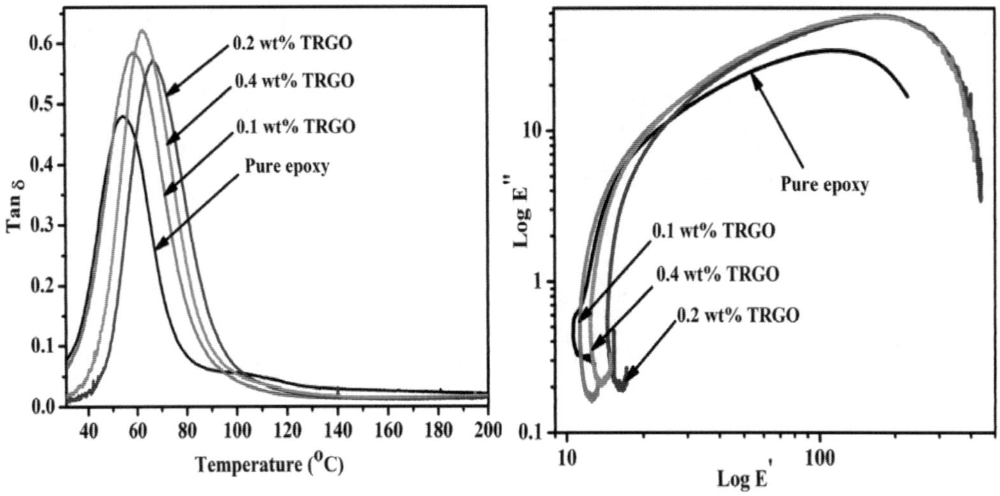

Figure 3. (a) tan δ vs. temperature and (b) Cole - Cole plot of the prepared CF/epoxy composites

The CF/epoxy composites containing 0.4 % of TRGO presented lower T_g value than 0.2 % of TRGO loaded composite. The decrease in T_g value referred the presence of voids close to the agglomerates in the composites. However, the incorporation of TRGO in the epoxy resin made a higher interfacial adhesion between CF and the epoxy resin and decreased the chain mobility of the epoxy resin and consequently increased the T_g. Thus, the T_g of the CF/epoxy composites varied with the TRGO content. The dispersion of TRGO in epoxy resin was analyzed by Cole-Cole plot analysis. The mechanical properties of the composites obtained by DMA study were expressed on Cole-Cole complex plane E = *f*(E') [12], when the properties were evaluated as a function of temperature and frequency. The appearance of a semi-circle in the Cole-Cole plot suggested homogeneous dispersion of TRGO in the epoxy matrix and a heterogeneous system exhibits an imperfect semi-circular curve [13] and it has been shown in Figure 3(b). From the Figure 3(b), it was observed that with increase of TRGO content, the curve shows more semi-circular compared to others i.e. more homogeneous.

Table 1 The Dynamic mechanical properties of the TRGO/CF/epoxy laminate

wt% of TRGO	T_g	E' (GPa)	E''(GPa)	tan δ
0	54	221	34	0.48
0.1	58	360	57	0.58
0.2	62	435	58	0.57
0.4	66	410	57.5	0.62

4. Conclusions

The DMA study of the TRGO/CF/epoxy was utilized to investigate the effect of TRGO nanofiller on the dynamic mechanical properties of the CF/epoxy composites. It was observed that the TRGO acted as a good filler material for increasing the adhesion between the CF and epoxy resin via better cross-linking with epoxy resin. The dynamic mechanical analysis of the TRGO/CF/epoxy composites showed ~ 96% enhancement in storage modulus and ~ 12 °C increments in glass transition temperature (T_g) with 0.2 wt% of TRGO incorporation compared to the base CF/epoxy composites. The fiber-matrix interaction was studied by Cole-Cole plot analysis. It confirmed the homogeneous dispersion of TRGO in the epoxy resin. The homogeneous dispersion of the TRGO in the epoxy matrix increased the overall enhancement of dynamic mechanical properties of the composites.

References

[1] Dave R S and Loos A C 2000 *Processing of Composites* Hanser Publishers.
[2] Geng Y et al 2008 *Compos. Part A* **39** 1876–1883.
[3] Wang L et al 2010 *Polym Eng Sci* **50** 2176–2181.
[4] Pathak AK et al 2016 *Compos Sci Technol* **135** 28-38.
[5] Trappe V et al 2012 *Polym. Test* **31** 654-659.
[6] Stark W 2013 *Polym. Test* **32** 231-239.
[7] Stark W et al 2009 *Polym. Test* **28** 561-566.
[8] Jaunich M. 2009 *Mater. Test. Mater. Compon. Techn. App* **51** 828-834.
[9] Kuila T et al 2012 *Carbon* **50** 914–921.
[10] Adak NC et al 2018 *Crystals* (Accepted).
[11] McCrum NG 1967 John Wiley and Sons.
[12] Jawaid M 2011 *Bio-Resources* **6** 2309-2322.
[13] Agarwal G 2013 *Tribo. Materi.Surf. & Interf.* **7** 150-160.

Estimation of Effective Directional Strength of Single Walled Wavy CNT Reinforced Nanocomposite

Krishnendu Bhowmik, Pranav Kumar, Niloy Khutia and Amit Roy Chowdhury

Department of Aerospace Engineering and Applied Mechanics, Indian Institute of Engineering Science and Technology, Shibpur, Howrah 711103, India

E-mail: krishnendub@aero.iiests.ac.in

Abstract. In this present work, single walled wavy carbon nanotube reinforced into composite has been studied to predict the effective directional strength of the nanocomposite. The effect of waviness on the overall Young's modulus of the composite has been analysed using three dimensional finite element model. Waviness pattern of carbon nanotube is considered as periodic cosine function. Both long (continuous) and short (discontinuous) carbon nanotubes are being idealized as solid annular tube. Short carbon nanotube is modelled with hemispherical cap at its both ends. Representative Volume Element models have been developed with different waviness, height fractions, volume fractions and modulus ratios of carbon nanotubes. Consequently a micromechanics based analytical model has been formulated to derive the effective reinforcing modulus of wavy carbon nanotubes. In these models wavy single walled wavy carbon nanotubes are considered to be aligned along the longitudinal axis of the Representative Volume Element model. Results obtained from finite element analyses are compared with analytical model and they are found in good agreement.

1. Introduction

Carbon nanotube (CNT) brought a new interest to use as reinforcement as Iijima [1] found a needle tube like structure made of carbon atoms. High specific strength of CNT is the potential candidate for application in industries like aerospace, automotive, electronics and infrastructure [2]. It has been reported that addition of only 1% weight of straight CNT in the matrix enhances the stiffness of composite by 36-42% [3]. Feng et al. [4] conducted several experiments with single walled carbon nanotube (SWCNT) reinforced in epoxy matrix with maximum weight fraction of 39.1%. It is observed from their experiments that the tensile strength and Young's modulus are increased by 183% and 408%, respectively compared to the epoxy. Fisher et al. [5] investigated the behaviour of waviness of nanotubes inside polymers in the form of wavy sinusoidal long nanotube with high aspect ratio using both experimental and numerical approaches. Three dimensional (3-D) finite element (FE) analysis of wavy particle is modelled into infinite matrix to numerically compute the dilute strain concentration tensor using Mori–Tanaka approach for aligned or randomly oriented particles. They observed that wavy characteristic reduces effective tensile modulus of the composite. Shao et al. [6] observed that the waviness of CNTs and interfacial bond between CNT and matrix are the major parameters that affect the efficiency of reinforcement. Joshi et al. [7] investigated the effect of wavy carbon nanotubes on elasticity and strength for nanocomposites using a 3-D representative volume element (RVE) for long as well as short wavy carbon nanotubes. Carbon nanotube is modelled as a continuous hollow tube with curvature in its geometry. It is being observed that the waviness significantly reduces the effectiveness of reinforced composite. Tsai et al. [8] investigated the effects of randomly distributed or partially aligned inclusion of wavy reinforced composite. Different configurations of waviness, aspect ratio are considered in their analyses. It is being concluded that the waviness of the inclusion has a greater effect on tensile modulus as well as shear modulus of the nanocomposite. Yazdchi and Salehi [9] used 3-D RVE model for a wavy CNT to visualize the stress transfer mechanism of SWCNT reinforced polymer composite and to predict the axial and interfacial shear stresses along wavy CNTs. Dastgerdi et al. [10] developed a micro mechanical model for wavy CNT based composite to investigate the overall stiffness of the composite. The nanotubes are taken as bow shaped solid short fibers with a circular cross-section. Paunikar and Kumar [11] studied the effect of waviness of nanotube in the form of cosine function using micromechanics approach and 3-D finite element analysis (FEA). Nanotubes are taken as solid cylindrical shape without any end cap. Alian et al. [12] reported a multiscale model to predict the effect of waviness and agglomeration of CNTs on the elastic properties of nanotube epoxy composites. Free vibration and bending behaviour of CNT reinforced composite plate under hygroscopic environment are investigated using shear deformation theory by Mehar and Panda [13]. Similarly, the bending behaviours of functionally graded sandwich spherical panel are studied under hygroscopic environment using first order shear deformation theory by Mahapatra et al. [14]. Kumar and Srinivas developed an analytical model for wavy CNT reinforced polymer composites to show that the elastic parameters are not only depending on the volume fraction of CNT but also on geometry and agglomeration of CNTs [15].

2. Methodology

In this current study, waviness pattern of CNT is being considered as $y=A\cos(2\pi x/\lambda)$; a periodic cosine function (Fig. 1), where A is the amplitude of a cosine function, x is the axial direction of the RVE, λ is the wavelength. Both long (continuous) and short (discontinuous) CNTs are being considered and idealized as solid annular tube. Short CNT is modelled with hemispherical cap at its both ends. In present study, parameters considered for the FE analysis are; waviness ratio, w (ratio of amplitude and wavelength, A/λ); modulus ratio, λ_1 (ratio of Young's modulus of CNT and matrix); particle aspect ratio, λ_2 (ratio of CNT height and diameter); particle height fraction, λ_3 (ratio of CNT height and RVE length); particle volume fraction, V_{CNT} (ratio of CNT volume and RVE volume).

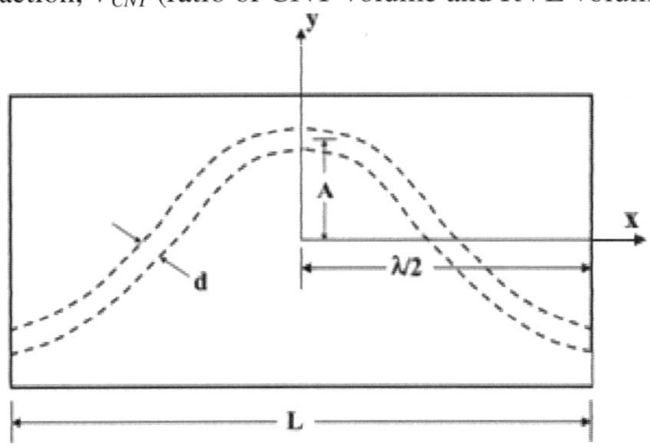

Figure 1. Schematic of typical rectangular RVE with wavy long CNT [11].

The orientation of the CNT with respect to the global co-ordinate system (x, y, z) varies throughout the length of the RVE. The principal (local) coordinate system at any location along the length of the RVE is assumed as (x_0, y_0, z_0), where axis x_0 is along the central path of the CNT. A general coordinate transformation from the principal to global axes can be described by Euler angles [11].

The angle of rotation γ, at any location along the length of RVE is calculated as

$$\gamma = -\tan^{-1}\left(\frac{dy}{dx}\right) = -\tan^{-1}\left(-\frac{2\pi A}{\lambda}\sin(\frac{2\pi x}{\lambda})\right) \quad (1)$$

The stress and strain vectors in the global and principal co-ordinate systems are

$$\{\sigma'\} = [T_1]\{\sigma\} \quad (2)$$

$$\{\varepsilon'\} = [T_2]\{\varepsilon\} \quad (3)$$

where, σ, ε are stress and strain in global and σ', ε' are stress and strain in principal co-ordinate systems; $[T_1]$ and $[T_2]$ are the transformation matrix given by

$$[T_1]=\begin{bmatrix} m^2 & n^2 & 0 & 0 & 0 & 2mn \\ n^2 & m^2 & 0 & 0 & 0 & -2mn \\ 0 & 0 & 1 & 0 & 0 & 0 \\ 0 & 0 & 0 & m & -n & 0 \\ 0 & 0 & 0 & n & m & 0 \\ -mn & mn & 0 & 0 & 0 & m^2-n^2 \end{bmatrix};\ [T_2]=\begin{bmatrix} m^2 & n^2 & 0 & 0 & 0 & mn \\ n^2 & m^2 & 0 & 0 & 0 & -mn \\ 0 & 0 & 1 & 0 & 0 & 0 \\ 0 & 0 & 0 & m & -n & 0 \\ 0 & 0 & 0 & n & m & 0 \\ -2mn & 2mn & 0 & 0 & 0 & m^2-n^2 \end{bmatrix} \quad (4)$$

where, m=cos γ and n=sin γ

The stress-strain relation in the principal coordinate direction at any point in the RVE

$$\{\varepsilon'\}=[S']\{\sigma\} \quad (5)$$

where, S' is the compliance matrix in principal cordinate system and their values are calculated as

$$[S']=\begin{bmatrix} 1/E_1 & -\nu_{12}/E_1 & -\nu_{12}/E_1 & 0 & 0 & 0 \\ -\nu_{12}/E_1 & 1/E_2 & -\nu_{23}/E_1 & 0 & 0 & 0 \\ -\nu_{12}/E_1 & -\nu_{23}/E_1 & 1/E_2 & 0 & 0 & 0 \\ 0 & 0 & 0 & 1/G_{23} & 0 & 0 \\ 0 & 0 & 0 & 0 & 1/G_{12} & 0 \\ 0 & 0 & 0 & 0 & 0 & 1/G_{12} \end{bmatrix}$$

The components of the compliance matrix of the composite are calculated using rule of mixture

$$E_1=v_f E_f+(1-v_f)E_m \quad (6)$$

$$E_2=\left(\frac{v_f}{E_f}+\frac{1-v_f}{E_m}\right)^{-1} \quad (7)$$

$$G_{12}=\left(\frac{v_f}{G_f}+\frac{1-v_f}{G_m}\right)^{-1} \quad (8)$$

$$\nu_{12}=v_f\nu_f+(1-v_f)\nu_m \quad (9)$$

$$\nu_{23}=v_f\nu_f+(1-v_f)\left(2\nu_m-\frac{\nu_{12}E_2}{E_1}\right) \quad (10)$$

$$G_{23}=\frac{E_2}{2(1+\nu_{23})} \quad (11)$$

Similarly, using principle of complimentary energy, the effective compliance matrix of composite is calculated as

$$[S_{RVE}]=\frac{1}{\lambda}\int_0^{\lambda}[T_2]\,[S'][T_1]^{-1}dx \quad (12)$$

For Short CNT, RVE can be considered as two parts (Fig. 2). Central part is composed of CNTs/matrix connected in parallel; end part consists of matrix only. Equivalent longitudinal Young's modulus of central part can be estimated using eqn. (12). The effective Young's modulus ($E_{Effective}$) of the composite in the longitudinal direction can be obtained using inverse Rule of Mixtures considering the compatibility of strains and equilibrium of stresses as given by eqn. (13).

$$E_{Effective}=\left[\frac{1}{E_m}\left(\frac{L_e}{L}\right)+\frac{1}{E_{center}}\left(\frac{L_c}{L}\right)\right]^{-1} \quad (13)$$

where, L_c, L_e and L are the length of central part, end part and total length (L_c+ L_e), respectively. E_{center} is taken as S_{RVE} in eqn. (13) and E_m is the Young's modulus of matrix.

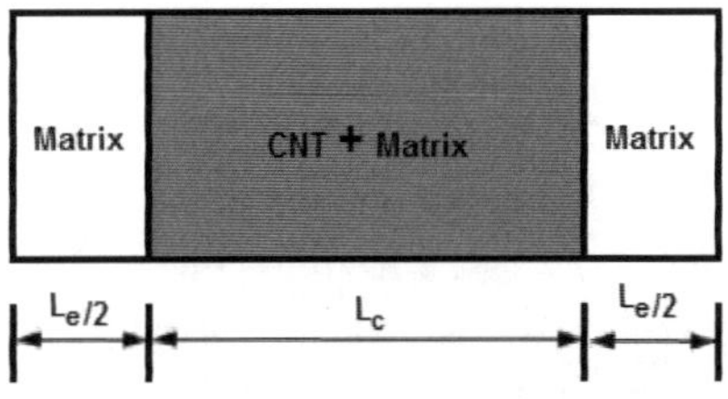

Figure 2. Typical representation of short CNT inside RVE model.

3. Finite Element Analysis

In this current study, rectangular based RVE is being considered for 3-D finite element (FE) analyses (ANSYS 14.0) for the composite. Both long (continuous) and short (discontinuous) CNTs are being idealized as solid annular tube. Nominal diameter and thickness of CNT (24,0) are being taken as 1.88 nm and 0.34 nm, respectively. Short CNT is modelled with hemispherical cap at its both ends. Five wavy CNTs are modelled with equal spacing of 15 nm in transverse direction (Fig. 3). Wavy CNTs are reinforced centrally inside the full RVE. Due to symmetry, 1/4th of the full RVE is being considered in FEA. Figures 3a and 3b represent typical schematic diagrams of full RVE and 1/4th RVE models, respectively. Two symmetry boundary conditions are imposed on two respective faces of RVE where out of plane translational degrees of freedom (DOF) are being restricted. Uniform displacement corresponding to 1% longitudinal strain is being applied in the FE model. Young's modulus and Poisson's ratio of CNT are taken as 1000 GPa and 0.3, respectively. Young's modulus of matrix is being varied to produce a wide range of modulus ratios (λ_1). However, Poisson's ratio of matrix is being taken as 0.3 for the entire analyses. Both particle and matrix are considered as linear and isotropic in nature. Wavy CNT is assumed to be perfectly bonded to the surrounding matrix. CNTs and Matrix are modelled using second order tetrahedron (solid186) elements with reduced integration. Typical number of nodes and elements are taken as 216270 and 131952 (waviness ratio, w=0.1 and CNT volume fraction of 2.0%), respectively. Longitudinal Young's modulus (E_c) of composite under axial load can be estimated using equation, $E_c = \frac{F}{A\varepsilon}$, where F is the longitudinal reaction force at support obtained from FEA, A is the cross-sectional area of RVE and ε is the average strain.

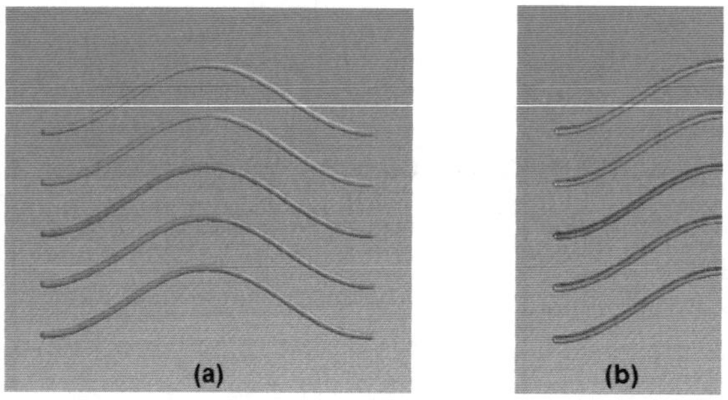

Figure 3. Typical rectangular RVE for short CNTs (w=0.1, λ_3=0.8) (a) full model and (b) quarter model.

4. Results and Discussions

4.1. Effective Young's modulus versus CNT volume fraction

The variations of normalised Young's modulus (ratio of Young's modulus of nanocomposite and matrix; E_c/E_m) versus CNT volume fraction (V_{CNT}) are shown in Figs. 4-5. CNT volume fractions are taken as 0.5%, 0.75%, 1.0%, 1.25%, 1.5%, 1.75% and 2.0%. The variations of the strength of the composite with the volume fraction for long CNTs (λ_3=1) are shown in Figs. 4a and 4b. Figure 4a shows the variation of E_c/E_m versus V_{CNT} for modulus ratio (λ_1) of 50 with different waviness ratios (w) of 0.1, 0.2, 0.3, 0.4 and 0.5. Values of E_c/E_m increase linearly with V_{CNT}. Further, Fig. 4a indicates that for same volume fraction and modulus ratio, higher waviness exhibits lower effective modulus of composite. It can be noted from Fig. 4a that the slope of curve is highest in case of w=0.1 compared with w=0.2, 0.3, 0.4 and 0.5. In case of V_{CNT}=0.5%, values of E_c/E_m are 1.120, 1.049, 1.030, 1.026 and 1.022 for w of 0.1, 0.2, 0.3, 0.4 and 0.5, respectively. Corresponding values are 1.464, 1.186, 1.108, 1.088 and 1.081 in case of V_{CNT}=2.0%. Sensitivities on overall directional elastic modulus of composites are less when waviness ratio (w)>0.3 and values are close to each other for w=0.4 and 0.5. Figure 4b illustrates the variation of E_c/E_m versus V_{CNT} with w=0.1 for modulus ratios (λ_1) of 5, 10, 20, 50, 100 and 200. Rate of change of E_c/E_m is more with higher value of modulus ratio. Minimum and maximum values of E_c/E_m obtained from FEA are 1.011 and 1.976, respectively, for V_{CNT}=0.5%, λ_1=5 and V_{CNT}=2.0%, λ_1=200. Figure 5a exhibits variation of E_c/E_m versus V_{CNT} for short CNT with height fraction (λ_3) of 0.8 and λ_1=50. Results for different waviness ratios (w) of 0.1, 0.2, 0.3, 0.4 and 0.5 are also plotted in Fig. 5a. The variations of E_c/E_m with V_{CNT} shown in Fig. 5a are similar to Fig. 4a, except the values are lower in case of short CNTs for same waviness and volume fraction. Effect of height factions on overall elastic response of composite are shown in Fig. 5b. Considering w=0.1 and λ_1=200, height fractions (λ_3) are taken as 0.6, 0.7, 0.8 and 0.9. Rate of enhancement of effective Young's modulus with V_{CNT} slightly decreases in case of λ_3=0.6. However, the slope of E_c/E_m with V_{CNT} remains almost constant for λ_3=0.9. Values of E_c/E_m for V_{CNT}=0.5% are 1.138 and 1.181 for λ_3=0.6 and 0.9, respectively. Corresponding values for V_{CNT}=2.0% are 1.448 and 1.679, respectively.

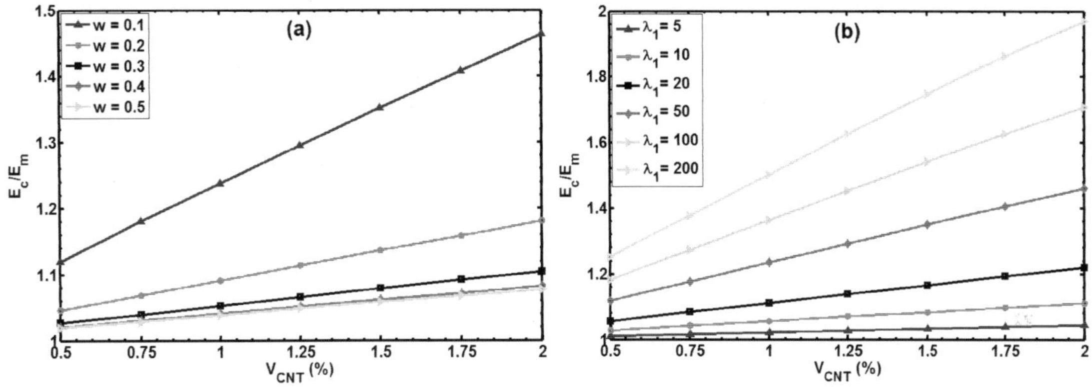

Figure 4. Young's modulus versus volume fraction for long CNTs (a) λ_1=50 and (b) w=0.1.

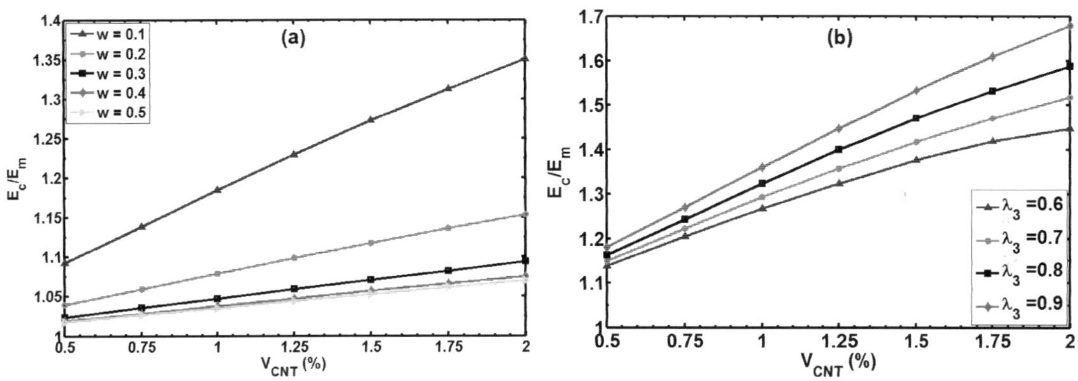

Figure 5. Young's modulus versus volume fraction for short CNTs (a) λ_1=50, λ_3=0.8 and (b) w=0.1 λ_1=200.

4.2. Effective Young's modulus versus CNT waviness

The effect of waviness is being illustrated in Figs. 6a and 6b, considering volume fraction of CNT as 2.0%. Results obtained from FEA and analytical solution based on micromechanics approach for different modulus ratios of 10, 20, 50 and 200 are also plotted in Figs. 6a and 6b. In case of long CNTs (Fig. 6a), with increase of waviness, the strength of the composite decreases sharply between w=0.1 and 0.2 but when w>0.3 (can be considered as critical value) E_c/E_m variation remains almost constant for same modulus ratio. Figure 6b shows E_c/E_m variation with waviness for short CNTs (λ_3=0.8). In case of analytical model, the overall modulus are being calculated using eqn. (12) for long CNTs; and using both eqn. (12) and eqn. (13) for short CNTs. FEA results under predicts the analytical solution for all cases. The maximum variations of FEA and analytical results for long and short CNTs are 7.2% and 11.6%, respectively in case of w=0.1 and λ_1=200. However, finite element results and analytical solutions are very close to each other when w>0.2. Figures 6a and 6b show similar trend in variation of normalized Young's modulus with waviness. The rate of reduction of E_c/E_m is more with higher modulus ratio as demonstrated in Figs. 6a and 6b. However, the results obtained from the analytical approach are comparatively closer to FEA solution in case of long CNTs than short CNTs.

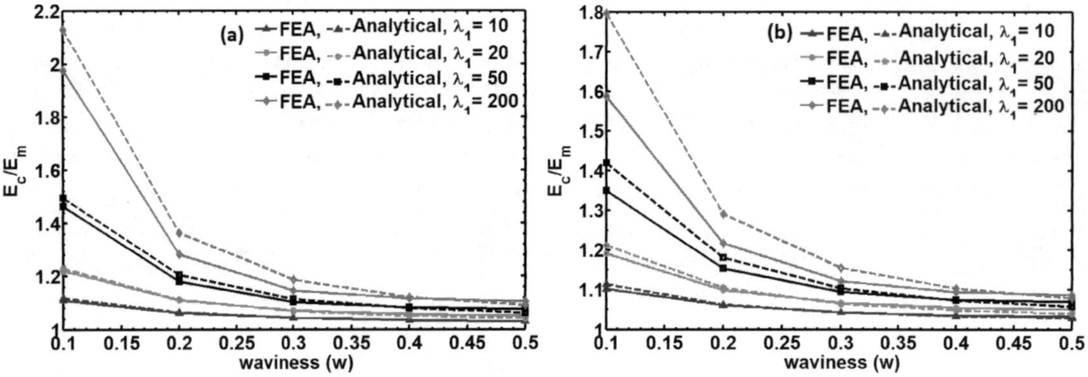

Figure 6. Young's modulus versus waviness with V_{CNT}=2.0% (a) long CNTs (b) short CNTs, λ_3=0.8.

5. Conclusions

Following conclusions can be drawn from the present analyses:

- Straight CNTs are more efficient than wavy CNTs.
- Effective Young's modulus of composite decreases with CNT waviness. However, sensitivity of modulus of composite becomes less significant with higher waviness.
- Long CNTs are more effective in enhancement of overall modulus of composite than short CNTs.
- Enhancement of overall elastic response of the composite increases with higher modulus ratios.
- Finite element analysis results are in good agreement with the results obtained from micromechanics approach.

References

[1] S. Iijima 1991 *Nature* **354** pp 56–58
[2] M. S. Dresselhaus, G. Dresselhaus and R. Saito 1995 *Carbon* **33 (7)** pp 883-891
[3] D. Qian, E. C. Dickey, R. Andrews and T. Rantell 2000 *Appl. Phy. Lett.* **76 (20)** pp 2868–2870
[4] Qing-Ping Feng, Xiao-Jun Shen, Jiao-Ping Yang, Shao-Yun Fu, Yiu-Wing Mai and Klaus Friedrich 2011 *Polymer* **52** pp 6037-6045
[5] F. T. Fisher, R. D. Bradshaw and L. C. Brinson 2003 *Compos. Sci. Technol.* **63** pp 1689–1703
[6] L. H. Shao, R. Y. Luo, S. L. Bai and J. Wang 2009 *Compos. Struct.* **87** pp 274–281
[7] U. A. Joshi, S. C. Sharma and S.P. Harsha 2011 *Physica E* **43** pp 1453–1460
[8] Chao-hsi Tsai, Chuck Zhang, David A. Jack, Richard Liang and Ben Wang 2011 *Composites Part B* **42** pp 62–70
[9] K.Yazdchi and M. Salehi 2011 *Composites Part A* **42** pp 1301–1309
[10] J. N. Dastgerdi, G. Marquis and M. Salimi 2013 *Compos. Sci. Technol.* **86** pp 164–169
[11] S. Paunikar and S. Kumar 2014 *Comput. Mater. Sci.* **95** pp 21–28
[12] A. R. Alian, S. El-Borgi and S. A. Meguid 2016 *Comput. Mater. Sci.* **117** pp 195–204
[13] K. Mehar and S. K. Panda 2016 *IOP Conf. Ser.: Mater. Sci. Eng.* **115** 012014
[14] T. R. Mahapatra, K. Mehar, S. K. Panda, S. Dewangan and S. Dash 2017 *IOP Conf. Ser.: Mater. Sci. Eng.* **178** 012031
[15] P. Kumar and J. Srinivas 2017 *IOP Conf. Ser.: Mater. Sci. Eng.* **178** 012016

IOP Conf. Series: Materials Science and Engineering **338** (2018) 012017 doi:10.1088/1757-899X/338/1/012017

Dynamic Response of Functionally Graded Carbon Nanotube Reinforced Sandwich Plate

Kulmani Mehar[1] and Subrata Kumar Panda[2]

[1,2] Department of Mechanical Engineering,
National Institute of Technology Rourkela, Odisha, India

Corresponding author Email: kulmanimehar@gmail.com

Abstract: In this article, the dynamic response of the carbon nanotube-reinforced functionally graded sandwich composite plate has been studied numerically with the help of finite element method. The face sheets of the sandwich composite plate are made of carbon nanotube-reinforced composite for two different grading patterns whereas the core phase is taken as isotropic material. The final properties of the structure are calculated using the rule of mixture. The geometrical model of the sandwich plate is developed and discretized suitably with the help of available shell element in ANSYS library. Subsequently, the corresponding numerical dynamic responses computed via batch input technique (parametric design language code in ANSYS) of ANSYS including Newmark's integration scheme. The stability of the sandwich structural numerical model is established through the proper convergence study. Further, the reliability of the sandwich model is checked by comparison study between present and available results from references. As a final point, some numerical problems have been solved to examine the effect of different design constraints (carbon nanotube distribution pattern, core to face thickness ratio, volume fractions of the nanotube, length to thickness ratio, aspect ratio and constraints at edges) on the time-responses of sandwich plate.

Keywords: Carbon nanotube; Dynamic response; Finite element method; ANSYS

1. Introduction

Functionally graded carbon nanotube-reinforced composite (FG-CNTRC) is a new kind of composite material. In functionally graded composite, material properties varies in one or more coordinate directions by some defined grading pattern. From last one decade, many researchers have been attracted to work on the FG-CNTRC structures. Because, CNT possesses extraordinary mechanical, electrical and thermal properties. In 2009, first time Shen [1] used the functionally graded concept to examine the nonlinear bending responses of the FG-CNTRC plate with functionally graded concept based on the higher-order shear deformation theory (HOSDT). Later, Zhu *et al.* [2] implemented the first-order shear deformation theory (FOSDT) to examine the natural frequency and mid-point displacement of the FG-CNTRC plate. Dastjerdi *et al.* [3] implemented a refine shear deformation theory to examine the free vibration behaviour of FG-CNTRC plate resting in an elastic medium. They found that increasing the CNT volume fraction within the matrix lead to increase the natural frequency. It is also found that the increasing of thickness ratio and aspect ratio lead to increase the fundamental frequency. Mehar *et al.* [4] computed the linear fundamental frequency of the FG-CNTRC plate using the HOSDT mid-plane kinematics and found the similar trend of results as discussed in [3]. Further, the concept of functionally graded concept implemented on the CNT reinforced sandwich structure [5–7]. It is well known that sandwich structure consists minimum three faces, two stiffer face sheets to wear the external load and one lightweight core to transfer the shearing force between both face sheets. The static and free vibration behaviours of the CNT reinforced sandwich plate are examined by Natarajan *et al.* [5] using shear deformation theories with different degree of freedom (DOF). The results indicate that increases in DOF of nodes lead to increase the flexibility of the plate. In another word central deflection is increases and the natural frequency decreases with increase the DOF of nodes. Wang and Shen [6] examined the nonlinear mechanical characteristics of the CNT-reinforced sandwich plate using HOSDT. The ratio of nonlinear to linear frequency increases with the higher value of amplitude ratio (ratio of maximum central deflection to thickness of the plate). Lei *et al.* [8] illustrated the dynamic behaviour of the FG-CNTRC plate using an element-free kp-Ritz method based the FOSDT kinematic model.

From the literature review, it is found that dynamic response of the CNT-reinforced sandwich plate with functionally graded concept is not presented in open sources. In the present paper, the dynamic response of the FG-CNT-reinforced sandwich plate using ANSYS environment based on the FOSDT has been examined. Extended ROM has implemented to examine the material properties of the CNT/epoxy composite and Newmark's time integration method to solve the final governing equation. Finally, effect of different design constraints on the dynamic response of the sandwich plate has been computed and illustrated in details.

2. Theory and formulation

In this paper, two different type CNT-reinforced sandwich plate are used namely, UD-CNT and FG-CNT. UD and FG represent for uniformly distributed and functionally graded, respectively. The geometrical dimensions of the sandwich plate are taken as length (a), width (b), the thickness of the core (h_c), the thickness of each face sheet (h_f) and the total thickness of the sandwich plate as ($h=h_c+2h_f$) as shown in Figure 1. In the face sheets, CNT volume fraction is assumed to be distributed using predefined grading patterns in Z-direction and align along the length of the plate. The effective volume fraction (V_{CNT}) with respect to Z- direction can be calculated using the following formulae:

For FG-CNT-reinforced sandwich

$$\left.\begin{aligned} V_{CNT} &= 2\left(\frac{h_1 - Z}{h_1 - h_0}\right)V^*_{CNT} && \text{for top bottom sheet} \\ V_{CNT} &= 0 && \text{for core} \\ V_{CNT} &= 2\left(\frac{Z - h_2}{h_3 - h_2}\right)V^*_{CNT} && \text{for top face sheet} \end{aligned}\right\} \quad (1)$$

For UD-CNT-reinforced sandwich

$$\left.\begin{aligned} V_{CNT} &= V^*_{CNT} && \text{for top and bottom face sheet} \\ V_{CNT} &= 0 && \text{for core} \end{aligned}\right\} \quad (2)$$

where, V^*_{CNT} is the total CNT volume fraction within each face sheet.

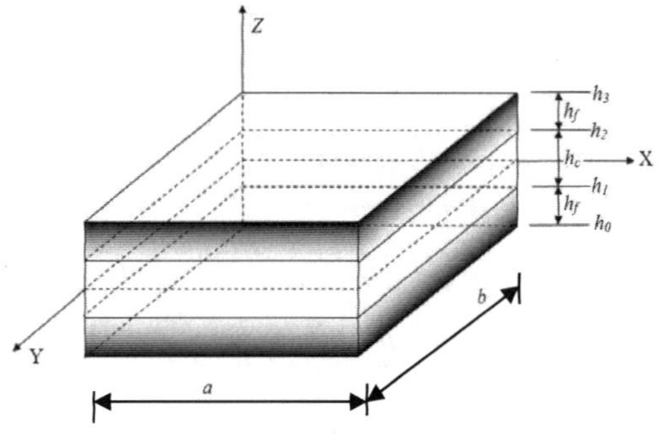

Figure 1. Geometry of CNT-reinforced sandwich plate.

The core of the sandwich is made of by the pure epoxy an isotropic homogeneous material. To the convenience, the final elastic and shear modulus of the face sheets are calculated by the extended rule of mixture (ROM) as follows [2,9]:

$$E_{11} = V_m E^m + \eta_1 V_{CNT} E_{11}^{CNT} \tag{3}$$

$$\frac{\eta_2}{E_{22}} = \frac{V_m}{E^m} + \frac{V_{CNT}}{E_{22}^{CNT}} \tag{4}$$

$$\frac{\eta_3}{G_{12}} = \frac{V_m}{G^m} + \frac{V_{CNT}}{G_{12}^{CNT}} \tag{5}$$

where, η_1, η_2 and η_3 are the effectiveness parameters of CNT within the composite. The values of effectiveness parameters taken as same as [2]. Similarly, the density and Poisson's ratio of the composite are calculated using the ROM [2].

To examine the dynamic behaviour of the CNT-reinforced sandwich plate, a simulation model is developed using ANSYS parametric design language (APDL) code. In ANSYS, various kind of elements are available for modelling of the sandwich structure. In the present study, shell 281 has been implemented to develop the sandwich plate model. Shell 281 possesses eight nodes on each element and each node possesses six DOF (three translations along the X, Y and Z-directions and three rotations about X, Y and Z-directions). It is based on the FOSDT and the kinematic model is as follows:

$$\left.\begin{aligned} u_x &= u_{0x} + Z\theta_x \\ u_y &= u_{0y} + Z\theta_y \\ u_z &= u_{0z} + Z\theta_z \end{aligned}\right\} \tag{6}$$

where, u_x, u_y and u_z are the displacement of any point along the X, Y and Z-directions, respectively. Similarly, θ_x, θ_y and θ_z are representing the rotation of mid-plane about X, Y and Z-directions, respectively. "0" in u_{0x}, u_{0y} and u_{0z} is represented for mid-plane.
The final governing equation for dynamic analysis can be written in the following form:

$$[M]\{\ddot{\delta}\} + [K]\{\delta\} = \{f\} \tag{7}$$

where, $[M]$ and $[K]$ are global mass and stiffness matrix, respectively. $\{\ddot{\delta}\}$, $\{\delta\}$ and $\{f\}$ are acceleration, displacement and force vector, respectively.

Finally, the dynamic response is computed by solving the final governing equation using Newmark's time integration method with time steps "Δt" and total time "T".

3. Result and discussion

In the present analysis, single-walled carbon nanotube and the epoxy are used as the reinforcement and matrix material, respectively. The material properties of the CNT are taken as same as [8] E_{11}^{CNT} =5.6466TPa, $E_{22}^{CNT} = E_{33}^{CNT}$ =7.08TPa, G_{12}^{CNT} =1.9445TPa, v_m=0.17 and ρ_m=1400kg/m^3. The material properties of the isotropic epoxy are considered as E_m=2.1GPa, v_m=0.34 and ρ_m=1150kg/m^3. In general material and geometrical properties are taken as V_{CNT}^* =0.17, a/h=10, a/b=1, h_c/h_f=2, q_0=0.1MPa, Δt=1µ-sec and all edges simply supported condition (SSSS), if not explained. To provide the restriction at free edges, SSSS type supported condition is used such that:

$u_{0y}=u_{0z}=\theta_y=\theta_z=0$ at X=0 and a
$u_{0x}=u_{0z}=\theta_x=\theta_z=0$ at Y=0 and b

Convergence and validation

In this section, the accuracy and reliability of the developed model have been checked by performing the convergence and validation studies. The convergence study is performed for different mesh size with other design parameters as FG-CNT, SSSS, V_{CNT}^* =0.11, a/h=10, a/b=1, h_c/h_f =2, q_0=0.1MPa and time step Δt=1µ-sec. Convergence study shows that developed model is converging well with mesh refinement at 10×10 mesh size as illustrated in Figure 2 and this size used for the further study.

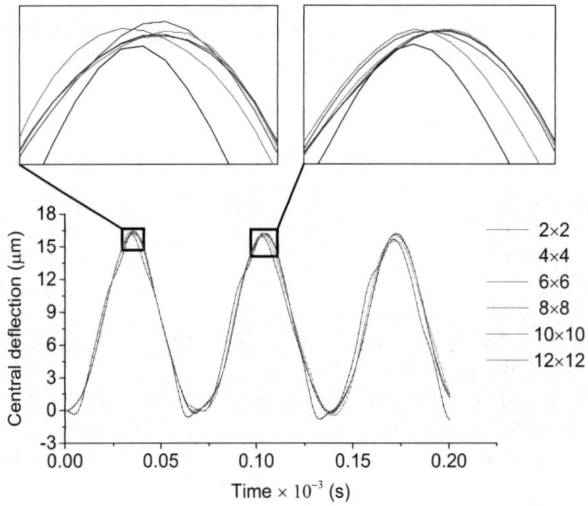

Figure 2. Convergence with mesh refinement.

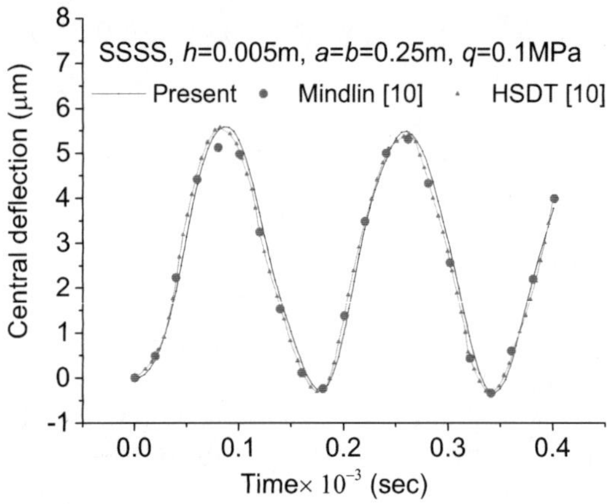

Figure 3. Comparison of dynamic response of isotropic composite plate.

To illustrate the reliability of developed model for dynamic analysis, two comparison studies have been performed. First comparison study conducted for isotropic plate by taking the design parameters as same as Kant *et al.* [10] such as E_2=2.1×10^{6}N/cm^2, E_1=25E_3=25E_2, G_{12}-=G_{13}=G_{23}=0.5E_2, ν_{12}= ν_{23}= ν_{13}=0.25, density =8×10^{-6} N sec^2/cm^4, h=5cm, a=b=25cm, q_0=10N/cm^2 and Δt=2μ-sec with all edges simply supported and illustrated in Figure 3.

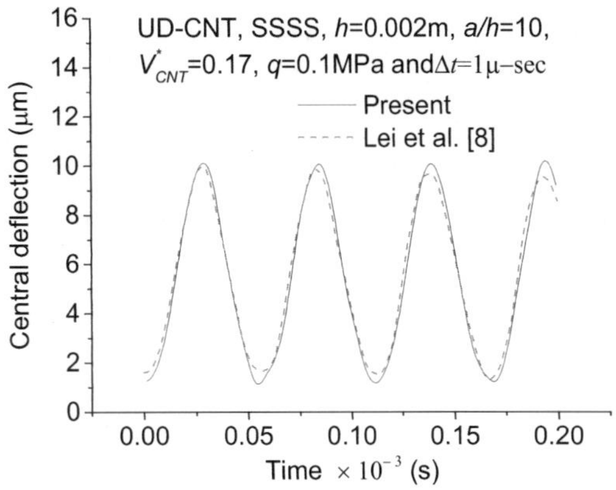

Figure 4. Comparison of dynamic response of UD-CNTRC plate.

The second comparison study performed for the UD-CNTRC plate using same material properties as same as [8] and presented in Figure 4. Both comparison studies show that the developed model is agreement well with previously published results.

Parametric study

In this section effect of different design parameters on the dynamic behaviour of the CNT-reinforced sandwich plate are examined as shown in Figures 5-8. Figure 5 illustrated that the central point transverse displacement of the FG-CNT-reinforced sandwich plate decreases with rising of CNT volume fraction within the composite. Because CNT possesses higher stiffness as compared to the matrix and the concentration of CNT within composite lead to increase the total stiffness of the composite.

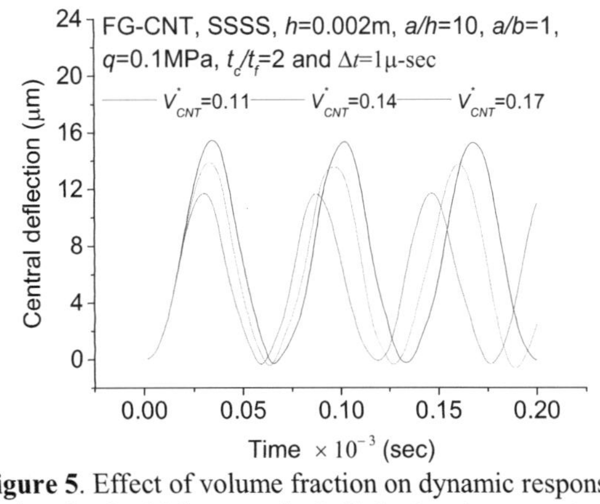

Figure 5. Effect of volume fraction on dynamic response.

Figure 6 illustrated that rise in thickness ratio (*a/h*) lead to increase the central point transverse displacement and cycle time of displacement with higher values of thickness ratio. Further, the influence of aspect ratio for three different values (*a/b*=1, 1.5 and 2) and effect of CNT grading within the matrix are examined as shown in Figure 7. From the figure, it is clear that the transverse displacement of the central point and cycle time of displacement decreases with higher value of aspect ratio (*a/b*). Figure 7 also illustrated that the central deflection and cycle time of the FG-CNT are less as compared to the UD-CNT-reinforced sandwich plate.

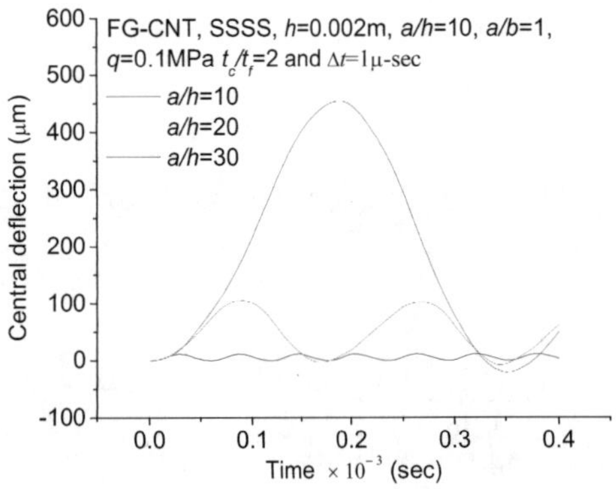

Figure 6. Effect of volume fraction on dynamic response.

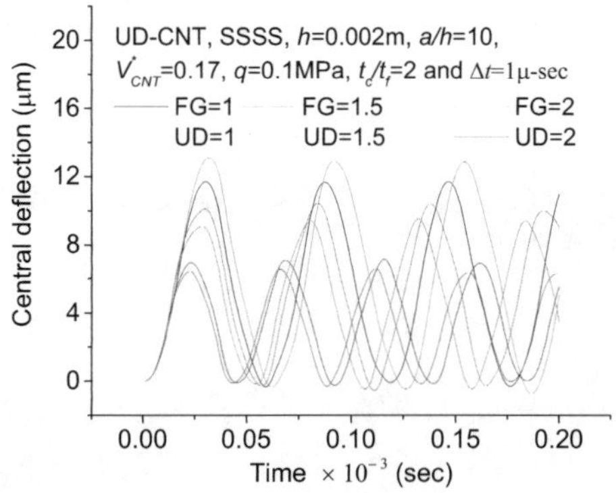

Figure 7. Effect of volume fraction on dynamic response.

4. Conclusion

In the present study, the dynamic responses of the FG-CNT-reinforced composite plate based on the FOSDT is examined for two different types of CNT grading patterns. The effective material properties are computed using the extended ROM and solved using Newmark's time integration method. Finally effect of different design constraints on the dynamic behaviour is computed and the major conclusions discussed in the following points.

- The transverse displacements of the central point of plate decreased for the higher values of aspect ratios and volume fractions of CNT within the composite, however, it follows a reverse trend for the thickness ratio.
- The cycle time of response decreases for the higher values of the aspect ratios and volume fractions of CNT within the composite, however, it shows a converse trend for the thickness ratio.
- The results reveal that the FG-CNT-reinforced sandwich plate is the stiffer configuration than that of the UD-CNT-reinforced sandwich plate.

Nomenclature

a, b and h are dimension of the plate in X, Y and Z-direction, respectively.
h_c and h_f= thickness of the core and face sheets, respectively
E_m, G_m and v_m = Young's modulus, shear modulus and Poisson's ratio of the epoxy, respectively.
E_{11}^{CNT}, E_{22}^{CNT} and E_{33}^{CNT} = Young's modulus of CNT in X, Y and Z-directions, respectively.
G_{12}^{CNT} and υ_{12}^{CNT} = Shear modulus and Poisson's ratio of the CNT
E_{11}, E_{22} and E_{33} = Young's modulus of the composite in X, Y and Z-directions, respectively.
V_m and V_{CNT} = total volume fraction of epoxy and CNT in face sheets

References

[1]. Shen H S 2009 *Compos. Struct.* **91** 9-19.
[2]. Zhu P, Lei Z X, Liew K M 2012 *Compos. Struct.* **94** 1450-1460.
[3]. Moradi-Dastjerdi R, Payganeh G, Malek-Mohammadi H, Dastjerdi R M, Payganeh G, Mohammadi H M 2015 *J. Solid. Mech.* **7** 158-172.
[4]. Mehar K, Panda S K, Dehengia A, Kar V R 2016 *J. Sandw. Struct. Mater.* **18** 151-173.
[5]. Natarajan S, Haboussi M, Manickam G 2014 *Compos. Struct.* **113** 197-207.
[6]. Wang Z X, Shen H S 2012 *Compos. Part B Eng.* **43** 411-421.
[7]. Mehar K, Panda S K 2017 *Polym. Compos.* doi:10.1002/pc.24266.
[8]. Lei Z X, Zhang L W, Liew K M 2015 *Int. J. Mech. Sci.* **99** 208-217.
[9]. Mehar K, Panda S K 2016 *IOP. Conf. Ser. Mater. Sci. Eng.* **115** 012014.
[10]. Kant T, Ravichandran R V, Pandya B N, Mallikarjuna B N 1988 *Compos. Struct.* **9** 319-342.

IOP Conf. Series: Materials Science and Engineering **338** (2018) 012018 doi:10.1088/1757-899X/338/1/012018

Electrodeposition of near stoichiometric $CuInSe_2$ thin films for photovoltaic applications

Ramkumar Chandran, Archana Mallik*

Electrometallurgy and Corrosion Laboratory, Department of Metallurgical & Materials Engineering, National Institute of Technology, Rourkela, Orissa, India
Corresponding author Email: archananitrkl@gmail.com

Abstract: This work investigates on the single step electrodeposition of quality $CuInSe_2$ (CIS) thin film absorber layer for photovoltaics applications. The electrodeposition was carried using an aqueous acidic solution with a *pH* of 2.25. The deposition was carried using a three electrode system in potentiostatic conditions for 50 minutes. The as-deposited and nitrogen (N_2) annealed films were characterized using XRD, FE-SEM and Raman spectroscopy. It has been observed that the SDS has the tendency to suppress the copper selenide (Cu_xSe) secondary phase which is detrimental to the device performance.

Keywords: *Semiconductor, Electrodeposition, Thin films, Solar cells*

1. Introduction

The ternary $CuInSe_2$ is an excellent photovoltaic material due to its high optical absorption coefficient ($\alpha=10^5$ cm^{-1}) [1]. When compared to first and third generation solar cells such as organic solar cells [2], Si-based solar cells [3] etc., photovoltaic devices (PV) based on CIS thin films are efficient as they can be easily fabricated over flexible substrates. CIS thin films are deposited by various techniques such as sputtering [4], spray coating [5], chemical synthesis [6–8]etc. Though most of the reported CISe/CIGSe high efficiency devices are fabricated by vacuum techniques, bottle necks such as high capital and material wastage is a major hurdle for industrialization. However, when compared to vacuum based techniques, electrodeposition is cheaper as it requires less equipment capital with minimized material wastage and is industrially scalable [9,10]. Several schemes in electrodeposition such as Cu/In binary stack deposition, CuInGa alloy deposition, Cu/In/Ga alloy deposition and further complicated schemes were also reported [11]. These layers can be either electroplated in conventional direct current mode (DC) or by using pulsed current mode (PC). These steps require selenization using Se powder or H_2Se to obtain the photovoltaic quality $CuInSe_2$ or $CuInGaSe_2$ thin films [12]. Most often, the compositional deviation is dependent on the thickness of the layers deposited in the stack. Also deposition of each layer followed by post-processing annealing and compositional optimizations are often time consuming. Hence, when compared to other deposition schemes and techniques, electrodeposition of CISe or CIGSe using single-step deposition techniques is highly advantageous where co-deposition of all the elements takes place by applying one constant potential. The stoichiometry of the CISe films can be tuned by varying the electrolyte composition. Till date most of the high efficiency CISe or CIGSe devices are often of Cu-poor composition. Even a slightest deviation [Cu/In $\geq$1] from the stoichiometry leads to the formation of detrimental Cu_2Se secondary phase. These highly conductive Cu_2Se phase contribute to the leakage current and affects the open circuit voltage (Voc) of the devices. Due to this reason, the [Cu/In] < 1 ratio is always preferable [13].

In the present work, we report the electrodeposition of $CuInSe_2$ thin films by single step electrodeposition technique. For improving the film adhesion, we have used sodium dodecyl sulfate (SDS) as surfactant. The advantage of using SDS is that they can limit the undesirable copper selenide binaries (Cu_2Se) phase and thus improves the film quality. The as-deposited and N_2 annealed CISe films were subjected to XRD, FE-SEM and Raman spectroscopy analysis.

2. Materials and methods

The deposition bath electrolyte was prepared by using chloride salts of copper chloride ($CuCl_2$) - 2 mM and indium chloride ($InCl_3$) – 9.375 mM along with Selenious acid (H_2SeO_3)- 5 mM for the

Selenium ions. The final *pH* of the bath was maintained at 2.25 using H_2SO_s. A well cleaned Fluorine doped tin oxide (FTO) coated glass substrates (2 x 3 cm^2) was used as the substrates and platinum foil as the counter electrode. All the electrochemical studies and deposition were performed in Biologics SP-150 and the deposition was carried out for 50 minutes at – 0.70 V/SCE. The as-deposited films were then annealed in N_2 atmosphere at 450 °C for 40 min. The phase, structure and morphology was analyzed by using XRD (Pan analytical with Cu source λKα1=1.54 Å), FESEM imaging/compositional studies (Nova NanoSEM 650) and Micro-Raman spectroscopy (HORIBA Jobin Yvon LabRamHR800L with high resolution) with Ar^+ polarized laser green line (λ=514.53 nm). To avoid crystallization effects, 2-5% of the laser beam intensity was used on the films with the acquisition time of 400 s.

3. Results and Discussion

The linear sweep voltammetry (LSV) behavior of Cu-In-Se ternary system is shown in fig.1. The potential was swept between OCP to -1.2 V/SCE at a scan rate of 10 mV/Sec. The reduction peak appearing at C1 can be assigned to the single step reduction of Cu^{2+} to Cu^+ followed by the reduction of Cu^+ to Cu(0) and subsequent formation of Cu-Se (Cu_3Se_2 or Cu_2Se) secondary phase at C2. As in scan #2, the reduction peak C2 is considered to be the formation of $CuInSe_2$ phase through the surface induced reactions which is indicated by the reduced intensity of the reduction peak. The Cu_3Se_2 phase further reduces to form H_2Se and Cu_2Se phase which initiates at the reduction peaks C3 and C4. The chemical reaction between the evolved H_2Se and In^{3+} ions takes place at C4 to form a In_2Se_3 phase which can chemically reacts with the CuSe phase to form $CuInSe_2$ phase [14–18]. The direct reduction of In^{3+} to In(0) at C5 and C6 along with hydrogen evolution reaction (HER) due to the protonation of H^+ ions. From the LSV studies, the suitable deposition potential for co-deposition of $CuInSe_2$ as a single phase can take place between C3 to C5. Hence, the potential was chosen to be -0.70 V/SCE and the deposition were carried for 50 min using chronoamperometry (CA) technique. The as-deposited CIS films were metallic bright in color indicating the presence of In-rich layer over the surface. The as-deposited and N_2 annealed (400 °C for 30 min) $CuInSe_2$ thin films were used for XRD, FE-SEM and Raman spectra characterization.

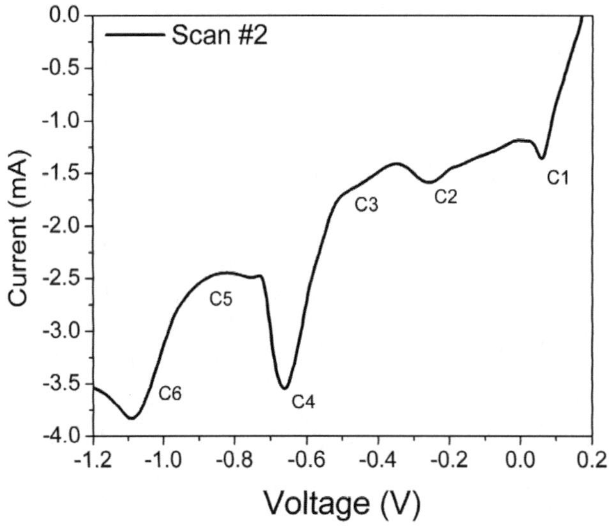

Figure 1. 1 LSV behavior of Cu-In-Se ternary system (Cu- 2mM, In-9.375 mM, Se-5 mM, SDS-1 mM , LiCl- 150 mM)

Table 1: Compositional studies of CIS thin films

Elements	Cu (at.wt %)	In (at.wt %)	Se (at.wt %)
As-deposited	25	20	55
Annealed	23	25	52

Fig.2 (a,b) shows the FE-SEM of as-deposited and annealed CIS thin films. After 50 min deposition, the CIS films with spherical particle morphology were homogenous and smooth. Such smooth films is a required during annealing as it can eliminate the formation of pinholes and porosity issues which are detrimental to the solar cell device performance. Moreover, the as-deposited CIS thin films exhibited good adherence to the FTO substrates. After annealing, the crystallization of the CIS thin films was improved as shown in fig.2b. In Fig.2c, the XRD of as-deposited and annealed CIS thin films were given. For the as-deposited CIS, the broad diffraction pattern corresponding to (220)/(204) planes indicates the presence of poor crystalline CIS precursor film. However, upon annealing, the improved crystallinity of the peaks indicates the formation of a single phase $CuInSe_2$ thin films. Well defined characteristic peaks at 17°, 26.6°, 35.5°, 44.1° and 52.4° , 70.9°, correspond to the diffraction of the (101), (112), (211) (204) ,(312) and (332) planes respectively. The diffraction peaks corresponding to annealed $CuInSe_2$ films were in good agreement with the JCPDS card No: 40-1487. Using the Scherrer's formula, the average crystallite size for as-deposited and annealed CIS films were 32 nm and 60 nm respectively, indicating the improvement in the crystallinity after annealing.

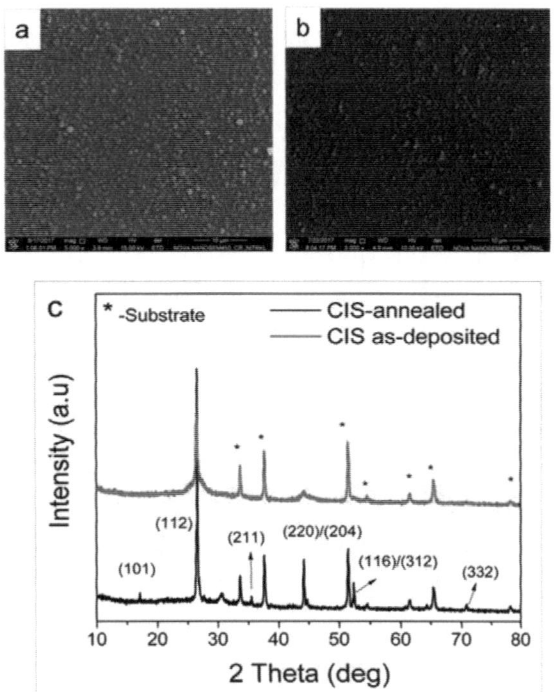

Figure.2 FE-SEM of as-deposited (a) and annealed (b) CIS thin films, c) XRD of as-deposited and annealed CIS films (*-FTO substrate peaks)

The compositional studies were carried by FE-SEM EDAX studies (table.1). From the compositional studies, it was found that the as-deposited CIS were initially rich in Cu and Se. In the as-deposited films, the nanocrytsalline In_2Se_3 phase might be dispersed in the precursor film .Due to their nano-crystalline nature; it is difficult to confirm their presence through XRD and Raman studies. The increase in indium content after annealing have been observed as the indium selenide binaries involves in alloy formation with Cu-Se to form a single $CuInSe_2$ chalcopyrite phase.

The deposits and annealed CIS films were subjected to micro-Raman studies to understand the phase formation and is shown in Fig.3. The Raman analysis was carried between the spectral regions of 100-600 cm^{-1}. The peak at 177 cm^{-1} is assigned to the characteristic A1 mode of the chalcopyrite $CuInSe_2$ phase. The presence of elemental selenium is confirmed by the peak at 240 cm^{-1} due to trigonal selenium. An appearance of a shoulder peak around 160 cm^{-1} indicates the Cu-poor $CuIn_3Se_5$ compounds known as ordered vacancy compounds (OVC). The presence of OVC indicates the Cu-poor composition of the as-deposited $CuInSe_2$ films [19]. The Raman peaks corresponding to the copper selenide binaries (260 cm^{-1}) were not observed due to the selection of high [Se^{4+} /Cu^{2+}] ions ratio used for electrodeposition. After annealing, the blue shift of the A1 mode from 177 to 176 cm^{-1} is probably due to a nanocrytsalline structure or to residual stress.

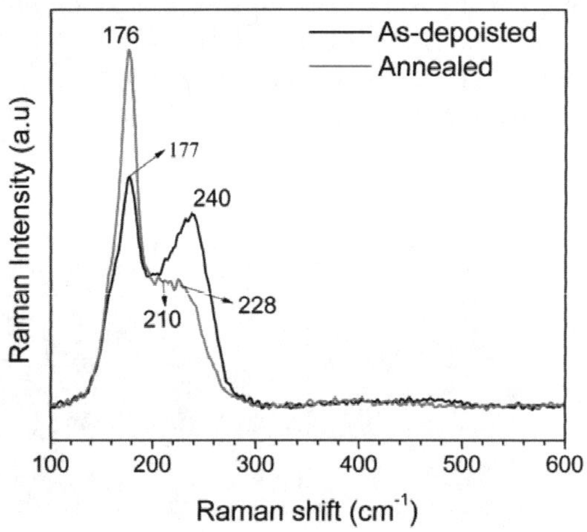

Figure.3 Raman spectra of as-deposited and annealed $CuInSe_2$ films

The disappreanece of Se (0) peak after annealing confirms that they have involved in the chalcopyrite phase formation followed by the appearance of peaks at 210 and 228 cm^{-1} which can be assigned to the B2, E mode of the chalcopyrite CIS films. The decrease in full width half maxima (FWHM) of A1 mode indicates the improved crystallinity of the annealed CIS films.

4. Conclusions

In the present study we have successfully electrodeposited CIS thin films. The as-deposited and annealed CIS films were stoichiometric with minimized CuSe secondary phase. The absence of copper selenide binaries are due to the high [Se^{4+}/Cu^{2+}] ion ratio bath used for the electrodeposition. Moreover, the use of SDS surfactant helps in inhibiting the copper selenide formation and improves the film compactness. The mentioned properties are confirmed by XRD, FE-SEM and Raman studies.

Acknowledgement

The Authors would like to thank NIT-RKL for providing the infrastructure and experimental facilities

References

[1] Siebentritt S 2017 *Current Opinion in Green and Sustainable Chemistry* **4** 1–7

[2] Aneesh P M, Ramkumar C, Reshmi Varma P C, Vivek K, Manoj Namboothiry A G 2015 *Organic Photonics and Photovoltaics* **3** 64-70

[3] Battaglia C, Cuevas A, De Wolf S 2016 *Energy Environ. Sci.* **9** 1552–76

[4] Jackson P, Hariskos D, Wuerz R, Kiowski O, Bauer A, Friedlmeier T M and Powalla M 2015 *phys. stat. sol. (RRL)* **9** 28–31

[5] Ye H, Park H S, Akhavan V a, Goodfellow B W, Panthani M G, Korgel B a and Bard A J 2011 *J. Phys. Chem. C* **115** 234–40

[6] Eluri R and Paul B 2012 *J. Nanopart. Res* **14** 1–14.

[7] Eluri R and Paul B 2012 *Mater. Lett.* **76** 36–9

[8] Graeser B K, Hages C J, Yang W C, Carter N J, Miskin C K, Stach E A and Agrawal R 2014 *Chem. Mater.* **26** 4060–3

[9] Bhattacharya R N and Fernandez A M 2003 *Sol. Energy Mater. Sol. Cells* **76** 331–7

[10] Dharmadasa I M, Burton R P and Simmonds M 2006 *Sol. Energy Mater. Sol. Cells* **90** 2191–200

[11] Başol B M, Pinarbaşi M, Aksu S, Freitag J, Gonzalez P, Johnson T, Matus Y, Metin B, Narasimhan M, Nayak D, Norsworthy G, Soltz D, Wang J, Wang T and Zolla H 2009 *34th IEEE Photovoltaic Specialists Conference (PVSC)*(Philadelphia, PA) 2310-2315

[12] Kyoung Kim W, Hanket G M and Shafarman W N 2011 *Sol. Energy Mater. Sol. Cells* **95** 235–8

[13] Siebentritt S, Gütay L, Regesch D, Aida Y and Deprédurand V 2013 *Sol. Energy Mater. Sol. Cells* **119** 18–25

[14] Calixto M E, Dobson K D, McCandless B E and Birkmire R W 2005 *MRS Proc.,* **865** F14.17

[15] Thouin L, Massaccesi S, Sanchez S and Vedel J 1994 *J. Electroanal. Chem.* **374** 81–8

[16] Chassaing E, Ramdani O, Grand P P, Guillemoles J F and Lincot D 2008 *Phys. Status Solidi C.* **5** 3445–8

[17] Chandran Ramkumar, Behera C K and Mallik A 2017 *Materials Today: Proc.,* **4** 12473–9

[18] Chandran Ramkumar, Rajneesh Pandey and Mallik A 2015 One step electrodeposition of CuInSe2 from an acidic bath: A reduction Co-deposition study *Mater. Lett.* **160** 275–7

[19] Ramdani O, Guillemoles J F F, Lincot D, Grand P P P, Chassaing E, Kerrec O and Rzepka E 2007 *Thin Solid Films* **515** 5909–12

IOP Conf. Series: Materials Science and Engineering **338** (2018) 012019 doi:10.1088/1757-899X/338/1/012019

Numerical Flexural Strength Analysis of Thermally Stressed Delaminated Composite Structure under Sinusoidal Loading

C K Hirwani[1], S Biswash[2], K Mehar[3] and S K Panda[4]

[1,2,3,4]Department of Mechanical Engineering, National Institute of Technology Rourkela Rourkela, Odisha, India.

Email[1]: chetanhirwani111@gmail.com

Abstract. In this article, we investigate the thermomechanical deflection characteristics of the debonded composite plate structure using an isoparametric type of higher-order finite element model. The current formulation is derived using higher-order kinematic theory and the displacement variables described as constant along the thickness direction whereas varying nonlinearly for the in-plane directions. The present mid-plane kinematic model mainly obsoletes the use of shear correction factor as in the other lower-order theories. The separation between the adjacent layers is modeled via the sub-laminate technique and the intermittent continuity conditions imposed to avoid the mathematical ill conditions. The governing equation of equilibrium of the damaged plate structure under the combined state of loading are obtained using the variational principle and solved numerically to compute the deflection values. Further, the convergence test has been performed by refining the numbers of elements and validated through comparing the present results with available published values. The usefulness of the proposed formulation has been discussed by solving the different kind of numerical examples including the size, location and position of delamination.

1. Introduction

With increasing application of the composite material in many weight-sensitive industries, the prediction of the structural behavior close to the actual one is very challenging and become a critical issue nowadays. Because when they exposed to the real-life situation, experience different direct (externally applied mechanical) and indirect (environmental loading, i.e., unlike temperature) loading. Additionally, the bonding between the laminas is not perfect as considered in the theoretical analysis, i.e., debonding always present due to different reasons (presence of foreign element during fabrication, matrix cracking, geometrical discontinuity, etc.). In view of the same (environmental exposer and delamination) the stiffness of the structure made using composite material reduces considerably. Therefore, the consideration of such factors or real-life situation that highly affects the structural integrity and the final performance, while prediction of structural responses is highly important.

Many works, presented in the past have considered such condition in their solution approach, few very important of them are discussed in following lines. The analytical solution approach used to analyze the bending behavior and strain energy release rate of the orthotropic debonded layered composite structure are presented by Davidson et al. [1] and Hamed et al. [2]. Further, Parhi et al. [3] used the first-order shear deformation theory (FSDT) to model the debonded composite plate structure under low-velocity impact and reported the bending as well as first-ply failure behavior. Bruno et al. [4] applied the fracture and contact mechanics approach with that of the FSDT and investigated the mixed-mode delamination. Similarly, the layerwise theory and the FSDT is used by Nanda [5] to obtain the transverse deformation behavior of the centrally located debonded shell panel. Szekre´nyes [6] developed an exact kinematic condition so-called system of exact kinematic conditions (SEKC) for the delamination modeling in layered composite structure and applied that in FSDT model of the plate structure.

Different analytical [7-8], as well as numerical method [9-10] based on different mid-plane kinematic (lower and higher-order shear deformation theory (HSDT)), are also presented to investigate the central deflection responses of the composite structure subjected to thermo-mechanical loading. Further, the finite element (FE) model derived using four variable kinematics is presented by the Tounsi and his co-author [11-12] to investigate the effect of thermo-mechanical loading on the central deflection behavior of the functionally graded plate. A simulation model of the layered composite plate under elevated thermal environment is developed by Das et al. [13] in FE based software ANSYS and analyze the stress behavior of the same. Han et al. [14] also investigated the thermomechanical behavior of the composite plate using a mathematical model developed based on higher-order zig-zag theory.

It is clear from the review of the literature that many solution methodologies to obtain the structural responses of the layered structure with and without considering the effect of debonding and environmental condition have already been reported in open literature. It is also clear that methodology based on analytical approach and FSDT are major in number and hardly a work is reported using HSDT. In order to fulfill such gap in the literature, current work aim's to develop a FE model of the delaminated composite structure subjected to mechanical (sinusoidally distributed) and thermal (uniform unlike temperature) load using HSDT mid-plane kinematics to analyze the structural response (Bending) of the same. The square-shaped delamination is assumed at the center of the laminate using the sub-laminate method. Effect of the environmental condition is incorporate via constitutive relation. The model's accuracy is checked via comparison study. Further, the parametric study has been performed to show the influence of the input parameters (geometrical and the material) on bending behavior of the debonded structure.

2. Mathematical Formulations

Figure 1 shows a layered composite plate structure consisting a delamination at centre mid-plane whose geometrical parameter is defined as length, a; breadth, b and height, h in ξ_1, ξ_2 and ζ direction, respectively. The laminate is made of n number of equally thick laminae oriented at an angle θ.

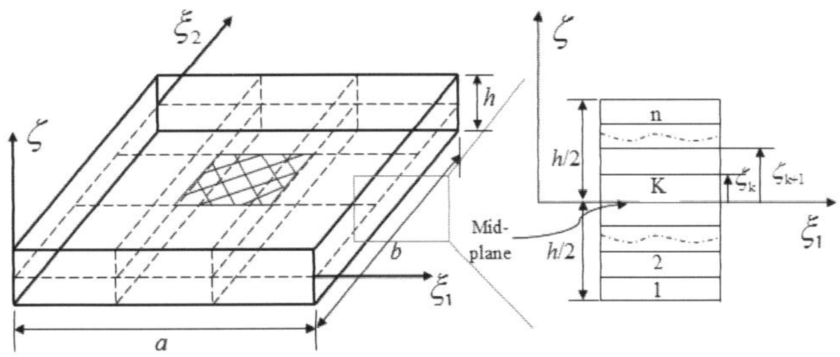

Figure 1. Geometry of the layered composite plate structure

The higher-order displacement kinematics (nine degrees of freedom) as presented in Reddy and Liu [15] is used for the current modeling purpose and shown as:

$$\left.\begin{aligned} \bar{u}(\xi_1,\xi_2,\zeta) &= u_0 + \zeta\phi_1 + \zeta^2\psi_1 + \zeta^3\theta_1 \\ \bar{v}(\xi_1,\xi_2,\zeta) &= v_0 + \zeta\phi_2 + \zeta^2\psi_2 + \zeta^3\theta_2 \\ \bar{w}(\xi_1,\xi_2,\zeta) &= w_0 \end{aligned}\right\} \quad (1)$$

where $\bar{u}, \bar{v}$ and $\bar{w}$ represents the displacement of any point along ξ_1, ξ_2 and ζ directions, respectively. The u_0, v_0 and w_0 denotes the mid-plane displacement whereas ϕ_1 and ϕ_2 are the rotation at ($\zeta = 0$) of the normal to mid-plane about ξ_2 and ξ_1, respectively. Rest of the terms $\phi_3, \psi_1, \psi_2, \theta_1$ and θ_2 are the higher order term of Taylor series expansion.

The laminate constitutive relations in terms of force and moments can be expressed as [16, 17]:

$$\{F_{ij}\} = [D]\{\{\varepsilon_{ij}\} - \{\alpha_{ij}\}\Delta T\} \quad (2)$$

where, $\{F_{ij}\},[D]$ and $\{\varepsilon_{ij}\}$ are denoted the forces and moment vector, elastic constant matrix, mid-plane strain and curvature matrices, respectively. Further, 'ΔT' is the temperature gradient and $\{\alpha_{ij}\}=\{\alpha_{\xi_1\xi_1}\ \alpha_{\xi_2\xi_2}\ 0\ 0\ \alpha_{\xi_1\xi_2}\}$'' coefficient of thermal expansion along the principal material directions.

The $\{F_{ij}\}$ and $[D]$ matrix can be explored as:

$$\{F_{ij}\}=\left[N_{\xi_1\xi_1}\ N_{\xi_2\xi_2}\ N_{\xi_2\zeta}\ N_{\xi_1\zeta}\ N_{\xi_1\xi_2}\ M_{\xi_1\xi_1}\ M_{\xi_2\xi_2}\ M_{\xi_2\zeta}\ M_{\xi_1\zeta}\ M_{\xi_1\xi_2}\right]^T \tag{3}$$

$$[D]=\sum_{k=1}^{n}\int_{\zeta_{k-1}}^{\zeta_k}\left(\bar{Q}_{ij}\right)_k\left(1,\zeta,\zeta^2...\zeta^6\right)d\zeta \tag{4}$$

$\left(\bar{Q}_{ij}\right)_k$ in eqation (4) is the transformed reduced stiffness matrix whose detail can be seen in [18]. Similarly, the necessary strain of Equation (2) is elaborated as following and detail can be seen in Mahapatra and Panda [19]:

$$\{\varepsilon_{ij}\}=\left[\varepsilon_{0\xi_1\xi_1}\ \varepsilon_{0\xi_2\xi_2}\ \gamma_{0\xi_2\zeta}\ \gamma_{0\xi_1\zeta}\ \gamma_{0\xi_1\xi_2}\ K_{\xi_1\xi_1}\ K_{\xi_2\xi_2}\ K_{\xi_2\zeta}\ K_{\xi_1\zeta}\ K_{\xi_1\xi_2}\right]^T \tag{5}$$

The displacement vector using FEM concept (Cook et al. [20]) can be represented as:

$$\{\delta\}=\sum_{i=1}^{9}[N_i]\{\delta_i\} \tag{6}$$

where, $[N_i]$ and $\{\delta_i\}$ are shape functions and the nodal displacement vector, respectively. The nodal displacement elaborated as:

Now, the mid-plane strain vector is expressed in terms of nodal displacement vector:

$$\{\bar{\varepsilon}\}=[B]\{\delta_i\} \tag{7}$$

where, $[B]$ matrix which establish a relation between strain and displacement.

The strain energy and the total work done can be presented as follows:

$$U=\frac{1}{2}\iint\left\{\sum_{k=1}^{n}\int_{\zeta_{k-1}}^{\zeta_k}\{\sigma_{ij}\}\{\varepsilon_{ij}\}d\zeta\right\}d\xi_1 d\xi_2 \tag{8}$$

$$W=\int_A\{\delta\}^T\{q^e\}dA \tag{9}$$

where, $\{q^e\}=\{q^e_m\}+\{q^e_{th}\}$ the combined form of thermal and mechanical load vector.

The integral form of the elemental stiffness matrix can be represented as:

$$[k]=\int_A\left(\sum_{k=1}^{n}\int_{\zeta_{k-1}}^{\zeta_k}[B]^T[D][B]d\zeta\right)dA \tag{10}$$

The mathematical modelling steps for the delamination i.e. displacement kinematics, stiffness matrix and intermittent continuity condition can be seen in Hirwani et al. [21].

The governing equation of the deflection analysis obtained by minimizing the total energy functional via variational principle and presented as:

$$\partial\Pi=\partial U-\partial W=0 \tag{11}$$

where, ∂ represents a variational symbol and Π is used to express the total potential energy functional.

Now, equation (8) and (9) is substituted into equation (11) to get the final equation and depicted as:

$$[K]\{\delta\}=\{q\} \tag{12}$$

The equation (12) is solved by applying different set of boundary condition as in Hirwani et al. [21]:

3. Result and discussion

Now, the above-discussed mathematical formulation is further converted into a MATLAB code using MATLAB-12. Four types of square delamination located at the center mid-plane of the laminate namely Dtype-1, Dtype-2, Dtype-3 and Dtype-4 are considered in this work. Dtype-1 is the healthy laminate, Dtype-2, Dtype-3 and Dtype-4 are the laminate having delamination with a side length of *a*/4, *a*/2 and 3*a*/4, respectively. The material property is used is given below. The external mechanical load is sinusoidally distributed on (ξ_1, ξ_2) plane as in equation below and the thermal load is considered uniform through the thickness as well as on (ξ_1, ξ_2) plane. The nondimensional form of the external load and final central deflection responses are presented in equations below. Now, the convergence and accuracy of the developed code have been checked and discussed in the next section.

$$q_m = q_0 \sin\left(\frac{\pi\xi_1}{a}\right)\sin\left(\frac{\pi\xi_2}{b}\right); \qquad Q = q_0 * E_{\xi_2} * (a/h)^4; \quad W = w/h$$

$$E_{\xi_1} = 181\ GPa; E_{\xi_2} = 10.3 GPa; G_{\xi_1\xi_2} = G_{\xi_1\xi_3} = 7.17 GPa; G_{\xi_2\xi_3} = 6.21 GPa; \upsilon_{\xi_1\xi_2} = \upsilon_{\xi_1\xi_3} = \upsilon_{\xi_2\xi_3} = 0.25$$

3.1 *Convergence and comparison study*

In order to check the consistency of the current model, the nondimensional central deflection responses of the simply supported, symmetric laminated composite plate (0º/90º/90º/0º) subjected to thermomechanical load is obtained. The responses are obtained for five different amplitude of mechanical load (*Q*=30, 60, 90, 120 and 150) and eight different element sizes. The geometrical parameter (*a*/*h*) and temperature gradient (ΔT) is taken as 60 and 50, respectively. The responses are plotted and presented in Figure 2. The figure indicates that the responses are showing very good convergence for different mesh sizes. Based on this analysis a mesh size of (8×8) is utilized for further investigation.

For the validation study, a numerical example of cylindrical shell panel has been chosen because of unavailability of the result of delaminated plate bending (best of author knowledge) as the present model is so general that it can easily be converted into a plate to shell. The eight-layer simply supported cross-ply cylindrical shell panel (laminated and Dtype-2 delaminated) subjected to uniformly distributed mechanical load has been analyzed. The nondimensional deflection values are evaluated via current model and compare with the values of Nanda [5] and tabulated in Table 1. All the input parameter are taken as same as given in the corresponding reference. The table indicates that the reference (layerwise FE model) underpredict the responses. It is due to the lower DOF, i.e., 8 (eight) of the reference model compare to the present model, i.e., 9 DOF.

Table 1. Validation study of nondimensional central deflection responses of the simply supported laminated/delaminated cylindrical shell panel subjected to uniformly distributed load

(ξ_1 / a)	Laminate		Delaminate (*a*/4)	
	Present	Nanda [5]	Present	Nanda [5]
0	0	0	0	0
0.125	0.2322	0.21934	0.26	0.22607
0.25	0.4201	0.40915	0.4668	0.41656
0.375	0.5395	0.52964	0.5952	0.54027
0.5	0.5803	0.57273	0.6413	0.58611
0.625	0.5398	0.53105	0.5946	0.54193
0.75	0.4205	0.40928	0.4578	0.41804
0.875	0.2324	0.22268	0.2508	0.22762
1	0	0	0	0

3.2 *Numerical illustrations*

3.2.1 *Effect of centrally located delamination size*

The influence of the size of debonding on nondimensional central deflection response of laminated composite plate is analyzed in this illustration. For the illustration, a simply supported antisymmetric cross-ply (0°/90°/0°/90°) laminate with different type of delamination (Dtype-1, Dtype-2, Dtype-3 and Dtype-4) under thermo-mechanical (*Q*=50, 100, 150, 200, 250 and ΔT=80ºC) load is considered. The nondimensional response is evaluated using currently developed model and presented in Figure 3. The deflection responses are lower for Dtype-1 and following the increasing trend with Dtype-2, Dtype-3 and Dtype-4. This indicates that with the increase in debonding sizes the overall stiffness of the laminate reduces considerably resulting the deflection response increases.

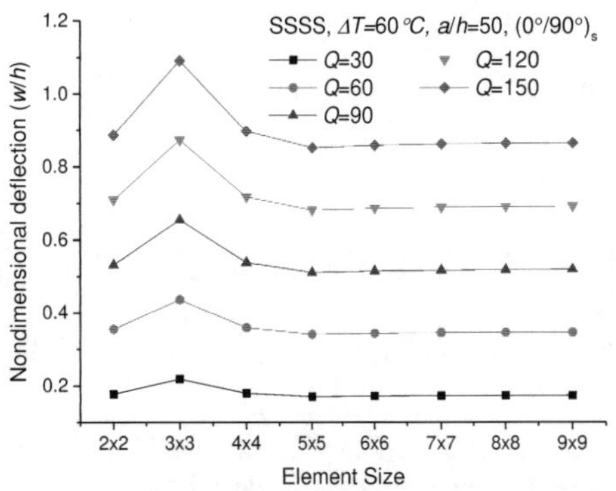

Figure 2. Convergence behavior of layered composite plate

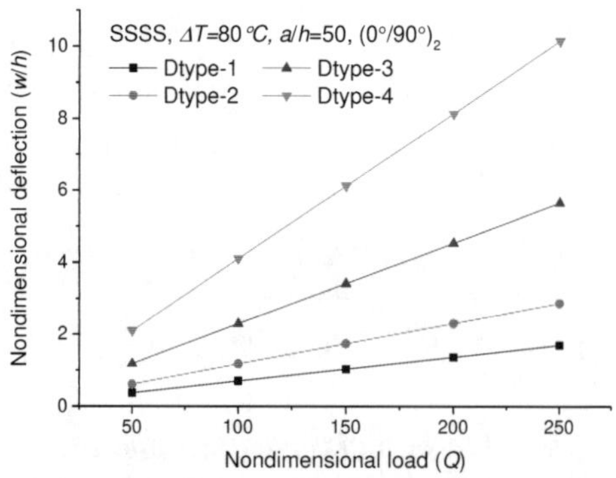

Figure 3. Nondimensional deflection responses of the antisymmetric delaminated composite plate

3.2.2 Influence of Young's modulus ratio

A clamped symmetric (0°/90°/90°/0°) laminate with delamination (Dtype-2) subjected to the combined effect of thermo-mechanical loading is analyzed for central deflection responses. The responses are obtained for five Young's modulus ratio (E_{ξ_1} / E_{ξ_1}) and five mechanical loading amplitude (*Q*=100, 200, 300, 400 and 500) and presented in Figure 4. The nondimensional response decreases with increase in Young's modulus ratio. In the current example, the transverse modulus is kept constant. Therefore longitudinal modulus increases with increase in Young's modulus ratio. The increase in longitudinal Young's modulus increases the overall stiffness and reduces the deflection responses.

3.2.3 Influence of aspect ratio

This example investigates the influence of aspect ratio on bending characteristic of the delaminated composite plate structure. For the investigation, cross-ply (0°/90°/90°/0°) delaminated (Dtype-2) composite plate (*b*/*h*=50) with simply supported edges under external mechanical and elevated temperature loading is considered. The nondimensional responses is examined for six aspect ratio (*a*/*b*=1, 1.4, 1.8, 2.2, 2.6 and 3) and five different load (*Q*=40, 80, 120, 160 and 200). The obtained responses are plotted and shown in Figure 5. It is evident from the figure that the nondimensional responses are increasing with the increase in (*a*/*b*) ratio. It is due to the reduction of the overall stiffness of the laminate with increasing the (*a*/*b*) ratio as *b* is kept constant and *a* is changing.

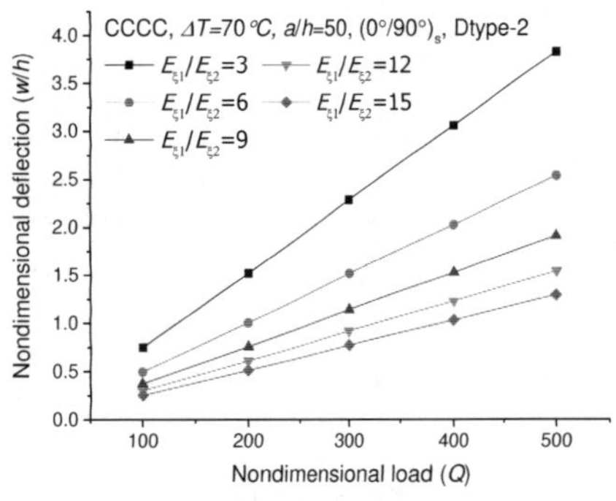

Figure 4. Effect of Young's modulus ratio on deflection responses of delaminated composite plate

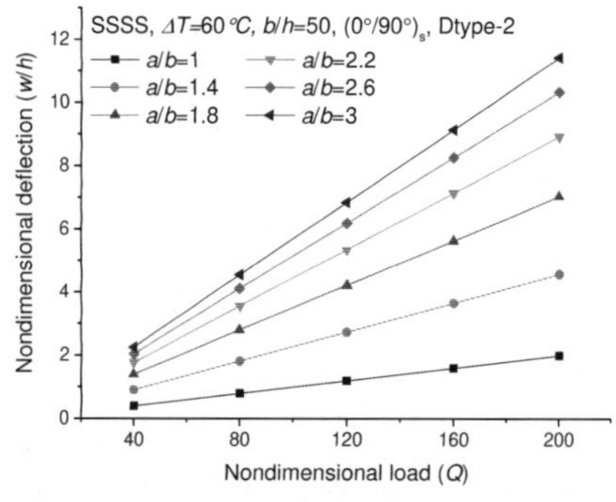

Figure 5. Effect of modulus ratio on deflection responses of delaminated composite plate

3.2.4 Influence of end conditions

To show the effect of boundary condition, eight-layer antisymmetric cross-ply $(0°/90°/0°/90°)_2$ delaminated (Dtype-3) composite plate structure is analyzed and central deflection responses are obtained for different boundary condition (CCCC, SCSC, CFCF, SSSS, CFFF). CCCC, SCSC, CFCF, SSSS and CFFF stand for all side clamped, two sides clamped two sides simply supported, two sides clamped two sides free and one side clamped and other three free, respectively. The plate geometry is considered as (*a*/*h*=70) under mechanical and thermal load (ΔT=70°C). The computed responses are shown in Table 2. The table shows the nondimensional values is lowest for the CCCC and increasing with SCSC, CFCF, SSSS and CFFF. The trend in response is due to the increase in constrants degrees of freedom with CCCC, SCSC, CFCF, SSSS and CFFF.

Table 2. Nondimensional central deflection responses of the eight-layer antisymmetric cross-ply delaminated plate.

Load (Q)	Boundary condition				
	CCCC	SCSC	CFCF	SSSS	CFFF
40	0.1074	0.1527	0.1578	0.7052	2.5515
80	0.233	0.3364	0.3453	1.4123	3.697
120	0.3586	0.52	0.5327	2.1193	4.8426
160	0.4842	0.7037	0.7202	2.8264	5.9881
200	0.6098	0.8873	0.9076	3.5334	7.1337

4. Conclusion

The deformation behavior of the laminated and delaminated composite plate structure subjected to sinusoidal mechanical and uniform thermal loading is investigated in the present investigation. The mathematical model of the plate structure is developed using higher-order mid-plane kinematic. For the debonding modeling, a sub-laminate method is employed and continuity between the nodes of the laminated and delaminated boundary is established using intermittent continuity condition. A nine noded element with nine 81 DOF per element is used to discretize the entire plate structure. The final bending equation is obtained from variational principle and solved via FE approach. The necessary model's consistency and accuracy have been checked via solving numerical examples. Finally, few parametric studies have been performed and based on that some conclusion has been drawn. The presence of delamination reduces the overall stiffness and as its size increases, the overall stiffness decreases. The deflection response increases with the increase in aspect ratio and reduces with the increase in Young's modulus and increasing constraint DOF.

References

[1] Davidson B D, Kruger R and Konig M 1995 *Composites Science and Technology* **54** 385-394.
[2] Hamed M A, Nosier A and Farrahi G H (2006) *Materials and Design* **27** 900–910.
[3] Parhi P K, Bhattacharyya S K and Sinha P K 2001 *Bulletin of Materials Science* **24** 143–149.
[4] Bruno D, Greco F, and Lonetti P 2005 *European Journal of Mechanics A/Solids* **24** 127–149.
[5] Nanda N 2014 *Acta Mechanica* **225** 2893–2901.
[6] Szekrényes A 2013 *International Journal of Mechanical Sciences* 7717–7729.
[7] Zenkour A M 2004 *Compos Struct* **65** 367–379.
[8] Zenkour A M, Allam M N M and Radwan A F 2013 *Int J Mech Mate. Des* **9** 239–251.
[9] Cho M and Oh J 2003 *Compos: Part B* **34** 67–82.
[10] Oh J and Cho 2004 M *Int J Solids Struct* **41** 1357–1375.
[11] Houari M S A, Benyoucef S, Mechab I, Tounsi A and Bedia E A A 2011 *J Thermal Stresses* **34** 315-334.
[12] Tlidji Y, Daouadji T H, Hadji L, Tounsi A, and Bedia E A A 2014 *J Thermal Stresses* **37** 852–869.
[13] Das R R, Singla A and Srivastava S 2016 *Procedia* **144** 1060–1066.
[14] Han J W, Kim J S and Cho M 2017 *Compos Part B* **122** 173-191.
[15] Reddy J N and Liu C F 1985 *Int J Eng Sci* **23** 319–330.
[16] Jones R M 1994 *Taylor & Francis* Philadelphia.
[17] Tsai S W and Hahn H T 1980 *Technomic* Westport CT.
[18] Reddy J N 2004 *CRC Press* Florida.
[19] Mahapatra T R and Panda S K 2015 *J Therm Stresses* **38** 39–68.
[20] Cook R D, Malkus D S and Plesha M E 2000 *John Willy and Sons (Asia) Pvt. Ltd* Singapore.
[21] Hirwani C K, Patil R K, Panda S K, Mahapatra S S, Mandal S K, Srivastava, L and Buragohain M K 2016 *Aerosp Sci Technol* **54** 353–370.

IOP Conf. Series: Materials Science and Engineering **338** (2018) 012020 doi:10.1088/1757-899X/338/1/012020

Lifetime Prediction of Nano-Silica based Glass Fibre/Epoxy composite by Time Temperature Superposition Principle

Abhijeet Anand*, Poulami Banerjee, Rajesh Kumar Prusty, Bankin Chandra Ray
Composite Materials Group, Metallurgical and Materials Engineering,
NIT Rourkela, India- 769008
Corresponding author Email : anand.abhiappu@gmail.com

Abstract: The incorporation of nano fillers in Fibre reinforced polymer (FRP) composites has been a source of experimentation for researchers. Addition of nano fillers has been found to improve mechanical, thermal as well as electrical properties of Glass fibre reinforced polymer (GFRP) composites. The in-plane mechanical properties of GFRP composite are mainly controlled by fibers and therefore exhibit good values. However, composite exhibits poor through-thickness properties, in which the matrix and interface are the dominant factors. Therefore, it is conducive to modify the matrix through dispersion of nano fillers. Creep is defined as the plastic deformation experienced by a material for a temperature at constant stress over a prolonged period of time. Determination of Master Curve using time-temperature superposition principle is conducive for predicting the lifetime of materials involved in naval and structural applications. This is because such materials remain in service for a prolonged time period before failure which is difficult to be kept marked. However, the failure analysis can be extrapolated from its behaviour in a shorter time at an elevated temperature as is done in master creep analysis. The present research work dealt with time-temperature analysis of 0.1% SiO_2-based GFRP composites fabricated through hand-layup method. Composition of 0.1% for SiO_2nano fillers with respect to the weight of the fibers was observed to provide optimized flexural properties. Time and temperature dependence of flexural properties of GFRP composites with and without nano SiO_2 was determined by conducting 3-point bend flexural creep tests over a range of temperature. Stepwise isothermal creep tests from room temperature (30°C) to the glass transition temperature T_g (120°C) were performed with an alternative creep/ relaxation period of 1 hour at each temperature. A constant stress of 40MPa was applied during the creep tests. The time-temperature superposition principle was followed while determining the Master Curve and cumulative damage law. The purpose of a Master Curve was to determine the variation of compliance with respect to increase in time and temperature of the specimen. The shift factors at any reference temperature were determined by Arrhenius activation energy method at a far lower temperature than T_g (Glass transition temperature) and by manual shift method at a temperature near T_g (Glass transition temperature).

1. Introduction

Studies to determine the short-term as well as the long-term durability of various types of fiber reinforced polymer (FRP) composites are being conducted widely. This is due to the fact that the composite structure may be prone to failure at longer durations of time under a constant stress. The viscoelastic behavior of the polymer matrix further leads to a gradual deterioration of the composite material under adverse environmental conditions and for longer durations if not for sudden failure. The glass transition temperature (T_g) exhibited by the polymer matrix is defined as the temperature above which the transition in the amorphous regions between the glassy and rubbery state occurs[1]. It is observed that due to drastic structural changes above the T_g, structural properties of the composites deteriorate significantly. Therefore, inference can be made that the supplication temperature of a composite be maintained below the glass transition temperature for satisfactory mechanical performance and creep resistance.

2. Materials and methods

2.1 Materials

The resin (Diglycidyl ether of Bisphenol A based epoxy) and curing agent (Triethylene tetra amine) used in this study were supplied by Atul Industries, India under the trade names Lapox L12 and K6 respectively. The glass fiber (3 K plain weave E-glass) having a filament diameter of about 15 μm, was purchased from Owens Corning, India. Areal weight of this fabric was 360 g/m2,

IOP Conf. Series: Materials Science and Engineering **338** (2018) 012020 doi:10.1088/1757-899X/338/1/012020

composed of 16 and 14 yarns/inch in warp and weft directions respectively. Silica having spherical structure with diameter in the range of 15–30 nm was received in the powder form from Sigma Aldrich Pvt. Limited.

2.2 Creep: Creep can be defined as the tendency of a material to deform permanently under the influence of mechanical stresses owing to long term exposure to high levels of stress[2]. For a perfectly elastic material (both linear and non-linear), strain(ε) is a time-independent property and is directly related with the applied stress (σ)

Polymers such as epoxy exhibit viscoelastic properties. They store elastic energy within themselves when deformed and utilize this energy to revert back to their original form after the deforming stress is removed. Above a certain temperature, they behave as liquids and exhibit viscosity. Here, their deformation becomes time-dependent.

Here, S (t) is termed as the creep compliance, which can be defined as the inverse of stiffness, but here it is a time-dependent function:

$$S(t) = \frac{\varepsilon(t)}{\sigma} \tag{1}$$

It takes a general polymeric material around tens of thousands of years to deform and is therefore impractical to be determined through long-term creep testing till the completion of the designated lifetime of the specimen. Thus, an accelerated creep testing on materials at an elevated temperature over a shorter period of time is conducted, and the results are extrapolated to ascertain the behaviour of materials at room temperature for an elongated period of time.

2.3 Time-Temperature Superposition model (TTSP) :

Based on TTSP, it is possible to generate short-term creep model at various temperatures and a master curve is generated accordingly. Master creep generation involves a shift parameter (a_T) that matches the creep behavior at a reference temperature [3]. According to the TTSP principle, the frequency function of a dynamic property (say compliance S) at one temperature T_{ref} is similar in shape to the same function in neighboring temperatures. Therefore, S (t) versus logarithmic frequency (or logarithmic response time) at one temperature can be horizontally shifted along the reduced time axis and made to overlap at the neighboring temperatures. This shift distance along the reduced time axis is called the shift factor (a_T) and is given as

$$a_T = \frac{t}{t_r} \tag{2}$$

Where t is the generated time steps for a temperature, t_r is the reduced time corresponding to the reference temperature T_{ref} [1].

In the TTS diagram, the curve at every temperature can be shifted at a horizontal distance to overlap the curve at the reference temperature (T_{ref}) at the neighboring temperature. In this way, for every reference temperature chosen there is a fully overlap curve, termed as the master curve.

3. Experimental Procedure

3.1 Specimen Preparation:

3.1.1 Preparation of SiO_2 nano filler modified epoxy resin:

169 g was taken and preheated at 120°C at 400rpm for half an hour. 0.1%weight of SiO_2 with respect to weight of 5-layers of glass fiber laminate was measured and was added to the epoxy resin, and stirred for 3 hours, followed by 1 hour of sonication after which the modified epoxy resin was placed in a vacuum chamber for 18 hours, in order to remove air bubbles.

3.1.2 Fabrication of the FRP composite with and without hybridized epoxy resin:

Fabrication of 5 layer laminate of bidirectional GFRP composite was conducted placing the epoxy resin (without/with nano-Silica) uniformly between each layer of glass fibers, one after the other. The composite was then cured in a hot press for 20 minutes to solidify the specimen. The finished sample was then kept at room temperature for 24 hours. Following this, specimens were cut and polished following ASTM B7264 standard. Polishing was followed by post curing in a hot air oven at 140°C for 6 hours, in order to facilitate cross-linking of the resin.

3.2 Testing conditions:

Determination of creep/relaxation isotherms were performed in a 3-point bending jig. Stepwise isothermal creep tests from room temperature (30°C) to glass transition temperature T_g (120°C) with an increment of 5°C at each step were performed with an alternative creep/relaxation period of 1 hour each at each temperature. A constant stress of 40MPa was applied during the creep tests and data points were taken after every 15 seconds. The T_g of the specimen was determined by conduction of DMA analysis through tan δ peak method determined at various frequencies.

3.3 Calculations:

Creep compliance is determined from the equation:

$$D_c\,(t) = \frac{4bh^3\delta(t)}{PL^3} \tag{3}$$

where $D_c(t)$ is creep compliance (1/MPa) at a given time t; P is the constant load (N) [4].

4. Result and Discussion

Determination of Flexural Creep at elevated temperatures:

It is highly impractical to determine the design lifetime of a material by conducting creep tests for the entire period. Therefore, prediction of durability is conducted using time-to-failure methods for a given load at elevated temperatures.

Figure 1 shows the creep compliance curve versus the actual time for the neat as well as Nano Silica modified GFRP specimens. The breaks between successive curves are an indication of the alternate recovery periods whereby data points have not been recorded.

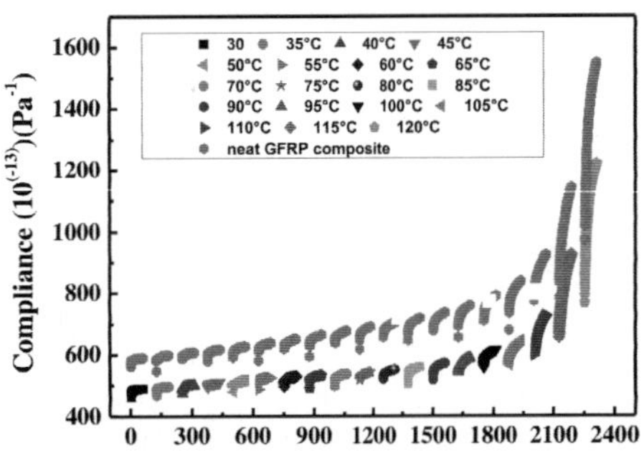

Figure 1: Creep Compliance vs Actual test time for nano modified GFRP and neat GFRP composite.

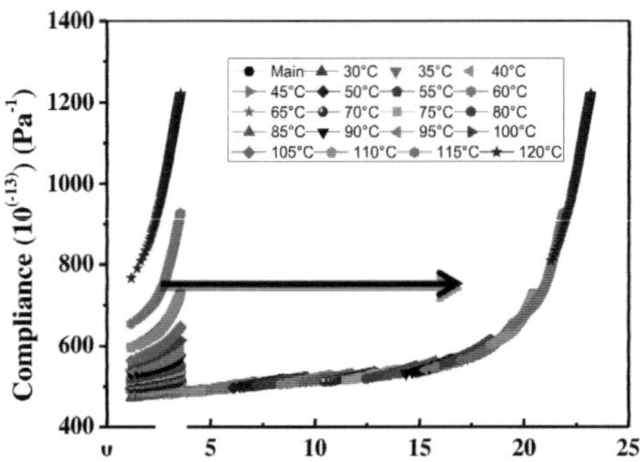

Figure 2.Unshifted creep compliance data and nano Silica Silica modified GFRP composite master curve for T_{ref} =30°C

where t is the actual test time while t_r is the reduced time.

The reduced time is nothing but an expansion of the time scale at an isothermal reference temperature while generation of the master curve, and is given as

$$\log(t_r) = \log(t) - \log(a_T) \tag{4}$$

The creep compliance is not shifted for the reference temperature T_{ref}, since log a_T is zero. For all the other temperatures, the creep compliance curve is shifted horizontally on the reduced time (t_r) scale as is determined by the shift factor. For temperatures greater than T_{ref}, the curve is shifted to the right, since a_T is negative. Similarly, for temperatures less than T_{ref}, the curve is shifted to the left, since in this case a_T is positive. In case of manual shift factor method, the shift factors are chosen by hand so that the isothermal curves overlap in the subsequent temperatures.

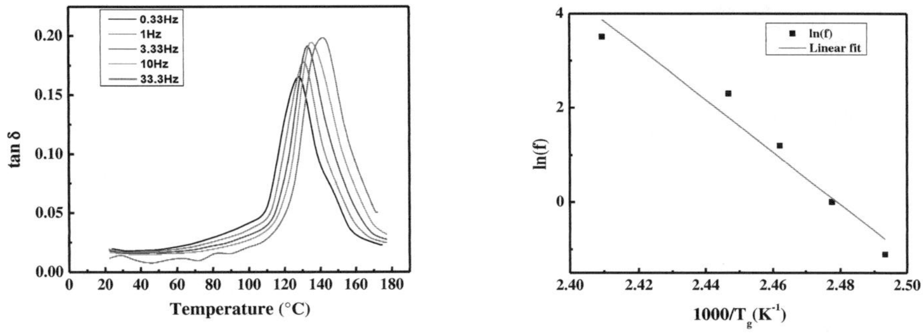

Figure 3(a). tan δ curve peak T_g (b) variation of T_g as obtained by tan delta peak for nano Silica modified GFRP composite

In case of activation energy method, it is possible to determine shift factors by the activation energy measured from the glass transition relaxation. The activation energy for glass transition can be measured from the slope of ln(f) vs $1/T_g$ graph as shown in Figure 3. Here, T_g is measured as each of the peak values of tanδ vs temperature graph for various assigned frequencies (0.33Hz, 1Hz, 3.33Hz, 10Hz and 33.3Hz). Graph shown in Figure 3(a) is obtained by conduction of dynamic mechanical analysis[5].

The equation for calculation of activity is given by

$$\Delta H = -R\frac{d(ln(f))}{d(\frac{1}{Tg})} = -(8.314\times10^{-3})(-55.60084\times10^{3}) = 462.06 \text{ KJ/mol} \quad (5)$$

The determination of activation energy in this manner makes the task of obtaining shift factors much easier as there is no requirement of generating the master curve in this case.

The shift factor is determined using time-temperature superposition model[1]. Using the constant activation energy technique, shift factor can be calculated by the following formula.

$$\text{Log}(a_T) = \frac{\Delta H}{R}\left(\frac{1}{T} - \frac{1}{T_{ref}}\right)\log e \quad (6)$$

This is derived from the Arrhenius relationship.

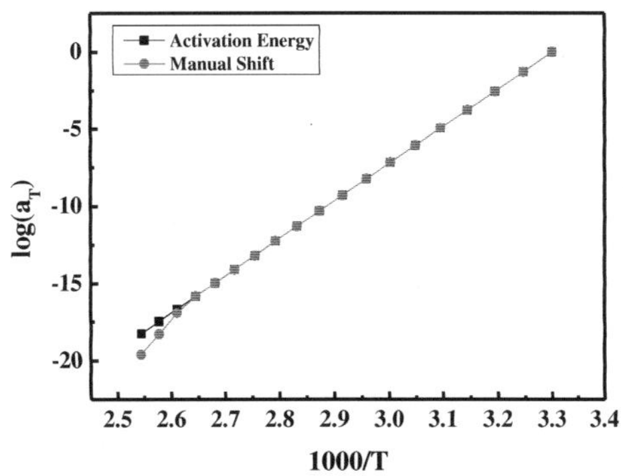

Figure 4. Shift factors versus 1/T for manual shift and constant activation energy estimation of for nano Silica modified GFRP

It has been observed that this equation is valid only up to a certain temperature below T_g, above which Willian-Landers-Ferry model (WLF) is applicable [6]. In case of nano Silica modified GFRP composite, the activation energy obtained is 462.06 KJ/mol.

As inferred from Figure 4ithad been observed that the manual shift factor and the constant activation energy shift factor coincide well upto 1000/T= 2.6, which is T= 111.6°C. Above this, the manually shifted data exhibits larger values of shift factor.

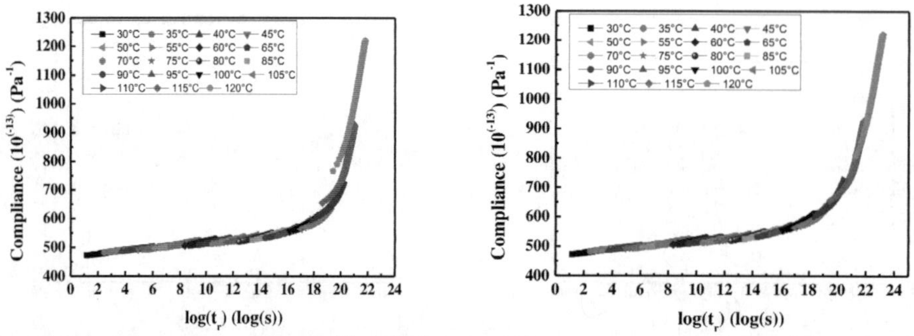

Figure 5. Master Curve for nano Silica modified GFRP at T_{ref} =30°C: (a) constant activation energy (b) manual shift

It can be well deduced that constant activation energy curve falls invalid near temperature just below T_g, i.e. 115°C. Below this temperature range, the compliance data line quite well up, and starts to get distorted above it as shown in Figure 5(a).

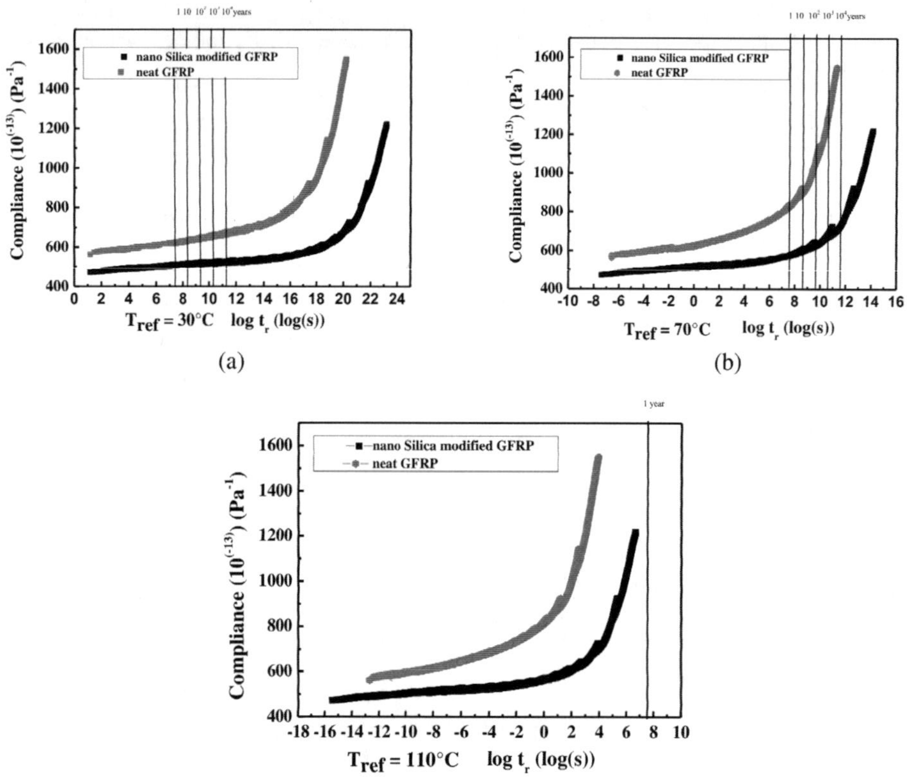

(c)

Figure 6. Master Creep Curve for nano Silica modified GFRP and neat GFRP at (a) T_{ref} : 30°C (b) T_{ref} : 70°C (c) T_{ref} : 110°C

The lifetime prediction of the SiO_2-modified GFRP composites have been done by using reference temperatures of 30°C, 70°C and 110°C along with a time sequence of 1, 10, 100, 1000 years and so on.

A comparative study of the master curves for the two specimens, first being the neat GFRP composite and the second being the nano-Silica modified GFRP composite indicates a similar yet distinguishable difference in compliance with respect to time in case of three reference temperatures (T_{ref} = 30°C, 70°C and 110°C) as shown in Figure 6.

5. Conclusions

The study of master creep analysis of a 0.1%SiO_2-modified Glass fiber/epoxy composite was conducted at various reference temperatures for a prolonged duration of time at a constant stress. The following conclusions were drawn:

- The constant activation energy method stands valid only up to a temperature far lower than T_g of the material after which the manual shift factor method provides more consistent master curve plot.
- A total of three master curves (T_{ref} = 30°C, 70°C and 110°C) were generated using each method of shift factor estimation. The adequate model gives an important curvature of the master curve, due to the stress sensibility exponent of the model. It was inferred that the flexural creep strain is both time and temperature dependent, and becomes more viscoelastic near T_g.
- A comparative study between neat GFRP and nano-Silica modified GFRP composites showed that nano-Silica modified GFRP composite has a better stiffness (lower compliance) as compared to neat GFRP composite and therefore is more resistant to deformation than the latter.
- Lifetime prediction at various reference temperatures indicate that the modified GFRP can be sustained at room temperature for a high period of time (around 10,000 years) without exhibiting any significant change in the compliance. At a reference temperature of 70°C, durability up to 10 years is observed beyond which the compliance increases (i.e. stiffness decreases) drastically. At an elevated temperature of around 110°C, both the composites fail to sustain even for 1 year.

6. Acknowledgement:

The authors would like to thank National Institute of Technology Rourkela, India and Council of Scientific and Industrial Research, New Delhi, India (No. 22(0735)/17/EMR-II) for their support and for providing the necessary facilities to carry out this study. The technical assistance received from Mr. Rajesh Patnaik is also highly appreciated.

References

[1] R. Li, “Time-temperature superposition method for glass transition temperature of plastic materials,” *Mater. Sci. Eng. A*, vol. 278, no. 1, pp. 36–45, Feb. 2000.

[2] S. Kumar Ghosh, R. K. Prusty, D. K. Rathore, and B. C. Ray, “Creep behaviour of graphite oxide nanoplates embedded glass fiber/epoxy composites: Emphasizing the role of temperature and stress,” *Compos. Part Appl. Sci. Manuf.*, vol. 102, no. Supplement C, pp. 166–177, Nov. 2017.

[3] B. Ponsot, D. Valentin, and A. R. Bunsell, “The effects of time, temperature and stress on the long-term behaviour of CFRP,” *Compos. Sci. Technol.*, vol. 35, no. 1, pp. 75–94, Jan. 1989.

[4] I. Ivaneiko, V. Toshchevikov, K. W. Stöckelhuber, M. Saphiannikova, and G. Heinrich, “Superposition approach to the dynamic-mechanical behaviour of reinforced rubbers,” *Polymer*, vol. 127, no. Supplement C, pp. 129–140, Oct. 2017.

[5] V. Fabre, G. Quandalle, S. Cantournet, and N. Billon, *Time-Temperature-Water Content equivalence on dynamic mechanical response of Polyamide 6,6*. 2014.

[6] J. D. Ferry, W. C. Child, R. Zand, D. M. Stern, M. L. Williams, and R. F. Landel, “Dynamic mechanical properties of polyethyl methacrylate,” *J. Colloid Sci.*, vol. 12, no. 1, pp. 53–67, Feb. 1957.

IOP Conf. Series: Materials Science and Engineering **338** (2018) 012021 doi:10.1088/1757-899X/338/1/012021

Experimental and numerical investigations on the temperature distribution in PVD AlTiN coated and uncoated Al_2O_3/TiCN mixed ceramic cutting tools in hard turning of AISI 52100 steel

Ch Sateesh Kumar[1*], Saroj Kumar Patel[2], Anshuman Das[3]
[1,2,3]Department of Mechanical Engineering, National Institute of Technology, Rourkela (Odisha)

chigulla51@gmail.com

Abstract. Temperature generation in cutting tools is one of the major causes of tool failure especially during hard machining where machining forces are quite high resulting in elevated temperatures. Thus, the present work investigates the temperature generation during hard machining of AISI 52100 steel (62 HRC hardness) with uncoated and PVD AlTiN coated Al_2O_3/TiCN mixed ceramic cutting tools. The experiments were performed on a heavy duty lathe machine with both coated and uncoated cutting tools under dry cutting environment. The temperature of the cutting zone was measured using an infrared thermometer and a finite element model has been adopted to predict the temperature distribution in cutting tools during machining for comparative assessment with the measured temperature. The experimental and numerical results revealed a significant reduction of cutting zone temperature during machining with PVD AlTiN coated cutting tools when compared to uncoated cutting tools during each experimental run. The main reason for decrease in temperature for AlTiN coated tools is the lower coefficient of friction offered by the coating material which allows the free flow of the chips on the rake surface when compared with uncoated cutting tools. Further, the superior wear behaviour of AlTiN coating resulted in reduction of cutting temperature.

1. Introduction

The finishing or semi-finishing machining operation of materials having hardness in the range of 45-70 HRC is termed as hard machining process. The advantages of hard machining is making it an imperative alternative to grinding process [1]. However, hard machining process is associated with elevated machining forces and temperature which demands excellent hot hardness and wear resistance for the employed cutting tools [2]. Thus, cutting tools manufactured from ceramics, mixed ceramics, polycrystalline cubic boron nitride (PCBN) or cubic boron nitride (CBN) that exhibit excellent thermal and chemical stability at elevated temperatures are employed for machining of hard materials [3]. Among these cutting tools, the popularity of alumina based mixed ceramic cutting tools in the industry and the research community is basically due to their economical nature. But the higher brittleness of these ceramic cutting tools make them more prone to brittle failure under impact or shock loads [4]. However, the progress made in the processing technology of ceramics [5] has generated the possibility of addition of certain ductile phases like TiCN and TiC to the brittle ceramic matrix to form a composite commonly termed as mixed ceramic or composite ceramic material. The addition of these phases impart additional fatigue and shock loading capacity, and wear resistance [6] to the ceramic cutting tools.

The performance of these tools can further be improved by deposition of hard coatings [7–11]. The literature apparently indicates that the AlTiN coated tools exhibit excellent thermal stability [12] which is a must condition in hard machining where large amount of heat is generated. Further, temperature generation in cutting tools is one of the major causes of tool failure especially during hard machining where machining forces are quite high resulting in elevated temperatures [1]. Thus, the present work investigates the temperature generation during hard machining of AISI 52100 steel (62 HRC hardness) with uncoated and PVD AlTiN coated Al_2O_3/TiCN mixed ceramic cutting tools. Further, a finite element model has been adopted to predict the temperature generated during the hard machining process for comparative assessment with the measured temperature.

2. Experimental methodology

AlTiN coating was deposited on to the Al_2O_3/TiCN (designation: SNGA120408) based mixed ceramic cutting tool using Oerlikon Balzers rapid coating machine (RCM) in cathode are evaporation mode. The detailed coating process has been elaborated by Kumar and Patel [2]. After deposition of coating on the ceramic substrates, the coated and uncoated cutting tools were subjected to continuous turning tests under dry cutting environment on a heavy duty lathe machine provided by HMT Ltd., India (see Fig. 1) by mounting the cutting tools on a tool holder with the specification as PSBNR2020K12. Each experiment was carried out 3 times to avoid machining errors. The machining parameters used for the turning tests are listed in Table 1. The temperature measurements were performed during each experimental run with the help of a non-contact infrared thermometer (Make: HTC, Model: IRX-65) that has the capability to measure the temperature of a 1 mm spot from a distance of 30 mm. The temperatures were recorded after every 2 min during each experimental run and an average was considered as the final cutting temperature.

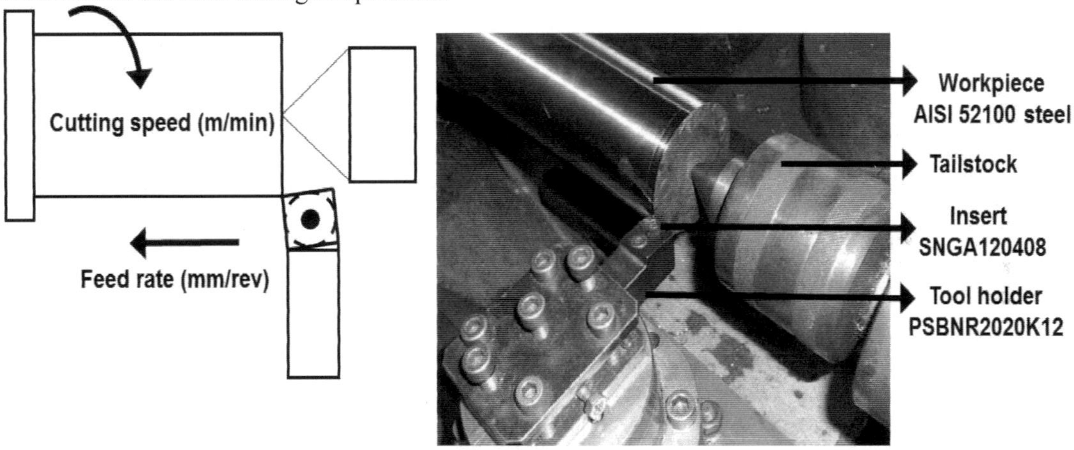

Fig. 1 Experimental setup for turning tests

Table 1 Machining parameters for turning tests

Experimental run	Cutting speed (m/min)	Feed rate (mm/rev)	Depth of cut (mm)
1	110	0.12	0.5
2	145	0.12	0.5
3	180	0.12	0.5

3. Finite element analysis

In the present work, Deform 3D finite element analysis and modelling software for the numerical analysis of the temperature generation during dry and hard turning of AISI 52100 steel. All the simulations were performed using Al_2O_3/TiCN based mixed ceramic inserts as cutting tools in coated and uncoated condition. A thickness of 3 μm was considered for the AlTiN coating. The separation of the chips has been defined using Lagrangian approach in the numerical modelling process. A linear simplified workpiece model as illustrated in Fig. 2 has been implemented for the present study. The tool has been considered as rigid whereas the workpiece was considered as visco-plastic. The cutting tool and the workpiece were meshed with 1,55,000 and 90,000 tetrahedral thermally coupled elements respectively. For the purpose of precisely investigating the cutting zone, a very fine mesh has been defined at the cutting zone for the cutting tool. The heat transfer coefficient and air convection on the free surface were assumed as per the considerations from Özel et al. [13]. Also, all the simulations were performed considering an initial temperature of 20 ^{0}C for both the cutting tool and the workpiece.

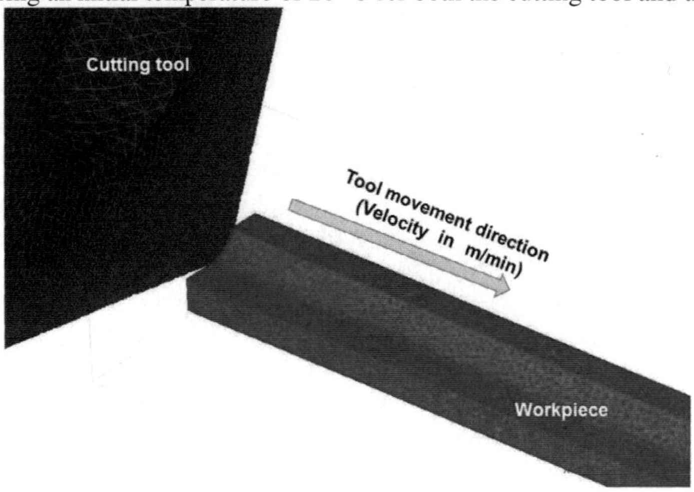

Fig. 2 Simplified linear finite element meshing model for turning simulations

The equivalent flow stress for the workpiece material has been defined with the help of Johnson-Cook (J-C) material flow stress model. J-C model applied by Kim et al.[3] has been implemented for the present numerical study. J-C model indicating the equivalent stress has been shown in Equation 1.

$$\bar{\sigma} = [A + B.\varepsilon^n]\left[1 + C.\ln\left(\frac{\dot{\varepsilon}}{\dot{\varepsilon}_0}\right)\right]\left[1 - \frac{T - T_{Room}}{T_{Melt} - T_{Room}}\right]^m \quad (1)$$

where A, B, n, C, ε, $\dot{\varepsilon}$, $\dot{\varepsilon}_0$, T_{Room}, T_{Melt} and m are material model constants. The values of model constants and temperature dependent material properties for workpiece are adopted as implemented by Kumar and Patel [1] .The model constants used for the present work are illustrated in Table 2

Table 2 J-C material constitutive model constants for AISI 52100 steel (62 HRC)

A (MPa)	B (MPa)	n	C	m	T_{Melt} (K)
774.78	134.46	0.371	0.0173	3.171	1760

The researchers have employed various friction models that are basically based on Coulomb's coefficient of friction or shear friction factors [14–16] but there is scarcity of studies that involve implementation of hybrid models. Thus, the present work involves implementation of a hybrid model which has been shown in Equation 2.

$$s_f = \begin{cases} \alpha\sigma_n \text{ ,} if \ \alpha\sigma_n < \nu\bar{\tau} \\ \nu\bar{\tau} \text{,} if \ \alpha\sigma_n \geq \nu\bar{\tau} \end{cases} \quad (2)$$

where s_f is the frictional shear stress, α is the Coulomb's coefficient of friction, σ_n is the normal stress, ν is the shear friction factor and $\bar{\tau}$ is the shear flow stress of workpiece material which is further expressed as

$$\bar{\tau} = \frac{\bar{\sigma}}{\sqrt{3}} \quad (3)$$

The values of shear friction factor and Coulomb's coefficient of friction have already been elaborated by Kumar and Patel [2]. The present model tries to explore the sliding and sticking zones of friction during the machining process.

4. Results and discussions

Fig. 3 represents the variation of cutting temperature with cutting speed for coated and uncoated cutting tools respectively. The dashed lines represent the corresponding numerically predicted cutting temperatures while machining of AISI 52100 steel at 62 HRC hardness under dry cutting environment. It was observed that the measured cutting temperatures were less as compared to the numerically predicted cutting temperatures. These lower readings during experimental procedure have accounted due the hindrance offered by the moving chips and thus, the exact cutting zone could not be explored. Therefore, the temperatures recorded were only of the top surface of the chips. However, owing to the high thermal conductivity of workpiece material and small thickness of the chips, there would not be much difference in the cutting zone temperature and the temperature of the top surface of the chips. This phenomenon has also been discussed by Kumar and Patel [1,2].

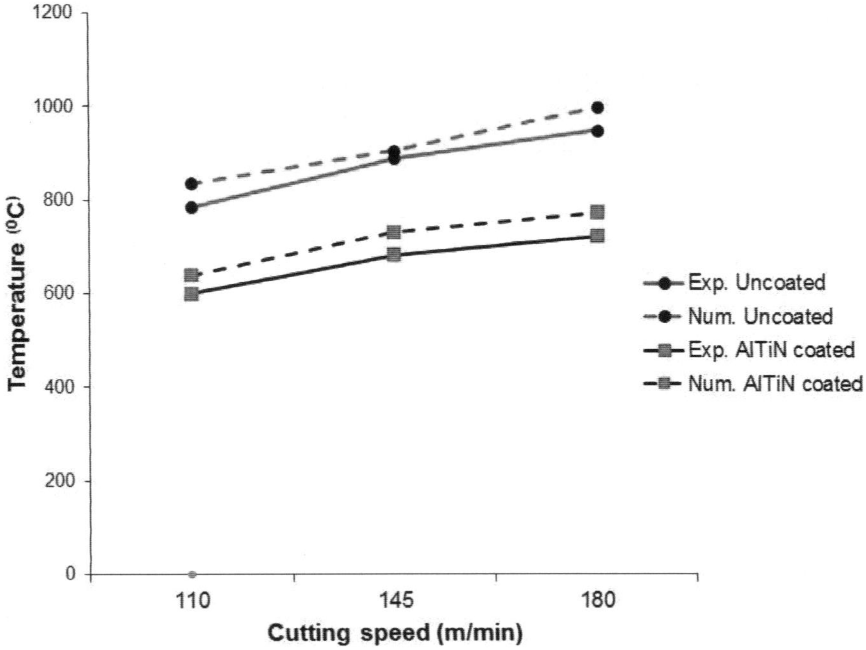

Fig. 3 Variation of temperature with cutting speed

The investigation of temperature generation revealed that the cutting zone temperature reduced drastically while machining with AlTiN coated tools when compared with uncoated cutting tools. Numerical results also reveal similar trend, though the temperatures are on the higher side due to the reasons mentioned above. However, the temperature gain by the workpiece is almost same when machining with coated and uncoated cutting tools which indicates that the energy required for material removal from a specimen remains almost constant (see Fig. 4). The reduction in temperature generation during machining with AlTiN coated tools is mainly due to the lower friction offered by the coating material as compared to uncoated cutting tool [2]. Further, higher hardness of AlTiN coating when compared with uncoated cutting tool, provides it with added wear resistance.

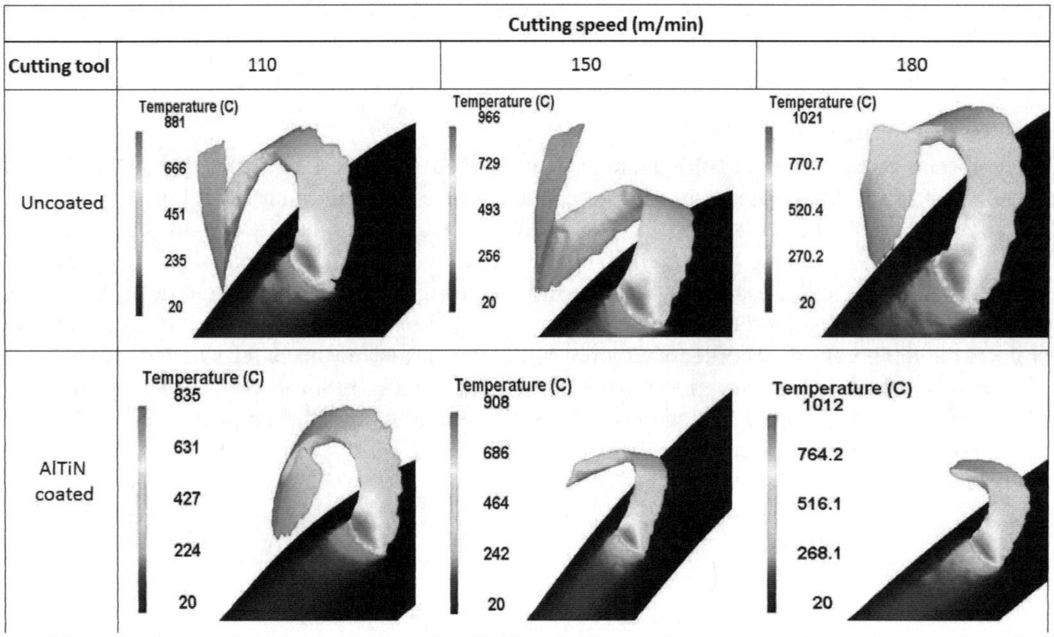

Fig. 4 Numerically predicted temperature distribution in the workpiece material during turning of AISI 52100 steel

5. Conclusions

The present work investigated the machining behavior of AlTiN coated and uncoated Al_2O_3/TiCN based mixed ceramic cutting tools during turning of hardened AISI 52100 steel (62 HRC). Deposition of AlTiN coating on the mixed ceramic substrate resulted in the reduction of cutting temperature. The lower coefficient of friction and superior wear resistance offered by the AlTiN coating led to the reduction of temperature generation during the machining process.

References

[1] C. Sateesh Kumar, S. Kumar Patel, Hard machining performance of PVD AlCrN coated Al2O3/TiCN ceramic inserts as a function of thin film thickness, Ceram. Int. 43 (2017) 13314–13329. doi:10.1016/j.ceramint.2017.07.030.

[2] C.S. Kumar, S.K. Patel, Experimental and numerical investigations on the effect of varying AlTiN coating thickness on hard machining performance of Al 2 O 3 -TiCN mixed ceramic inserts, Surf. Coat.Technol. 309 (2017) 266–281. doi:10.1016/j.surfcoat.2016.11.080.

[3] D.M. Kim, V. Bajpai, B.H. Kim, H.W. Park, Finite element modeling of hard turning process via a micro-textured tool, Int. J. Adv. Manuf. Technol. 78 (2015) 1393–1405. doi:10.1007/s00170-014-6747-x.

[4] Q. Xu, J. Zhao, X. Ai, Fabrication and cutting performance of Ti(C,N)-based cermet tools used for machining of high-strength steels, Ceram. Int. 43 (2017) 6286–6294. doi:10.1016/j.ceramint.2017.02.034.

[5] H. Yang, F. Shang, L. Gao, Microstructure and mechanical properties of gas pressure sintered Al 2 O 3 / TiCN composite, Ceram. Int. 33 (2007) 1521–1524. doi:10.1016/j.ceramint.2006.07.001.

[6] K. Aslantas, T.I. Ucun, A. çicek, Tool life and wear mechanism of coated and uncoated Al 2O3/TiCN mixed ceramic tools in turning hardened alloy steel, Wear. 274–275 (2012) 442–451. doi:10.1016/j.wear.2011.11.010.

[7] A. Mkaddem, A. Ben Soussia, M. El Mansori, Wear resistance of CVD and PVD multilayer coatings when dry cutting fiber reinforced polymers (FRP), Wear. 302 (2013) 946–954.

doi:10.1016/j.wear.2013.03.017.

[8] A. Devillez, F. Schneider, S. Dominiak, D. Dudzinski, D. Larrouquere, Cutting forces and wear in dry machining of Inconel 718 with coated carbide tools, Wear. 262 (2007) 931–942. doi:10.1016/j.wear.2006.10.009.

[9] S. Chinchanikar, S.K. Choudhury, Machining of hardened steel - Experimental investigations, performance modeling and cooling techniques: A review, Int. J. Mach. Tools Manuf. 89 (2015) 95–109. doi:10.1016/j.ijmachtools.2014.11.002.

[10] E.O. Ezugwu, C.I. Okeke, Tool life and wear mechanisms of TiN coated tools in an intermittent cutting operation, J. Mater. Process. Technol. 116 (2001) 10–15. doi:10.1016/S0924-0136(01)00852-4.

[11] Q. An, C. Wang, J. Xu, P. Liu, M. Chen, Experimental investigation on hard milling of high strength steel using PVD-AlTiN coated cemented carbide tool, Int. J. Refract. Met. Hard Mater. 43 (2014) 94–101. doi:10.1016/j.ijrmhm.2013.11.007.

[12] Y.P. Feng, L. Zhang, R.X. Ke, Q.L. Wan, Z. Wang, Z.H. Lu, Thermal stability and oxidation behavior of AlTiN, AlCrN and AlCrSiWN coatings, Int. J. Refract. Met. Hard Mater. 43 (2014) 241–249. doi:10.1016/j.ijrmhm.2013.11.018.

[13] T. Özel, M. Sima, A.K. Srivastava, B. Kaftanoglu, Investigations on the effects of multi-layered coated inserts in machining Ti-6Al-4V alloy with experiments and finite element simulations, CIRP Ann. - Manuf. Technol. 59 (2010) 77–82. doi:10.1016/j.cirp.2010.03.055.

[14] P.J. Arrazola, T. Özel, Investigations on the effects of friction modeling in finite element simulation of machining, Int. J. Mech. Sci. 52 (2010) 31–42. doi:10.1016/j.ijmecsci.2009.10.001.

[15] C. Maranhão, J. Paulo Davim, Finite element modelling of machining of AISI 316 steel: Numerical simulation and experimental validation, Simul. Model. Pract. Theory. 18 (2010) 139–156. doi:10.1016/j.simpat.2009.10.001.

[16] D.M. Kim, I. Lee, S.K. Kim, B.H. Kim, H.W. Park, Influence of a micropatterned insert on characteristics of the tool-workpiece interface in a hard turning process, J. Mater. Process. Technol. 229 (2016) 160–171. doi:10.1016/j.jmatprotec.2015.09.018.

Study of effective utilization of iron ore sinter through arc plasma

***[1]Biswajit Swain, [1]S K Samal, [1]M K Mohanty, [1]A Behera, [1]S C Mishra,**
[1]Metallurgical and Materials Engineering Department, N.I.T Rourkela
Email: enggbiswajit92@gmail.com

Abstract
Generation of fines is common in mining, sizing, and beneficiation and also in high-temperature metallurgical processes as the disintegration of agglomerate/compact occurs. Extraction of metallic iron from ore fines is one of the challenging aspects of iron making industries as the liberation of fines blocks, the charge burden porosity and hence hinders the reduction rate. Along with size factor, mineral composition plays a vital role in the extraction process; particularly silica. As silica has the very high tendency towards iron oxide, at comparatively low temperature, the activity of silica should be suppressed to prevent silicate phases. Adjustment of such conditions is controlled by addition of lime, but sometimes excessive slag generation increases the cost of production. In the present work, carbothermic reduction of partially reduced iron bearing pellets has been melted through 20 KW DC arc plasma furnace, and a comparative study has been made for considering different slag chemistry approaches. Pellets as aforementioned are made available from Patnaik Steel and Alloys Ltd, Odisha, having high silica content ore fines (of about 8.6%) as obtained from the chemical analysis. X-Ray analysis and optical image analyzer result of sinter thus obtained reveal that fayalite phase has major fractional value. Smelting works were done for sinter with/without adjustment of slag chemistry, where argon and nitrogen were used as plasma forming gases. A range of recovery rates (between 87-94%) is achieved by charge composition, ionizing gases, and smelting duration. It is observed that use of nitrogen as plasma forming gas increases the recovery rate than that of using only argon plasma; due to high energy flux of nitrogen which increases the enthalpy due to its diatomicity. A maximum recovery rate of about 94% is achieved for process duration of 13minutes utilizing nitrogen plasma. Smelting of charge with the addition of hydrated lime targeting melilite as final slag resulted in the formation of metallic iron as confirmed from XRD and XRF analyses. In the other hand, ferrosilicon is liberated in the metallic parts where smelting of charge was done without adjustment of slag chemistry. Both metal and slag thus obtained are characterized by XRD, XRF, microhardness and wet chemical analysis suitably.

Keywords: siliceous iron ore, arc plasma furnace, argon and nitrogen, recovery.

1. Introduction
With increasing global demand for iron ore, the production has increased by initiating steps to utilize the low-grade iron ores, fines and slimes [1]. Generation of fines is common in mining, sizing and also in high-temperature metallurgical processes as the disintegration of agglomerate/compact occurs. These fines being of low grade, cannot be used in the blast furnace directly and if used then it reduces process efficiency [2, 3]. However, this difficulty can be overcome with the application of plasma technology as it possesses several advantageous features [4, 5 and 6].

Along with size factor, mineral composition plays a vital role in the extraction process; particularly silica. As silica has the very high tendency towards iron oxide, at comparatively low temperature, the activity of silica should be suppressed to prevent several iron-silicate phases. Silica hinders the reduction reaction rate by lowering surface area of pellet samples due to the formation of liquid fayalite phase at comparatively low temperature [7, 8, 9, 10 and 11]. In the thermal plasma process, high density of ionic charges makes uniform heat transfer to the charge material the reactions are completed in short duration [12]. Ferrosilicon is a ferroalloy which is used as a deoxidizer in steel industries [13]. In the present work an attempt has been made for the comparative study of smelting of sinter obtained from siliceous iron ore; with/ without adjustment of slag chemistry.

2. Materials and experiments

2.1. Materials

Iron ore sinter was collected from Pattanaik steels & Alloys Ltd, Odisha, India for our present study. The chemical analysis of sinter was done through wet chemical analysis and is listed in table 1 given below. It is clear from the compositional analysis that content of all gangue oxides contributes about 18.23 wt. % in which only silica content is 8.6wt. %.

Table 1. Chemical analysis of sinter.

Oxides	Al_2O_3	SiO_2	MgO	CaO	TiO_2	Fe_t	Others
In Wt.%	7.2	8.6	0.63	1.4	0.4	72.2	9.57

2.1.1. Elemental analysis of sinter:

Elemental analysis of sinter was done by EDS analysis and is given in table 2.

Table 2. Elemental analysis of sinter.

Elements	Fe	Al	Mg	Si	Ca	Ti	O	C
In Wt. %	63.2	3.79	0.38	3.99	1.01	0.53	18.65	8.44

The electron image and EDS spectra of powdered sinter sample are shown in figure 1. In the EDS spectra, the major peaks identified are Fe, Al, and Si. Besides oxygen, other elements are found to be in trace amount.

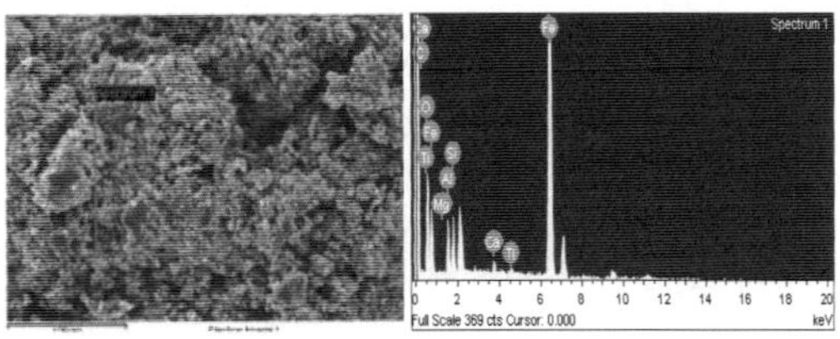

Figure 1. SEM of sinter with EDS spectra.

2.1.2. Optical micrograph of sinter:

Further, the polished surface of sinter sample was observed by using a Zeusius light emission electron microscope under a constant magnification of 200X, and area-fraction analysis was obtained by using Axio vision software of version 4.8 as shown in figure 2.

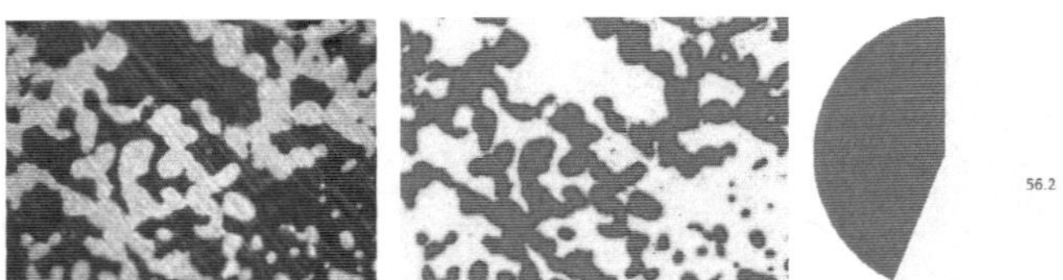

Figure 2. Optical micrograph of sinter with image analysis

Two distinct phases viz. bright and dull phases are seen to be widely distributed along with the very small fraction of dark/black phases. Bright phases are mostly interconnected and surrounded by dull phases which signifies the possibility of formation and flow of slag phases over metallic phases. From area-fraction analysis, it is found that bright phases have lesser fractional value by about 13%.

2.1.3. *x-ray analysis of sinter:*

Further X-Ray analysis of sinter was done to detect the presence of phases formed in the sinter by using a Philips X-Ray Diffractometer with Cu-Kα (1.54 A^0) and is shown in figure 3.

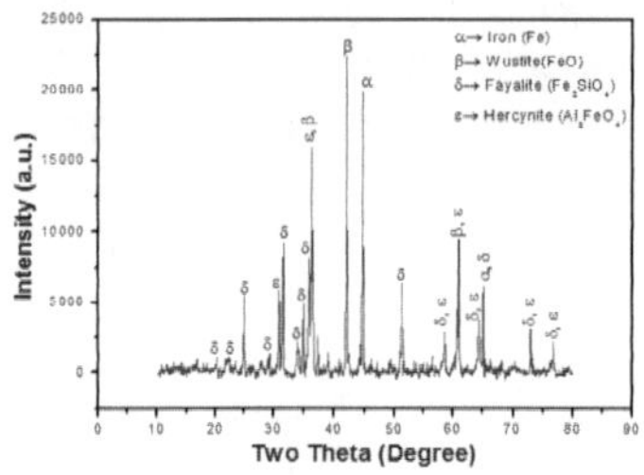

Figure 3. X-ray diffractogram of sinter

Wustite (FeO), fayalite (Fe2SiO4), iron (Fe) and hercynite (Al2FeO4) are found to be the major phases present in the XRD pattern as shown above. Bright and dark phases seen in optical micrograph are thus confirmed to be metallic and Fe-silicate/Fe-aluminium-silicate, i.e., fayalite and hercynite phases respectively. The existence of any other major phases is not detected in X-ray diffractogram as other compounds are in trace amount which is evident form chemical analysis results. Fayalite is a stable phase that not only consumes more reductant and energy to reduce but also requires a higher reducing atmosphere, i.e., high working temperature. The requirement of high energy flux at elevated temperature can be fulfilled by thermal plasma and hence implemented for the smelting of sinter.

2.2 Experiments

A 20KW DC extended arc plasma arc furnace was employed for the smelting experiments. Specification of the above mentioned set up is illustrated in table 3 as given below. The generation of plasma was made through arcing mode and is identical as followed and described by S K Samal et al. [4]. The composition of charge feed along with some of the critical smelting parameters are listed in table 4.

Table 3. Specification of plasma setup.

Crucible (graphite)	Outer diameter- 145 mm Wall thickness- 15 mm Height- 300mm
Hollow tapered rod (graphite)	Length- 400 mm Inner dia- 5 mm Outer dia- 35 mm

Table 4. Important operating parameters.

Sample No.	Plasma forming gas used	Gas flow rate (in LPM)	Process duration (min.)
1	Ar	2.5	7.0
2	Ar	2.5	13
3	Ar+ N_2	2.5	7+2
4	N_2	2.5	7
5	Ar	2.5	13
6	N_2	2.5	9

*Ar-argon, N_2-Nitrogen

First four samples were smelted as such with the addition of coke in 25% excess of stoichiometry. Voltage and current values as maintained throughout the process are 50V and 250A respectively. The weight of charge was in the range of 300-500gm. Further, two more smelting works, i.e., samples 5 and 6, were done with the addition of 8.6% hydrated lime as flux after complete melting of charge. In both cases, petroleum coke was used as reducing agent.

3. Results and discussion

3.1 Recovery:

Recovery rate is the ratio of the weight of the metallic fraction of smelted product to the weight of metal present in the charge feed before smelting. The of the recovery rate of each sample w.r.t smelting duration is shown in Figure3.1.

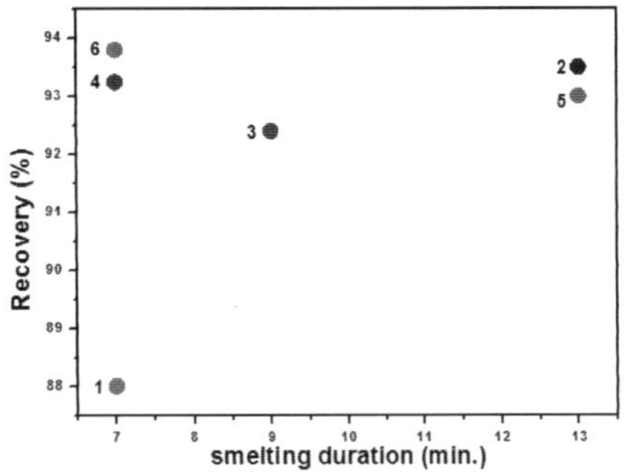

Figure 4. Percentage of recovery of samples smelted by argon and nitrogen plasma

Points representing recovery rates of nitrogen plasma smelted samples lie above to that of argon plasma as evident from figure 4. The highest recovery rate of about 93.82 % was achieved for sample smelted with an addition of lime by nitrogen gas.

2.2. SEM study of final product:

Further well-polished surfaces of metallic fractions of smelted products (metal) were observed in scanning electron microscope and are shown in figure 3.2. EDS analysis at different appearing phases was done and is given by Wt% in table 5.

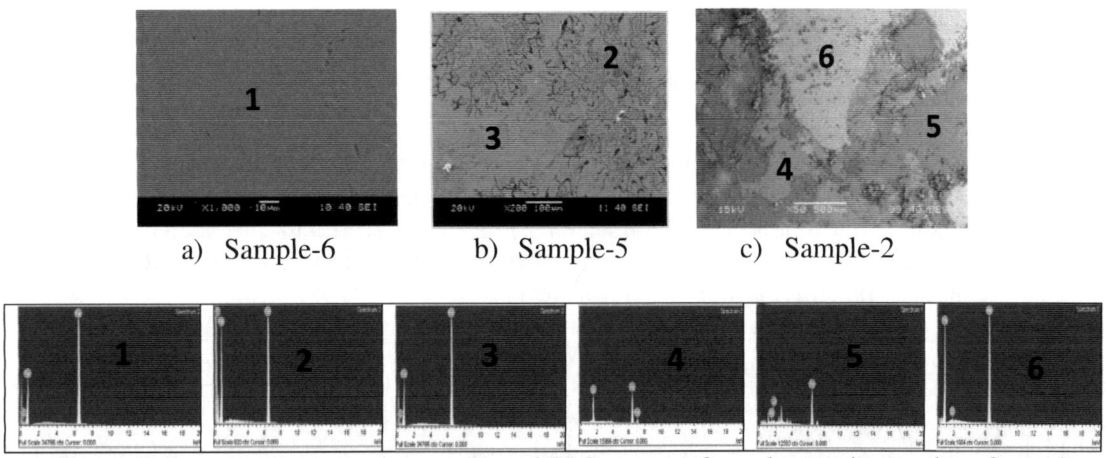

Figure 5. Scanning electron micrographs and EDS spectra of metals at various region of samples.

In micrograph of sample-5, i.e., smelted in argon plasma for 13minutes with adjustment of slag chemistry, elongated flakes are seen to be present randomly in between another bright phase as shown in figure 5(b). Samples smelted in nitrogen plasma with slag-adjustment for the comparatively lower duration of 9minutes exhibited a single phase which is confirmed as Fe from EDS analysis. Almost similar type of microstructure, i.e., dark to bright phases is observed in samples 1 to 4 as shown in figure 5(c). Elemental analysis (EDS) on three separate looking phases shows that silicon content in dark, dull and bright phases are in decreasing order. For dark phases, silicon content varies in the range of 5.35 to 0.99 % while for bright phases it varies from 0.32 to 0.15%. Dull phases contain silicon intermediate to that of bright and dark phases.

Table 5. Elemental analysis of metals at various region

Points	Fe	Si	C
1	96.03	0.18	3.79
2	90.52	0.50	4.23
3	85.45	0.70	9.56
4	88.74	8.75	3.50
5	90.46	3.02	6.40
6	99.63	0.37	0.45

2.3. X-Ray analysis:

XRD patterns of metallic fractions of all smelting products are shown in figures 6. Metallic parts as obtained from the smelting of sinter by Argon plasma for 7 min is detected as ferrosilicon (Fe_3Si). Here time for smelting is less, so recovery rate is lowest, i.e., 88%. Further smelting of sinter by Argon plasma gas for 13 min resulted in the formation of Fe and Fe_3Si phases. Although no metal carbide phase is identified in the XRD pattern, carbon was found to be present in fewer percentages as obtained from EDS analysis results. Here smelting duration is comparatively more, so recovery rate is enhanced.

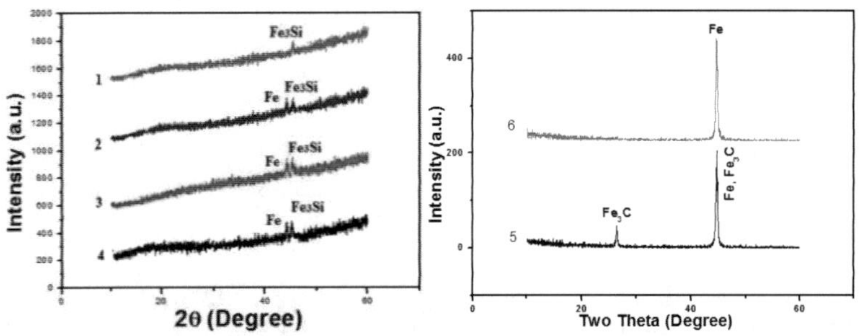

Figure 6. X-ray diffractograms of metallic fractions of all smelted products

Metallic yield is enhanced by nitrogen plasma smelting of charge over argon plasma where both Fe and Fe_3Si phases were detected in the respective XRD patterns. Although recovery rate is a function of smelting period, samples smelted with the application of nitrogen plasma has shown higher recovery than that of argon plasma in lesser operation length. The principal reason for increased recovery is di-atomicity of nitrogen that increases enthalpy of the system which in turn increases reducibility.

Further, the lime addition was made after complete melting of charge targeting melilite as final slag with employment of argon and nitrogen plasma. The objective of flux addition was to suppress the activity of silica to improve reducibility. Although recovery rate in both cases is comparatively similar, the formation of phases is different as confirmed from X-Ray analysis results. Along with Fe, cohenite (Fe_3C) phase is detected in case of sample smelted through argon plasma whereas single metallic iron phase is identified in case of nitrogen plasma smelted product.

2.4. Microhardness:

Another characterization technique hardness test as separately appearing phases was applied for confirmation of phases. A LECO microhardness tester LM248AT was implemented for hardness testing where load and dwell time were 100gf and 10 seconds respectively. Three distinct hardness values (average) viz. 618.5, 524.7, 493.2 HV with a maximum deviation of ±5 were obtained; is due to presence of iron carbide, metallic iron and ferrosilicon phases respectively depending on microstructure as shown in figure 5.

From the above results, it is seen that from minerals having high silica content, both metallic iron, as well as ferrosilicon alloy, can be produced. For extraction of iron it requires fluxing agents which in turn increases the final slag volume; hence heating of additional slag volume consumes more energy. On the contrary; initial silica in the ore minerals can be considered for value addition if smelted as such for producing ferrosilicon alloy. In the latter case additional material (flux) cost can be eliminated; hence can be considered as economic in comparison to the other case.

3. Conclusions

1. Ferro-silicon alloy and hot metal (molten iron) can be produced by plasma smelting of siliceous iron ore sinter by varying various plasma forming gases.
2. Recovery rate is a function of plasma forming gas viz. argon, nitrogen or mixture of gases and smelting duration.
3. A maximum recovery rate of about 93 percentages is achieved with employment of argon and nitrogen plasmas for durations of 13 and 7 minutes respectively.
4. XRD patterns of metal show presence of the three distinct phases viz. Fe, Fe_3C, and Fe_3Si per operating conditions.
5. Dissimilar type of microstructures is observed for samples subject to different operating conditions.
6. Silicon content at distinct phases in the alloy thus produced varies from 0.18 to 8.75 %.
7. Flux addition helped in suppressing the activity of silica; hence silicon picks up in the metal is reduced to a significant extent.
8. As both production of iron and ferrosilicon is feasible through the arc plasma, conversion of lean grades into ferrosilicon alloy puts more importance.

References

1. Roy S 2009 *Open Mineral Processing Journal 2* pp 17-30
2. Srivastava M P, Sutaone A T, Ghosh S K and Raju K S 1997
3. Gupta S, French D, Sakurovs R, Grigore M, Sun H, Cham T, Hilding T, Hallin M, Lindblom B and Sahajwalla V 2008 *Progress in Energy and Combustion Science 34*(2) pp 155-197
4. Samal S K, Sindhoora L P, Mishra S C. and Mishra B 2015 In *IOP Conference Series: Materials Science and Engineering* Vol 75 No. 1 p 012030
5. Gonterman J R and Weinstein M A 2010 In *Fiberglass and Glass Technology* pp 431-451
6. Mishra S C, Bhattacharjee S and Jena P K SG iron production by plasma smelting of blue dust.
7. Dwarapudi S, Ghosh T K, Shankar A, Tathavadkar V, Bhattacharjee D and Venugopal R 2011 *International Journal of Mineral Processing 99*(1-4) pp 43-53
8. A.K. Biswas 1984 SBA Publication, Calcutta
9. Bogdandy L von, Engell H J 1971 Springer-Verlag New York
10. Park H and Sahajwalla V 2014 *ISIJ international 54*(1) pp 49-55
11. El-geassy A A, Nasr M I, Khedr K S and AbdeL-Halim K S 2004 *ISIJ international 44*(3) pp 462-469
12. Dutta S K and Ghosh A 1993 *ISIJ international 33*(11) pp 1168-1173
13. Park J H and Kang Y B 2006 Metallurgical and Materials Transactions B 37(5) pp 791-797

Mechanical and Tribological Characteristics of the AMC, Prepared by P/M Route along with Thermo-Mechanical Treatment

Sambit Kumar Mohapatra[1], Kalipada Maity[2], Subrat Kumar Bhuyan[3*] and Mantra Prasad Satpathy[1]

[1]School of Mechanical Engineering, KIIT University, Bhubaneswar, Odisha 751024, India.
[2]Department of Mechanical Engineering, NIT Rourkela, Odisha 769008, India.
[3]Department of Metallurgical and materials Engineering, NIT Rourkela, Odisha 769008, India.

*E-mail: subratbhuyan2014@gmail.com

Abstract. Thermo mechanical treatments have the ameliorated impacts on the mechanical and tribological properties of powder metallurgy components. In this investigation an aluminium matrix composite (AMC) {Al (92) + Mg (5) + Gr (1) + Ti (2)} has been prepared by following powder metallurgy technique, with double axial compaction and ulterior sintering. Secondary thermo-mechanical treatment i.e. hot extrusion through mathematical contoured cosine profiled die was considered. The die causes minimum velocity relative differences across the extrusion exit cross-section, which provides smooth material flow. Comparative result analysis for the mechanical and tribological characteristics of the specimen before and after extrusion was concentrated. Extrusion engenders significant amount of improvements of the properties those are attributed to excellent bond strength and uniform density distribution due to high compressive stress. Oxidative and delaminated wear mechanisms were found predominating type. To furnish the suitable explanation scanning electron microscopies have been performed for the wear surfaces.

Keywords: AMC, Extrusion, Powder Metallurgy, Wear.

1. Introduction

Light weight high performance materials have the substantial implications in the automotive, aerospace, hand hold tools and sports sectors. Aluminium matrix composites (AMCs) in the above segment of materials was been severely commercialized to fulfill the advanced needs [1]. Among different methods of manufacturing, powder metallurgy (P/M) route suits good for its own advantages like apparently infinite compositional possibilities and uniform distribution of the proportions [2, 3].

A significant amount of investigations have been acquitted to investigate the results of different ceramic reinforcements in different aluminium series of powder matrix [4, 5]. Aluminium possesses poor wettability with ceramic reinforcements so as to improve bond strength as well as distribution, the P/M component is provided a secondary thermo-mechanical treatment. These secondary treatments like forging, extrusion [6] etc. helps to improve its mechanical as well as tribological properties [5, 7]. Addition of graphite improves the tribological property by minimizing interface friction with compromising hardness, flexural strength [8-10].

This research area has influenced the investigators to tailor different properties of the composite by altering the compositions with their proportions and types of manufacturing process [11]. During preparation of the specimen particles get oxidise around its surface. This oxide layer breaks due to the higher shear stress, induced by the implementation of secondary treatments which leads to the formation of strong microstructural bond [12]. For extrusion of P/M specimen, shear faced dies does not perform well in concern to flow characteristics and surface properties. Hence a mathematical contoured die is used for the extrusion operation.

Here in this research one composition has been selected for manufacturing the component by P/M route and subsequent secondary thermo-mechanical treatment: extrusion. The improvement of properties after extrusion has been investigated.

2. Methodology followed

2. 1. Specimen preparation

The composition selected for the specimen preparation is in weight percentages of Al (92) + Mg (5) + Gr (1) + Ti (2). The composition was blent uniformly in a centrifugal blender. At 200 RPM, the stainless balls and powder (in a weight ratio of 10:1) were revolved for 10 hours. Apparent density and tap density were checked before the cold compaction, conducted by double axial compaction method. The blend was compressed at 275 MPa in a cylindrical die with low rate of pressure rise and accorded for ten minutes of dwelling period. Green specimens were sintered in an argon atmospheric tubular furnace. Dwell periods of 20, 30, 90 minutes at 110℃ to evaporate water vapour, 450 ℃ to gasify zinc stearate lubricant, 590℃ to form the metallic bond were maintained respectively.

2. 2. Secondary processing

Sintered specimens of 10 mm diameter were extruded through a mathematical contoured die profile. The contour plot is shown in Figure 1. The cylindrical specimens were extruded at an operating temperature range of 400-450°C, to square cross-section. For composite extrusion shear faced die are not suitable, therefore a specially designed cosine profiled die was implemented to minimize the surface defects. Figure 1 shows the cross sectional change of the die as cosine profile, presented in one quarter-circle. By considering the generated coordinates of the die profile, three dimensional die has been manufactured indigenously.

2. 3. Mechanical and tribological characterization

Different densities like apparent and tap density, green and sintered density, theoretical and extruded densities were measured. By the relation mentioned in Equation 1, theoretical densities were checked.

$$\rho_{Theoritical} = \sum(\rho_i \times m_i) \quad (1)$$

Where ρ_i and m_i are the density of each constituent and the mass fraction of each constituent respectively

The solid body's densities were measured by adopting Archimedes' principle mentioned in Equation 2.

$$\rho_{sample} = \frac{W_A \times \rho_f}{W_A - W_f} \quad (2)$$

Where W_A represents the mass of the solid specimen measured at atmospheric condition, W_f and ρ_f are the mass and density of the fluid chosen respectively.

Vicker's hardness number in micro scale was determined the both sintered as well as extruded specimen. To determine the number, 50 grams of load through a pyramidal diamond with 136° of face angle was applied on the polished surface for a dwell period of 15 seconds.

Transverse rupture strength (TRS) of both the specimen were determined by 3-point bend test. Ram rate of 2 mm/min for the span of 30 mm specimen was maintained during the test in the universal testing machine Instron -5979.

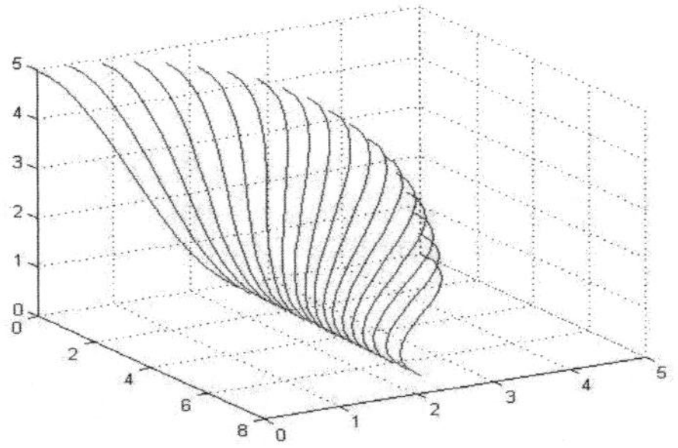

Figure 1. Cosine profile for round to square die one quadrant.

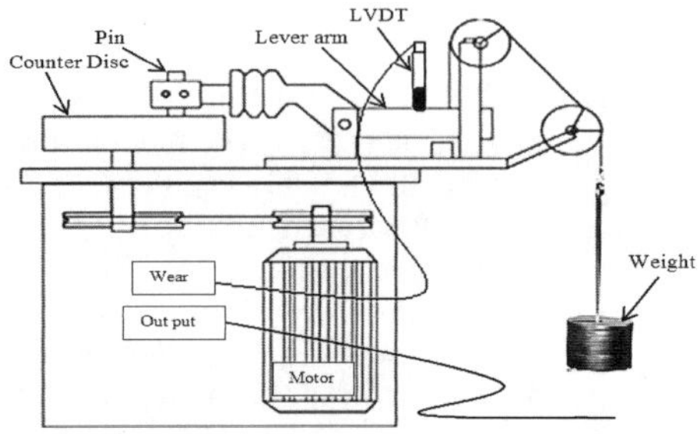

Figure 2. Schematic of wear testing apparatus.

The prepared specimens were subjected to two-body pin-on-disc type wear testing setup for the investigation of dry sliding wear characteristics. The cylindrical pins of diameter Ø 10mm and height 25 mm with two parallel surface were prepared. Specimen holder carries the stationary pin vertical to the horizontal counter disc surface. The EN-31 counter disc (60 HRC) has a surface roughness in average of 2 μm. The schematic of the wear testing setup is presented in Figure 2. Each experiment was conducted for a time duration of ten minutes. The pin (specimen) was weighed with an average accuracy of 0.1 mg for determining the weight loss due to wear.

3. Results and discussion

3. 1. Physical characterisation

Physical attributes of the powder particles has the influence on flowability, hence strength and tribological performance get affected. Among different popular techniques like sieve analysis, LASER diffraction and microscopy along with image analysis the latter one is used to determine Particle size. Scanning electron microscopy (SEM) images shown in figure 3 depict the shape. The selective informations of different powders are detailed in Table 1.

Table 1 Physical attributes of the selected powders.

Powder type	Supplier	Average size	Particle shape	Purity / Assay (%)
Al	Loba Chemie	45 μm	Spherical or sub-rounded	98.0
Mg		140 μm	Flakey	99.0
Graphite		20 μm	Rounded and Flakey	98.0
Ti		85 μm	Extreme angular	98.0

Different densities like bulk density or apparent density and tap density were directly recorded. Nearly 30-35% enhancement of density occurs due to tapping, depicted in the Figure 4, and indicates good flow ability of the blend. Good consolidations of blended particles in green specimens were observed at 275 MPa of compaction pressure. Effect of sintering in terms of shrinkage or growth was ascertained by the non-dimensional densification parameter presented in Equation 3.

$$\text{Densification parameter} = \frac{(\text{sintered density} - \text{green density})}{(\text{theoretical density} - \text{green density})} \quad (3)$$

Shrinkage and swelling or growth were affirmed by Positive and negative densification parameter respectively. The specimens here possess swelling behavior. Thermo-mechanical treatment (extrusion) causes improved density which can be calculated by the relation illustrated in Equation 4.

$$\text{Density improvement due to extrusion} = \frac{\text{extruded density - sintered density}}{\text{sintered density.}} \quad (4)$$

50% dimensional reduction causes 15-20% of density improvement here, still lesser than the theoretical density, discernible in Figure 4. Densification factor, improved density, and porosity are tabulated in Table 2.

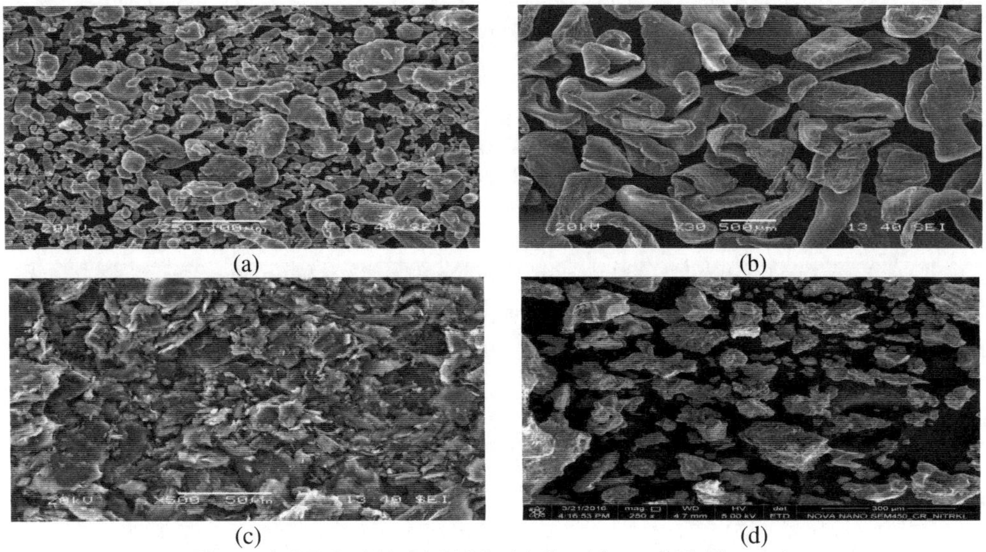
(a) (b) (c) (d)

Figure 3. SEM of (a) Al (b) Mg (c) Graphite and (d) Ti powder.

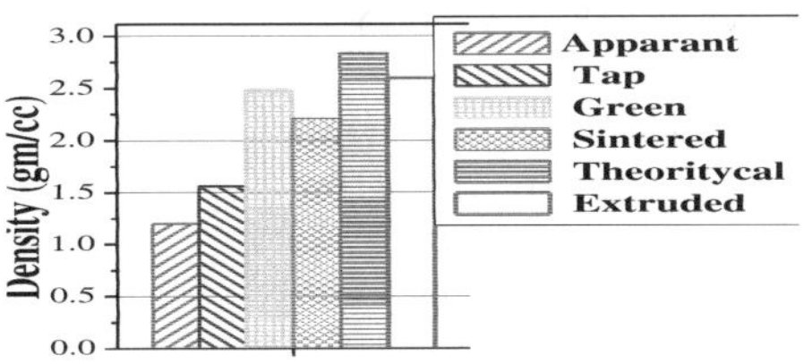

Figure 4. Density plot.

Table 2 Porosity analysis.

Parameters	Data
Calculated densification factor	-0.742
Improvement of density after extrusion (in %)	17.35
Porosity of the sintered specimen (in %)	22
Porosity after extrusion (in %)	8.53

3. 2. Mechanical characterisation

Considering the ram rate of 3 mm/min compression test of the specimen was carried out at room temperature condition in the Instron-setec series UTM. The average ultimate stress found is 381MPa depicted from the stress versus strain plot conducted for three different tests shown in Figure 5.

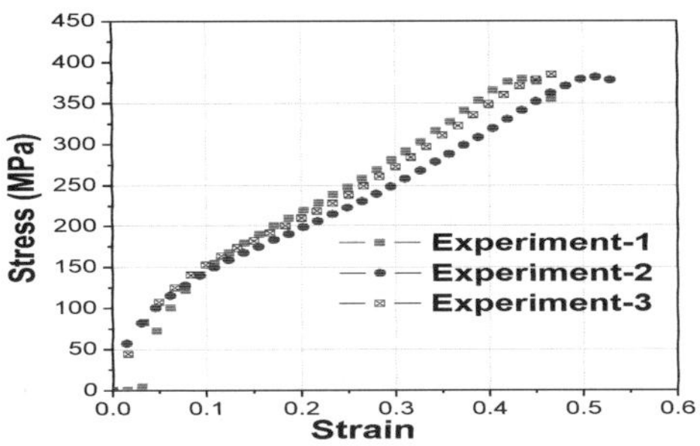

Figure 5. Stress versus strain.

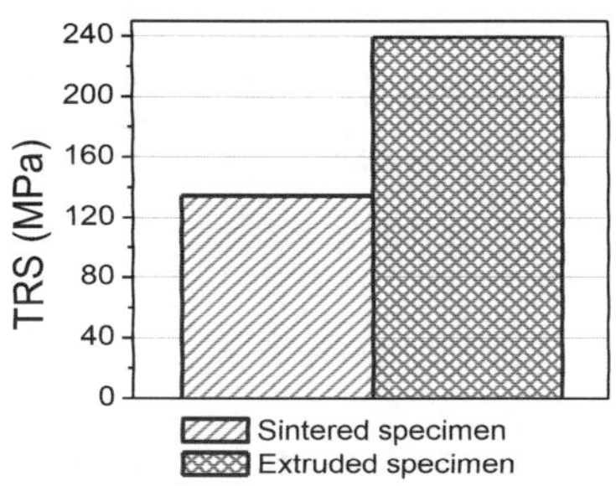

Figure 6. Transverse rupture strength.

The average hardness found from ten readings taken at different locations of both sintered as well as extruded specimen were 80 HRV and 110 HRV respectively. The hot working causes the storage of large thermal and internal stresses that establish improved properties. In loosely packed sintered specimen the TRS is very low as shown in Figure 6, which could be attributed to the stress concentration around the large titanium particles as well as presence of porosity.

3. 3. Microstructural analysis

Scanning electron micrographs of the thermo-mechanical treated specimen at different magnifications are presented in Figure 7. Secondary operation causes the improved density by minimizing porosity and creates a well dispersion in the matrix phase. The irregular and angular shape of Ti is very suitable for the improved bond during plastic deformation. Distribution of Ti and graphite powders in the matrix is very fine and appears to be homogeneous.

Figure 7. Microstructure of the specimen after thermo mechanical treatment.

3. 4. Wear analysis

Two-body dry sliding wear test was conducted for the specimen before as well as after extrusion. By considering three variables such as normal load, track diameter and revolution per minute of the disc in three levels as in Table 3, wear rate in terms of mass loss was analysed through L_9 orthogonal array. The mass loss in a run is converted to volume loss per sliding distance and the same is plotted in Figure 8 for two types of specimen comparatively. It was observed the wear rate is reduced by the implication of secondary treatments to the P/M specimens. The improved bond strength avoids the erosion of the pin material and consequent three body wear [10].

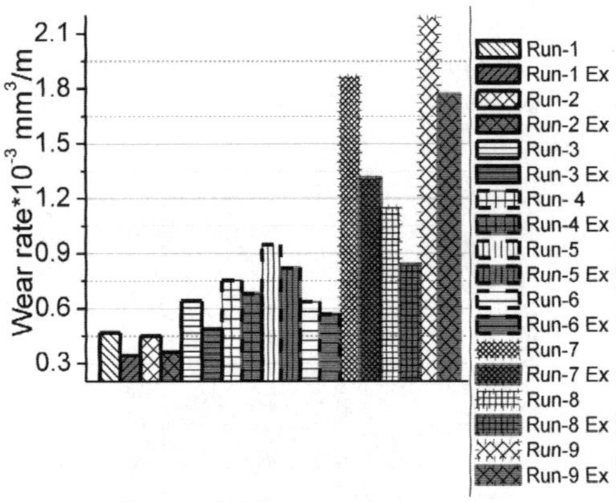

Figure 8. Wear rate for L_9 array.

Table 3 L_9 orthogonal array adopted for wear characterization.

Run	Normal Load (N)	Track Diameter (mm)	RPM of the counter disc
1	20	50	200
2	20	70	400
3	20	90	600
4	40	50	400
5	40	70	600
6	40	90	200
7	60	50	600
8	60	70	200
9	60	90	400

FESEM images of arbitrarily chosen weared surfaces were captured. Figure 9 (a) run-2 (b) run-7 (c) run-2 for extruded sample and (d) run-7 for extruded sample is presented below. It was observed delamination as well as combination of three such as abrasion, delamination and adhesion is the predominating mechanism of wear for high velocity and heavy loading conditions. Shear deformation of the surface occurs by the continuous sliding over the crack. Presence of graphite as a solid lubricant at the tribo-interface minimizes friction and wear. Presence of porosity causes severe erosion, crack formation and fracture of the specimen.

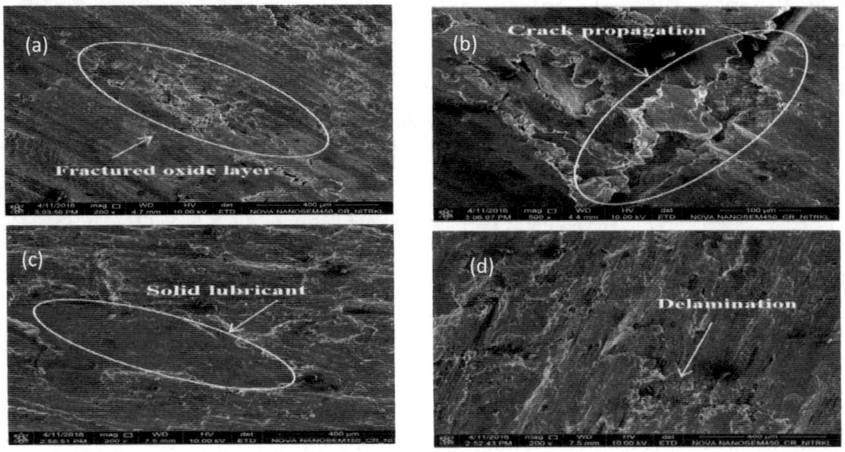

Figure 9. Wear morphology.

4. Conclusion

Effect of thermo-mechanical treatment on the P/M specimen was analysed successfully. There observed a significant enhancement of mechanical as well as tribological properties due to the thermo-mechanical treatment considered as secondary operation. Very less amount of surface defects due to smooth flow characteristics caused by cosine die extrusion made the specimen stronger. Presence of fine graphite particles ameliorates the tribological properties by compromising with strength marginally. A motley type of wear i.e. oxidative, adhesive, delamination as well as abrasion is predominant in the AMC surfaces.

5. References

[1] Fogagnolo JB, Velasco F, Robert MH and Torralba JM 2003 *Materials Science and Engineering* **A 342(1)** 131-43.
[2] Torralba J, Da Costa C and Velasco F 2003 *Journal of Materials Processing Technology* **133(1)** 203-06.
[3] Min KH, Kang SP, Kim D-G and Do Kim Y 2005 *Journal of alloys and compounds* **400(1)** 150-53.
[4] El-Kady O and Fathy A 2014 *Materials & Design* **54** 348-53.
[5] Taleghani MJ, Navas ER and Torralba JM 2014 *Materials & Design* **55** 674-82.
[6] Adamiak M, Fogagnolo JB, Ruiz-Navas EM, Dobrzański LA and Torralba JM 2004 *Journal of Materials Processing Technology* **155-156** 2002-06.
[7] Abdollahi A, Alizadeh A and Baharvandi HR 2014 *Materials & Design* **55** 471-81
[8] Ravindran P, Manisekar K, Narayanasamy P, Selvakumar N and Narayanasamy R 2012 *Materials & Design* 39 42-54.
[9] Ravindran P, Manisekar K, Narayanasamy R and Narayanasamy P 2013 *Ceramics International* **39(2)** 1169-82.
[10] Baradeswaran A and Perumal AE 2014 *Composites Part B Engineering* **56** 472-76.
[11] Yigezu BS, Jha PK and Mahapatra MM 2013 *Materials and Manufacturing Processes* **28(9)** 969-79.
[12] Bykov YV, Rybakov K and Semenov V 2001 *Journal of Physics D Applied Physics* **34(13)** 55.

Parametric Investigation on Microstructure and Mechanical Properties of Ultrasonic spot welded Aluminium to Copper sheets

Mantra Prasad Satpathy[1], Kasinath Das Mohapatra*, Ananda Kumar Sahoo[2] and Susanta Kumar Sahoo[3]

[1]School of Mechanical Engineering, KIIT University, Bhubaneswar, 751024, Odisha
[*23]Department of Mechanical Engineering, National Institute of Technology Rourkela, Rourkela 769008, India

[*]Corresponding Author E-mail: kitu.kasinath1@gmail.com

Abstract. Ultrasonic welding is one of the promising solid state welding methods which have been widely used to join highly conductive materials like aluminum and copper. Despite these applications in the automotive field, other industries also have a strong interest to adopt this process for joining of various advanced alloys. In some of its applications, poor weld strength and sticking of the workpiece to the tool are issues. Thus, an attempt has been taken in the present study to overcome these issues by performing experiments with a suitable range of weld parameters. The major objectives of this study are to obtain a good joint strength with a reduced sticking phenomenon and microstructure of Al-Cu weld coupons. The results uncovered the mechanical strength of the joint increased up to 0.34 sec of weld time and afterward, it gradually decreased. Meantime, the plastic deformation in the weld zone enhanced the formation of an intermetallic layer of 1.5 μm thick, and it is composed of mainly Al_2Cu compound. The temperature evolved during the welding process is also measured by thermocouples to show its relationship with the plastic deformation. The present work exemplifies a finer understanding of the failure behavior of joints and provides an insight of ultrasonic welding towards the improvement in the quality of weld.

1. Introduction

The invention of Ultrasonic metal welding (USMW) was in the 1950s. Meanwhile, this technique has been continuously applied in various sectors such as an automobile, electrical and electronics, manufacturing and aerospace industries. Due to the high thermal and electrical conductivity nature, aluminum and copper are used in electric power industries to transmit power efficiently. Due to the friction, an interface temperature is developed between the sheets, and that is capable of softening the overlapping zone. Thus, the plastic deformation occurs in the weld area, and the two materials are joined [1]. Zhang et al. [2] accomplished various tests by taking different ultrasonic energy values in order to measure the temperature at the weld center. He concluded that the softening of the material is achieved when the peak temperature is above 500°C, to boost the joining of aluminium sheets. Recently, USMW is tried to join thin sheets of aluminium with other lightweight materials in the structural automotive and aerospace application [3]. This process is also being used to accumulate secondary aerospace structures such as helicopter access doors: These doors can sustain loads of about 5 to 10 times the design load without the weld failure in air load tests as described by American welding society [4]. Meantime, there is a necessity to understand the influences of various welding parameters on the mechanical strength of welds during the USMW process. Most of the literatures [5] reported about the ability of the USMW process to weld metals like titanium, steel, nickel, copper and their dissimilar combinations with other non-metallic materials. Kodama [6] performed experiments using soft materials like aluminium, copper, brass and gold. The study also investigated the various parameters like specimen roughness, hardness, dimensions, coefficient of friction and anvil geometry on the bond strength. The results showed that the good strength could be obtained when the amplitude of vibration was reduced with the increase of frequency of vibration. Hetrick et al. [7] also demonstrated the ultrasonic welding process as a low heat producing process. For this reason, the issues like part distortion and the degradation of Al alloy property can be eliminated. Barnes and Pashby [8] reported that USMW could be used in aerospace and automotive sectors for joining of lightweight structural components.

A large number of studies have examined the joint strength by considering one or two of the parameters as constant. However, a good result can be obtained by varying these control parameters simultaneously to study the joint strength and quality. The majority of studies had also not considered the effect of dissimilar material combination and the effect intermetallic compounds (IMCs) on the weld strength. Therefore, in this study, an attempt has been taken to weld the dissimilar materials such as AA1100 aluminium alloy with UNS C10100 copper along with the investigations of the effects of various process parameters like weld time, weld pressure and vibrational amplitude on T-peel failure and tensile shear loads along with the generation of temperature during joining.

2. Experimentation

In this study, the welding experiments have been conducted between commercially available 0.45 mm thick sheets of aluminium (AA1100) and copper (UNS C10100) (further it is abbreviated as 0.45Al-0.45Cu combination). Sheets were sheared to a rectangular shape of 100 mm length and 20 mm wide with an overlapped area of 20 × 20 mm^2 for the lap welding. Meantime, the sheet hardness also has a vital role in deciding the quality of the joint [9]. As the top specimen (i.e., on the sonotrode side) is more subjected to the impact of ultrasonic energy, thus, a softer material, i.e., AA1100 (45 HV) is selected for all the experiments. Moreover, it can be more plastically deformed than the relatively harder material, i.e., UNS C10100 (72 HV) (i.e., on the anvil side). The configuration is shown in Figure 1. The mechanical and thermal properties of weld specimen are presented in Table 1.

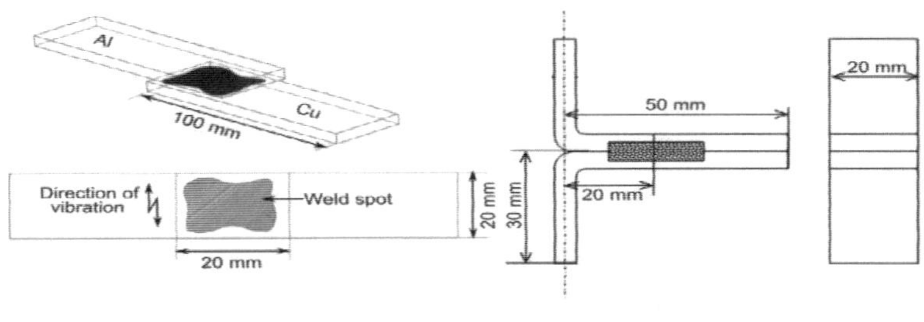

Figure 1. Schematic diagrams for testing of weld coupons.

Table 1. Thermal and mechanical properties of weld materials

Materials	AA1100	UNS C10100
Ultimate tensile strength (Pa)	135.5E6	302.1E6
Brinell Hardness (BHN)	32	70
Yield strength (Pa)	115.2E6	251.5E6
Thermal Conductivity (Wm-1°C-1)	220	391

The welds are produced by a Telsonic® ultrasonic lateral drive spot welder (Figure 2) with a flat serrated weld tip size of 11 mm × 9 mm on time control mode. For achieving a good weld, there is a need to select proper process parameters with suitable operation levels in such a way that the required quantity of ultrasonic energy can be transmitted to the weld zone. Therefore, from various literature studies and screening tests, three process parameters like weld time (WT), vibration amplitude (A) and weld pressure (WP) have been selected. In the current analysis, WP, WT and A are varied in three, seven and four levels respectively. These factors with their levels are represented in Table 2. The full factorial design of experiment has been employed in this study to record the influence of each process parameters on the outputs like T-peel failure and tensile shear loads. Thus, a total of 84 numbers of experiments are executed using this design. For better accuracy of results, each experiment has been replicated six times, and a total of 504 samples are prepared. Out of which 252 samples each are employed for tensile shear and T-peel tests.

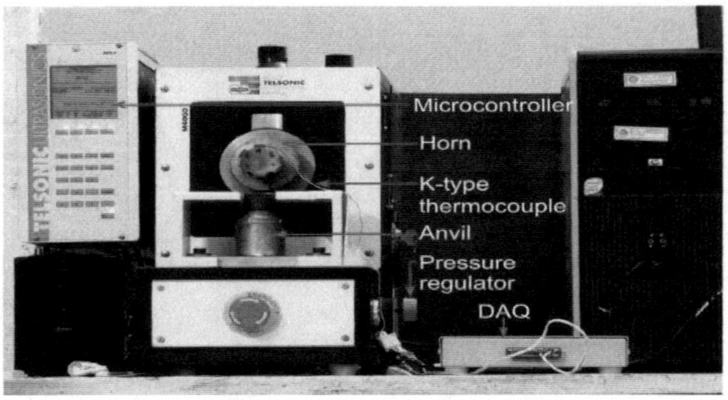

Figure 2. Lateral drive ultrasonic spot welder.

Prior to welding, the face of weld strips are cleaned by swabbing with acetone and was dried by hot air to get rid of any contaminants and oxides. The interface temperature was analyzed by embedding 0.1 mm K-type thermocouples near to the periphery of the weld spot. The implanted position of the thermocouple is illustrated in Figure 3. This temperature measurement is essential to provide an insight to predict the formation of intermetallic phases. The T-peel failure load and tensile shear tests were carried out using the fully computerized universal testing machine (UTM) INSTRON® 1195. This UTM is equipped with 10 kN load cell with a constant crosshead speed of 2 mm/min to avoid any type dynamic stress effect on the joint.

Table 2. Experimental domain for Ultrasonic metal welding of 0.45Al-0.45Cu weld coupons

	Amplitude (μm)	Weld time (Sec)	Weld pressure (MPa)
Level 1	47	0.26	0.26
Level 2	54	0.28	0.34
Level 3	60	0.30	0.42
Level 4	68	0.34	...
Level 5	...	0.36	...
Level 6	...	0.38	...
Level 7	...	0.41	...

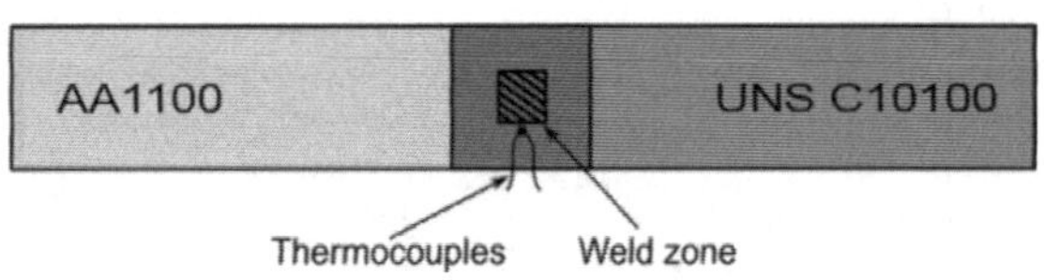

Figure 3. Weld samples with positions of thermocouples for temperature measurement.

The metallographic specimens were prepared from the weld coupons by cutting in the transverse cross-sections and followed by polishing with different grades of emery paper and diamond paste. The microstructural examinations were performed using scanning electron microscope (SEM) (JEOL®, JSM-6480LV).

3. Results and Discussion

3.1. Interface temperature analysis

Figure 4 represents the peak temperatures at each condition of process parameters. It can be noticed that the temperatures rise very quickly up to 0.28 sec for all the vibration amplitudes and then these increasing rates gradually slow down. The maximum temperature of 292.78 °C is observed at the weld time of 0.41 sec, vibration amplitude of 47 μm and weld pressure of 0.34 MPa. However, this interface temperature value is increased to 303.83 °C, 314.39 °C and 328.98 °C with the increase in vibration amplitude to 54 μm, 60 μm, and 68 μm respectively. It can be observed that most of the frictional works are performed at the weld interface, and this is originated from the scrubbing motion of the sheets. Thus, increasing in amplitude value means the increasing the degree of scrubbing motion and finally, it leads to rise in temperature.

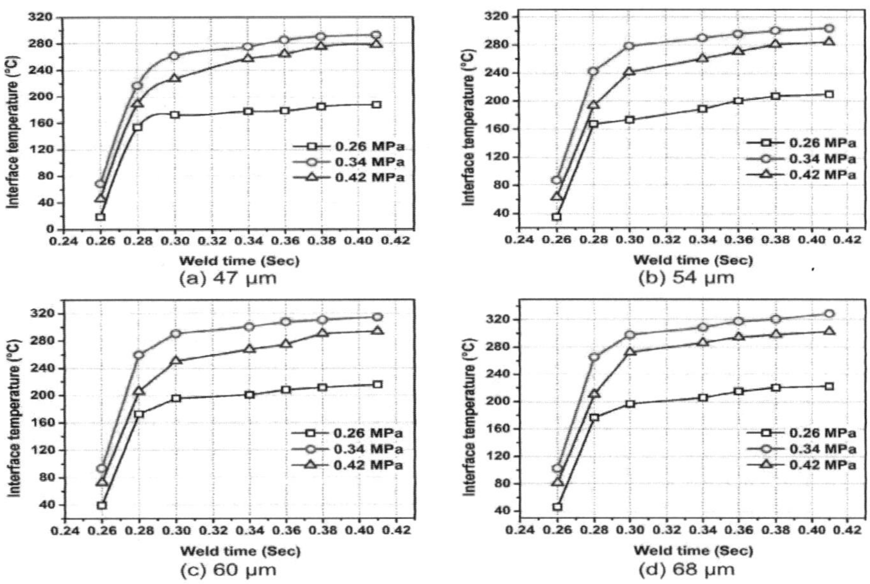

Figure 4. Weld interface temperatures for 0.45Al-0.45Cu weld coupons at different vibrational amplitudes.

3.2. Analysis of weld area

Figure 5 illustrates the graph between different input parameters with weld area for the weld coupons of 0.45Al-0.45Cu. Similar to previous analysis, these tests are also conducted using four amplitude settings comprising of 47 μm, 54 μm, 60 μm and 68 μm. The weld area instigates to rise and reaches the peak value of 67.19 mm^2 at maximum weld pressure of 0.34 MPa, vibration amplitude of 47 μm and weld time of 0.41 sec. Similarly, the weld area exhibit higher values of 69.63 mm^2, 72.52 mm^2 and 79.19 mm^2 respectively when the vibration amplitude values are increased to 54 μm, 60 μm, and 68 μm respectively. This nature of the graph is obvious because the highest weld time results in less oxidation and a higher rate of heat generation during welding. Furthermore, on the increase of vibration amplitude, the rate of plastic deformation in the mating surfaces rises, and it leads to the dense distribution of micro-bonds in the interface.

3.3. Fractographic analysis

The microscopic photograph of 0.45Al-0.45Cu is illustrated in Figure 6. From these figures, it can be observed that no appreciable deformation takes place in the copper materials. Meanwhile, it is believed that the inter-metallics formed between the weld coupons affect the joint strength. From this figure, some micro cracks are noticed on the aluminium side at the higher level of weld time due to its higher ductility and lower hardness properties. Furthermore, the inter-granular cracking is the other reason for reducing the joint strength. Once again, when the weld pressure is too high, then the relative motion between the sheets disappears, and the sticking may happen. Likewise, the weld time also plays a crucial role in getting good weld strength. With less weld time, the interatomic diffusion will not be complete, and it results in lower weld strength with voids in the interface region. From the weld quality studies, it is confirmed that the strength of a joint depends on the specific values of input parameters.

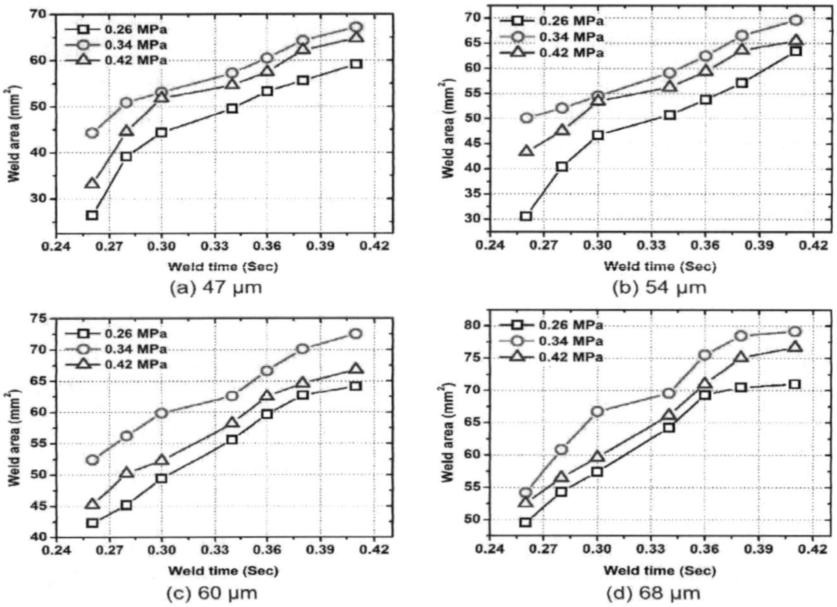

Figure 5. Weld areas of 0.45Al-0.45Cu weld coupons at different vibration amplitudes.

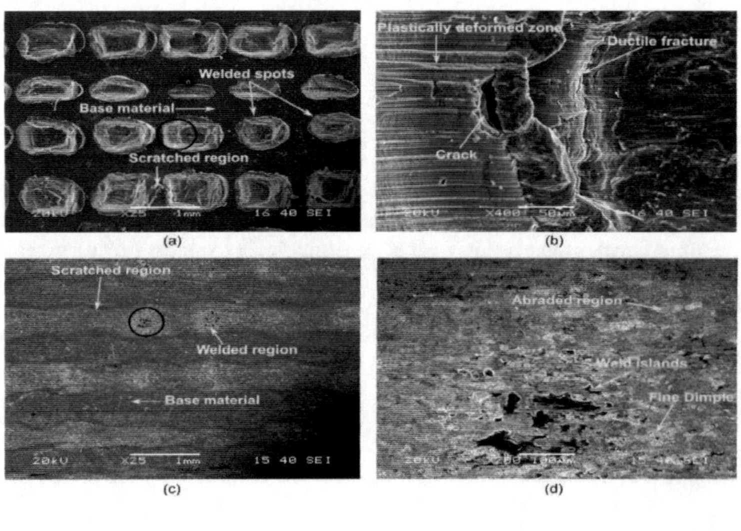

Figure 6. Fractured surfaces of ultrasonic welded specimens (a) Al side, (b) enlarged portion of figure "a" (c) Cu side and (d) enlarged portion of figure c.

4. Conclusions

The present study deals with the extensive analysis on every aspect of USMW of Al-Cu dissimilar materials. This study includes the exploration of input parameters of the welding process on the outputs like T-peel strength failure loads, tensile shear and weld area with providing information temperature during the process. The following conclusions can be extracted from this research:

- At various input parameter, AA1100 and UNS C10100 sheets are welded successfully.
- The interface temperature of 328.98 °C achieved during welding is the highest among all the observed results. It is obtained at weld pressure of 0.34 MPa, vibration amplitude of 68 μm and weld time of 0.41 sec. Initially, the temperature profile suddenly increases up to 0.28 sec, but as the relative motion between the sheets vanishes, the temperature comes to a steady state. Due to the sudden increase of interface temperature, the oxide film on the weld surface ruptures and extensive plastic deformation occurs.
- The maximum weld area of 79.19 mm^2 is obtained at a weld time of 0.41 sec, vibration amplitude of 68 μm, and weld pressure of 0.34 Mpa. It is because of denser micro-bonds formation at the highest vibration amplitude.
- From the fractographic study of weld surfaces, it is observed that for Al-Cu weld specimens, the number of scratch marks increases and it leads to the formation of cracks around the weld zone.

References

[1] Janaki Ram G D, Robinson C, Yang Y and Stucker B E 2007 Use of Ultrasonic Consolidation for Fabrication of Multi-Material Structures *Rapid Prototyp. J.* **13**(4): 226–235.

[2] Zhang C Y, Chen D L, and Luo A A 2014 Joining 5754 Automotive Aluminum Alloy 2-Mm-Thick Sheets Using Ultrasonic Spot Welding *Weld. J.* **93**(4): 131-138.

[3] Komiyama K, Sasaki T and Watanabe Y 2016 Effect of Tool Edge Geometry in Ultrasonic Welding *J. Mater. Process. Technol.* **229**: 714–721.

[4] O'brien R L 1997 *Jefferson's Welding Encyclopedia*, American Welding Society.

[5] Seo J S, Jang H S and Park D S 2015 Ultrasonic Welding of Ni and Cu Sheets *Mater. Manuf. Process.* **30**(9): 1069–1073.

[6] Kodama M 1989 Ultrasonic Welding of Non-Ferrous Metals *Weld. Int.* **3**(10): 853–860.

[7] Hetrick E, Jahn R, Reatherford L, Skogsmo J, Ward S, Wilkosz D, DEVINE J, Graff K and Gehrin R 2005 Ultrasonic Spot Welding: A New Tool for Aluminum Joining *Weld. J.* **84**(2): 26–30.

[8] Barnes T A and Pashby I R 2000 Joining Techniques for Aluminium Spaceframes Used in Automobiles: Part I—solid and Liquid Phase Welding *J. Mater. Process. Technol.* **99**(1): 62–71.

[9] Yang J W, Cao B, He X C and Luo H S 2014 Microstructure Evolution and Mechanical Properties of Cu-Al Joints by Ultrasonic Welding *Sci. Technol. Weld. Join.* **19**(6): 500–504.

DFT Studies on Interaction between Lanthanum and Hydroxyamide

Anindita Pati*[1], T. K. Kundu[1] and Snehanshu Pal[2]
[1]Department of Metallurgical and Materials Engineering, Indian Institute of Technology Kharagpur, Kharagpur, West Bengal, India-721302.
[2]Department of Metallurgical and Materials Engineering, National Institute of Technology Rourkela, Rourkela, Odisha, India-769008.

* Email:patianindita14@gmail.com

Abstract: Extraction and separation of individual rare earth elements has been a challenge as they are chemically very similar. Solvent extraction is the most suitable way for extraction of rare earth elements. Acidic, basic, neutral, chelating are the major classes of extractants for solvent extraction of rare earth elements. The coordination complex of chelating extractants is very selective with positively charged metal ion. Hence they are widely used. Hydroxyamide is capable of forming chelates with metal cations. In this present study interactions of hydroxyamide ligand with lanthanum have been investigated using density functional theory (DFT). Two different functional such as ωB97XD and B3LYP are applied along with 6-31+G(d,p) basis set for carbon, nitrogen, hydrogen and SDD basis set for lanthanum. Stability of formed complexes has been evaluated based on calculated interaction energies and solvation energies. Frontier orbital (highest occupied molecular orbital or HOMO and lowest unoccupied molecular orbital or LUMO) energies of the molecule have also been calculated. Electronegativity, chemical hardness, chemical softness and chemical potential are also determined for these complexes to get an idea about the reactivity. From the partial charge distribution it is seen that oxygen atoms in hydroxyamide have higher negative charge. The double bonded oxygen atom present in the hydroxyamide structure has higher electron density and so it forms bond with lanthanum but the singly bonded oxygen atom in the hydroxyamide structure is weaker donor atom and so it is less available for interaction with lanthanum.
Keywords: Solvent extraction, Lanthanum, Hydroxyamide, Density Functional Theory

1. Introduction

Rare earth elements (REEs) includes lanthanides which represents elements with atomic number 57-71 on the periodic table. Scandium and yttrium with atomic number 21 and 39 respectively are also included because of their similar properties [1-2]. There are two types of REEs namely heavy rare earth elements and light rare earth elements. The light REEs are lanthanum, cerium, praseodymium, neodymium, and samarium. REEs are in great demand in aerospace, defense, health care and transportation [3-4]. There has been increasing demand of rare earth elements in throw-away technologies like phones, tablets, laptops, televisions, hybrid and electric vehicles, wind turbines, cameras and solar cells, as well as high-power magnets, fluorescent lamps, catalytic converters and metal alloys [5-6]. Two important ores for REEs are bastnaesite and monazite. Both the ores consist of mainly light REEs but in case of monazite the fraction of heavy rare element present is more than that of bastnaesite. Very low concentrations of REEs are found in ore and mineral deposits [7-8]. Recycling of REEs is very difficult as they are present in very less quantity in consumer products. The products are designed in such a way that it is difficult to separate the REEs from other components. Once separated they are needed to be purified and mixed with other components to create usable products. Research has been for recycling of the REEs from end-of-life (EOL) products and E-scrap. From end- of life neodymium magnets, neodymium and dysprosium can be recycled. Extraction of REE has been a challenge because of their chemical similarity [9-11]. $1s^2 2s^2 2p^6 3s^2 3p^6 3d^{10} 4s^2 4p^6 4d^{10} 4f^{0-14} 5s^2 5p^6 5d^{0-1} 6s^2$ is the electronic configuration of lanthanides. The $6s^2$ configuration is common in all lanthanides. But the subshell 4f and 5d vary in their occupancy because of their closeness in energy [12-14]. In case of Gadolinium, ytterbium and lutetium the electron enters subshell 5d orbitals instead 4f. Lanthanides have half-filled or completely filled or empty 4f subshell [15-16]. Due to their low charge density, they don't have the tendency to form complexes. The order in which lanthanides have the tendency to form complexes is $Ln^{4+} > Ln^{3+} > Ln^{2+}$. Solvent extraction is one of the suitable method for the extraction of REEs. It is a hydrometallurgical method used for the recovery of metal ions from their aqueous solutions [17-18]. Before solvent extraction, in order to produce REE rich ore concentrate, beneficiation of rare earth ore is done which is followed by leaching of these concentrates with inorganic acids like HCl, HNO_3, and H_2SO_4 etc. [19]. The leachate which is the aqueous phase is mixed with the organic phase which contains the extractant. The extractant is designed to selectively transfer the desired ions to the organic phase [20].

2. Computational details

The density functional calculations are carried out using Gaussian 09 suite of programs. Two different functional such as ωB97XD and B3LYP are applied along with 6-31+G (d, p) basis set for carbon, nitrogen, hydrogen and SDD basis set for lanthanum. ωB97XD is parameterized hybrid long range corrected (LC) B97 Generalised Gradient Approximation (GGA) exchange correlational (XC) functional with dispersion. B3LYP is hybrid GGA XC functional. B3LYP functional is among one of the Becke's non-local gradient-corrected three parameter exchange functional with two correlation functionals developed by Lee-Yang-Parr [9].

This exchange correlational functional is given by:

$$E_{XC} = 0.2 * E_X(HF) + 0.8 * E_X(LSDA) + 0.72 * \Delta E_X(B88) + 0.81 * E_C(LYP) + 0.19 * E_C(VWN) \quad (1)$$

Where E_{XC} is the exchange correlational functional for B3LYP which contains 20% non-local Fock exchange, 80% LSDA, 72% of the gradient corrections of the Becke88 exchange functional, 81% of LYP and 19% of the Vosko-Wilk-Nusair correlation functional III respectively. The LC functional i.e. ωB97XD yield better results in case of covalent systems than the hybrid functionals and also contain dispersion corrections. Its performance is similar to the DFT-D scheme but its computational cost is much higher.

6-31G (d) basis set is for non-hydrogen atom and includes d polarizations functions. 6-31G (d, p) basis set includes d and p polarization functions for hydrogen atoms. 6-31+G (d, p) basis set is 6-31G basis set along with s and p polarization functions for non-hydrogen atoms. SDD basis set is Effective core potential (ECP) basis set. ECP basis set are used for two main reasons i.e. reducing computational cost and includes the relativistic effects in deep core electrons [11]. The interaction energy i.e. ΔE value for the complex is obtained from the mathematical formula is given below:

$$\Delta E = E_{complex} - (E_{ligand} + E_{lanthanum}) \quad (2)$$

$\Delta E_{complex}$, is the optimized energy of the hydroxyamide ligand with lanthanum, ΔE_{ligand} is the optimized energy of the hydroxyamide ligand and $\Delta E_{lanthanum}$ is the optimized energy of the pure lanthanum atom. The solvation energy for the complex is obtained from the mathematical formula given below:

$$\Delta E_{solvation} = E_{complex\ in\ solution\ phase} - E_{complex\ in\ gas\ phase} \quad (3)$$

From the frontier orbital (highest occupied molecular orbital or HOMO and lowest unoccupied molecular orbital or LUMO) energies of the molecule we can calculate different chemical parameters like electronegativity, chemical hardness and chemical softness which gives insight into reactivity of the complex formed. From HOMO and LUMO energies we can calculate the ionization potential IP=-HOMO and electron affinity EA=-LUMO. Then from IP and EA values we get the values for the above mentioned chemical parameters as given below:

$$\text{Electronegativity } (\chi) = \left(\frac{\text{IP+EA}}{2}\right) \quad (4)$$

$$\text{Chemical hardness } (\eta) = \frac{\text{IP}-\text{EA}}{2} \quad (5)$$

$$\text{Chemical softness (S)} = \frac{1}{\eta} \tag{6}$$

3. Results and Discussion

Optimized structure of hydroxyamide ligand with lanthanum is presented in Figure 1. The hydroxyamide ligand and lanthanum are also optimized separately. B3LYP and ωB97XD functionals are used for optimization of hydroxyamide ligand and lanthanum individually and also for the complex but 6-31+G (d, p) basis set is used for hydroxyamide ligand and SDD basis set for lanthanum. The optimized energies of hydroxyamide ligand and lanthanum are -203108.928 and -272475.675 Kcal/mol respectively. The optimized energy of the complex using B3LYP functional is -475796 Kcal/mol. The optimized energy of the complex using ωB97XD functional is -475736 Kcal/mol. From the above calculations it can be seen that the complex shows lowest energy configuration in case of B3LYP functional. Negative value of interaction energy indicates favourable interaction within them. The optimized energy of hydroxyamide ligand and complex are -203117 Kcal/mol, -476359 Kcal/mol respectively in the presence of water as solvent. The solvation energy as calculated using equation 3 is -563 Kcal/mol. It can be seen that in the presence of solvent the extraction efficiency is higher as the complex is found to have lowest energy configuration in the presence of solvent. Table 1 represents the interaction and solvation energy of the complex formed. Table 2 represents the bond lengths and bond angles of the complex.

Table 1: Interaction and Solvation energy of the complex

Complex (Kcal/mol) Optimized energy	Hydroxyamide Ligand Optimized energy (Kcal/mol)	Lanthanum (Kcal/mol) Optimized energy	Interaction energy (Kcal/mol)	Functional
-475796	-203109	-272476	-211	B3LYP
-475736	-203109	-272476	-151	ωB97XD
-476359	-203117	-272476		B3LYP

Table 2: Bond Length and Bond Angle of complex

Bond Name	Bond Length (A^0)	Bond Angle Name	Bond Angle
O1 - C2	1.363	H11-N10-C2	127.921
O1 - La14	2.060	N10-C2-C3	125.305
C2 - C3	1.476	N10-C2-O1	111.221
C2 - N10	1.332	O1-C2-C3	123.461
C3 - C4	1.587	La14-O1-C2	140.806
C3 - H5	1.097	C2-C3-H5	110.771
C3 - H6	1.103	C3-C4-H7	111.575
C4 - H7	1.094	H5-C3-C4	110.016
C4 - H8	1.095	H9-C4-H7	110.692
C4 - H9	1.094		
N10 - H11	1.029		
O12 - H13	0.997		

*The symbols used for bond angle name in Table 2 are indicated in Figure 1

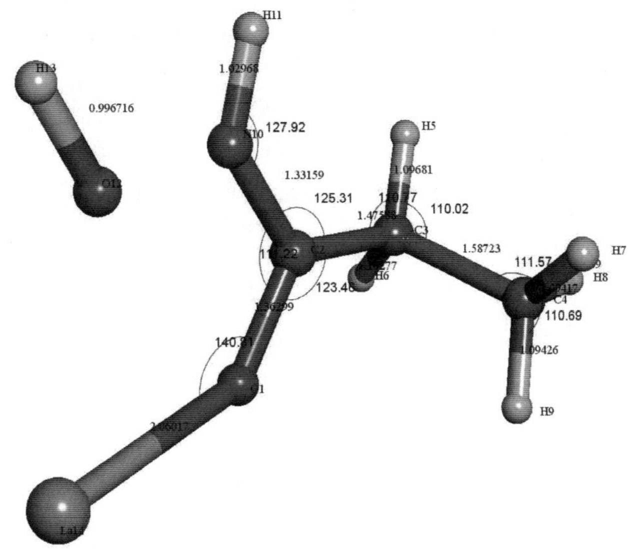

Figure 1: Optimized structure of complex with bond lengths and bond angles

Table 3: Mulliken Charge Distribution of the complex

Atom Label	Muliken Charge Distribution before addition of La	Muliken Charge Distribution after addition of La
1O	-0.400	-0.355
2C	0.556	0.589
3C	-0.478	-0.453
4C	-0.499	-0.550
5H	0.233	0.292
6H	0.194	0.314
7H	0.193	0.248
8H	0.171	0.321
9H	0.189	0.232
10N	-0.408	-0.170
11H	0.331	0.471
12O	-0.472	-0.518
13H	0.391	0.514
14 La	3	2.066

Table 3 represents the mulliken charge distribution. The lanthanum ion charge in the free state is +3. It can be seen that positive charge of the lanthanum ion decreases in complex formed with hydroxyamide, which shows electron transfer from ligand to the lanthanum ion and coordination bond forms. After complex formation lanthanum has a partial charge of +2.066 as can be seen from the charge distribution diagram. From the partial charge distribution it is seen that oxygen atoms in hydroxyamide have higher negative charge. The double bonded oxygen atom present in the hydroxyamide structure has higher electron density and so it forms bond with lanthanum but the singly bonded oxygen atom in the hydroxyamide structure is weaker donor atom and so it is less available for interaction with lanthanum.

Potential energy: To model interactions between atoms in a chemical system potential energy surface is used. It is used to represent atomic arrangements of a system by using atomic positions as variable as shown in Figure 3. Most stable nuclear configuration i.e. global minimum and the path travelled by atoms during the chemical reaction can be determined from potential energy surface diagram. The minima is at about 2.2 A^0.

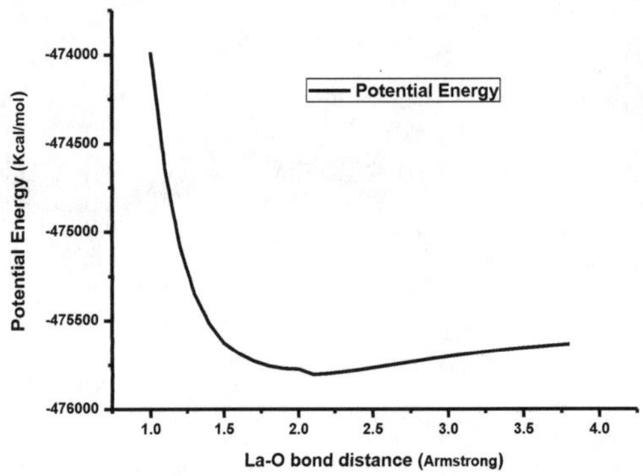

Figure 2: Potential energy surface

Frontier orbitals, highest occupied molecular orbital's energy is found to be -20.555 eV and that of lowest unoccupied molecular orbital's energy of the complex is found to be -17.079 eV. The lanthanum contributes for the LUMO shapes of the complex and the hydroxyamide ligand contributes for the HOMO shapes of the complex. The overlap of HOMO and LUMO give rise to chemical bond formation between La-O. The difference between HOMO LUMO energy is found to be 3.475712 eV as shown in Figure 8. From the HOMO LUMO energy the electron affinity and ionization potential calculated are 17.08 and 20.56 eV respectively. From the equations 4, 5, 6 the electronegativity, chemical hardness and chemical softness as calculated are 18.82, 1.74 and 0.58 eV respectively. Electron affinity is the energy released when an electron is added to a neutral atom. Ionization potential is the energy required to remove an electron from an atom. Electronegativity is the ability of an atom to attract an electron towards itself during bond formation. Chemical hardness is the energy gap between frontier orbitals and it helps to determine the stability of the complex formed. Table 4 represents the properties of the complex. Figure 3 represents the HOMO LUMO shapes of the hydroxyamide ligand and the HOMO LUMO shapes of the lanthanum (+3) ion.

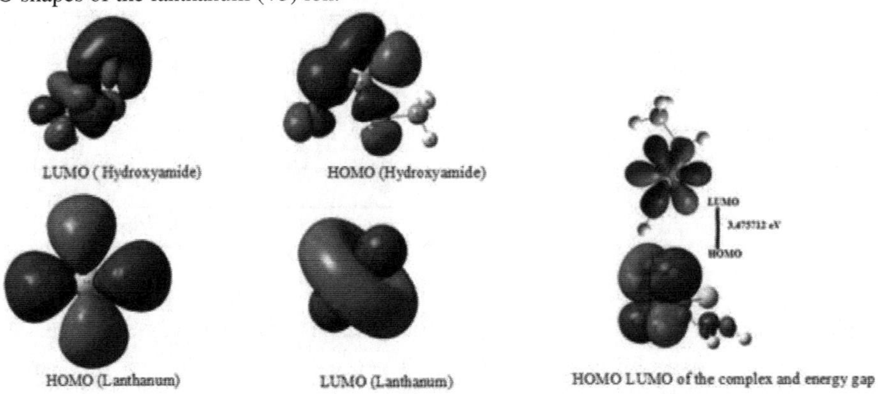

Figure 3: Representation of HOMO and LUMO

Table 4: Properties of the complex from HOMO LUMO energy

HOMO (eV)	LUMO (eV)	HOMO LUMO energy gap (eV)	Electron Affinity (eV)	Ionization Potential (eV)	Electronegativity (eV)	Chemical hardness (eV)	Chemical Softness (eV)
-20.56	-17.08	3.475712	17.08	20.56	18.82	1.74	0.58

4. Conclusion

Hydroxyamide forms a chelating bond with lanthanum. From the interaction energy calculations it can be seen that the complex shows lowest energy configuration in case of B3LYP functional then ωB97XD functional and hence complex is more stable in case of B3LYP functional. The solvation energy of the complex shows that it is more stable in the solvent phase then gas phase. The partial charge distribution shows that oxygen atoms in hydroxyamide have higher negative charge. Due to the higher electron density of the double bonded oxygen atom present in the hydroxyamide structure than the single bonded oxygen atom in the hydroxyamide structure and as single bonded oxygen atom is a weaker donor atom, lanthanum forms chelating bond with double bonded oxygen atom. From the potential energy curve the lowest configuration energy is found to be at 2.2 (La-O bond) A^0.

References

[1] Feng Xie, Ting An Zhang, David Dreisinger, Fiona Doyle, 2014, Minerals Engineering, **56**, 10–28.

[2] Koen Binnemans , Peter Tom Jones , Bart Blanpain , Tom Van Gerven , Yongxiang Yang, Allan Walton , Matthias Buchert, 2013 , Journal of Cleaner Production, **51** , 1-22.

[3] Adam Jordens, Ying Ping Cheng, Kristian E. Waters, 2013, Minerals Engineering, **41**, 97–114.

[4] Adam Jordens; The beneficiation of rare earth element-bearing minerals; Thesis, 2016.

[5] Rita Kakkar, Rajni Grover, Pragya Gahlot, 2006, Journal of Molecular Structure, THEOCHEM, **767**, 175–184.

[6] Sarvaramini, D. Azizi, F. Larachi Applied Surface Science, **387**,986–995.

[7] M. Atanassova, 2007, Russian Journal of Inorganic Chemistry, **52**, 1304–1311.

[8] Dongbei Wu, Qian Zhang, Borong Bao, 2007, Hydrometallurgy, **88**, 210–215.

[9] Miaomiao Tian , Naizhong Song , Dingding Wang , Xinjun Quan , Qiong Jia , Wuping Liao, Li Lin, 2012, Hydrometallurgy,**111–112**, 109–113.

[10] TIAN Miaomiao, JIA Qiong, LIAO Wuping, 2013, JOURNAL OF RARE EARTHS, **31**, 604.

[11] T.S. Urbanski, P. Fornari, C. Abbruzzese, 1996, Hydrometallurgy, **40**,169-179.

[12] Rajesh B. Gujar, Seraj A. Ansari, Arijit Sengupta, Mallekav S. Murali, Prasanta K. Mohapatra, 2016, Inorganic Chemistry Communications, **73**, 72–76.

[13] Alok Rout, Venkatesan K. A Antony M. P, 2016, J Radioanal Nucl Chem, **310**, 91–97.

[14] Billy W. McCann, Nuwan De Silva, Theresa L. Windus, Mark S. Gordon, Bruce A. Moyer, Vyacheslav S. Bryantsev,and Benjamin P. Hay, 2016, Inorg. Chem., **55**, 5787−5803.

[15] Jason M. Keith and Enrique R. Batista, 2012, Inorg. Chem., **51**, 13−15.

[16] Qiong Jia , Shanshan Tong, Zhiying Li ,Weihong Zhou, Hongfei Li , Shulan Meng, 2009, Separation and Purification Technology,**64**, 345–350.

[17] Renata D. Abreu, Carlos A. Morais, 2014, Minerals Engineering, **61**, 82–87.

[18] Tayebeh Hosseinnejad, Saeed Dehghanpour, and Sara Basiri_nasab, 2014, Russian Journal of Physical Chemistry A, **88**, 2004-2011.

[19] P.K.Mohapatra, D.R.Raut, M.Iqbal, J.Huskens, W.Verboom, 2014 Applied Radiation and Isotopes, **85**, 133–138.

[20] Arijit Sengupta, Sk. Musharaf Ali, K.T. Shenoy, 2016, Polyhedron, **117**, 612–622.

IOP Conf. Series: Materials Science and Engineering **338** (2018) 012026 doi:10.1088/1757-899X/338/1/012026

Flexible nano-GFO/PVDF piezoelectric-polymer nano-composite films for mechanical energy harvesting

Monali Mishra[1,3], Amritendu Roy[2], Sukalyan Dash[3], Somdutta Mukherjee[1]

[1]Colloids and Materials Chemistry Department, CSIR-IMMT Bhubaneswar-751013, India
[2]School of Minerals, Metallurgical and Materials Engineering, Indian Institute of Technology Bhubaneswar, Bhubaneswar-752050, India
[3]Department of Chemistry, VSSUT Burla, Burla-768018, India

Email: msomdutta@gmail.com

Abstract: Owing to the persistent quest of renewable energy technology, piezoelectric energy harvesters are gathering considerable research interest due to their potential in driving microelectronic devices with small power requirement. Electrical energy (milli to microwatt range) is generated from mechanical counterparts such as vibrations of machines, human motion, flowing water etc. based on the principles of piezoelectricity. Flexible high piezoelectric constant (d33) ceramic/polymer composites are crucial components for fabricating these energy harvesters. The polymer composites composed of gallium ferrite nanoparticles and polyvinylidene fluoride (PVDF) as the matrix have been synthesized by solvent casting method. First, 8 wt. % PVDF was dissolved in DMF and then different compositions of $GaFeO_3$ or GFO (10, 20, 30 wt. %) (with respect to PVDF only) nanocomposites were synthesized. The phase of the synthesized nanocomposites were studied by X- Ray diffraction which shows that with the increase in the GFO concentration, the intensity of diffraction peaks of PVDF steadily decreased and GFO peaks became increasingly sharp. As the concentration of GFO increases in the PVDF polymer matrix, band gap is also increased albeit to a small extent. The maximum measured output voltage and current during mechanical pressing and releasing conditions were found to be ~ 3.5 volt and 4 nA, respectively in 30 wt % GFO-PVDF composite, comparable to the available literature.

1. Introduction:

In recent days the dimension and power consumption of electronic devices have reduced significantly to make them portable and convenient to use in daily requirements. Piezoelectricity based energy harvesters that convert mechanical energy to electrical energy are perfect for powering those electronic devices where various mechanical vibrations originating from human movement or movement of any machine can be utilized. This can crumb the present dependency on batteries, give more freedom to mobile consumers and minimize the environmental issue related to battery disposal.

Piezoelectric materials with large piezoelectric constants (d_{33}) are natural choices for making composites. For this purpose various piezoelectric ceramics viz. ~ barium titanate ($BaTiO_3$), Lead zirconium titanate (PZT), zinc oxide (ZnO), lead magnesium niobate – lead titanate (PMN-PT) have been used for their large d_{33} value. However, application prospect of these piezoelectric ceramics are limited due to their poor mechanical property, *i.e.*, brittleness. Further, piezoceramics experience fatigue crack when exposed to cyclic loading. Such unfavourable mechanical characteristics inspires the search for alternative materials with superior properties.

Polyvinylidene difluoride (PVDF) is a piezoelectric polymer which is highly flexible. PVDF thus has attracted significant research attention due to its favourable mechanical property and ferroelectric, pyroelectric and piezoelectric properties. It has a piezoelectric coefficient of 12-23 pC/N.[1] A series of studies have been performed investigating the piezoelectric based energy harvesting behaviour of PVDF. Though, the pure PVDF based devices showed generation of current in a few hundred of nAmp range the voltage generated was restricted to ~ 0.5 V resulting in a very low power output.[2, 3] In this context, piezoelectric ceramics are blended with the PVDF polymer matrix to improve output current and voltage characteristics. In the present study we have chosen gallium ferrite, $GaFeO_3$ or GFO nanopowder, a piezoelectric material with piezoelectric coefficient ~ 4 pC/N with appreciable conductivity. [4]

2. Experimental details:

2.1 Synthesis of Gallium Ferrite or GFO nanopowders:

GFO nanopowders, were synthesized using modified Pechini method.[5] gallium nitrate hydrate ($Ga(NO_3)_3 .xH_2O$) (sigma Aldrich ≥ 99.9%) and iron nitrate nonahydrate ($Fe(NO_3)_3.9H_2O$) (sigma Aldrich ≥ 98%) were taken as the precursor materials.

2.2 Synthesis of GFO-PVDF nano-composites:

GFO/PVDF nano-composite was synthesized by solvent casting method. For the preparation of composites, we used N, N-Dimethyl formamide (DMF), Polyvinylidiene difluorie (Sigma Aldrich), surfactant Span-80 and as synthesized GFO nano-powder. Initially, GFO nanopowder (10 wt.%, 20 wt.%, 30 wt. % with respect to PVDF) was mixed with ethanol and water for 1 h to achieve OH order in the surface of GFO nanopowder.[6] The mixture was then dried at 80° C. Surfactant Span-80 and solvent acetone was added to the above and stirred overnight. After removal of acetone 8 wt. % PVDF dissolved in DMF was mixed with the above GFO nano-powders. The solution was then poured into a petri dish and heated at 100° C to remove all residual solvent present in it.

2.3 Characterization

To perform the structural characterization of the samples, x-ray diffraction of synthesized nano-powders and polymeric composites were performed using Rigaku diffractometer with CuKα radiation. Morphological evolution of GFO/PVDF composites was examined by scanning electron microscopy (SEM). Room temperature Fourier transform infrared spectra (FTIR) were also studied. The output voltage and the current generated by the composites were measured by using a Kiethley 6517B electrometer.

3. Results and Discussion

3.1 X-ray diffraction analysis

Fig. 1 shows the XRD patterns of PVDF and GFO-PVDF composites synthesized with various weight fractions of GFO. PVDF exists in three crystalline phases (i.e. α, β, γ- phase). There are two intense peaks observed at 2θ ~ 17.7°, 19.9° and 26° assigned to α-phase.[7-9] The peak position corresponding to 35.6° and 38.3° could be indexed to β- phase and α-phase of PVDF, respectively.[8] The XRD pattern of GFO nanopowder over 2θ ~ 20° - 80° indexed with JCPDS card no. 76-1005 is shown in the inset of Fig. 1. From the figure, orthorhombic $Pc2_1n$ symmetry of GFO is evident without presence of any secondary phase. Fig. 1 shows that with increase in GFO concentrations in PVDF matrix, the diffraction peak intensities of PVDF gradually decrease and the peaks of GFO became prominant. It is also clear that, the naure and positions of GFO peak remain almost unaltered with change in GFO weight percent. So we conclude that structure of GFO remains un-affected upon blending with PVDFmatrix.

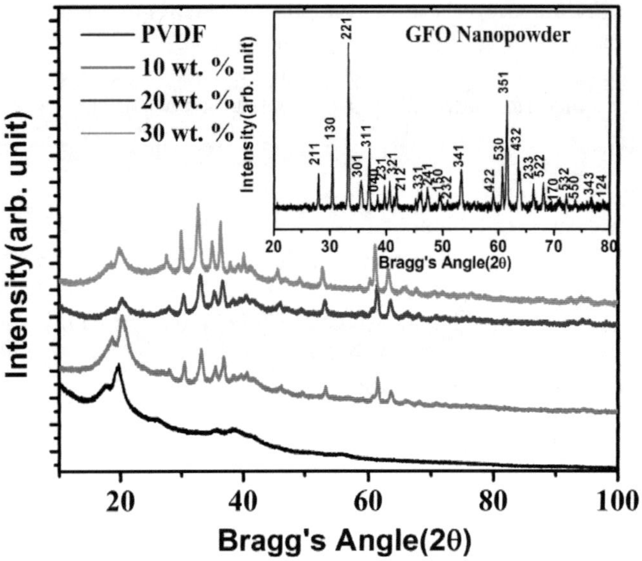

Fig. 1 Room temperature x-ray diffraction patterns of PVDF and GFO-PVDF nano-composite films with various weight percentage of GFO. Inset shows XRD pattern of GFO nano-powder indexed with JCPDS card no. 76-1005.

3.2 Microstructural analysis

The microstructures of the samples were examined by scanning electron microscopy (SEM). The SEM micrographs of pure PVDF, 10-30 wt. % GFO-PVDF composites are shown in Fig. 2(a-d), respectively. From Fig. 2(a), it was seen that pure PVDF film is clean, neat and uniform without any pores but it changes as the concentration of GFO nanoparticles changes in the polymer matrix. The micrographs reveal the uniform distribution of GFO nanopowders in the PVDF polymer matrix with no pores. In 10 wt. % of GFO-PVDF composite grains can be seen. As the concentration of GFO increases, packing of particles became denser. So we can say that there is excellent compatibility between PVDF and GFO nanoparticles. Inset of Fig. 2(d) shows the optical image of the 30 wt. % GFO-PVDF composite film and it was found that the composite film was quite flexible.

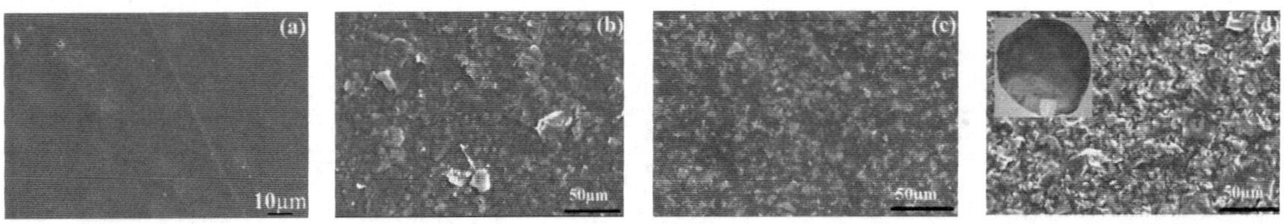

Fig. 2 (a) SEM image of pure PVDF. (b-d) SEM images of 10-30 wt. % of GFO-PVDF composite films, respectively. Inset of Fig. 2(d) shows optical image of 30 wt. % GFO-PVDF composite film.

3.3 FTIR analysis

Phase evolution of GFO-PVDF nanocomposite was further studied using FTIR analysis. Fig. 3 plots the FTIR spectra of pure PVDF and GFO-PVDF composites with different weight percentages of GFO in the spectral range of 650-2000 cm^{-1}. As discussed earlier, PVDF exists in α, β, γ phases. Bands at 759 (CH_2 stretching) and 971 cm^{-1} are related to α-phase. The β phase is observed at 833, 871, 1169 and 1279 cm^{-1}.[6] The 680 cm^{-1} band is due to head to head and tail to tail arrangements. The band at 833 cm^{-1} is due to CH_2 rocking, CF_2 stretching, skeletal C-C structure and stretching in β/γ phase. The band at 1067 cm^{-1} is assigned to CH_2 wagging. Similarly, the band at 1169 cm^{-1} is assigned to symmetrical stretching of CF_2 groups and β - phase, 1279 cm^{-1} is assigned to CF_2 symmetric stretching and β- phase, 1405 cm^{-1} is marked to CF_2 symmetric stretching and α- phase.[8] Most of the vibrational bands of β and γ phases appear at the same frequencies due to the presence of tail to tail structure in the chain configuration of PVDF.[10] The α- phase becomes quite weak with gradual increase in GFO concentration. Most of the PVDF peaks are present in FTIR spectra. A few vibrational frequencies below 800 cm^{-1} disappeared which suggests interaction between PVDF and GFO during formation of PVDF-GFO nano-composite. PVDF's original structure is changed and some groups in PVDF are fixed in GFO. As the concentration of GFO increases, some vibrational bands of PVDF gradually decreases.[11]

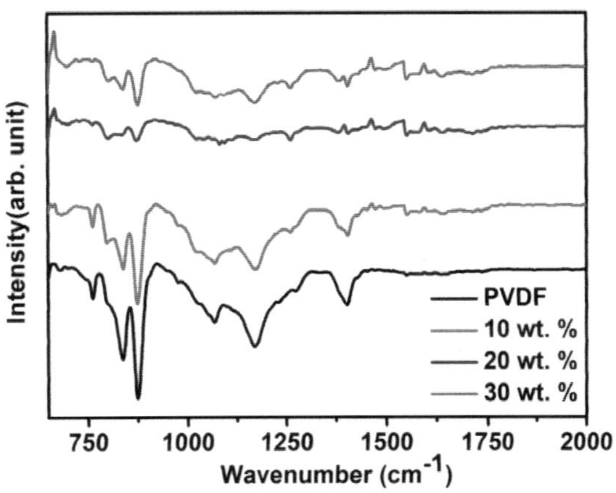

Fig. 3 FTIR spectra of GFO-PVDF nanocomposites and PVDF.

3.4 Electrical characterization

The above synthesized nano GFO-PVDF free standing flexible composite films were subjected to energy harvesting studies. The composite films were cut into 1.5 × 1.5 cm^2 dimension and silver paste was applied on both the sides of the films to act as electrodes. Two thin connecting wires were placed on both sides of the film and the whole system was laminated to make the final device. Fig. 4 shows a representative image of the final device. The output voltage and current were measured by manually finger tapping on the device. Fig. 5(a-c) shows the output voltage generated by 10-30 wt. % composite films. The 10 wt. % GFO-PVDF composite generates an output voltage of 1.5 V (Fig. 5(a)). With increasing GFO content the output voltage increased to 3 V and 4 V, respectively for 20 and 30 wt. % of GFO. Fig. 5(d-f) shows the output current generated by 10-30 wt% GFO-PVDF composite films. Fig. 5(d) shows that 10 wt. % composite film generates an output current of 1.2×10^{-11} Amp. With increase in GFO content in polymer matrix the output current increased to 6×10^{-10} Amp for 20 wt. % and 4×10^{-9} Amp for 30 wt. % of GFO.

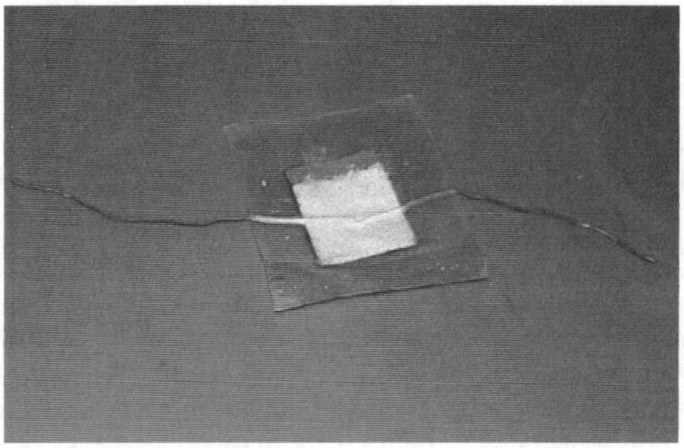

Fig. 4 Optical image of flexible composite device with 30 wt. % of GFO.

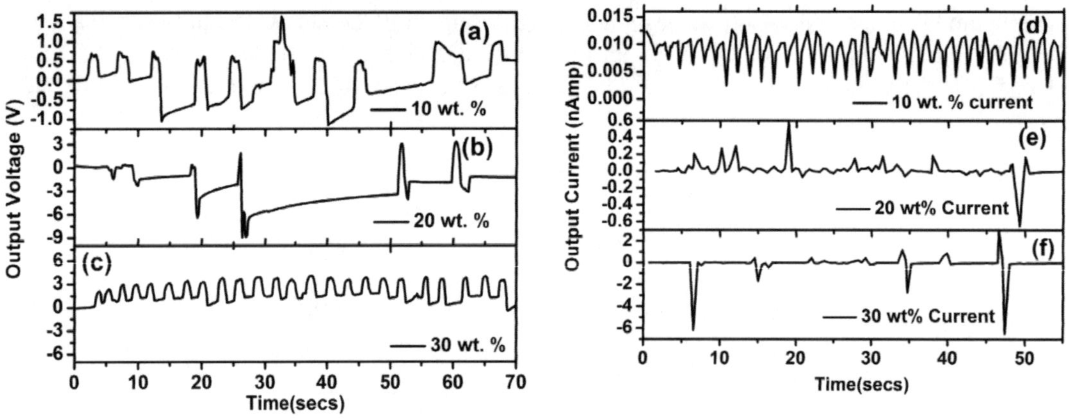

Fig. 5 Output voltage signal of (a) 10 wt. %, (b) 20 wt. %, (c) 30 wt. % GFO-PVDF composite films, respectively. Output current signal of (d) 10 wt. %, (e) 20 wt. %, (f) 30 wt.% GFO-PVDF composite films, respectively.

The voltage generated with increasing GFO content is at par the PVDF based energy harvesting devices.[3, 12] However, the current output of our films is low compared to recent reports on PVDF based nano-composite energy harvesters.[3, 12] The output current could be improved by modifying the composite materials. Multi-walled carbon nanotubes (CNT), reduced grapheme oxide (RGO) could be incorporated as conductive fillers to improve the current output.[13, 14]

4. Conclusion

Using solvent casting method thick films of GFO-PVDF composites are produced. XRD and FTIR analysis confirms the presence of α, β, γ phase in PVDF. SEM reveals the uniform structure of PVDF with no pores and the uniform dispersion of GFO in polymer matrix. The investigation on output voltage and current reveals that 30 wt. % GFO-PVDF composite film shows maximum voltage of 4 V and maximum current of 4 nA, respectively. This output voltage is comparable with other reported flexible piezoelectric composites. The output current characteristics could be improved by modifying the synthesis of composite material. Hence, flexible nano GFO-PVDF composites could be a promising candidate for piezoelectric energy harvesting.

IOP Conf. Series: Materials Science and Engineering **338** (2018) 012026 doi:10.1088/1757-899X/338/1/012026

Acknowledgement:

Authors acknowledge support from grant no. IFA-13/MS-03 by Department of Science and Technology, India under INSPIRE faculty award program with. A. Roy acknowledges financial support from SERB, Govt. of India through project no. YSS/2014/000287.

References

1. Li, H., C. Tian, and Z.D. Deng, *Energy harvesting from low frequency applications using piezoelectric materials.* Applied Physics Reviews, 2014. 1(4): p. 041301.
2. Nunes-Pereira, J., et al., *Energy harvesting performance of piezoelectric electrospun polymer fibers and polymer/ceramic composites.* Sensors and Actuators A: Physical, 2013. 196(Supplement C): p. 55-62.
3. Nunes-Pereira, J., et al., *Energy harvesting performance of BaTiO3/poly(vinylidene fluoride–trifluoroethylene) spin coated nanocomposites.* Composites Part B: Engineering, 2015. 72(Supplement C): p. 130-136.
4. Remeika, J.P., *GaFeO3: A Ferromagnetic-Piezoelectric Compound.* Journal of Applied Physics, 1960. 31(5): p. S263-S264.
5. Kavita, S., et al., *Magnetic and 57 Fe Mössbauer study of magneto-electric GaFeO 3 prepared by the sol–gel route.* Journal of Physics: Condensed Matter, 2013. 25(7): p. 076002.
6. Luo, B., et al., *Fabrication, characterization, properties and theoretical analysis of ceramic/PVDF composite flexible films with high dielectric constant and low dielectric loss.* Journal of Materials Chemistry A, 2014. 2(2): p. 510-519.
7. Sita, M.D., et al., *Structural changes in PVDF fibers due to electrospinning and its effect on biological function.* Biomedical Materials, 2013. 8(4): p. 045007.
8. Dash, S., R.N.P. Choudhary, and M.N. Goswami, *Enhanced dielectric and ferroelectric properties of PVDF-BiFeO3 composites in 0–3 connectivity.* Journal of Alloys and Compounds, 2017. 715: p. 29-36.
9. Tan, K.S., et al., *Pyroelectricity enhancement of PVDF nanocomposite thin films doped with ZnO nanoparticles.* Smart Materials and Structures, 2014. 23(12): p. 125006.
10. Mandal, D., K. Henkel, and D. Schmeißer, *The electroactive β-phase formation in Poly(vinylidene fluoride) by gold nanoparticles doping.* Materials Letters, 2012. 73: p. 123-125.
11. Yang, C., H.-s. Song, and D.-b. Liu, *Effect of coupling agents on the dielectric properties of CaCu3Ti4O12/PVDF composites.* Composites Part B: Engineering, 2013. 50: p. 180-186.
12. Karan, S.K., D. Mandal, and B.B. Khatua, *Self-powered flexible Fe-doped RGO/PVDF nanocomposite: an excellent material for a piezoelectric energy harvester.* Nanoscale, 2015. 7(24): p. 10655-10666.
13. Bhavanasi, V., et al., *Enhanced Piezoelectric Energy Harvesting Performance of Flexible PVDF-TrFE Bilayer Films with Graphene Oxide.* ACS Applied Materials & Interfaces, 2016. 8(1): p. 521-529.
14. Das, S., et al., *Electrical and mechanical behavior of PMN-PT/CNT based polymer composite film for energy harvesting.* Applied Surface Science, 2018. 428(Supplement C): p. 356-363.

Influence of thermo-mechanical treatment in ferritic phase field on microstructure and mechanical properties of reduced activation ferritic-martensitic steel

Prakash[1], J. Vanaja[2], K. Laha[3] and G. V. S. Nageswara Rao[1*]
[1]Department of Metallurgical and Materials Engineering
National Institute of Technology, Warangal-506021, India
[2]Creep Studies Section, Indira Gandhi Centre for Atomic Research, Kalpakkam-603102, India.
[3]AUSC Mission Directorate, Noida-201305, Uttar Pradesh, India
[*]Corresponding Author. E-mail address: nr_gvs@yahoo.com;
Phone: +91 870 2462521

Abstract: The present study focuses on the evaluation of microstructure and mechanical properties of reduced activation ferritic-martensitic (RAFM) steel (9Cr-1W-0.06Ta) subjected to thermo-mechanical treatment (TMT) in ferritic phase field. The results obtained were compared with the steel in conventional normalised plus tempered (N+T) condition. The microstructure of the steel in N+T and TMT conditions was assessed by optical and scanning electron microscopes. Hardness, tensile and creep studies were carried out and the results were correlated with the microstructural studies. While the TMT processed steel resulted in coarser prior austenite grains and exhibited ferritic microstructure with large distribution of fine $M_{23}C_6$ and MX precipitates, the N+T steel reveals tempered martensitic structure with finer prior austenitic grains with coarser $M_{23}C_6$ and MX precipitates. Although ferritic structure is present in TMT processed steel, it exhibits better tensile and creep rupture strengths than N+T steel due to the presence of increased dislocation density and finer distribution of precipitates.
***Keywords*:** Reduced activation ferritic-martensitic steel; Thermo-mechanical treatment; Electron microscopy; Tensile and creep strengths.

1. Introduction

Reduced activation ferritic-martensitic (RAFM) steels having 9Cr-1W-0.22V-0.06Ta as the major alloying elements are potential candidate materials for the test blanket module (TBM) compared to austenitic steels due to their attractive properties, for example, low thermal expansion coefficient, high thermal conductivity and favourable void swelling resistance [1]. The steel in normalized and tempered condition consists of tempered martensite with laths, blocks, and packets inside the prior-austenite grains. The laths, blocks and packets contain high concentration of transformation induced dislocation network, decorated with fine intra-granular MX type (where M=V, Ta; X=C, N) precipitates, and chromium-rich $M_{23}C_6$ carbides along the boundaries [2-4]. The microstructure of the ferritic-martensitic steel depends on the heat treatment temperatures especially those inducing phase transformation [5]. The precipitates coarsen on thermal aging and also during creep exposure at elevated temperatures by Ostwald ripening mechanism. The coarsening rate of $M_{23}C_6$ carbides is much faster than the MX carbo-nitrides [6]. The coarsening of $M_{23}C_6$ carbides present along the prior austenitic grain boundaries promote faster microstructural recovery in the vicinity of grain boundaries [7], thus resulting in increase in minimum creep rate, early onset of tertiary stage of creep deformation and hence reduce the creep rupture life. To increase creep rupture strength of the steel, it is desirable either to (i) decrease coarsening kinetics of $M_{23}C_6$ precipitates or (ii) eliminate the presence of unstable $M_{23}C_6$ carbides from martensitic microstructure and promote the precipitation of very fine and thermally more stable MX carbo-nitrides not only within the matrix but also along the prior austenitic grain boundaries.

In the present work, an attempt has been made to understand the effect of thermo-mechanical treatment on microstructure and mechanical properties of RAFM steel. The tempered martensitic structure of RAFM steel in normalized and tempered condition is transformed into ferritic (α) phase by prolonged soaking. The steel in ferritic phase field was subjected to thermo-mechanical treatment by warm rolling. The microstructural stability, tensile and creep properties of thermo-mechanically processed steel in ferritic phase condition were compared with those obtained in normalized and tempered condition.

2. Experimentation

2.1 Materials and Methods

The RAFM steel (9Cr - 1W - 0.06Ta), in the form of plate, was received in normalised and tempered condition. The normalizing was done at 1253 K for 30 min followed by tempering at 1038 K for 90 min. The major alloying elements (in wt.%) present in the steel are given in **Table 1.** The received plate was subjected to thermo-mechanical treatment (TMT) by warm rolling. Prior to TMT, the steel was austenitized at 1423 K for 10 min followed by air cooling up to 973 K and then holding for 2 h at the same temperature in a furnace to transform in to fully ferritic phase. Subsequently, warm rolling was carried out at 973 K using a 200 ton capacity, two-high rolling mill (Make: DEMAG) with a roll diameter of 300 mm. During warm rolling, a reduction of 25 % in thickness was imparted in five passes. This was followed by tempering at 1038 K for 150 min. Representative samples of steel from normalized and tempered condition and TMT condition were cut and polished to determine their microstructures using optical and electron microscopes. The samples were immersion etched with Villella's reagent (1 g picric acid and 5 ml hydrochloric acid in 100 ml ethanol). Hereafter, the steel in normalized and tempered condition is referred to as "N+T" and the steel with 25 % reduction in thickness by warm rolling followed by tempering is referred to as "FF-TMT25". A Leica metallurgical microscope (Model: DMi8A) equipped with image analysis software (Leica Application Suite (LAS), version 4.8) was used for obtaining optical microstructures of steel in N+T and FF-TMT25 conditions. High magnification images of microstructural features of steel in N+T and FF-TMT25 conditions were obtained by using scanning electron microscope (SEM) (Make: TESCAN, Model: VEGA 3 LMU). The hardness of the steel in N+T and FF-TMT25 conditions was determined by using Vickers macro hardness tester (Make: Matsuzawa, Model: VMT-X) using 10 kgf force and 15 s dwell time. At least ten measurements were made in each condition and their average and standard deviation values were reported. Threaded cylindrical tensile specimens were prepared from N+T and FF-TMT25 steel plates whose gauge length is parallel to the rolling direction. Tensile properties were evaluated in the temperature range of 300 - 923 K and at nominal strain rate of 3×10^{-4} s^{-1}. The creep rupture life of the steel in N+T and FF-TMT25 conditions was assessed by carrying out constant load creep tests at 823 K in ambient air at 240 MPa.

Table 1. Chemical composition of RAFM steel (in wt. %) (Major elements)

Element	C	Cr	W	V	Ta	Mn	N	Fe
wt.%	0.08	9.04	1.01	0.22	0.06	0.55	0.0226	Rest

3. Results and Discussion

3.1 Microstructural analysis

Figures 1a and 1b shows the optical microstructures of steel in N+T and FF-TMT25 conditions respectively. The steel in N+T condition revealed tempered martensitic structure (Figure 1a) and that in FF-TMT25 condition revealed fully ferritic phase (Figure 1b). The existence of ferrite phase was also confirmed through macro hardness measurement and is discussed in the next session. The average prior austenitic grain (PAG) size of steel in N+T and FF-TMT25 conditions was found to be 15±2 μm and 150±20 μm respectively. As the PAG size in FF-TMT25 condition is very large, a lower magnification image of the steel is shown in Figure 1b to cover more PAGs. Coarser PAG size by TMT was also reported by Sakthivel et al. [8] for modified 9Cr-1Mo steel (T91 steel). The coarsening of PAGs in FF-TMT25 steel is due to the reaustenitising at high temperature where most of the precipitates dissolved.

SEM images of RAFM steel in N+T and FF-TMT25 conditions are shown in Figures 2a and 2b respectively. In N+T condition the PAGs are seen clearly and are decorated by $M_{23}C_6$ precipitates along the boundaries, whereas in FF-TMT25 condition the PAG structure is refined and the precipitates are formed along the sub grain boundaries too and found occasionally within the grains. Based on the previous studies [4], the larger precipitates are of $M_{23}C_6$ and the finer one are of MX type. It can also be clearly seen that the number density of precipitates in TMT steel is higher than those in the N+T condition. This can be attributed to the fact that the warm rolling of steel develops higher dislocation density. These dislocations serve as preferential nucleation sites for the precipitation of $M_{23}C_6$ and MX precipitates. Thus resulting in the formation of large number of precipitates in TMT steel compared to N+T steel. Further, it can be observed that the sizes of $M_{23}C_6$ and MX precipitates in FF-TMT25 condition is finer than those observed in N+T condition. While the size of $M_{23}C_6$ and MX precipitates are in the range of 50-160 and 20-30 nm respectively in N+T condition, their size was refined and found to be in the range of 30-120 and 5-30 in TMT condition.

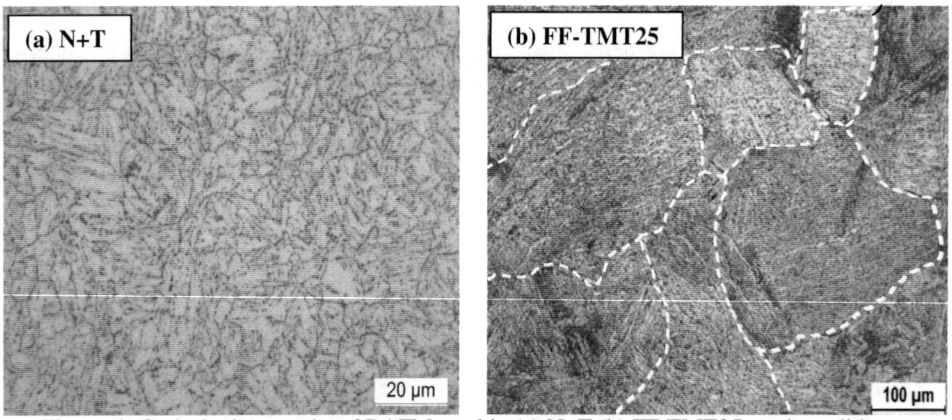

Figure 1. Optical micrographs of RAFM steel in (a) N+T (b) FF-TMT25 steel condition.

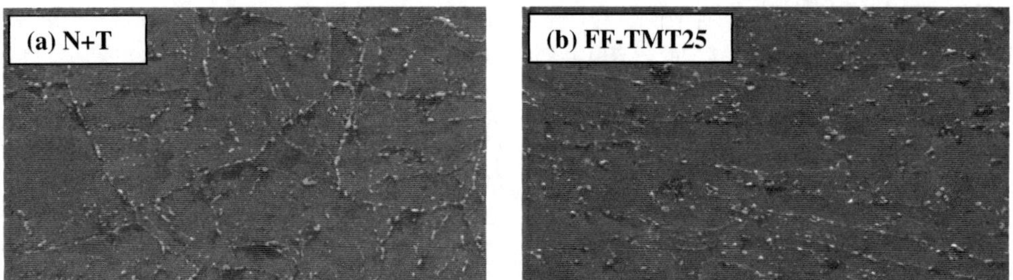

Figure 2. Scanning electron microscope images of RAFM steel in (a) N+T (b) FF-TMT25 steel condition.

3.2 Hardness

The Vickers hardness values of the steel in N+T, TMT in ferritic phase field condition (without tempering), and TMT processed plus tempered (FF-TMT25) conditions are shown in Figure 3. The hardness of steel in N+T condition is found to be 206±2 HV, which is increased to 409±2 HV upon TMT in ferritic phase field. This sudden rise in hardness can be attributed to the development of high density of dislocation structure resulted due to deformation and the existence of large number of fine precipitates [8-10]. A drastic reduction in hardness can be observed after tempering (i.e. in FF-TMT25 condition), which could be due to the presence of fully ferritic phase and annihilation of dislocations.

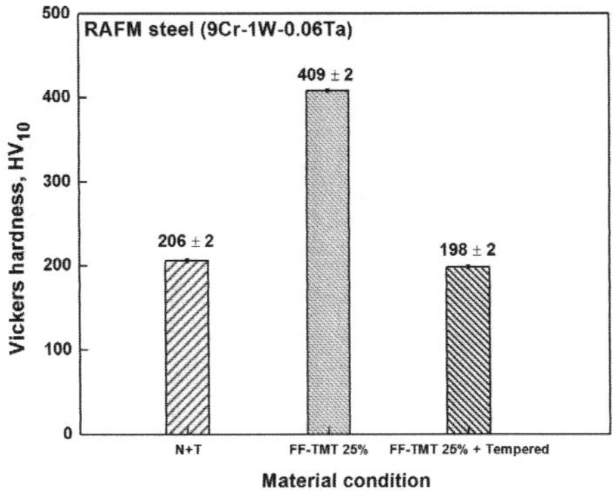

Figure 3. Macro Vickers hardness of RAFM steel in different conditions.

3.3 Tensile analysis

Engineering stress-strain curves of RAFM steel in N+T and FF-TMT25 conditions tested in the temperature range of 300 K - 923 K at strain rate of 3×10^{-4} s^{-1} are shown in Figure 4. The curves show that the steel in both conditions follow monotonic behavior at all test temperatures. Further, there is no evidence of serrated flow in both conditions of the steel, which commonly noticed in ferritic steels in the intermediate temperatures [11, 12]. While the strain to failure is reduced as the temperature is increased from 300 K to 773 K, and it is increased beyond 773 K in both N+T and FF-TMT25 conditions.

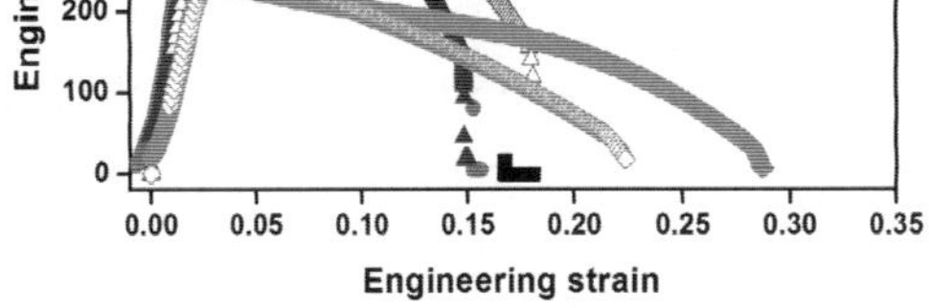

Figure 4. Comparison of engineering stress - engineering strain plots of RAFM steel in N+T and FF-TMT25 conditions at selected temperatures.

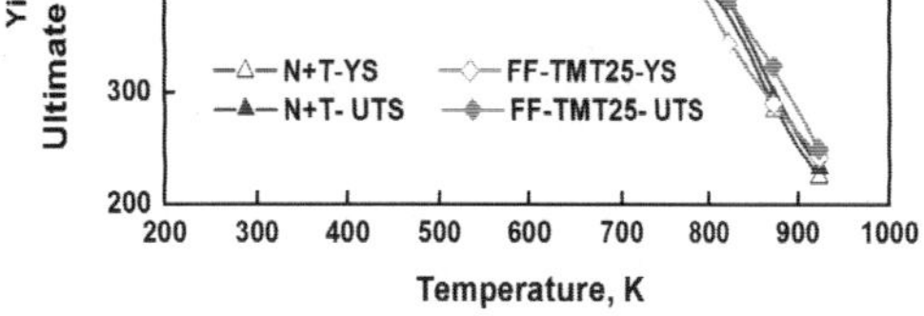

Figure 5. Variation of tensile strengths with test temperature.

The variation of yield strength (YS) and ultimate tensile strength (UTS) of RAFM steel in N+T and FF-TMT25 conditions with temperature is shown in Figure 5. The YS and UTS were found to decrease with increase in the test temperature. TMT of RAFM steel in ferritic phase field is found to impart slight improvement in YS and UTS than the steel in N+T condition. However, plateau region is observed in the YS in the intermediate temperature range (473 - 573 K) for N+T condition and similar feature is observed at (623 K-723 K) for FF-TMT25 condition, for the UTS plots the plateau region for N+T condition is found for temperature range of (473 K- 523 K) and no plateau region found for FF-TMT25 condition. A rapid decrease in YS and UTS was observed beyond 773 K for both conditions of steel. The yield stress of FF-TMT25 steel was higher than the N+T steel up to 573K beyond which close fluctuations are observed in both conditions. Higher UTS values are observed in TMT condition, except at 523 - 573 K, than N+T condition. On a whole slight improvement in YS and UTS are observed in FF-TMT25 condition than the N+T condition. This slight increase in tensile strengths of FF-TMT25 steel over N+T condition at higher temperatures can be attributed to the presence of fine and homogeneous distribution of precipitates, which assisted in delaying the dynamic recovery of the dislocation structure; where the dominance of dynamic recovery has been observed in the steel generally at higher test temperatures. This behaviour can be correlated with the microstructural features discussed in Figures 2a and 2b.

Figure 6 shows the percentage elongation and percentage reduction in area of RAFM steel in N+T and FF-TMT25 conditions. Percentage elongation is lower at room temperature for FF-TMT25 condition than N+T condition and small dip is observed at 523 K- 773 K in both conditions with closer values; higher elongations are observed for N+T condition beyond 773 K. Percentage reduction in area is lower for FF-TMT25 condition than the N+T condition over all the temperature range. Slight decrease at intermediate temperatures and rapid increase at higher temperatures for both conditions of the steel are observed. The appearance of mild plateau in YS and UTS as well as dip in elongation curves in the intermediate temperature range is an indication of the occurrence of dynamic strain aging (DSA) [11-13].

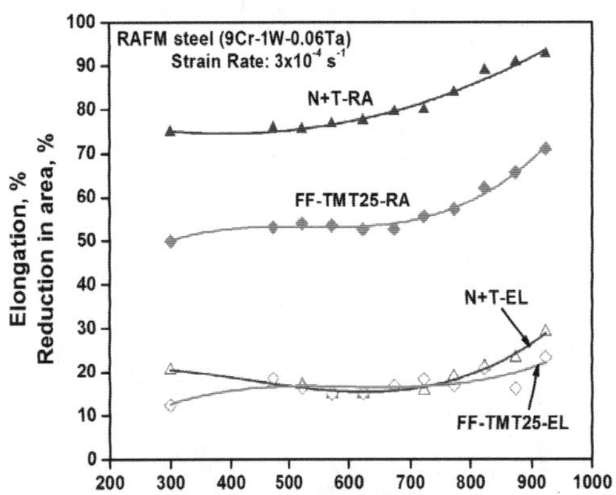

Figure 6. Variation of percentage elongation and percentage reduction in area with temperature.

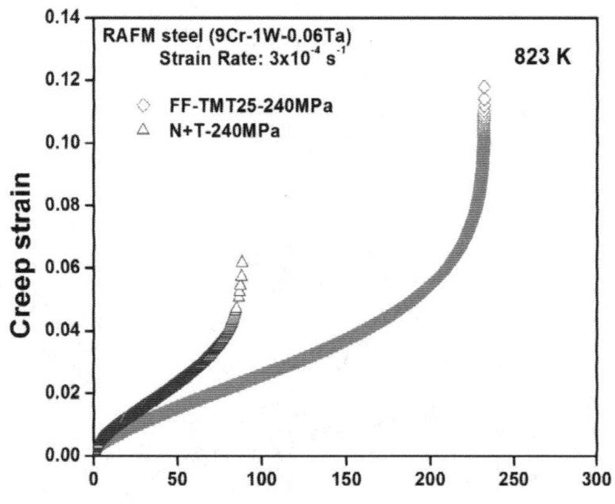

Figure 7. Comparison of creep curves of RAFM steel in N+T and FF-TMT25 conditions tested at 240 MPa and 823 K.

3.4 Creep properties

The creep curves of RAFM steel in N+T and FF-TMT25 conditions at stress level of 240 MPa and temperature of 823 K are compared in Figure 7. From these curves, it is evident that TMT processing of steel in ferritic phase field has significantly lowered the creep deformation and greatly improved the rupture life. The reduced creep deformation and improved rupture life can be attributed to the presence of large number density of nano sized $M_{23}C_6$ and MX precipitates. The presence of these nano size precipitates significantly delays the onset of tertiary stage of creep, there by leading to the enhancement of rupture life than the same steel in N+T condition.

4. Conclusions

In the present work, the effect of TMT in ferritic phase field on tensile properties and creep rupture life of RAFM steel was studied and the data obtained was compared with the steel in N+T condition. The following are the salient conclusions:

(i) The microstructure of the RAFM steel was refined through TMT. Coarser prior austenite grain sizes, finer sub-grains, dense and homogeneous distribution of precipitates were achieved in the TMT steel as compared to the N+T steel. The number density of $M_{23}C_6$ and MX type precipitates are more in FF-TMT25 condition than N+T condition.

(ii) Vickers hardness value of 198 HV_{10} at the end of TMT followed by tempering proved that the transformation of tempered martensitic structure (in N+T condition) to ferritic phase (in FF-TMT25 condition).

(iii) TMT of RAFM steel in ferritic phase field was found to impart slight improvement in YS and UTS than the steel in N+T condition at all test temperatures and at strain rate of 3×10^{-4} s^{-1}.

(iv) Creep tests performed at 240 MPa and 823 K on N+T and FF-TMT25 steels showed significant improvement in rupture life of TMT steel over N+T steel.

References

[1] Zinkle S J and Busby J T 2009 *Mater. Today* **12** pp12-19.
[2] Jayaram R and Klueh R L 1998 *Metall. Mater. Trans.* A **29** pp1551-1558.
[3] Klueh R L, Alexander D J and Sokolov M A 2002 *J. Nucl. Mater.***304** pp139-152.
[4] Fernandez P, Lancha A M, Lapena J and Hernandez-M 2001 *Fusion Eng. Des.* **58–59** pp787-792.
[5] Saroja S, Dasgupta A, Divakar R, Raju S, Mohandas E, Vijayalakshi M, Bhanu Sankara Rao K and Raj B 2011 *J. Nucl. Mater.* **409 (2)** pp131-39.
[6] Ravikirana, Mythili R, Raju S, Saroja S, Jayakumar T, Rajendrakumar E *Mater. Charact.* **84** pp196-204.
[7] Chandravathi K S, Sasmal C S, Laha K, Parameswaran P, Nandagopal M, Vijayanand V D, Mathew M D, Jayakumar T and Rajendra Kumar E 2013 *J. Nucl. Mater.***435** pp 128-136.
[8] Sakthivel T, Shruti P, Parameswaran P, Nageswara Rao G V S, Laha K and Srinivasa Rao T 2016 *Trans Indian Inst Met* **47** pp204-209.
[9] Benjamin A. Shassere, Yamamoto Y, Suresh Babu S 2016 *Metall. Mater. Trans.* A **47A** 2190.
[10] Matthias Nohrer, Walter Mayer, Sophie Primig, Sabine Zamberger, Ernst Kozeschnik, and Harald Leitner 2014 *Metall. Mater. Trans.* A **45A** pp4210-4219.
[11] Vanaja J, Laha K, Shiju Sam, Nandagopal M, Panneer Selvi S, Mathew M D, Jayakumar T, Rajendra Kumar E 2012 *J. Nucl. Mater.***424** pp116-122.
[12] Vanaja J, Laha K, Nandagopal M, Shiju Sam, Mathew M D, Jayakumar T, and Rajendra Kumar E, 2013 *J. Nucl. Mater.* **433** pp412-418.
[13] Palaparti D P, Choudhary B K, Isaac Samuel E, Srinivasan V S and Mathew M D 2012 *Mater. Sci. Eng.* A **538** pp110-117.

The influence of void and porosity on deformation behaviour of nanocrystalline Ni under tensile followed by compressive loading

Md. Meraj[1], Shradha Nayak[2], Kumar Krishanjeet[1], Snehanshu Pal[1,*]

[1]Metallurgical and Materials Engineering Department, National Institute of Technology Rourkela, Rourkela 769008, Odisha, India

[2]Metallurgical and Materials Engineering Department, Veer Surendra Sai University of Technology Burla, Sambalpur 768018, Odisha, India

Corresponding author Email : snehanshu.pal@gmail.com

Abstract: In this paper, we present a lucid understanding about the deformation behaviour of nanocrystalline (NC) Ni with and without defects subjected to tensile followed by compressive loading using molecular dynamic (MD) simulations. The embedded atom method (EAM) potential have been incorporated in the simulation for three kinds of samples-i.e. for NC Ni (without any defect), porous NC Ni and NC Ni containing a centrally located void. All the three samples, which have been prepared by implementing the Voronoi method and using Atom Eye software, consist of 16 uniform grains. The total number of atoms present in NC Ni, porous NC Ni and NC Ni containing a void are 107021, 105968 and 107012 respectively. The stress-strain response of NC Ni under tensile followed by compressive loading are simulated at a high strain rate of 10^7 s^{-1} and at a constant temperature of 300K. The stress-strain curves for the NC Ni with and without defects have been plotted for three different types of loading: (a) tensile loading (b) compressive loading (c) forward tensile loading followed by reverse compressive loading. Prominent change in yield strength of the NC Ni is observed due to the introduction of defects. For tensile followed by compressive loading (during forward loading), the yield point for NC Ni with void is lesser than the yield point of NC Ni and porous NC Ni. The saw tooth shape or serration portion of the stress-strain curve is mainly due to three characteristic phenomena, dislocation generation and its movement, dislocation pile-up at the junctions, and dislocation annihilation. Both twins and stacking faults are observed due to plastic deformation as the deformation mechanism progresses. The dislocation density, number of clusters and number of vacancy of the NC sample with and without defects are plotted against the strain developed in the sample. It is seen that introduction of defects brings about change in mechanical properties of the NC Ni. The crystalline nature of NC Ni is found to decrease with introduction of defects. The findings of this work can thus be extended in bringing a whole new insight related to the deformation behaviour and properties of Nano-materials during cyclic deformation at various temperatures.

Keywords: Molecular dynamics; deformation behaviour; stress-strain curve; Nanocrystalline; Ni

1. Introduction

Nanocrystalline (NC) materials are those polycrystalline materials which have a grain size that is less than 100 nm in dimension [1, 2]. They have been one of the prime areas of research in the past few years for the scientific and technological community. Reducing the average grain size of material enhances its properties [3]. The NC materials are well known for their distinct electrical, magnetic, mechanical and optical properties as compared to their bulk counterparts [4]. NC Ni can be utilized for wear and corrosion resistant coating. The nano-scale grain boundaries (GB) and defects such as stacking faults, point defects and porosity present in NC material caused by severe plastic deformation are crucial for determining the mechanical properties of the material [5]. NC materials undergo plastic deformation through mechanisms that are distinctive of their coarse grained counterparts [6]. The NC Ni, having an average grain size of 20-30nm, has a suitable face centered cubic (FCC) model for studying the mechanism of plastic deformation [7]. In-situ TEM deformation reveal that the dislocations originate from the grain boundaries and spread to the neighbouring grain boundaries where they get absorbed. This is also observed in case of an in-situ X-ray diffraction study of NC Ni undergoing tensile test at room temperature [8]. Reversible peak broadening is observed, signifying that large occurrence of defects such as twins or stacking faults are not found [9]. The role of interface structure on deformation was observed by molecular dynamics (MD) simulations [10-12]. The mechanical behaviour of NC FCC is observed to alter if the grain growth is stress driven [13, 14]. The fraction of grain boundaries in NC is more due to less average grain size, and this causes hindrance in dislocation motion. The strength of NC materials is reduced if they contain imperfections. Imperfections like pores, flaws, voids and other impurities are not always recognised and often lead to data misinterpretation while computing the reports [15]. Modelling techniques are hence used to provide a wider and deeper understanding of the material [16]. MD simulation provides a link between theory and experiments and is a widely favoured technique for observing material property of NC material [17]. Elder et al. [18] proposed the phase field model that is considered as a bridge between atomistic simulation and analytical solutions [19]. It provides a way to analyse complex many body system where analytical solution doesn't apply and experimental techniques are difficult to apply or simply don't exist [17]. MD is a powerful tool that can provide detailed microstructural evolution during simulation process [20]. Kim et al. has reported that introduction of porosity in a material decreases the elastic modulus of the material [21]. Wu et al. has reported that in NC Ni, the mechanism of the plastic deformation behaviour varies from one grain to another and at times even from place to place inside the grain. The concentration of plastic deformation is more in the regions near the grain boundaries. The authors have also reported that their results exhibit the inhomogeneous nature of deformation pattern inside the NC grain [7]. Constant time dependent deformation studies have been done which state that creep deformation of porous nano crystalline Ni is notably governed by grain-boundary diffusion [27-30]. However, there has been no evidence of studies containing NC Ni with or without defect subjected to tensile followed by compressive loading till date. In this work, we are going to elucidate and present a lucid understanding of the influence of void and porosity on the deformation behaviour of NC Ni under tensile followed by compressive loading using MD simulation.

2. Computational Details

NC Ni samples consisting of 16 uniform gains, each of 4 nm average grain size have been prepared by Voronoi method [16] using Atom eye software [22] as shown in **Fig.1**. The samples have random orientation and are inside three dimensional boxes of size 10 nm × 10 nm × 10 nm. In order to prevent grain growth, randomly oriented uniform grains are taken. The total number of atoms in NC Ni is 107,021. The porous NC Ni with percentage of porosity around one percent has 105,968 atoms in it. The NC Ni having a void of ~0.3 nm diameter in the centre consists of 107,012 atoms. The energy minimization of the sample has been carried out at 300K employing conjugate gradient algorithm. The time step chosen for the simulation is 0.002 ps. Studies via MD simulation on NC Ni, porous NC Ni and NC Ni with void have been executed using Large Scale Atomic/Molecular Massively Parallel Simulator (LAMMPS). Open visualization tool software OVITO has been used to find the atomic positions and visualize the extracted dislocation. The EAM (embedded atom method) potential for pure Ni system is incorporated in the simulation [23].

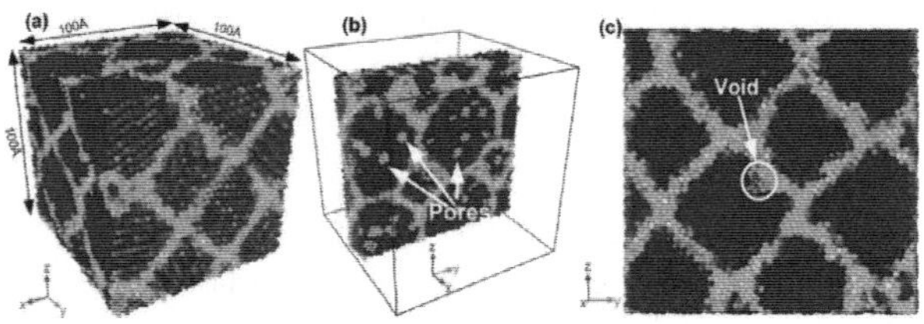

Fig. 1: Atomic snapshots of (a) NC Ni (b) porous NC Ni and (c) NC Ni with void.

3. Result and Discussion

The stress-strain curves for NC Ni, porous NC Ni and NC Ni containing a void for tensile, compressive and tensile followed by compressive loading are shown in **Fig.2 (a-c)**. In case of tensile loading, the yield point for NC Ni is ~1.9GPa after which the plastic zone is developed. The yield points of porous NC Ni and the one with void are 1.82 GPa and ~1.6 GPa respectively, i.e., lesser than that of NC Ni. The stress-strain ratio for the NC Ni with void in this curve shows the least value in the range of 0.09 to 0.2 strain. In case of compressive loading, the yield point of NC Ni is more than that of porous NC Ni and NC Ni with void. The stress-strain ratio for the porous NC Ni in this curve shows the most value in the range of 0.07-0.16 strain and the least value for NC Ni with void is in between the ranges of 0.09-0.11 strain and 0.13-0.2 strain. **Fig.2 (c)** shows the stress-strain curves for NC Ni, porous NC Ni and NC Ni with a void for tensile followed by compressive loading after a pre-strain of 0.20. During forward loading, the yield point for NC Ni with void is smaller than that of the other two samples. However in case of reverse loading, the yield point of NC Ni is more than that of porous NC Ni, which in turn is more than the yield point of NC Ni with void. The serration of the stress-strain curve is governed by three essential phenomena: generation of dislocation and its movement, dislocation pile-up at the junctions, and dislocation annihilation. It is observed that the Ultimate Tensile Strength (UTS) of NC Ni is more than that of porous NC Ni and NC Ni with void.

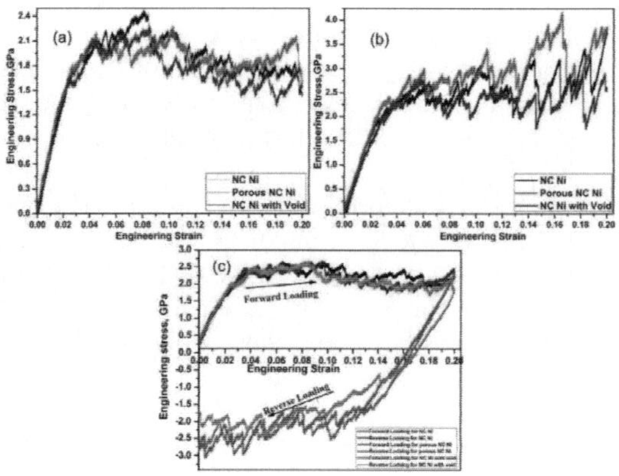

Fig.2: Stress-strain curves for NC Ni, porous NC Ni and NC Ni with void for (a) Tensile loading (b) Compressive loading (c) Tensile followed by Compressive loading.

The dislocation pile-up starts from the yield point hindering the movement of dislocations temporarily. Once the yield stress is reached, gliding of dislocation starts to occur and eventually the dislocations annihilate. This causes the stress to drop suddenly after the yield point. The reoccurrence of the process is observed upon subjecting the sample to further loading (either tensile or compressive). It can also be seen that the dislocation density reaches a maximum when dislocation pileup occurs (further explained in the dislocation density vs. strain curve section). These frequent fluctuations in the stress-strain curves are directly linked to strain hardening and softening of the material. The atomic configuration snapshots coloured according to common neighbourhood analysis (CNA) values for various position of deformation path for tensile followed by compressive

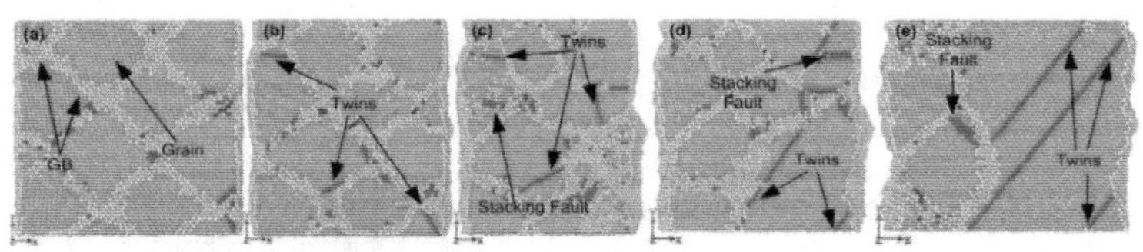

Fig.3: Atomic snapshot for forward followed by reverse loading of NC Ni at strain rates of (a) 0.0002 (b) 0.1 (c) 0.2 (d) 0.1 after a pre-strain of 0.2 (e) 0 after a pre-strain of 0.2

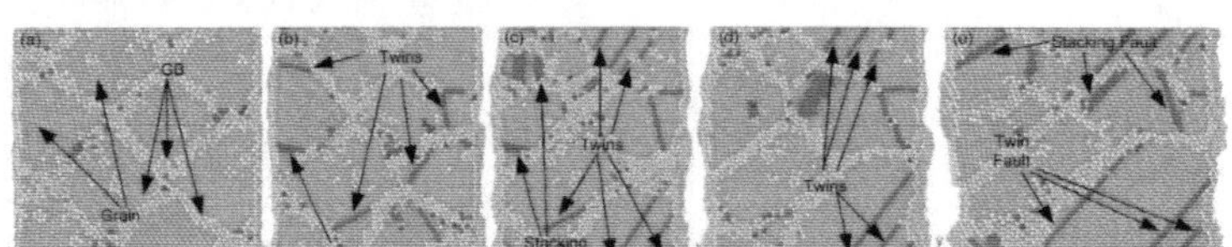

Fig.4: Atomic snapshot for forward followed by reverse loading of porous NC Ni at strain rates of (a) 0.0002 (b) 0.1 (c) 0.2 (d) 0.1 after a pre-strain of 0.2 (e) 0 after a pre-strain of 0.2

loading are taken and presented in **Fig. (3-5)**. Twins and stacking faults are not observed until a strain of 0.1 is reached. During forward loading, production of twins induces production of serrations in the stress-strain curve [24]. This fact is supported by **Fig.2 (c)** as we can see that serrations are present at a strain of 0.1. Stacking fault can also be understood as sub microscopic twins with thickness which is almost equal to atomic thickness [24]. However we can observe both twins and stacking fault for strains of 0.1 and 0.2, this is because stacking fault in most metals is produced by plastic deformation [24]. Twins play a major role in plastic deformation of materials. They are also responsible for varied mechanical behaviour of materials [5]. The NC FCC metals are deformed more easily by twinning. This is especially true for metals like Ni which have high stacking fault energy [24, 25]. From **Fig. (3-5)** it can be observed that at strain of 0.2 during forward loading, the number of dislocations for porous NC Ni is more than that of NC Ni with void which in turn is greater than that of NC Ni. It happens so because dislocation cores present in the sample act as a potential source of dislocations. Parallel twins are found on the surface of NC Ni. These formed twins glide across the sample as deformation progresses. FCC metals like Ni contain deformation twins which are usually formed due to the glide of partial dislocations [26]. As the applied stress becomes more than the critical resolved shear stress, gliding occurs and subsequently, dislocation annihilation is observed. It is because of this that a drop in the applied stress is observed after the yield point.

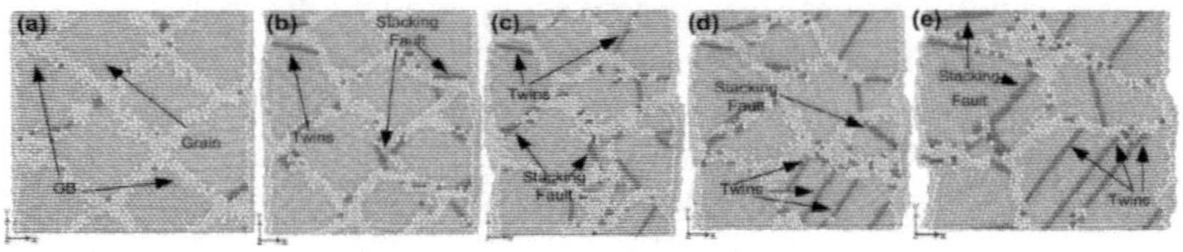

Fig.5: Atomic snapshot for forward followed by reverse loading of NC Ni with void at strain rates of (a) 0.0002 (b) 0.1 (c) 0.2 (d) 0.1 after a pre-strain of 0.2 (e) 0 after a pre-strain of 0.2

The calculated dislocation density with the progression of deformation subjected to tensile, compressive, and tensile followed by compressive loading conditions have been plotted in **Fig. 6**. It is observed in **Fig. 6 (a)** that more dislocations are produced in NC Ni as compared to porous NC Ni and NC Ni with void when subjected to tensile loading. In **Fig.6 (b)**, the general trend of the plot is an increase followed by a decrease, as can be seen for the cases of NC Ni and NC Ni with void present in it. However the dislocation density for porous NC Ni continuously increases with increase in strain. The general trend for forward loading in **Fig.6 (c)** as can be seen for the cases of NC Ni and porous NC Ni is an initial increase and then decrease of dislocation density with increasing strain. Contrary to this, the dislocation density for forward loading of NC Ni with void continues to increase along with strain. For the same plot, in case of reverse loading, the general trend for the dislocation density is to decrease and then increase, as can be seen in case of porous NC Ni and the one with void. It is observed from **Fig.6 (c)** that the dislocation density decreases on the initiation of reverse loading at a strain of 0.20 due to dislocation annihilation. During reverse loading the dislocation density of NC Ni decreases, while the dislocation density of porous NC Ni and NC Ni with void decreases and then increases. During reverse loading, the dislocation density of porous NC Ni increases after reaching a minimum at 0.16 strain and that of NC Ni with void starts increasing at 0.5 strain. These change in behaviour of dislocation density during tensile, compressive, and tensile followed by compressive loading can be attributed to introduction of voids and porosity in the samples.

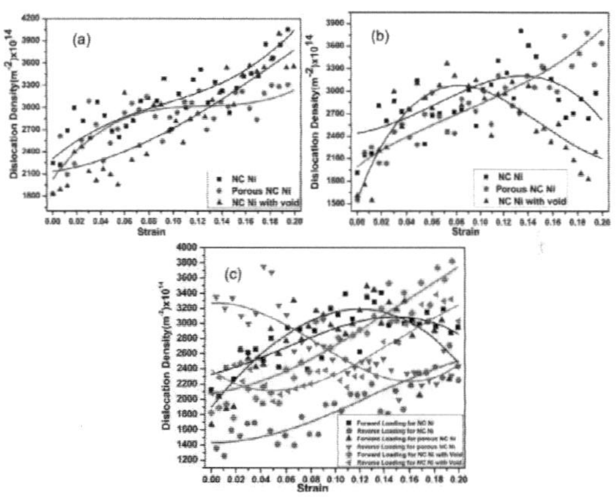

Fig.6: Dislocation density vs. strain curves for NC Ni, porous NC Ni and NC Ni with void for (a) Tensile loading (b) Compressive loading (c) Tensile followed by Compressive loading.

IOP Conf. Series: Materials Science and Engineering **338** (2018) 012028 doi:10.1088/1757-899X/338/1/012028

Cluster analysis has been conducted by taking a cut off radius of 2.5Å, since the first peak of radial distribution function plot is around 2.5 Å. The number of clusters vs. strain plots are given in **Fig.7**. The number of clusters along with strain for NC Ni, porous NC Ni, and NC Ni with void under tensile loading is given by the **Fig.7 (a)**. Initially the number of clusters increases with the strain but during 0.07-0.08 strain, number of clusters tend to decrease slightly. It is observed that there is an increase in cluster number when the strain is around 0.18. **Fig.7 (b)** gives the plot between the number of clusters and strain for NC Ni, porous NC Ni, and NC Ni with void during compressive loading. As observed in the plot on compressive loading, the number of clusters decreases with increasing strain. It can be inferred that decrease in crystallinity and cluster formation take place simultaneously. This can be supported by the number of clusters vs. strain plot for NC Ni, porous NC Ni and NC Ni with void under forward followed by reverse loading given in **Fig.7 (c)**. Both percentage amorphization and number of clusters are more in case of forward loading than in reverse loading. The number of vacancies vs. strain plots for NC Ni, porous NC Ni and NC Ni with a void for tensile, compressive and tensile followed by compressive loading are shown in **Fig.8**. Number of vacancies for all the three plots show similar behaviour. In **Fig.8 (c)**, it can be seen that vacancies generated during forward loading do not follow the same path during reverse loading. It can be inferred that vacancy is not just a function of stress and strain; it also involves some other parameters such as deformation. On comparing the plots of number of vacancy vs. strain for tensile loading and compressive loading, it may be observed that more number of vacancies is generated during compressive loading.

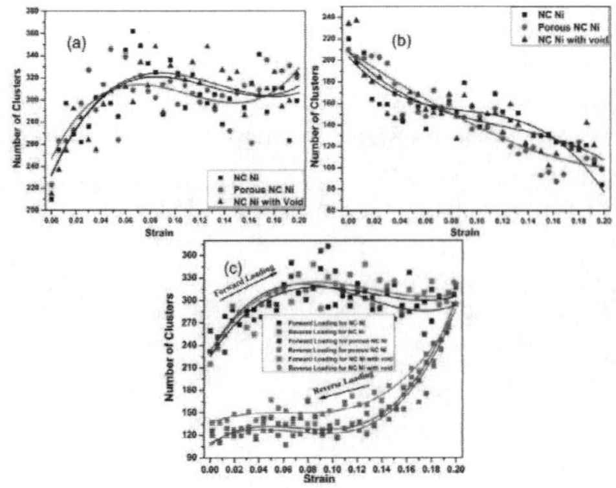

Fig. 7: Number of clusters vs. strain curves for NC Ni, porous NC Ni and NC Ni with void for (a) Tensile loading (b) Compressive loading (c) Tensile followed by Compressive loading.

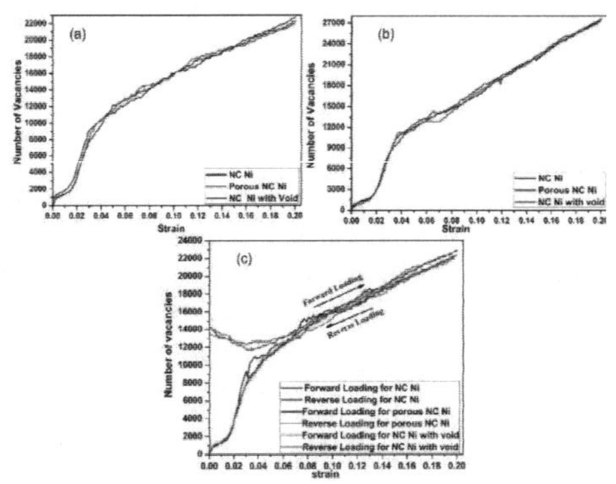

Fig. 8: Number of clusters vs. strain curves for NC Ni, porous NC Ni and NC Ni with void for (a) Tensile loading (b) Compressive loading (c) Tensile followed by Compressive loading.

4. Conclusion

MD simulations have been performed for NC Ni, porous NC Ni and NC Ni with void in order to study the deformation behaviour of NC Ni with and without defects under tensile followed by compressive loading. It is observed that the defects introduced decreases the yield strength of the NC Ni. Defects like stacking fault and twins are generated at the surface due to plastic deformation of the simulated samples. These change in behaviour of dislocation density during tensile, compressive, and tensile followed by compressive loading can be attributed to the introduction of voids and porosity in the samples. The number of clusters seems to be greater during forward loading. As for vacancies, compressive loading (reverse loading) condition contains the maximum number of vacancies. It is also seen that the crystalline nature of NC Ni decreases with introduction of defects. The summary of this investigation can be extended to bring perceptivity on the deformation behaviour and the properties of nano-materials during cyclic deformation at various temperatures.

Acknowledgements

The authors want to express their gratitude towards the computer centre of NIT Rourkela for providing the high performance computing (HPC) facility required to carry out this study.

References

[1] Chokshi AH, Rosen A, Karch J, Gleiter H, 1989, Scr. Metall., **23**, 1679
[2] Suryanarayana C, 1995, Int. Mater. Rev., **40**, 41
[3] Tucker GJ, Foiles SM, 2013, Mater. Sci. Eng.**571**, 207
[4] Meraj M, Yedla N, Pal S, 2016, Mater. Lett. , **169**, 265
[5] Paul SK, Kumar S, Tarafder S, 2017, Eng. Fract. Mech., **176**, 257
[6] Kumar KS, Van Swygenhoven H, Suresh S, 2003, Acta Mater. , **51**(19), 5743
[7] Wu X, Ma E, Zhu YT, 2007, J. Mater. Sci., **42**(5), 1427
[8] Brandstetter S, Budrovic Z, Van Petegem S, Schmitt B, Stergar E, Derlet PM, Van Swygenhoven H, 2005, Appl. Phys. Lett., **87**(23), 231910
[9] Budrovic Z, Van Swygenhoven H, Derlet PM, Van Petegem S, Schmitt B, 2004, Sci., **304**(5668), 273
[10] Cahn JW, Mishin Y, Suzuki A, 2006, Acta Mater. , **54**(19), 4953
[11] Horstemeyer MF, Baskes MI, Prantil VC, Philliber J, Vonderheide S, 2003, Modell. Simul. Mater. Sci. Eng., **11**(3), 265
[12] Tucker GJ, Zimmerman J, McDowell DL, 2009, Modell. Simul. Mater. Sci. Eng., **18**(1), 015002
[13] Gianola DS, Van Petegem S, Legros M, Brandstetter S, Van Swygenhoven H, Hemker KJ, 2006, Acta Mater., **54**(8), 2253
[14] Holm EA, Foiles SM, 2010, Sci., **328**(5982), 1138
[15] Sanders PG, Youngdahl CJ, Weertman JR, 1997, Mater. Sci. Eng., **234**, 77
[16] Chen D, 1995, Comput. Mater. Sci., **3**(3), 327
[17] Lyubartsev AP, Laaksonen A, 2000, Comput. Phys. Commun., **128**(3), 565
[18] Elder KR, Katakowski M, Haataja M, Grant M, 2002, Phys. Rev. Lett., **88**(24), 245701
[19] Yamanaka A, Takaki T, Tomita Y, 2008, Mater. Sci. Eng., **491**(1), 378
[20] Nie K, Wu WP, Zhang XL, Yang SM, 2017, J. Mater. Sci., **52**(4), 2180
[21] Kim HS, Bush MB, 1999, Nanostruct. Mater. , **11**(3), 361
[22] Li J, 2003, Modell. Simul. Mater. Sci. Eng., **11**(2), 173
[23] Mendelev MI, Kramer MJ, Hao SG, Ho KM, Wang CZ, 2012, Philos. Mag., **92**(35), 4454
[24] Dieter GE, Bacon DJ, 1986, Mechanical metallurgy, Vol. 3, New York: McGraw-Hill.
[25] Zhu YT, Liao XZ, Wu XL, 2012, Prog. Mater. Sci., **57**(1), 1
[26] Arzaghi M, Beausir B, Toth LS, 2009, Acta Mater. , **57**(8), 2440
[27] Meraj M, Pal S, 2017, Appl. Phys. A, **123**(2), 138
[28] Meraj M, Pal S, 2016, Trans. Indian Inst. Met., **69**(2), 277
[29] Meraj M, Pal S, 2017, Metall. Res. Technol., **114**(1), 107
[30] Pal S, Meraj M, Deng C, 2017, Comput. Mater. Sci., **126**, 382

IOP Conf. Series: Materials Science and Engineering **338** (2018) 012029 doi:10.1088/1757-899X/338/1/012029

Mechanical Properties and Wear Characteristics Al-ZrO2-SiCp and Graphite Hybrid Metal Matrix Composites

S.K.Nayak*[1], T.Mahanta[2], J.K.Sahoo[3] and A.Mishra[4]
[1,2,4]Department of Mechanical Engineering,
Indira Gandhi Institute of Technology, Sarang, Dhenkanal (ODISHA), INDIA-759146.
[3]Department of Metallurgical & Materials Engineering,
Indira Gandhi Institute of Technology, Sarang, Dhenkanal (ODISHA), INDIA-759146

Email: nayak.sandip200@gmail.com

Abstract: Development of Aluminum Metal Matrix Co mposites (AMMCs) has been one of the major requirements in engineering applicat ions due to their excellent mechanical properties, light weight and high strength. In the present investigation, Stir casting technique has been used for fabrication of co mposites, taking Alu miniu m as parent metal, Silicon Carbide (SiCp) of 7 vol. % of 220 mesh size and 1.75 vol. % of graphite as reinforcements. The Zirconia content was varied as 2.75, 4.5 and 6 vol. % to fabricate three d ifferent types of hybrid composites. The tensile strength and hardness were measured in UTM and Vickers hardness tester respectively and the wear characteristics were studied in a pin on disc friction monitor under dry sliding condition against steel counter face. The tensile strength was found to be 90 MPa, 120 MPa, 130 MPa and hardness 80.25 VHN, 103.22 VHN, 103.77 VHN for 2.75, 4.5 and 6vol. % of Zirconia respectively. Fro m the above investigation, it is recommended that composition with Al, 7 %-SiCp , 1.75 % -Gr and 6 vol %-ZrO2 showed better mechanical p roperties i.e . h igh tensile strength (130MPa) and reasonably good hardness (103.77 VHN) . The co mposite with Al, 7 % - SiCp, 1.75 % -Gr and 6 %-ZrO2 is good for short run frictional applicat ion and the composite with Al, 7 %- SiCp, 1.75 % -Gr and 4.5 %-ZrO2 may be used for long run frictional applicat ions after testing.

Key words: Composites of Al-SiCp-Gr-ZrO2, Strength, Hardness and Wear

1. Introduction

Available monolithic materials have limitations with respect to achievable combinations of strength, stiffness, and density. In order to overcome these shortages and to meet the increasing engineering demands of modern technology, metal matrix composites are gaining importance. The areas of applications of composite materials are growing rapidly and have even found new markets [1-3]. The current objective is to make them durable in tough conditions to replace other materials. Basically, Composites are the combination of two or more dissimilar materials (different in composition, characteristics and sometimes inform) which exist in different phases and are insoluble in each other. One or more discontinuous phases are, therefore, binded in a continuous phase to form a composite. Among the phases one is the matrix phase which contains the reinforcements in order to improve its own properties. The aluminium composites are widely used in the fields of Automotive Industry (Body, Brake pads, Drive shafts, Fuel tanks, Hoods, Spoiler, clutches), Aerospace (Nose, doors, struts, fairings, out board and inboard flaps etc), Sports (Tennis, Bicycles, Badminton), Constructions, marine industries and bio-medical applications etc.[4-6]. Therefore in the present investigation the aluminium matrix reinforced with different volume % of SiC_p, Graphite and Zirconia has been considered. These reinforcement materials have been chosen because of their better mechanical as well as tribological properties. Zirconia is one of the important ceramic which is used as a biomaterial that has a bright future because of its high mechanical strength and fracture toughness. Graphite and Silicon carbide have low densities. Silicon carbide has high strength, high hardness and excellent thermal shock resistance. Both graphite and Zirconia have high wear resistance also [7-8]. In this investigation the mechanical as well as wear behaviour of these hybrid composites are evaluated in order to distinguish the best material for suitable engineering applications

2. Experimental Details

In the present investigation Aluminium is considered as the matrix material and Zirconia, Silicon Carbide and Graphite were used as reinforcement materials. All the reinforcements used are in particulate form. Graphite and Silicon Carbide used are of 71 μm and Zirconia particles are of size 40μm at an average. The amount of reinforcement particle Silicon Carbide and Graphite are kept constant i.e. 7 and 1.75 vol. % respectively and the vol. % of Zirconia is varied as 2.75, 4.5 and 6 vol. % respectively.

Ingots of pure aluminium (900gms) are first melted in a muffle furnace in clay crucible at 850°C and after that the pre heated reinforcements are added to the liquid aluminium and the temperature is maintained for 20 minutes. The mixture is poured into a sand mould and left for 24hrs at room temperature to solidify in order to get the desired castings.

2.1 Characterization

2.1.1 Tensile Test

Tensile test samples (Fig.1) have been prepared by machining the cast samples to the desired sizes as per ASTM E8 standard. The tensile strength of the specimens was evaluated in the Universal Testing Machine (Fig-2) of model BSUT- 40-JD and having capacity 400kN.

2.1.2 Hardness test

For the hardness test three specimen are prepared (shown in fig.3) with varying vol% of ZrO_2 (2.75, 4.5 and 6 vol. %) and the test was conducted in a micro vicker's hardness testing machine by considering a load of 1 kgf and dwell time of 15 sec. in the test the The square based diamond pyramid was used as indenter. The hardness values were measured in three locations over the sample and the average length of the diagonals of the indentation were calculated. Then the Vickers Hardness Number (VHN) of the samples was obtained.

.

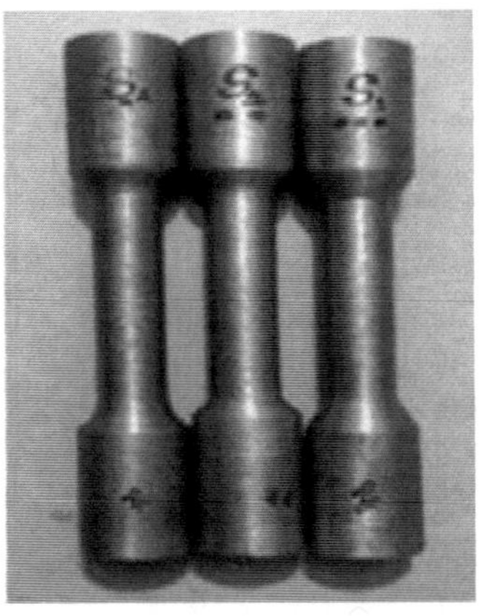

Fig.1 Standard specimens for tensile testing

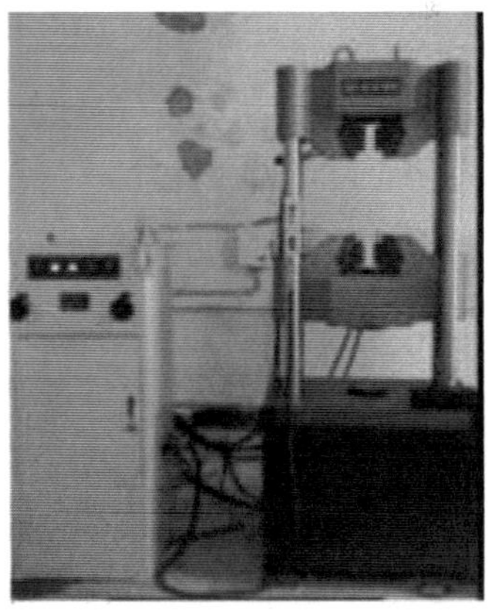

Fig.2UTM used for tensile testing

IOP Conf. Series: Materials Science and Engineering **338** (2018) 012029 doi:10.1088/1757-899X/338/1/012029

Fig. 3 Specimens used for hardness testing

2.1.3 Wear Testing

The wear test of the specimens was carried out in the dry sliding condition in a pin-on –disc tester (Fig.5).The short run wear tests were carried out to find out the co-efficient of friction and wear behaviour. The wear tests were conducted in two phases, in the first phase the pins (fig.4) were made to slide on the steel disc at 1, 1.5, 2 and 2.5 m/s respectively considering a constant load 5 Kg and constant time 4min and in the second phase the pins were made to slide on the pins for 1, 2, 3 and 4 minutes respectively considering a constant load 5 Kg and constant speed 2m/s.

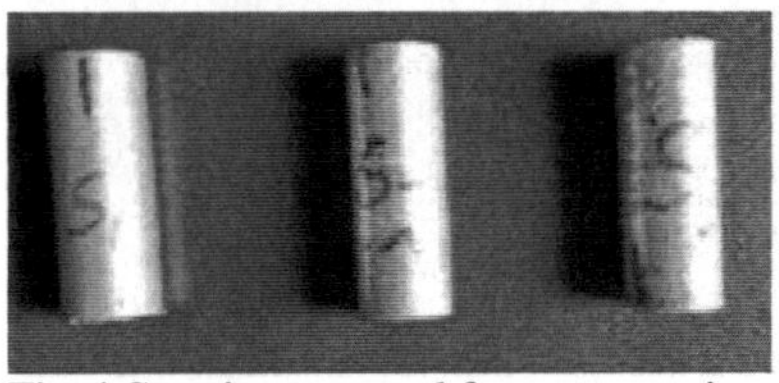

Fig.4 Specimens used for wear testing

Fig.5 Pin on Disc apparatus

For short run wear tests the pins of the composites were prepared by machining the cast samples to 10 mm in diameter and 32 mm length (Fig.4).

3. Results and discussion

3.1 Tensile Test

The results obtained from the tensile test were represented in the graphical form below (Fig.6). From the result it is clear that the composite consisting of 2.75 vol.% of Zirconia shows highest tensile strength (132.008 MPa) where as the composite with 4.5vol.% of Zirconia shows lower tensile strength (91.263MPa).

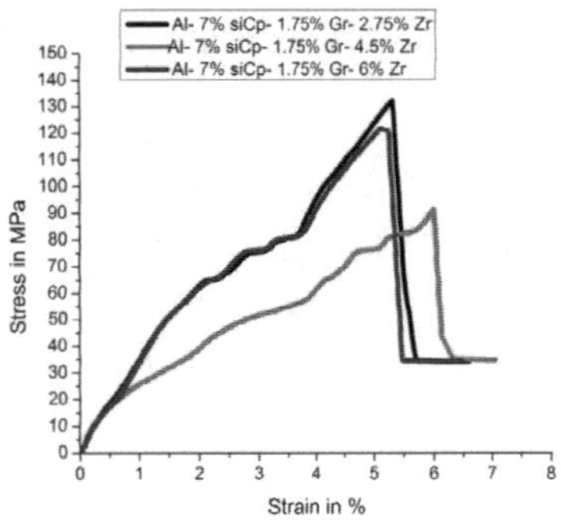

Fig- 6 Tensile Stress Strain curve of the composites

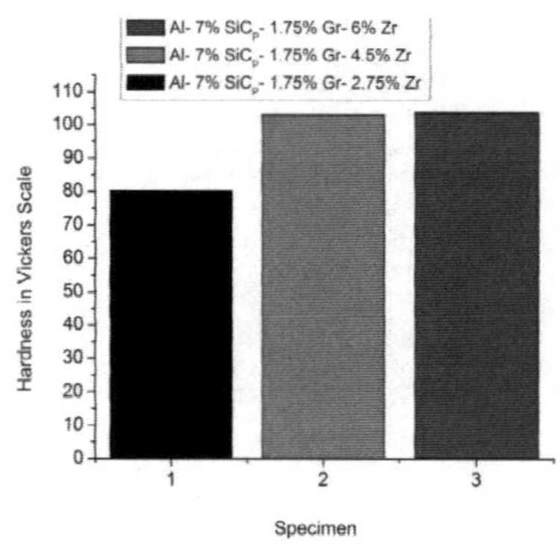

Fig.7 Hardness of the composites

IOP Conf. Series: Materials Science and Engineering **338** (2018) 012029 doi:10.1088/1757-899X/338/1/012029

3.2 Hardness test results

The hardness of the composites obtained from the micro Vickers hardness tester are represented in a bar graph as shown in Fig.7.From the results it is observed that the composite with 6 vol.% Zirconia is the hardest among the three and the hardness value is 103.77 and the composite with 2.75 vol.% Zirconium exhibits least hardness i.e 80.25 in the Vickers scale.

3.3 Analysis of wear test results

3.3.1 Co- efficient of friction (COF)

The C.O.F test of the composites is under the short run test category. In this test the composites are allowed to slide for a constant time period 1 minute with varying the load from 1 Kg to 4 Kg with an increment of 1. The coefficient of friction of the composites obtained at different load conditions are represented in Table 1

Table 1 Coefficient of friction of composites

Load(Kg)	Co- efficient of friction		
	Specimen-1 2.75% ZrO_2	Specimen-2 4.5% ZrO_2	Specimen-3 6% ZrO_2
1	0.67	0.385	0.49
2	0.48	0.32	0.524
3	0.38	0.28	0.53
4	0.26	0.279	0.47

The graphical representation of the above data is shown in Fig.8. From the obtained data it is seen that the C.O.F of specimen-1 and 2 decreases with the increase in load whereas the C.O.F of specimen-3 first increases and then decreases with the increase in load.

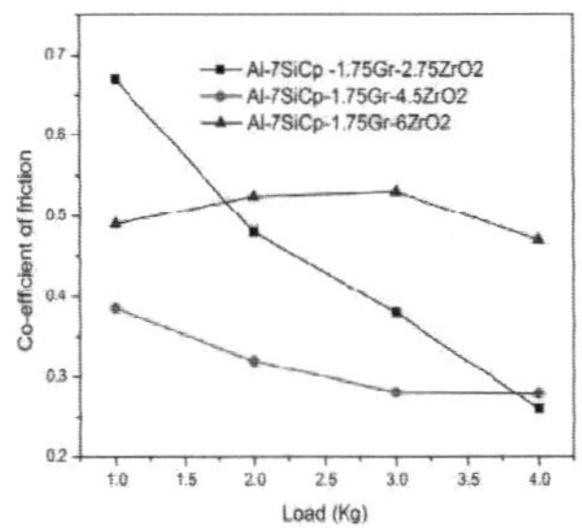

Fig.8 Co-efficient of friction vs. load

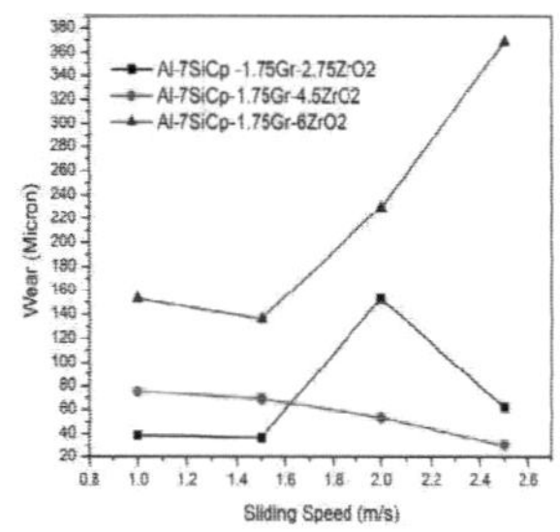

Fig.9 wear vs. sliding speed

3.3.2 Wear vs. Sliding speed

The test is carried out by changing the sliding speed of the composites. The other parameters like time (4 minute) and load (5 Kg) are kept constant. The different sliding speeds considered are 1, 1.5, 2 and 2.5 m/s respectively. The wear data obtained with respect to the sliding speed of the composites are represented in Table 2.

Tab.2 Wear of composites with respect to sliding speed

Slidin g spee d (m/s)	Wear (micron)		
	Specimen-1 2.75% ZrO_2	Specimen-2 4.5% ZrO_2	Specimen-3 6% ZrO_2
1	38.769	75.126	152.858
1.5	36.807	68.894	136.102
2	153.00 7	53.472	229.471
2.5	61.843	30.203	368.644

From the obtained data we can conclude that the wear of Specimen-1 and 3 first decreases and then increases with the increase in sliding speed where as the wear of specimen-2 decreases with the increase in sliding speed. However the wear of the specimen -2 is comparatively lesser than the other two which is shown in fig.9.

3.3.3 Wear vs. Time

The time of sliding of the specimen were varied from 1 minute to 4 minute. The other parameters like sliding speed (2m/s) and load (5Kg) are kept constant. The wear data obtained with respect to time of sliding of the composites are represented in Table 3.

Table 3. Wear of composite with respect to time

Time in minute	Wear (micron)		
	Specimen-1 2.75% ZrO_2	Specimen-2 4.5% ZrO_2	Specimen-3 6 %ZrO_2
1	13	24	17
2	15	18	28
3	17	23	35
4	34	28	48

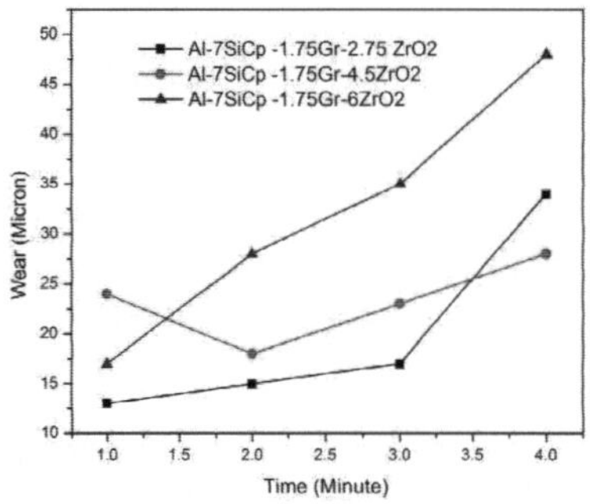

Fig.10 wear vs. time

From the above obtained data and the graph it is clear that in each case the wear increases with the increase in time, however the wear of specimen -3 is higher than the other two. Probably the wear will stabilize after running for a longer period.

4. Conclusions

From the present investigations the conclusions obtained are the followings:

1) The composite with 2.75 vol. % of Zirconia shows highest tensile strength (132.008 MPa) whereas the composite with 4.5vol. % of Zirconia shows lowest tensile strength (91.263MPa).
2) Addition of Zirconia to the composite increases the hardness of the composite. The highest hardness value obtained is 103.77 in Vickers scale which is obtained from the composite of 6 vol. % Zirconia.
3) The specimen-3 shows better C.O.F in all load conditions for which this may be used for frictional applications more effectively than the others.
4) Specimen-3 also exhibits highest wear with the increase in both the sliding speed as well as the time whereas the wear of specimen-2 changes within a certain range without fluctuating. Thus the specimen-2 can be proposed as a better composite than the other two for short run wear applications.

References

1. Scudino,S., Liu,G., Prashanth,K.G., Bartusch,B., Surreddi,K.B., Murty,B.S. and Eckert,J. "Mechanical properties of Al-based metal matrix composites reinforced with ZrO_2 based glassy particles" Acta Materialia Vol.57, 2009 pp.2029-2039.
2. Wang,J., Li,z. Fan,G., Pan,H., Chen,Z. and Zhang,D. "Reinforcement with graphene nanosheets in Aluminium matrix composites" Acta Materalia Inc. Published by Elsevier, Vol. 66, Issue 8, April 2012, pp.594-597.
3. Rana,R.S., Purohit,R. and Das,S. "Influences of alloying elements on the microstructure and mechanical properties of Aluminium alloys and Aluminium alloy composites" International Journal of Scientific and Research Publications, Vol. 2, Issue 6, June 2012,pp.1-7.
4. Sucitharan,S.K. , Senthilkumar,P., Shivalingappa,D. and Rino,J.J. "Wear behaviour of Al6063-Zircon sand metal matrix composite" IOSR Journal of Engineering ,Vol.3, Issue2,2013,pp.24-28.
5. Sudindra,S., Kumar,A.C. "Studies on Al6061/Al_2O_3 and Graphite hybrid metal matrix composites" International Journal of Metallurgical & Materials Science and Engineering, Vol. 3(3),2013, pp.35-42.
6. Bodunrin,M.O., Alaneme,K.K, and Chown,L.H. "Aluminium matrix hybrid composites: a review of reinforcement philosophies; mechanical, corrosion and tribological characteristics" Journal Of Materials Research And Technology vol. 4, 2015, pp.434-445.
7. Mariyappan,M., Sarangapani,P. ,Perumal,G. ,Sathiyamoorthi, A. "Improvement in tribological behaviour of Aluminium 356 hybrid metal matrix composites" Journal of Chemical and Pharmaceutical Sciences Special Issue 9: April 2015, pp.98-102.
8. Mishra, A., Nayak, S.K. "Mechanical behaviour and wear characteristics of Al-Cu- SiC_P hybrid composites", SSRG International Journal of Mechanical Engineering, Vol.4, issue-5, 2017, pp.39-46.

IOP Conf. Series: Materials Science and Engineering **338** (2018) 012030 doi:10.1088/1757-899X/338/1/012030

Numerical simulation of residual stress in laser based additive manufacturing process

Bibhu Kalyan Panda and Seshadev Sahoo*
Department of Mechanical Engineering, Faculty of Engineering and Technology, Siksha 'O' Anusandhan, Bhubaneswar, Odisha, India

Email: seshadevsahoo@soa.ac.in

Abstract: Minimizing the residual stress build-up in metal-based additive manufacturing plays a pivotal role in selecting a particular material and technique for making an industrial part. In beam-based additive manufacturing, although a great deal of effort has been made to minimize the residual stresses, it is still elusive how to do so by simply optimizing the processing parameters, such as beam size, beam power, and scan speed. Amid different types of additive manufacturing processes, Direct Metal Laser Sintering (DMLS) process uses a high-power laser to melt and sinter layers of metal powder. The rapid solidification and heat transfer on powder bed endows a high cooling rate which leads to the build-up of residual stresses, that will affect the mechanical properties of the build parts. In the present work, the authors develop a numerical thermo-mechanical model for the measurement of residual stress in the AlSi10Mg build samples by using finite element method. Transient temperature distribution in the powder bed was assessed using the coupled thermal to structural model. Subsequently, the residual stresses were estimated with varying laser power. From the simulation result, it found that the melt pool dimensions increase with increasing the laser power and the magnitude of residual stresses in the built part increases.

1. Introduction

Direct metal laser sintering (DMLS) is a laser based additive manufacturing technology received much attention nowadays due to its ability to produce complex shapes directly from the powder materials by using computer aided design (CAD) data. It creates metallic build parts through the layer-by-layer addition of materials. The key benefits of this process are freedom of design of any shape, minimal of material wastage, elimination of production steps and less expense over conventional manufacturing processes [1-3]. In this process a high intensity laser beam interacts with the powder layer which creates melt pool on a layer of spread powder. The melt pool rapidly solidifies and then another layer of powders is spread on top of the solidified layer and the process continues to add a new layer successively to build the parts [4]. The powder layer in DMLS process undergoes rapid melting and solidification during sintering, so a large thermal gradient will develop in the part which leads to build up residual stresses. The resulting residual stress influences the material properties and the distortion of the final product such that its geometry deviates from the required build dimensions and degrades the mechanical strength of the build part. For these reasons, tremendous efforts have been placed to understand and control residual stress formation during additive manufacturing.

The physical process associated with direct metal laser sintering process includes heat transfer and sintering of powders. Recently, many investigations have been carried out to develop computer models for the laser sintering process. Labudovic et al. [5] developed a transient thermal model for direct laser metal powder deposition process by using ANSYS. The author investigates the temperature profiles, fusion zone dimension and residual stresses in the powder bed from the simulated model and found that, the residual stress increases as the number of layer increase. Temperature distribution and residual stresses during laser sintering of hot-work tool steel powders with respect to laser power and scanning rate were investigated by Ibraheem et al. [6] using finite element approach. Mercelis and Kruth [7] investigated theoretically and experimentally the residual stress in selective laser melting and selective laser sintering. From their investigation they found that, high value of stress exists in the built part which connected to the substrate and the parts that are removed from the base plate contain much lower stress levels. Also, they found that built part suffered deformation during removal from the substrate.

Thermal behavior of the molten pool during selective laser melting of TiC/AlSi10Mg nanocomposites was simulated by Yuan and Gu [8]. From their simulation results, they found that, the convective heat transfer and dimension of the molten pool was significantly affected by Marangoni convection. A Finite Element Method (FEM) calculation of single-scan electron beam melting of Ti6Al4V was carried out by Vastola et al. [9] to understand how residual stresses are related to the process parameters, including beam size, beam power, scan speed, and bed preheating temperature. They found that the powder bed preheating temperature have the largest quantitative impact on the residual stress. Zhao et al. [10] developed a finite element model for predicting the thermo-mechanical response of Ti-6Al-4V built part using a laser sintering system by using COMSOL platform. From the simulation results, the authors studied the behavior of the melt pool size, temperature history and change of the residual stresses on the powder bed with the change of laser power and scan speed. The result of the simulation provides a better understanding of the complex thermo-mechanical mechanisms of laser sintering additive manufacturing processes. In the present work, a thermo-mechanical model is developed for the investigation of residual stress in the AlSi10Mg build parts during direct metal laser sintering. The developed model was used to estimate the transient temperature dispersal in the powder bed. Subsequently, the residual stresses were investigated at different laser power.

2. Numerical Modelling

The numerical method used to investigate the residual stress in DMLS process is a coupled thermo-mechanical transient model. In this coupled model, thermal analysis is carried out first, and the results of thermal analysis are used as a thermal load input for the stress analysis. The governing equation for heat transfer analysis is

$$\rho C\frac{\partial T}{\partial t} = \frac{\partial T}{\partial x}\left(K\frac{\partial T}{\partial x}\right) + \frac{\partial}{\partial y}\left(K\frac{\partial T}{\partial y}\right) + \frac{\partial}{\partial z}\left(K\frac{\partial T}{\partial z}\right) + Q \quad (1)$$

Where ρ is the material density (kg/m3);

C is the specific heat capacity (J/kg K);

T is the temperature;

t is the interaction time;

K is thermal conductivity (W/mK);

Q = (x, y, z, t) is the volumetric heat generation (W/m^3).

Heat transfer due to convection and radiation is calculated by Newton's law of cooling and Stefan-Boltzmann law which are given below

$$q=h(T-T_0) \quad (2)$$

$$q=\sigma\epsilon(T^4 - {T_0}^4) \quad (3)$$

Where h is the convection heat transfer coefficient (W/m^2K), T_0 is the ambient temperature (300K), σ is the Stefan Boltzmann constant (5.670×10^{-8} W/m^2K^4) and ϵ is the emissivity.

In the present model, the incident laser energy is considered as a heat flux boundary condition for the heat transfer analysis, in which the moving laser power source considered as the Gaussian heat source model. Based on Gaussian heat source model, the input heat flux is calculated by using the correlation [11]

$$q=\frac{2AP}{\pi\omega^2}e^{\left(-\frac{2r^2}{\omega^2}\right)} \quad (4)$$

Where A is the laser absorptance of the powder system, P is the laser power, ω represents radius of the Gaussian laser beam, and r is the distance of a point on the surface of the powder bed measured from the laser beam center.

To calculate the distribution of stress, the elastic finite element analysis simulation is used. The governing equation for stress is

$$\mathrm{div}\sigma = 0 \quad (5)$$

Whereσ is the stress. As the problem is elasto-plastic and involves thermal strains. Thermal strain can be represented as

$$\varepsilon^{th} = \alpha_e \Delta T = \alpha_e (T - T_{ref}) \quad (6)$$

where, T_{ref} is the reference temperature.
So, effective stress can be written as

$$\sigma_{eff} = \sqrt{\sigma_1{}^2 + \sigma_2{}^2 + \sigma_3{}^2} + 2\upsilon(\sigma_1\sigma_2 + \sigma_1\sigma_3 + \sigma_2\sigma_3) \quad (7)$$

Where σ_1, σ_2 and σ_3 are the three principal stresses, υ= poission's ratio
Initially, the powder bed layer is assumed to be at room temperature i.e., T (x, y, z, t) = To (x, y, z) = 300 K, at t=0. Also, the input laser energy from the moving laser beam at initial stage is assumed to be zero. When the sintering process starts, the rate of absorption of laser energy by the powder material also increases. Thus, heat flux is generated, and heat transfer takes place due to the temperature difference. From each surface of the powder bed, heat loss takes place by convection and radiation mode. But the bottom of the surface is considered as adiabatic. There is no loss of heat from the bottom surface of the substrate.

3. Simulation Method

The governing equations for heat transfer and mechanical analysis were discretized and solved by finite element method using ANSYS 17.0. The dimension of the computational domain is 10×1.5×1 mm^3 for powder bed and for substrate is 12.5×3×2 mm^3. Figure 1 shows the schematic diagram of the process and Figure 1 shows the computational domain used for simulation. Tetrahedral mesh structure is obtained for powder bed with fine meshing and hexahedral mesh for steel substrate is obtained with medium mesh. The total number of elements and nodes for transient thermal modelling are 20535 and 47500, respectively. Similarly, for stress simulation medium mesh are applied to both powder bed and substrate and total number of elements and nodes are 3850 and 19219, respectively.

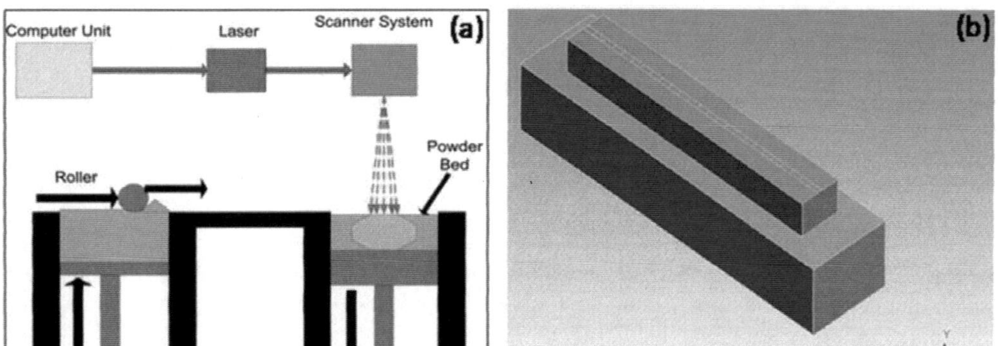

Figure 1. (a) Schematic diagram of the DMLS process (b) Computational domain used for simulation

The powder bed is considered as AlSi10Mg alloy and the substrate is made up of structural steel. The thermos-physical properties of AlSi10Mg and structural steel are given in Table 1 and Table 2.

Table 1. Thermo-physical properties of AiSi10Mg [12]

parameter	values
Laser absorptivity	0.9
Thermal conductivity (W/mK)	185
Density (Kg/m^3)	2650
Specific heat (J/Kg°C)	910
Young's modulus (MPa)	71000
Poisson's ratio	0.33

Table 2. Thermo-physical properties of structural steel [13]

Properties of steel	values
Thermal conductivity (W/mK)	60.5
Density (Kg/m^3)	7850
Specific heat (J/Kg°C)	434

4. Results and Discussions

In the present investigation, a transient thermos-mechanical model has developed based on finite element approach. The model used to investigate the residual stress on the built parts during direct metal laser sintering of AlSi10Mg alloy. The model considers conduction, convection and radiation mode of heat transfer in the powder bed as well as in the substrate. First thermal simulations are carried out and the results from the thermal simulation can be considered as the input for stress analysis. Simulations are carried out for sintering of AlSi10Mg powders in DMLS process by varying laser power from 70 watts to 100 watts, when the scan speed is kept constant i.e., 100 mm/s.

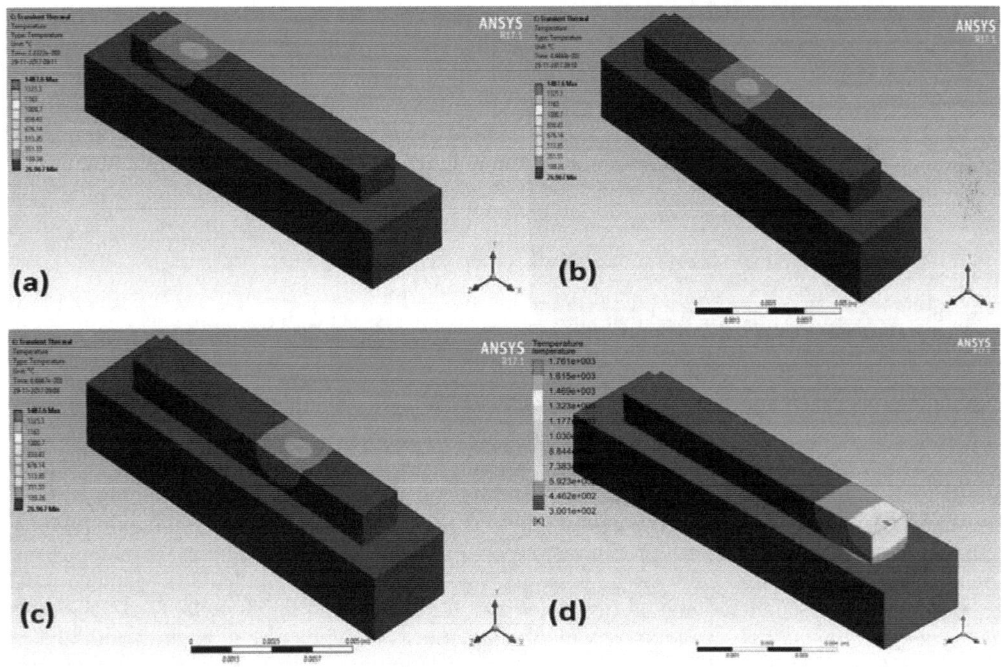

Figure 2. Thermal profile of AlSi10Mg built part at different time step (a) time=0.022 sec (b) time=0.044 sec (c) time=0.066 sec and (d) time=0.1 sec.

Figure 2 shows the thermal profile of AlSi10Mg built part at different time step, when the laser power is 100 watts and the scan speed is 100 mm/s. The temperature distribution in the powder bed is distinguished by color contour. In the thermal profile, the temperature contour of 830 K defines the solid-liquid interface. The temperature is higher than the melting temperature in the melt pool region than in the other regions. The melt pool dimensions are quite small and slightly larger than the laser spot size. As the DMLS process is a rapid manufacturing process, melting and solidification will occur within few seconds. So, the powders will melt and fuse rapidly, then cool down and reheat again when the laser beam passes on top of them again. Due to rapid heating and cooling, residual stresses are expected to develop in the fused part after it cools down to room temperature which is shown in Figure 3.

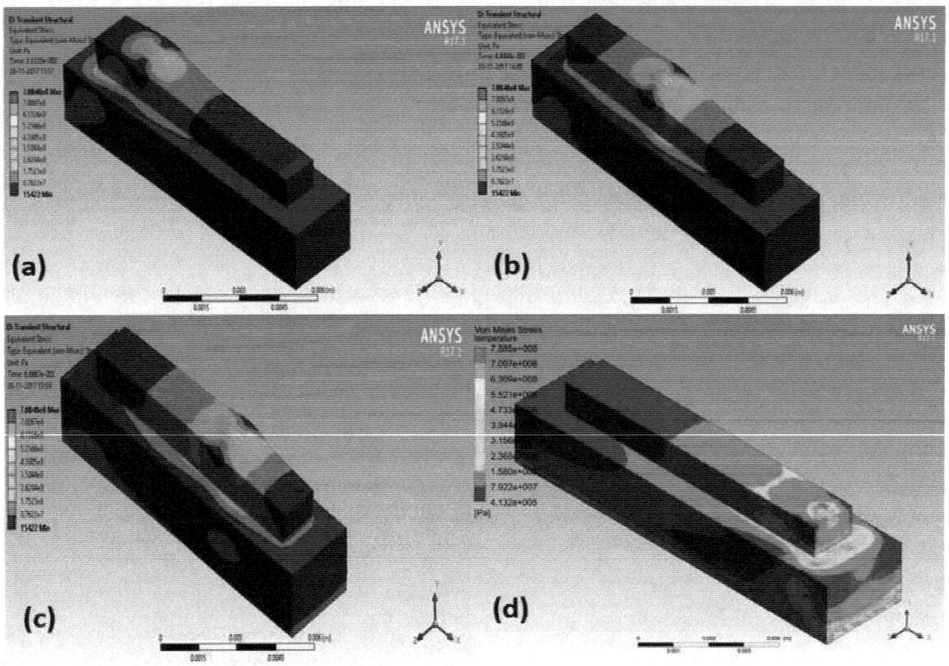

Figure 3. Residual stress profile at different time step (a) time = 0.022 sec (b) time = 0.044 sec(c) time = 0.066 sec and (d) time = 0.1 sec.

From the stress profile it is observed that, the magnitude of stress is maximum at the melt pool region. Also, there is some deformation occurs at the maximum stress region due to rapid cooling of built part. As the applied laser power increases the stress also increases. After 0.1sec it was observed that the stress at the start point of the powder layer is minimum and maximum stress only remains at the end position. It is because when the laser spot comes from 1st position to 2nd position, the temperature at the 2nd position is quiet high as compared to the 1st position so the 1st position has already solidified. This is because due to the temperature gradient from 2nd position the accumulated stress gets released from 1st position.

Figure 4 shows the graph between laser power and residual stress. It is observed that with increase in laser power, the magnitude of residual stress increases. As the laser power increases from 70 watts to 100 watts, the temperature in the molten pool increases. So, a large temperature gradient generates on the surface of the powder bed. Due to rapid melting and solidification, the amount of residual stress accumulation increases with increase in laser power.

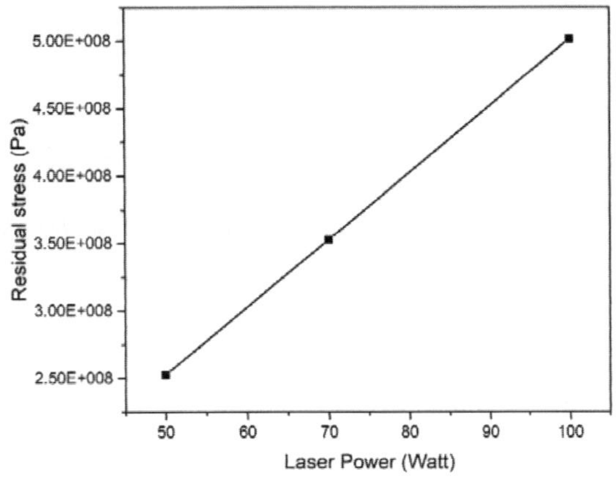

Figure 4. Residual stress vs. laser power

5. Conclusion

A finite element model is built for predicting the thermo-mechanical response of a AlSi10Mg alloy-built part manufactured with a direct metal laser sintering process. ANSYS 17.0 is used to model the coupled heat transfer and mechanical analysis. The thermal behavior, melt pool and residual stress are modelled using a 3-D thermo-mechanical analysis. From the simulation result, it is found that with increase in laser power the melt pool dimension increases. Also, the magnitude of residual stress in the built part increases significantly.

Reference

[1] Tuck C and Hague R. 2006, *International Journal of Mass Customisation.* **1** 360.
[2] Kruth JP, Wang X, Laoui T and Froyen L. 2003, *Assembly Automation.* **23** 357.
[3] Paul R and Anand Sam. 2012, *Journal of Manufacturing System.* **31** 429.
[4] Fogagnolo JB, Velasco F, Robert MH and Torralba JM. 2003, *Mater. Sci. Eng. A.* **342** 131.
[5] Labudovic M, D Hu and R kovacevic. 2003, *Journal of Materials Science.* **38** 35.
[6] Ameer KI, Brian D and Phillip JW. 2003, *Mat. Res. Soc. Symp. Proc.* **758**.
[7] Peter M and Kruth JP. 2006. *Rapid Prototyping Journal.* **12** 254.
[8] Pengpeng Y and Dongdong G. 2015, *J. Phys. D: Appl. Phys.* **48** 035.
[9] Vastola G, Zhang G, Pei QX and Zhang YW. 2016, *Additive Manufacturing.* **12** 231.
[10] Xinran Z, Iyer A, Promoppatum P and Yao CS. 2017, *Additive Manufacturing.* **14** 126.
[11] Roberts IA and Wang CJ. 2009, *International Journal of Machine Tools and Manufacture.* **49** 916.
[12] Nandy J, Sarangi H and Sahoo S. 2017, *IOP Conf. Series: Materials Science and Engineering.* **178** 1.
[13] Bogaard RH, Desai PD, Li HH and Ho CY. 1993, *Acta Thermochimica.* **218** 373.

IOP Conf. Series: Materials Science and Engineering **338** (2018) 012031 doi:10.1088/1757-899X/338/1/012031

Characterization of the Acoustic Radiation Properties of Laminated and Sandwich Composite Panels in Thermal Environment

Nitin Sharma[1], Trupti Ranjan Mahapatra[2,*], Subrata Kumar Panda[3] and Pruthwiraj Sahu[5]

[1,4]KIIT (Deemed to be University), Bhubaneswar: 751024, Odisha, India.
[2]Department of Production Engineering, Veer Surendra Sai University of Technology (VSSUT), Burla: 768018, Odisha, India.
[3]Department of Mechanical Engineering, NIT, Rourkela: 769008, Odisha, India

*E-mail: trmahapatra_pe@vssut.ac.in

Abstract. In this article, the acoustic radiation characteristics of laminated and sandwich composite spherical panels subjected to harmonic point excitation under thermal environment are investigated. The finite element (FE) simulation model of the vibrating panel structure is developed in ANSYS using ANSYS parametric design language (APDL) code. Initially, the critical buckling temperatures of the considered structures are obtained and the temperature loads are assorted accordingly. Then, the modal analysis of the thermally stressed panels is performed and the thermo-elastic free vibration responses so obtained are validated with the benchmark solutions. Subsequently, an indirect boundary element (BE) method is utilized to conduct a coupled FE-BE analysis to compute the sound radiation properties of panel structure. The agreement of the present sound power responses with the existing results available in the published literature establishes the validity of the proposed scheme. Finally, the current standardised scheme is extended to solve several numerical examples to bring out the influence of various parameters on the thermo-acoustic characteristics of laminated composite panels.

1. Introduction

Laminated and sandwich composite structures find extensive applications in high performance engineering industries largely owing to superior dynamic characteristics and higher-strength/lighter-weight properties in comparison to the conventional materials. When subjected to mechanical loading, the structures generate sound that propagates in to the surrounding medium and interacts with the other structures in its path and thus acts like an external load on those structures. For thin structures, consideration of such acoustic loading is a vital design parameter. Also, in real-time applications, the structures are often exposed to hostile environment. The acoustic radiation emitted by the structures is consequently affected by factors like temperature, moisture, dynamic loading and at times all of these combined together. Therefore, it is imperative to study and analyse the acoustic radiation characteristics of laminated and sandwich composite structures subjected to complex loading. The numerical prediction of the acoustic responses becomes more critical when the structures are curved (which is usually the case). In this regard, several mathematical models (theories) have been used to model to the mid-plane kinematics of the structures. However, the first-order shear deformation theory (FSDT) and the higher-order shear deformation theory (HSDT) are commonly used due their ease of implementation alongside the reasonable accuracy [1-3]. The acoustic radiation characteristics of isotropic and laminated composite structures with baffled and un-baffled configurations have been the subject of intense investigation in the past. Conventionally, the vibration behaviour is obtained using finite element method (FEM) and the free vibration responses are then utilized to obtain the acoustic responses by performing vibroacoustic analysis using coupled finite and boundary element (FEM-BEM) methods [4,5]. Acoustic radiation responses of layered composite flat panels subjected to thermal loading [6] and hygroscopic loading [7] have been studied in past. Further, the commercially available software tools have been exhaustively utilized to study the acoustic radiation emitted by vibrating isotropic flat panels [8], laminated composite flat panels [9,10] in ambient as well as elevated thermal environment. Further, several researchers have employed a coupled FE-BE technique to study the acoustic responses of cylindrical shells [11] in elevated thermal environment based on the thermo-elasticity theory [12] and the FSDT [13]. In addition, the acoustic radiation responses from orthotropic sandwich panels have been studied using the equivalent non-classical theory [14] and piecewise low order shear deformation theory [15]. The viscoelastic sandwich flat panels subjected to transient harmonic excitation have also been analysed [16] for their vibroacoustic responses under the influence of elevated thermal loading.

From this brief review of the literature, it is very much evident that the studies reporting the sound radiation characteristics of doubly curved panels under the influence of elevated thermal environment are scarce. Therefore, this article focuses on the acoustic responses of unbaffled laminated and sandwich composite spherical shell panels acted upon by concentrated harmonic load in an elevated temperature environment. The structural model of the panels is developed in ANSYS using ANSYS parametric design language code (APDL). The addition of temperature load generates thermal stresses in the panels and modal analysis is performed on this pre-stressed state. The mode shapes and the modal frequencies obtained from the modal analysis are then transferred to LMS Virtual Lab commercial package and an indirect BEM is used to compute the acoustic radiation characteristics of the panels. The effect of elevated temperature load on the natural frequency and the radiated sound power, the influence of varying elastic moduli on the radiation efficiency and the influence of lamination scheme and the core-to-face thickness ratio on the radiated sound power radiated of the laminated composite and sandwich composite spherical shell panels in an elevated thermal environment is investigated and the useful inferences are deliberated in detail.

2. Mathematical formulation

Figure 1 illustrates the stacking sequence and geometry of the orthotropic layered spherical shell panels analyzed in this study. A simulation model has been developed in ANSYS environment using the ANSYS Parametric Design Language (APDL) code to compute the free vibration responses of the laminated composite and sandwich composite spherical shell panels. For discretization purpose, the Shell281 element from the ANSYS element repository has been chosen. This element is eight-noded with a total 48 degrees of freedom. The FSDT based kinematic model for the spherical shell panel is defined and conceded as:

$$[u \quad v \quad w]^T = [u_0 + z\theta_x \quad v_0 + z\theta_y \quad w_0 + z\theta_z]^T \tag{1}$$

where, u, v and w are the displacements of any point on k^{th} layer at time t along the x, y and z coordinate axes, respectively; u_0, v_0 and w_0 represent the displacements of a point lying on the mid-plane; θ_x and θ_y depict the rotations of mid-surface normal w.r.t the y and x-axes, respectively. The variation of displacement is assumed to be linear along the z direction and is accounted by θ_z. The modal analysis is performed using this simulation model and the mode shapes and the corresponding natural frequencies of the structure are computed to be subsequently used as the input to the routine for computing the acoustic radiation response. The panel is acted upon by a uniform thermal load (ΔT). The effects due to the thermal load are introduced via the geometry stiffness matrix (K_Δ). The equivalent stiffness of the panels can be written as: $[K_{eq}] = [K] - [K_\Delta]$ (2)

where, $[K]$ is the linear stiffness matrix. First, a static analysis is carried out so that the thermal stresses are generated in the structure. This state is referred to as the pre-stressed state. Then, the natural frequency of the vibrating structure under the influence of thermal loads can be obtained by solving the eigenvalue equation: $([K_{eq}]-\omega^2[M])\{\Phi\}=0$ (3)

where, $[M]$ is the mass matrix, ω is the natural frequency of vibration, and $\{\Phi\}$ is the corresponding mode shape vector. The thermal load is chosen so as to keep the temperature below the critical buckling temperature of the structure. The mode shapes are imported in LMS Virtual Lab environment where an indirect BE technique is utilized to account for the sound radiated in the surrounding medium by the vibrating shell panel. The Helmholtz wave equation, which is the fundamental governing equation for the acoustic problems, is written as: $\nabla^2 p + k^2 p = 0$ (4)

where, p is the sound pressure in the surroundings, $k=\omega_e/c$ is the wave number. Further, ω_e is the frequency of excitation and c is the sonic speed in the surrounding fluid medium. The Eq. (4) is solved in accordance with the Neumann boundary constraints employed on the complete surface of the laminated composite spherical shell panel. The coupled vibroacoustic response of the structure is obtained by solving the coupled equation of the system [17]:

$$\begin{pmatrix} K_{eq}+i\omega_e\Theta-\omega_e^2 M & C \\ C^T & -\frac{1}{\rho\omega_e^2}\bar{D}(\omega_e) \end{pmatrix}\begin{Bmatrix} U \\ \mu \end{Bmatrix}=\begin{Bmatrix} F \\ 0 \end{Bmatrix} \quad (5)$$

For the individual terms in Eq. (5), [17] can be referred. Using the values of U and μ obtained by solving Eq. (4) the acoustic response indicators such as the sound power radiated by the panels and the corresponding radiation efficiency can be obtained by following the steps given in [17].

3. Results and Discussion

The acoustic radiation characteristics of the spherical laminated and sandwich composite shell panels are obtained using ANSYS and LMS Virtual.Lab commercial packages. The structure (spherical shell) is modelled using finite elements in ANSYS. An eight-noded isoparametric element (Shell281) is utilized for discretizing the panel model. The modal analysis is performed under the influence of thermal load and the results (*.rst* file) are exported to LMS Virtual Lab environment where an indirect BE technique is applied to obtain the acoustic radiation responses. All the thermal loads are barred by an upper limit equal to the critical buckling temperature of the spherical panels. The panels are subjected to point harmonic excitation of 1N at the central node. Firstly, the validity of the present model is confirmed. Subsequently, influence of various parameters on the acoustic radiation responses of spherical shell panels is investigated. The geometrical parameters of the shell panel are taken as: a=0.5m, b=0.4m, h=0.01m, R/a=10 ($R_1=R_2=R$) and t_c/t_f=18 unless specified otherwise. For computational purpose the properties of core and the face material are taken as Core [14]: $E_c=7\ GPa$, $\nu_c=0.3$, $\rho_c=1000\ kg/m^3$, $\alpha=1.8\times10^{-5}/°C$; Face [6]: $E_{f,1}=132\ GPa$, $E_{f,2}=10.3\ GPa$, $G_{f,12}=G_{f,13}=6.5\ GPa$, $G_{f,23}=3.91\ GPa$, $\nu_{f,12}=\nu_{f,13}=\nu_{f,23}=0.25$, $\rho_f=1570\ kg/m^3$, $\alpha_1=1.2\times10^{-6}/°C$, $\alpha_2=2.4\times10^{-5}/°C$. A constant modal damping ratio of 0.01 is assumed throughout the analysis. The properties of the laminated composite are the same as the aforementioned properties of the face material. The sandwich panel is considered to have (0°/90°/c/90°/0°) lay-up scheme, where c denotes the core. The details of the support conditions (SSSS, CCCC, HHHH and SCSC) used in the analysis can be seen in [18].

3.1Validation of sound power

The sound power radiated by a five layered (0°/90°/0°/90°/0°) all sides clamped layered composite flat panel subjected to a uniform temperature load of 80°C is computed and compared with the results of Li et al. [6] in Figure 2 to establish the validity of the present scheme. It is evident that the current model predicts responses those agree closely with the reference data. The differences observed are attributed to the fact that the reference utilized NASTRAN simulation model in conjunction with VA.One BE commercial ware whereas ANSYS and LMS Virtual.Lab model is utilized in the present work. It is to be noted that the present simulation model yields smaller sound radiation values as compared to the reference indicating that the present model leads to stiffer structures.

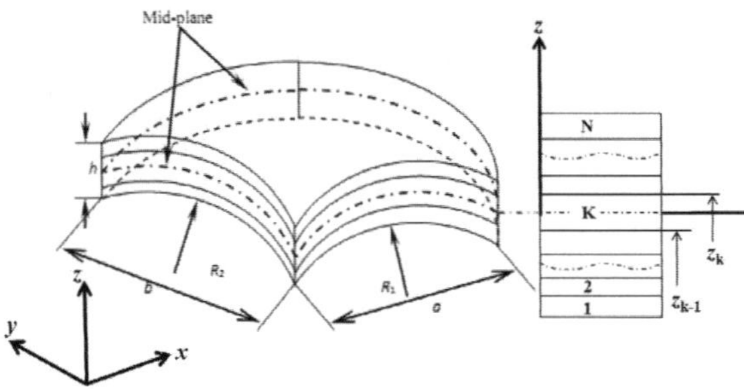

Figure 1. Geometry and stacking sequence of layered composite spherical shell panel.

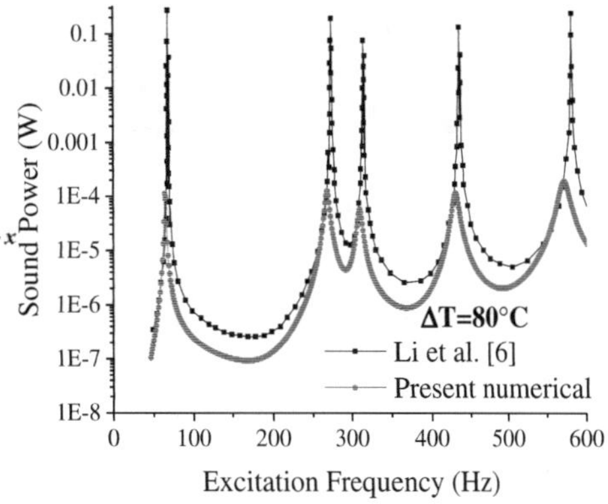

Figure 2. Comparison of sound power values of clamped laminated composite flat panel.

3.2. Numerical illustrations

Now, several numerical examples are discussed to characterize the acoustic radiation behaviour of spherical panels in a thermal environment. Figure 3(a) and (b) shows the variation of first eight natural frequencies with temperature (ΔT=0°, 30°C, 60°C, 90°C and 120°C) for all sides hinged laminated ([0°/90°]$_s$) and sandwich composite panels, respectively. It is evident that increase in temperature causes both the laminated as well as the sandwich composite panels to become increasingly soft and as a result the frequency decreases monotonically. However, the first natural frequency is smaller for the sandwich panel. Further, the influence of increasing temperature load on the sound power radiated by clamped laminated ([0°/90°]$_2$) and sandwich composite spherical shell panels is investigated. It is clear (from Figure 4) that the peaks in the radiation patterns shift to lower frequencies with increasing temperature. This is attributed to the decrease in the stiffness caused by softening of the panel with increasing temperature. However, the difference is more visible in the case of the composite sandwich due to the presence of the soft core.

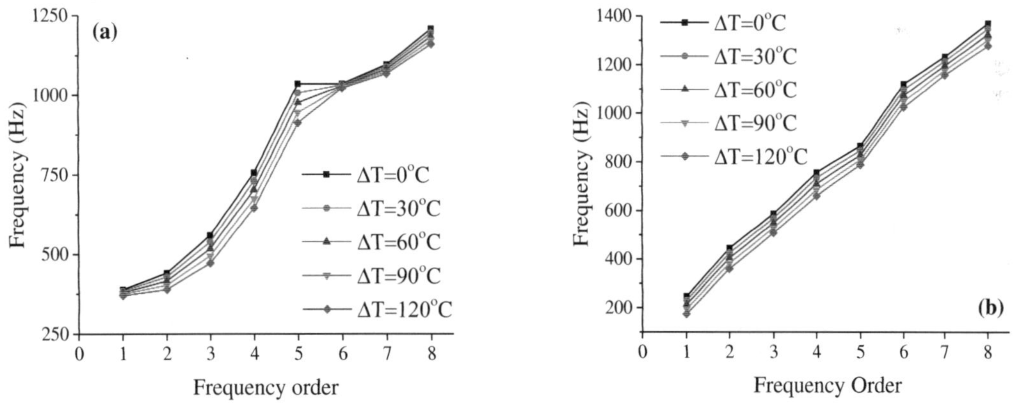

Figure 3. Effect of uniform thermal load on natural frequencies: (a) Laminated composite, (b) Sandwich composite spherical shell panel.

The influence of modular ratio (E_1/E_2) and the core-to-face modular ratio ($E_c/E_{f,1}$) on the radiation efficiency of laminated ([45°/-45°]$_s$) and sandwich spherical panels subjected to a uniform temperature load ΔT=60° under SCSC support condition is investigated and presented in Figure 5. Interestingly, the laminated composite panels radiate equally for different values of the modular ratio indicating the independence of the acoustic radiation characteristics on it. On the other hand, the core-to-face modular ratio has a significant influence on the radiation efficiency of sandwich panels. It is to be noted that E_c is varied keeping $E_{f,1}$ constant. As E_c increases, the panels become stiffer and consequently the fluctuations in the radiation efficiency pattern decrease. However, the panels tend to radiate more efficiently for increasing $E_c/E_{f,1}$ ratio values.

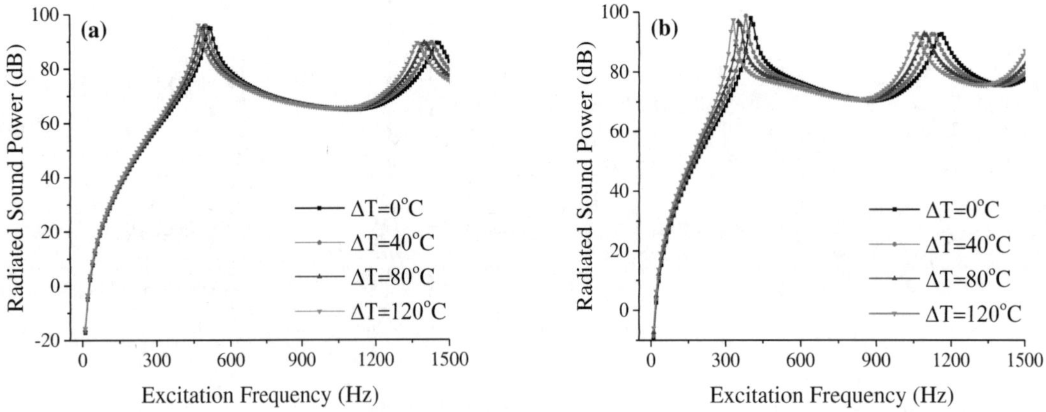

Figure 4. Effect of uniform thermal load on the sound power radiated by: (a) Laminated composite, (b) Sandwich composite spherical shell panel.

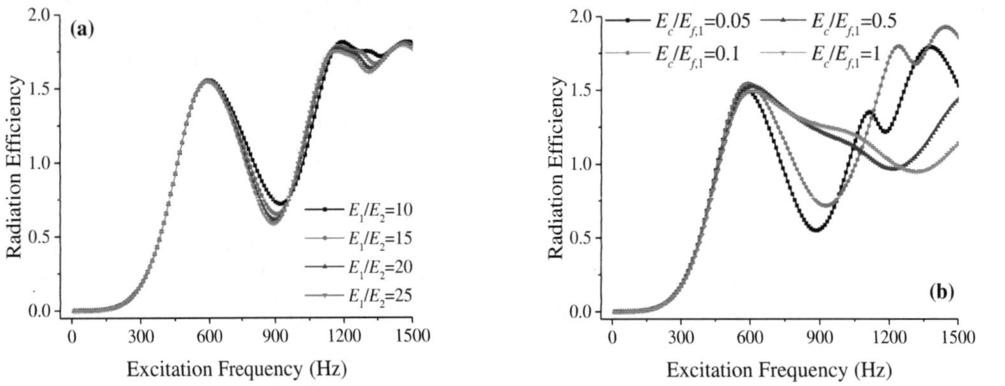

Figure 5. Variation of radiation efficiency of spherical shell panels for ΔT=60°C with modular ratio: (a) Laminated composite, (b) Sandwich composite.

The influence of varying fibre angle on the sound power radiated by clamped laminated composite spherical shell panel under an elevated thermal environment (ΔT=60°C) is investigated and depicted in Figure 6. The lamination scheme is chosen as symmetric angle ply ($[\theta/-\theta]_s$), where the fibre angle θ=15°, 30°, 45°, 60° and 90°. It can be observed that the first resonant peaks in the sound radiation pattern shift towards right along the frequency axis. On the contrary, the second resonant peaks shift to lower frequencies with increasing fibre angle. The θ=90° case radiates the most from (525–900) Hz range. Finally, the influence of varying core-to-face thickness ratio (t_c/t_f= 3, 6, 10 and 20) on the sound power radiated by hinged sandwich composite spherical panels is investigated and shown in Figure 7. The thickness of the core and the face is varied keeping the total thickness h of the panel constant. It is evident that as t_c/t_f increases, the core becomes thicker and the stiffness of the panels decreases. As a result, the resonance peaks cascade to lower frequencies. Consequently, the sound power radiated by the panel increases with increasing core-to-face thickness ratio.

4. Conclusion

The sound radiation characteristics of layered and sandwich composite spherical shell panels in an elevated thermal environment have been investigated. The free vibration responses obtained using FEM through ANSYS are transferred to LMS Virtual Lab environment in which an indirect BE technique is utilized to compute the vibroacoustic responses. The present sound power level is compared to and is found to have a good conformance with the results available in the published literature. The natural frequencies of the spherical laminated composite as well as the sandwich composite shell panels are found to increase with increasing temperature load. The composite sandwich panels are much more influenced by variation in temperature as compared to the laminated composite shell panels. The radiation efficiency of the laminated composite shell panels is least influenced by the modular ratio. The sound power radiated by laminated composite shell panels generally increase with increasing fibre angle. The panels with a thicker core radiate more sound compared to the panels with a thinner core.

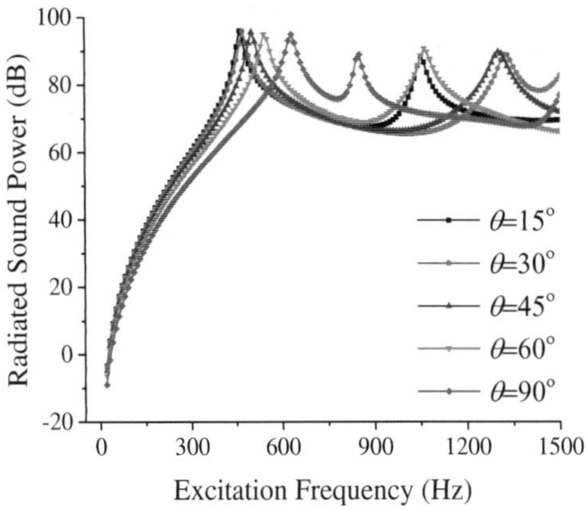

Figure 6. Effect of fibre angle on sound power radiated by laminated composite spherical shell panel (ΔT=60°C).

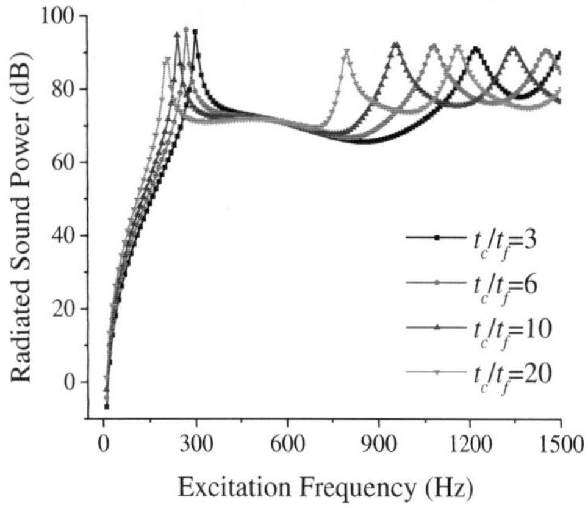

Figure 7. Effect of core-to-face thickness ratio on the sound power radiated by sandwich composite spherical shell panel (ΔT=60°C).

Reference

[1] Yin S, Hale JS, Yu T, Bui T Q, Bordas S P A 2014 *Composite Strucutres* **118** 121–38

[2] Mahapatra T R, Kar V R, Panda S K 2015 *Journal of Sandwich Structures and Materials* **17(5)** 511–545

[3] Attia A, Tounsi A, Bedia E A A, Mahmoud S R 2015 Steel and Composite Structures **18(1)** 187–212

[4] Täger O, Dannemann M, Hufenbach W A 2015 *Journal of Sound and Vibration* **342** 57–74

[5] Nowak Ł J, Zieliński T G 2015 *Journal of Vibration and Acoustics* **137(5)** 51001

[6] Li X, Yu K, Han J, Song H, Zhao R 2016 *International Journal of Mechanical Sciences* **119** 370–82

[7] Zhao X, Geng Q, Li Y 2013 *Journal of the Acoustical Society of America* **133** 1433–42

[8] Jeyaraj P, Padmanabhan C, Ganesan N 2008 *Journal of Vibration and Acoustics* **130(5)** 51005

[9] Geng Q, Li Y 2014 *Journal of Vibration and Control* 1–11

[10] Jeyaraj P, Ganesan N, Padmanabhan C 2009 *Journal of Sound and Vibration* **320(1–2)** 322–38

[11] Jeyaraj P, Padmanabhan C, Ganesan N 2011 *International Journal of Applied Mechanics* **3(3)** 525–41

[12] Geng Q, Li Y *International Journal of Applied Mechanics* **4(3)** 1250028–1:16

[13] Jeyaraj P 2010 *European Journal of Mechanics - A/Solids* **29(6)** 1088–94

[14] Liu Y, Li Y 2013 *Shock and Vibration* **20(5)** 1011–30

[15] Li X, Yu K 2015 *Composite Strucutres* **131** 1040–9

[16] Jeyaraj P, Padmanabhan C, Ganesan N 2011 *Journal of Sandwich Structures and Materials* **13(5)** 509–37

[17] Atalla N, Sgard F 2015 *Finite element and boundary methods in structural acoustics and vibration* Boca Raton, Florida, USA: CRC Press, Taylor and Francis Group

[18] Sharma N, Mahapatra T R, Panda S K 2017 *Latin American Journal of Solids and Structures*, **14(8)** 1547–1568

Development and Sliding Wear Response of Epoxy Composites Filled with Coal Mine Overburden Material

Prithika Das[*1], **Alok Satapathy**[2], **M. K. Mishra**[1]

[1] Department of Mining Engineering, [2]Department of Mechanical Engineering National Institute of Technology, Rourkela, Orissa, India

Corresponding author Email: das.prithikarkl@gmail.com

Abstract: The paper reports on development and characterization of epoxy based composites filled with micro-sized mine overburden material. Coal mine overburden material is typically highly heterogeneous and is considered as waste material. For excavating each ton of coal, roughly 5 tons of overburden materials are removed and is dumped nearby occupying large space. Gainful utilization of this waste is a major challenge. In the present work, this material is used as filler materials in making a new class of epoxy matrix composites. Composites with different weight proportions of fillers (0, 10, 20, 30 and 40) wt. % are prepared by hand layup technique. Compression tests are performed as per corresponding ASTM standards to assess the compressive strength of these composites. Further, dry sliding tests are performed following ASTM G99 standards using a pin on disk machine. A design of experiment approach based on Taguchi's L_{16} orthogonal arrays is adopted. Tests are performed at different sliding velocities for multiple sliding distances under varying normal loads. Specific wear rates of the composites under different test conditions are obtained. The analysis of the test results revealed that the filler content and the sliding velocity are the most predominant control factors affecting the wear rate. This work thus, opens up a new avenue for the value added utilization of coal mine overburden material.

Keywords: *Epoxy, taguchi method, coal mine overburden, polymer composites*

1. Introduction

India produces ample amount of coal every year. Maximum number of thermal plants as well as household activities depends upon coal. So, through opencast mining process, coal is extracted from the environment. For excavating each ton of coal, near about five to six tons of overburden material are excavated. After digging the coal, these mine overburden materials are left over lands. These soils are occupying useful lands and destroying its fertility. Overburden materials are combined with both coarse and fine grains with equal amounts. Since, it is having fine grains and is typically loose, so due to wind they are highly prone to flow and damages the surrounding which leads to health issues. So, utilization, recycle and reuse of mine overburden material are necessary. Earlier some trials had been made to discover reuse techniques for a huge range of waste substances along with blast furnace and steel slag, construction waste, fly ash, red mud and so forth.

Polymers are found in numerous applications including aerospace, vehicles, household appliances, constructing materials and many others. Due to weight sensitiveness, polymers are used, but their high cost is a drawback. Therefore, to reduce the cost, industrial wastes are used. Proper mixing of matrix and reinforcement will provide a better composite material. The addition of filler material to polymer matrix increases its resistance to moisture and improves its mechanical characteristics.

Although many works have been focused on polymer composites filled with particulates in the past, reports on the use of industrial solid wastes as particulate fillers have been rare. Only a few studies on use of industrial wastes like red mud, fly ash, copper slag etc. in polymer composites have been reported so far. Mechanical properties of epoxy and fly ash composites are studied by Chaowasakoo and Sombatsompop [1]. Compressive strength of fly ash epoxy composites are reported by Srivastava and Pawar [2].Patnaik et al. [3, 4] explained the fabrication of the glass-fibre reinforced fly ash filled polyester composites and showed the erosion wear using Taguchi's orthogonal array. Incorporation of solid waste and E-glass fiber with polyester has been showed by Mahapatra [5]. Gangilet al. [6] produced homogenous and functionally graded vinyl ester composites by mixing the particulate-filled CBPD and short Kevlar fibre. The physio-mechanical polypropylene composites filled with red mud studied by Zhang et al. [7].

Similarly, from copper extraction industries copper slag are generated during matter smelting and refining of copper [8]. Biswas and Satapathy [9] reported the use copper slag in polymers. Reports are also available on characterization and sliding wear behaviour of epoxy composites filled with steel industry wastes like blast furnace slag [10] and LD sludge [11].

Although a number of studies have been reported on the use of various solid wastes as fillers in polymer composites, till date there has not been any work reported on composites using overburden material produced from coal mines as filler. This work explores the possibility of developing such a composite and studies the effect of mine overburden material on the sliding wear behaviour.

2. Materials and methods

2.1 Material used

The overburden material is collected from the Bharatpur Mines, Talcher, Odisha. This coal belongs to Gondwana Series. Freshly dumped overburden materials in the dumping yard were collected in the gunny bag, then; the mouth is sealed and was packed into another poly bag and sealed. Couples of overburden material in multiple gunny bags are approximately transported to the laboratory. The sealing was to preserve the original moisture content. At the time of experimentation, gunny bags were opened and material was taken out. It was crushed and sieved to 0-2 microns grain sizes.

2.2 Composite fabrication

These overburden materials are collected and crushed properly. Then, these particles are sieved to get a size of range 70-100 micrometer. The epoxy (LY 556) and hardener (HY 951) is blended within the ratio of 10:1 by using weight. The overburden material and epoxy are then mixed together in numerous proportions to prepare the composites. The combination is very well combined until uniform dough is formed. Before pouring the dough inside the test tube, it is coated with wax. After that, dough is poured inside the test tube and then kept at room temperature for 24 hours for curing. When the composites are prepared, the test tubes are broken and the composites came out in easy manner. Likewise four samples are made with different compositions (0, 10, 20, 30, 40 wt. %) of overburden material.

2.3 Compressive test

The cylindrical shaped composite samples of required dimension of length 3 cm and diameter 1.2 cm are used for compressive test as per ASTM D695, ISO 604 so as to get the aspect ratio from 2-2.5. The specimen is kept in between two platens of the compression testing machine. A compressive load is applied to the sample slowly till the failure takes place.

2.4 Sliding wear test

In dry sliding test, by using the pin on disc machine, the wear test was done by ASTM G99 standard. The disc which is made up of hardened steel was rotating while the sample was kept constant. Samples are fitted to the holder and sliding rpm is set, track radius and normal load are set before starting the test. In a pin-on disc test, the sample is stationary and the disc is rotating with a certain

IOP Conf. Series: Materials Science and Engineering **338** (2018) 012032 doi:10.1088/1757-899X/338/1/012032

speed and a lever mechanism is used to apply load. A series of tests using Taguchi's L_{16} orthogonal array are conducted with four sliding velocities of 32, 63, 94 and 126 cm/sec under four different normal loadings of 20, 30, 40 and 50 N, respectively. After each trial, both the specimen and the disc are cleaned with acetone so as to get perfect readings. The sample is tested before and after wearing every time.

$$W_S = \frac{\Delta m}{\rho \times t \times V_s \times f_n}$$

Where, Δm the mass loss in the test (g), ρ is the density (g/mm^3), t is the test duration (s), V_s is the sliding velocity (m/s), f_n is the normal load (N) and W_s is the specific wear rate (mm^3/Nm).

2.5 Taguchi experimental design

Design of Experiment (DOE) is an effective device used for interpreting and modeling the influence of control factors on overall performance output. In design of experiment, selection of control factors is very much essential. Numerous parameters like sliding velocity, normal load, sliding distance and filler content are there to study the L_{16} orthogonal array. The use of Taguchi L_{16} reduced the number of experiments from 4^4=256 conventional runs to simply 16 runs. It saves the cost as well as time. The Signal to Noise (S/N) ratio for minimum wear rate can be expressed as "smaller is better" and can be calculated in logarithmic transformation as shown below.

$$\frac{S}{N} = -10 \log \frac{1}{n} \Sigma y^2$$

Where, y is the observed data and n is the number of observations.

Table 1: Control factors chosen for sliding wear test

Control Factors	1	2	3	4	Units
Sliding Velocity	32	63	94	126	cm/sec
Normal Load	20	30	40	50	N
Sliding Distance	500	1000	1500	2000	m
Filler Content	10	20	30	40	wt. (%)

3. Results and Discussion

3.1 Compressive strength

For applications like automobile, designing of hauled roads and ceramic parts, the unconfined compression test is generally used. It is often used as an index to measure the strength enhancement of materials due to treatment. After performing the compression test, the tested sample is shown in figure 1. Similarly in figure 2, the stress-strain diagram of 30 wt. % and 40 wt. % of epoxy-overburden material composites are shown. Table 2 and table 3 as shows the test results for composites filled with epoxy and overburden material of 30 wt. % and 40 wt. %, respectively. These tables also present the values of Young's modulus, Poisson's ratio, Modulus of rigidity and Bulk modulus of the composites. It is observed that the compressive strength of the epoxy-overburden material composite increases with increase in overburden material is mixed with epoxy resin. All the samples in unconfined compressive loading situations exhibited shear type failure. Except some samples, all samples failed by means of shear which reflects the mixed impact of sample and system traits.

IOP Conf. Series: Materials Science and Engineering **338** (2018) 012032 doi:10.1088/1757-899X/338/1/012032

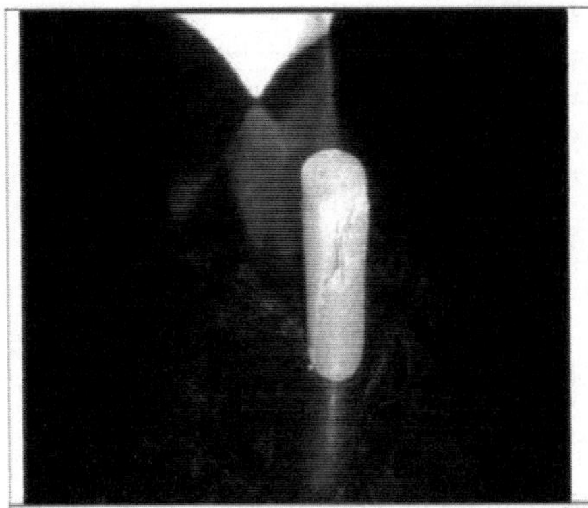

Figure 1. Cross-Sectional View of the sample after UCS test

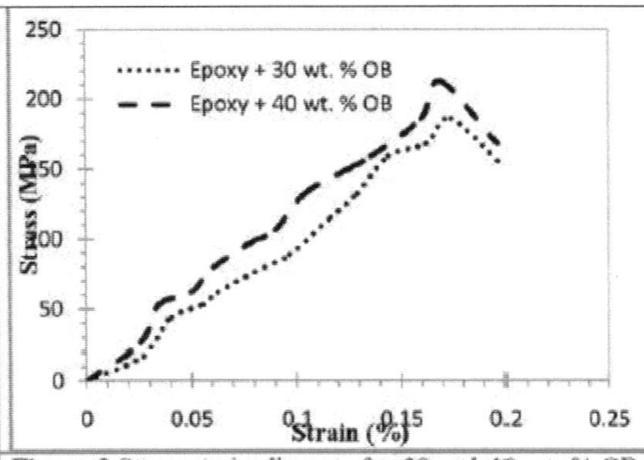

Figure 2. Stress-strain diagram for 30 and 40 wt. % OB filled epoxy composite

Table 2: Mechanical properties of epoxy composites filled with 30 wt. % OB

P(N)	A(mm2)	dL(mm)	L(mm)	dD(mm)	D(mm)	σ (Mpa)	dL/L(strain)	dD/D	E	μ	G	K
2000	113.04	0.81	30	0.11	12	17.69285	0.027	0.009167	655.2908	0.339506	244.6016	680.4943
5000	113.04	1.21	30	0.17	12	44.23213	0.040333	0.014167	1096.664	0.35124	405.7994	1228.67
6000	113.04	1.67	30	0.25	12	53.07856	0.055667	0.020833	953.507	0.374251	346.9187	1263.775
7000	113.04	1.88	30	0.33	12	61.92498	0.062667	0.0275	988.1646	0.43883	343.3918	2692.391
9000	113.04	2.55	30	0.36	12	79.61783	0.085	0.03	936.6804	0.352941	346.1645	1061.571
10000	113.04	2.9	30	0.42	12	88.46426	0.096667	0.035	915.1475	0.362069	335.9402	1105.803
13000	113.04	3.49	30	0.48	12	115.0035	0.116333	0.04	988.5691	0.34384	367.8152	1055.078
15000	113.04	3.88	30	0.57	12	132.6964	0.129333	0.0475	1026.003	0.367268	375.2019	1288.314
18000	113.04	4.32	30	0.66	12	159.2357	0.144	0.055	1105.803	0.381944	400.0896	1561.134
19000	113.04	4.86	30	0.73	12	168.0821	0.162	0.060833	1037.544	0.375514	377.1476	1389.108
21000	113.04	5.23	30	0.84	12	185.7749	0.174333	0.07	1065.631	0.40153	380.167	1803.64
17000	113.04	6.01	30	0.9	12	150.3892	0.200333	0.075	750.6951	0.374376	273.104	995.9552

Table 3: Mechanical properties of epoxy composites filled with 40 wt. % OB

P(N)	A(mm2)	dL(mm)	L(mm)	dD(mm)	D(mm)	σ (Mpa)	dL/L(strain)	dD/D	E	μ	G	K
3000	113.04	0.77	30	0.14	12	26.53928	0.025667	0.011667	1033.998	0.454545	355.4368	3791.325
6000	113.04	1.03	30	0.18	12	53.07856	0.034333	0.015	1545.977	0.436893	537.9583	4082.966
7000	113.04	1.49	30	0.29	12	61.92498	0.049667	0.024167	1246.812	0.486577	419.3565	15481.25
9000	113.04	1.82	30	0.35	12	79.61783	0.060667	0.029167	1312.382	0.480769	443.1419	11373.98
11000	113.04	2.36	30	0.39	12	97.31069	0.078667	0.0325	1237	0.413136	437.6792	2373.431
12000	113.04	2.7	30	0.45	12	106.1571	0.09	0.0375	1179.523	0.416667	416.3024	2359.047
15000	113.04	3.12	30	0.52	12	132.6964	0.104	0.043333	1275.927	0.416667	450.3271	2551.854
17000	113.04	3.77	30	0.59	12	150.3892	0.125667	0.049167	1196.731	0.391247	430.0932	1834.015
18000	113.04	4.08	30	0.68	12	159.2357	0.136	0.056667	1170.851	0.416667	413.2414	2341.701
21000	113.04	4.79	30	0.72	12	185.7749	0.159667	0.06	1163.517	0.375783	422.8565	1561.134

24000	113.04	5.08	30	0.79	12	212.3142	0.169333	0.065833	1253.824	0.38878	451.4122	1878.887
19000	113.04	5.91	30	0.85	12	168.0821	0.197	0.070833	853.2086	0.35956	313.7811	1012.543

3.2 Sliding wear results

The specific wear rates are determined on the basis of test results and the corresponding S/N ratios for all the 16 test runs are presented in table 4. By using the MINITAB 14 software, the graphs are drawn in figure 3 and from there, the lowest specific wear is observed. From this experiment, it is discovered that the specific wear rate is influenced more due to filler content as compared to all other 3 factors like sliding velocity, sliding distance and normal load.

Table 4: Taguchi experimental design for sliding wear test

Test Run	Sliding Velocity (cm/sec)	Load (N)	Sliding Distance (m)	Filler Content (wt. %)	Specific Wear Rate (mm^3/Nm)	S/N Ratio
1	32	20	500	10	9.739	-19.7703
2	32	30	1000	20	7.807	-17.8497
3	32	40	1500	30	5.761	-15.2100
4	32	50	2000	40	4.576	-13.2097
5	63	20	1000	30	7.023	-16.9305
6	63	30	500	40	5.424	-14.6864
7	63	40	2000	10	10.304	-20.2601
8	63	50	1500	20	6.661	-16.4708
9	94	20	1500	40	4.674	-13.3938
10	94	30	2000	30	6.276	-15.9537
11	94	40	500	20	8.753	-18.8431
12	94	50	1000	10	13.736	-22.7572
13	126	20	2000	20	10.130	-20.1122
14	126	30	1500	10	14.158	-23.0200
15	126	40	1000	40	5.521	-14.8404
16	126	50	500	30	6.509	-16.2703

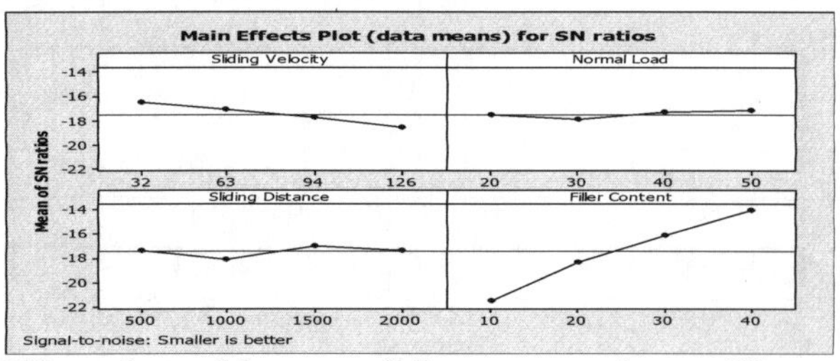

Figure 3. Effects of control factors on sliding wear rate of epoxy-OB composites

4. Conclusions

- The work establishes the use of overburden material, a solid waste in the coal mining sites as a potential filler material in the making of polymer composites.
- It shows that with the enhancement of filler content of coal mine overburden material compressive strength also increases.
- Using of overburden material has resulted in substantial improvement in the wear resistance of epoxy resin.
- Through Taguchi experimental design, dry sliding wear can be measured successfully.
- The analysis reveals that overburden material content and the sliding velocity are the most predominant control factors affecting the wear rates of the composites.
- In future, this study can be extended by mixing this waste with other industrial waste to produce better strength and to fabricate new composites.

References

1. Chaowasakoo T and SombatsompopN 2007 *Compos. Sci. Technol.* **67**2282.
2. Srivastava V K and Pawar A G Compos. *Sci. Technol.***66**3021.
3. Patnaik A Satapathy A Mahapatra S S and Dash R R 2009*J. Reinf. Plast. Compos.***28** 513.
4. Patnaik A Satapathy A Mahapatra S S and Dash R R 2008*Int. Polym. Proc.***13**192.
5. Mahapatra S S2010 J. *Eng. Tribol.* **224** 157.
6. Gangil B Patnaik A Kumar A and Biswas S*J. Eng. Tribol.*DOI: 1350650112460363.
7. Zhang Y Zhang A ZhenZ Lv F Chu P K and Ji J 2011*J. Compos. Mater.*DOI: 10.1177/0021998311401937.
8. Biswas S and Satapathy A 2010*Waste Manage. Res.***28** 615.
9. Biswas S and Satapathy A 2009*Mater. Des.***30** 2841.
10. Padhi P K and Satapathy A2014 *Int. Polym. Proc.***29** 233.
11. Purohit A and Satapathy A 2016 *J.Compo. Mat.***51** 899.

Study on effect of tool electrodes on surface finish during electrical discharge machining of Nitinol

Anshuman Kumar Sahu, Suman Chatterjee, Praveen Kumar Nayak, Siba Sankar Mahapatra
Department of Mechanical Engineering,
National Institute of Technology Rourkela, Odisha, India

Corresponding author Email : anshuman.sahu123@gmail.com

Abstract: Electrical discharge machining (EDM) is a non-traditional machining process which is widely used in machining of difficult-to-machine materials. EDM process can produce complex and intrinsic shaped component made of difficult-to-machine materials, largely applied in aerospace, biomedical, die and mold making industries. To meet the required applications, the EDMed components need to possess high accuracy and excellent surface finish. In this work, EDM process is performed using Nitinol as work piece material and AlSiMg prepared by selective laser sintering (SLS) as tool electrode along with conventional copper and graphite electrodes. The SLS is a rapid prototyping (RP) method to produce complex metallic parts by additive manufacturing (AM) process. Experiments have been carried out varying different process parameters like open circuit voltage (V), discharge current (Ip), duty cycle (τ), pulse-on-time (Ton) and tool material. The surface roughness parameter like average roughness (Ra), maximum height of the profile (Rt) and average height of the profile (Rz) are measured using surface roughness measuring instrument (Talysurf). To reduce the number of experiments, design of experiment (DOE) approach like Taguchi's L_{27} orthogonal array has been chosen. The surface properties of the EDM specimen are optimized by desirability function approach and the best parametric setting is reported for the EDM process. Type of tool happens to be the most significant parameter followed by interaction of tool type and duty cycle, duty cycle, discharge current and voltage. Better surface finish of EDMed specimen can be obtained with low value of voltage (V), discharge current (Ip), duty cycle (τ) and pulse on time (Ton) along with the use of AlSiMg RP electrode.

1. Introduction

Electro-discharge machining (EDM) is a mostly used non-traditional machining process due to its capability to produce complex, intrinsic cavity with excellent surface finish for applications in aerospace, biomedical, automobile, tool and die industries. In the EDM process, the work piece and the tool electrodes are immersed inside a dielectric medium. When the voltage is applied to the system, spark is generated in the electrode gap with generation of very high temperature (around 10000^0C). High temperature causes to melt and vaporize tiny amount of material from the work piece surface and leads to the removal of material from the work piece surface. EDM process is used for machining of hard-to-machine materials with reasonably accurate dimensions and surface finish to meet the requirement of the industries [1]. During EDM process, maximum time is devoted to produce the tool electrode which needs the use of conventional or non-conventional machining processes. Development of the tool electrode not only consumes the overall time required for the EDM process but also increases the cost of production of the final product. In order to reduce the manufacturing time of electrodes, rapid prototyping (RP) process like selective laser sintering (SLS) process can be used. By using the RP process, the preparation time for tool electrodes is reduced by 50%, which reduced the overall production cost of final product [2, 3].

Different type of EDM electrodes like bronze-nickel with copper phosphite by SLS, epoxy with silver paint and copper coating by thin coated stereolithography (SL) model, steel and phosphate taking polyester as binder by SLS were prepared by RP process and used in EDM [2-4]. Complex steel mold was prepared by EDM process by taking tool electrode of graphite prepared by RP process with the use of abrading tool of corundum and SiC [5]. Similarly, mold from steel work piece was prepared by EDM process by taking copper electrode prepared by electroforming process and graphite electrode prepared by abrading process [6]. From these works, it is found that rapid tooling (RT) technology is a rapid and effective technique to manufacture mold and RT has high prospective for reduction in time and cost.

In this work, EDM performance of AlSiMg electrode prepared by SLS process along with conventional copper and graphite electrodes have been studied during the machining of Nitinol. Here performance of the AlSiMg composite electrode is investigated by varying different EDM process parameters to enhance the surface properties of the machined speciment and compare to the performance of conventional copper and graphite tool electrodes. Nitinol is a nickel-titanium alloy used in aerospace, automobile, biomedical and chemical industries due to its high strength and corrosion resistance in a wide range of temperature variations. To meet the required applications in the field of chemical, aerospace, biomedical, die and mold making industries, the EDM specimen must be prepared with high precision and excellent surface finish. To examine the surface roughness characteristics (average roughness (Ra), average height of the profile (Rz) and maximum height of the profile (Rt)) of the machined surface during electrical discharge machining, different EDM input parameters like voltage (V), discharge current (Ip), duty cycle (τ) and pulse-on-time (Ton) have been varied along with different types of tool electrodes i.e. AlSiMg RP electrode, copper and graphite electrodes.

2. Materials and methods

In this work, AlSiMg RP electrode prepared by SLS process, copper and graphite are used as tool electrode for the EDM of Nitinol as work piece material and commercial EDM 30 oil as dielectric fluid. The chemical composition of Nitinol in weight percentage of the elemental component is given in Table 1.

Table 1. Chemical composition of Nitinol

Composition	Ni	Cr	Cu	Fe	Nb	Co	O	C	Ti
Weight percentage	55.64	<0.01	<0.01	<0.012	<0.01	<0.01	<0.05	<0.02	Balance

Table 2. Machining parameters with their levels

Parameters	Unit	Level-1	Level-2	Level-3
A-Voltage (V)	V	20	25	30
B-Discharge current (Ip)	A	20	25	30
C-Duty cycle (τ)	%	67	75	83
D- Pulse-on-time (Ton)	μs	100	200	300
E-Tool type	-	AlSiMg RP	Graphite	Copper

To study the effect of different types of tools like AlSiMg RP electrode, copper and graphite electrodes on the surface roughness of the machined surfaces, the EDM process parameters like voltage (V), discharge current (Ip), duty cycle (τ) and pulse on time (Ton) are varied during the EDM process. The values of the input parameters with different levels are listed in Table 2. To reduce the number of experiments, design of experiment (DOE) methodology like Taguchi's L_{27} orthogonal array is considered. The EDM process of work piece material Nitinol by taking different tool electrodes has been performed in a EDM (model: ELECTRA EMS 5535). The surface quality parameters like average roughness (Ra), average height of the profile (Rz) and maximum height of the profile (Rt) are measured by the use of surface roughness measurement machine (Taylor-Hobson-PNEUNO-Suetronic 3+) [7,8]. Here, the surface roughness parameters are measured three times and the average value for each EDM machined specimens is considered for further study. The surface roughness values are optimized by desirability approach (DA) and best optimal setting is obtained to get better surface finish of the machined specimens.

Average roughness (Ra) is the arithmetic mean of the absolute values of the heights of the contour along the estimated length. The average roughness (Ra) is estimated by the formula as given in Eqn. (1).

Average roughness, $$Ra = \frac{1}{L}\int y(x)\,dx = \frac{\sum_{i=1}^{i=n} y_i}{n} \quad (1)$$

Maximum height of the profile (Rt) is the perpendicular distance amongst the uppermost and bottommost points of the contour inside the estimated length. The maximum height of the profile (Rt) is calculated as in Eqn. (2).

Maximum height of the profile, $$Rt = Max(y_i) \quad (2)$$

Average height of the profile (Rz) is the mean of the consecutive values of maximum height of the profile (Rt) calculated over the estimated length as mentioned in Eqn. (3).

Average height of the profile, $Rz = \frac{\sum_{i=1}^{n} R_{ti}}{n}$ (3)

where L = sampler length, y = profile curve, x = profile direction, y_i = height of the profile. The surface roughness values are measured within L=0.8mm having cut off length Lc=0.4mm and N=2.

2.1. Desirability function approach

The desirability function approach is a optimization technique used to solve multi objective optimization problems. The elementary conception of desirability is to transform multi objective optimization problem into single objective. Here, each response variable y_i is converted into individual desirability value d_i, which value ranges between 0 to 1. If the characteristic of the output response is unacceptable, then desirability value is 0 and for optimal acceptable characteristic the desirability value is 1. After that composite or overall desirability is then calculated from the individual desirability. The highest value of the overall desirability indicates the optimum parametric setting [9-11]. The following steps are followed to evaluate the individual desirability index (d_i) and overall desirability index (d_o).

1. Calculation the individual desirability index (d_i).
   (i) Nominal is the best.

$$d_i = \begin{cases} (\frac{y_i - y_{min}}{T - y_{min}})^s, y_{min} \le y_j \le T, s \ge 0 \\ (\frac{y_i - y_{max}}{T - y_{max}})^s, T \le y_j \le y_{max}, s \ge 0 \\ 0, \quad \text{otherwise} \end{cases} \quad (4)$$

   (ii) Larger is the better.

$$d_i = \begin{cases} 0, \quad y_i \le y_{min} \\ (\frac{y_i - y_{min}}{y_{max} - y_{min}})^r, y_{min} \le y_j \le y_{max}, r \ge 0 \\ 1, \quad y_j \ge y_{min} \end{cases} \quad (5)$$

   (iii) Smaller is the better.

$$d_i = \begin{cases} 1, \quad y_i \le y_{min} \\ (\frac{y_i - y_{max}}{y_{min} - y_{max}})^r, y_{min} \le y_j \le y_{max}, r \ge 0 \\ 0, \quad y_j \ge y_{min} \end{cases} \quad (6)$$

   where y_j= observed value from the experiment.
   y_{max}= maximum observed value
   y_{min}= minimum observed value
   T = Target value

2. Calculate the composite or overall desirability.

$$d_o = (d_1^{w_1} * d_2^{w_2} * \dots\dots\dots d_i^{w_i})^{1/w} \quad (7)$$

   where $w_1 + w_2 + \cdots w_i = w$

3. Result and Discussion

The experiment of EDM is performed for the Nitinol work piece and the surface roughness parameters were measured as described in section 2. The surface roughness parameters like average roughness (Ra), average height of the profile (Rz) and maximum height of the profile (Rt) are measured for three numbers of times and the average value is consider for further analysis.

3.1 Effect of parameters on Ra

The main effect plot for average roughness (Ra) is presented in Figure 1. From the figure it is found that with increase in EDM parameters like voltage (V), discharge current (Ip), duty cycle (τ) and pulse on time (Ton), the average surface roughness (Ra) of the machined surface increased. This is due to the regeneration of more energy during sparking, which leads to more crater depth with increased average surface roughness (Ra). The surface of specimen machined with RP electrode is better with respect to average surface roughness (Ra) followed by surface machined with copper and graphite electrodes.

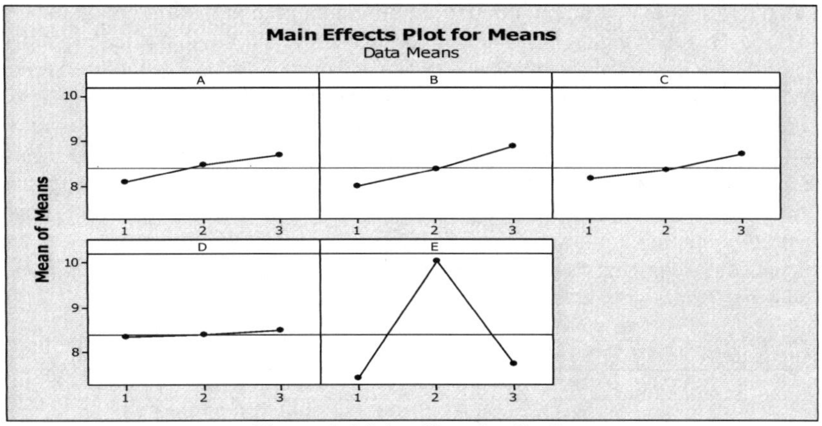

Figure 1. Main effect plot for Ra

3.2 Effect of parameters on Rt

The main effect plot for maximum height of the profile (Rt) is presented in Figure 2. From the figure it is found that the similar trend like average roughness (Ra) is also found in case of maximum height of the profile (Rt).

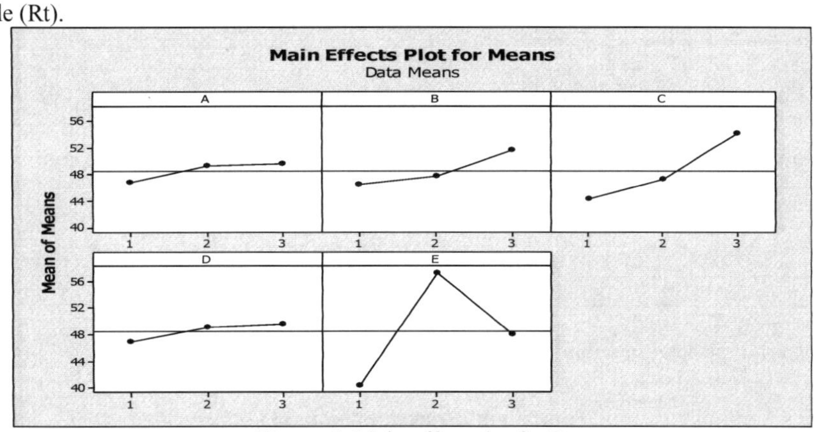

Figure 2. Main effect plot for Rt

3.3 Effect of parameters on Rz

The main effect plot for average height of the profile (Rz) is presented in Figure 3. From the figure it is found that the similar trend like average roughness (Ra) is also found in case of average height of the profile (Rz).

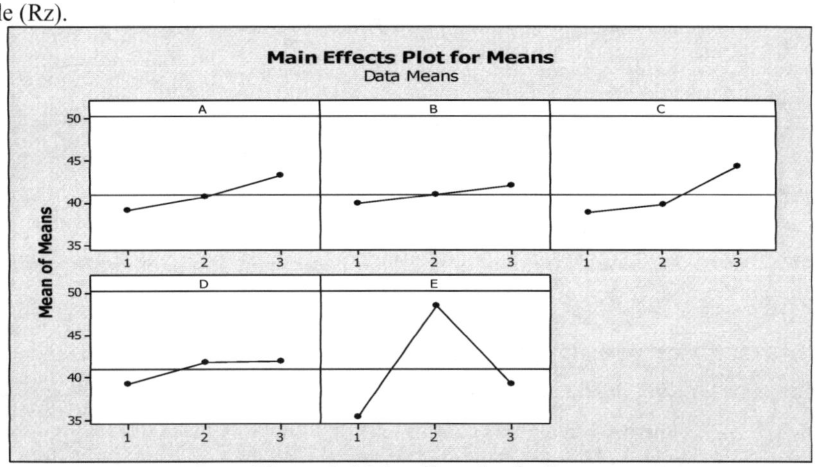

Figure 3. Main effect plot for Rz

3.4 Optimization by desirability function approach

To study all the responses concurrently, multi response optimization method i.e. desirability function approach is used. By following the procedure of desirability function approach as explain in Eqns. (4) to (7) [section 2], the individual desirability index (d_i) and overall desirability index (d_o) are calculated. The ANOVA for the overall desirability index (d_o) is tabulated in Table 3 with $R^2 = 97.5\%$. The main effect plot for the overall desirability index (d_o) are shown in Figure 4. The ANOVA is generate by the help of MINITAB 16. From the ANOVA table, the type of tool is found to be the most significant parameter followed by interaction of tool type and duty cycle, duty cycle, discharge current and voltage with percentage contribution of 61.98, 15.87, 10.94, 3.79 and 2.79 respectively. Better surface finish of EDM specimen produced with lower value of voltage (V), discharge current (Ip), duty cycle (τ) and pulse on time (Ton) along with the use of AlSiMg RP electrode during EDM. The optimal parametric setting is shown in response table for means of overall desirability index (Table 4).

Table 3. ANOVA for overall desirability index (d_o)

Parameters	DF	SS	MS	F	P	% Cont.
A	2	0.05266	0.026331	4.48	0.049*	2.79
B	2	0.07142	0.035709	6.08	0.025*	3.79
C	2	0.20629	0.103146	17.56	0.001*	10.94
D	2	0.01528	0.007639	1.30	0.324	0.81
E	2	1.16919	0.584594	99.52	0.000*	61.98
B*E	4	0.02517	0.006292	1.07	0.431	1.33
C*E	4	0.29944	0.074860	12.74	0.002*	15.87
Error	8	0.04699	0.005874			2.49
Total	26	1.88644				100

*Significant parameters at 95% confidence interval

Table 4. Response table for means of overall desirability index

Level	A	B	C	D	E
1	0.6701*	0.6722*	0.6955*	0.6443*	0.8179*
2	0.6027	0.6171	0.6491	0.6037	0.3269
3	0.5631	0.5465	0.4913	0.5878	0.6911
Delta	0.1070	0.1257	0.2042	0.0565	0.4909
Rank	4	3	2	5	1

*Optimal level of parametric setting

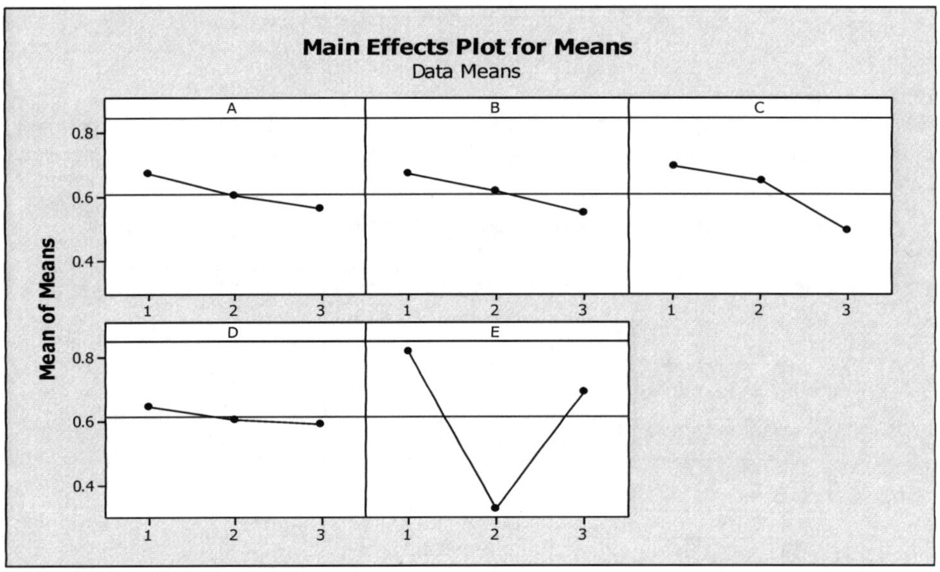

Figure 4. Main effect plot for overall desirability index (d_o)

3.5 Surface crack density of machined surface

The scanning electron micrograph of the machined surface by three different types of tool electrodes are taken and the surface crack density (SCD) of the machined surface are measured by pdf xchange viewer software. It is found that by the use of AlSiMg RP electrode less surface crack density is found followed by copper and graphite electrode.

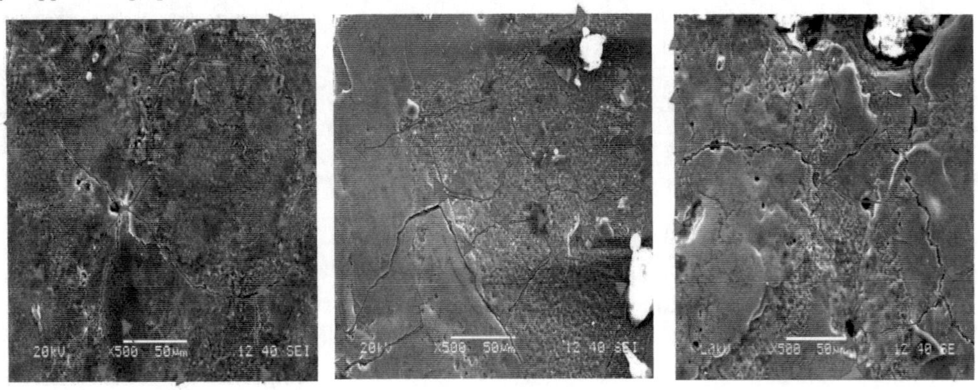

(a) SCD=0.02547μm/μm^2 (b) SCD=0.026526μm/μm^2 (c) SCD=0.029684μm/μm^2

Figure 5. Surface crack density of machined surface by different tool electrodes at parametric setting V=20V, Ip=10A, τ=67%, Ton=100μs (a) AlSiMg RP, (b) Copper, (c) Graphite

4. Conclusion

In this work, the performance of AlSiMg electrode prepared via SLS process has been studied and compared with the performance of conventional copper and graphite electrodes. It is found that better surface finish specimens were manufactured by EDM process by taking AlSiMg SLS electrodes followed by copper and graphite electrodes respectively. By the use of desirability function approach, all the responses were converted into single response i.e. the overall desirability index (d_o) and the optimal parametric setting has been obtained for the EDM process for getting better surface finish that are V=20V, Ip=20A, τ=67%, Ton=100μs and AlSiMg SLS tool electrode. Therefore, to get better surface finish, open circuit voltage (V), discharge current (Ip), duty cycle (τ) and pulse on time (Ton) should be minimized and AlSiMg electrode prepared via selective laser sintering (SLS) process must be used.

References

[1] Sahu A K, Mohanty P P, Sahoo S K 2017 *Advances in 3D Printing & Additive Manufacturing Technologies.*

[2] Durr H, Pilz R, Eleser N S 1999 *Computers in Industry* **39.**

[3] Arthur A, Dickens P M, Cobb R C 1996 *Rapid Prototyping Journal* **2** 1.

[4] Zhao J, Li Y, Zhang J, Yu C, Zhang Y 2003 *Journal of Materials Processing Technology* **138**.

[5] Tang Y, Hong J, Zhou H, Lu B 2005 *Journal of Materials Processing Technology* **166**.

[6] Ding Y, Lan H, Hong J, Wu D 2004 *Robotics and Computer-Integrated Manufacturing* **20**.

[7] Sahu A K, Mahapatra S S, Chatterjee S 2017 *International Conference on Materials, Manufacturing and Modelling(ICMMM), VIT University, Vellore, India.*

[8] Pradhan M K, Das R, Biswas C K 2009 *Proc. IMechE Part B: J. Engineering Manufacture* **223**.

[9] Datta S, Bandyopadhyay A, Pal P K 2006 *International Journal for Manufacturing Science & Production.*

[10] Karande P, Gauri S K, Chakraborty S 2013 *Materials and Design* **45**.

[11] Singaravel B, Selvaraj T 2016 *Journal of Advanced Manufacturing Systems* **15** 1.

IOP Conf. Series: Materials Science and Engineering **338** (2018) 012034 doi:10.1088/1757-899X/338/1/012034

Synthesis of Graphene-wrapped Silicon Nanoparticles by Ultra-sonication

P G Kale[1], Mihir Kumar Sahoo[2], Nishit Kumar Sadual[3] and S K Behera[4]

[1,2]Department of Electrical Engineering, NIT Rourkela, Odisha, India
[3,4]Department of Ceramic Engineering, NIT Rourkela, Odisha, India

pareshkale@nitrkl.ac.in

Abstract. Composite electrodes composed of silicon nanoparticles (SiNPs) in accommodation with graphene oxide (GO) synthesized by the ultra-sonication method can be used as a Li-ion battery anode. SiNPs wrapped with graphene is found to be used as an advanced material for Li-ion anode battery because of the high charge storage capacity, low cost, large surface area, cyclic stability, and reliability in comparison to the conventional planner Si particle. The SiNPs coated with graphene can accommodate the volume expansion problem of Si during lithiation without any fracture. The graphene-wrapped SiNPs morphology and dimension were characterized by high-resolution transmission electron microscopy (HRTEM).

1. Introduction

Graphene is a two-dimensional (2D) sheet, consists of one layer carbon (C) atoms with honeycomb lattice arrangement, and graphene oxide (GO) is the oxidation of graphite. The graphene and graphene-based devices are advanced research material due to its large surface area, higher electric and thermal conductivity, and stability. In recent times, research on graphene composites with silicon, for application in energy storage, and conversion device is on the rise.

Doped Si materials enhance conductivity at a low mass density making it a good candidate for stable electrodes for electrochemical devices, e.g. electrochemical supercapacitors, and micro fuel cells. The graphene coating over porous Si (PS) passivates surface charge traps and improves electrode-electrolyte electrochemical interface [1] improving energy density up to 40 times compared to Si material. Su et al. demonstrated that the conductivity of the direct-graphene coated PS was ten times better than that of the direct-graphene coated flat Si [2, 3]. The enhanced conductivity is attributed to the presence of suitable quality graphene layer. The formation of which can further be improved by using gold (Au) catalysts. Graphene and graphene-based composites are used for high-performance Li-ion batteries (LIB) because of their excellent electronic and mechanical properties. The amorphous Si (a-Si) NPs and graphene composite prepared by Hummers method and thermal decomposition of silane gas (SiH_4) helps in reducing induced strain and stress occurring during volume expansion [4]. As reported, hierarchical nano graphene shell indicates elastic, sponge-like aspects, which minimizes the volume expansion of problem of Si during lithiation process. Additionally, graphene provides low diffusion barriers and stipulates enough free space to accommodate the fast migration of ions, which results in long-term cycling stability, and rate performance [5]. The composition of PS and reduced-graphene-oxide (rGO), fabricated by steam etching of SiNPs and rGO aerogel nano-sheets with nano-holes, establish a porous structure to condense SiNPs, that reduces the transfer distance of Li-ions which further controls the aggregation and destruction of Si particles [6]. The composite electrode shows high specific capacity and excellent cycling stability (1004 mAh/g at 50 mA/g up to 100 cycles). The graphene-based materials adsorb hydrogen through physical absorption (or physisorption) compared to graphite, which is due to better electrical conductivity, larger surface area, and higher chemical tolerance. PS coated with Pd NPs followed by GO exhibits hydrogen storage capabilities up to 2.1 wt% hydrogen [7]. The use of composites of GO and PS is extended to hydrogen gas sensor [8].

Chemical inertness of pristine Si due to the weak reaction between Si and hydrocarbons imposes a significant restriction of direct graphene growth. Hence, application of surface engineering techniques for the growth of graphene becomes a critical process when Si is used as a substrate. The graphene-wrapped SiNPs composite, used as an anode for LIB, increase the capacity of the battery because of the large surface area. This paper reports the fabrication of graphene-wrapped SiNPs by the ultra-sonication method.

2. Materials and preparation method

Si (100) wafers (p-type, B-doped, 0.01–0.02 Ω-cm, 250 μm thick) were used to fabricate PS by electrochemical anodization technique using a solution consisting of 1:2 volume mixture of aqueous 48% hydrofluoric acid (HF) and acetic acid (CH_3COOH) [9]. An anodization current density of 30 mA/cm^2 for 5, 10, and 20 minutes helps in the production the PS samples. After the anodization, the PS film was lifted up from its parent wafer and crushed.

GO was prepared by the modified Hummers method. In a three-necked round bottom flask, two gram of fine graphite powder and 46 ml of concentrated H_2SO_4 were taken and stirred for 30 mins.The temperature of the reaction was kept below 10^0 C by keeping an ice bath below the flask. Six gram of $KMnO_4$, which serves as an oxidant, was then added and the mixture was again stirred for 30 minutes. The ice bath was then removed, and the mixture was vigorously stirred for one hour. The temperature rose to 35^0 C in this period. Controlled addition of water up to 92 ml was done to prevent the explosion of the reactants. The mixture was stirred for one hour with the temperature maintained at 98^0 C. Finally 50 ml of 30% H_2O_2 was added to oxidize the unreacted graphite. Ultrasonic treatment achieved the formation of GO from graphite oxide. The dissolved GO (in DI water with GO: DI:: 1:2 ratio) was mixed with freestanding PS film and was put in the ultra-sonication for nine hours to form graphene-wrapped SiNPs [10]. The colloidal was dried at 60^0 C in an oven to obtain The SiNPs-graphene composite in dry powder form.

3. Result and Discussion

The morphology of the composite, as shown in **Figure 1** (a), indicates a curly flower structure of graphene composites wrapped with SiNPs, obtained from the ultra-sonication method. The FESEM image of the multilayered composite, as shown in **Figure 1** (b), shows the layered structure of the composite, where well-separated GO films fully encapsulate si particles. The TEM characterization techniques show the internal structure and crystal orientation of Si/graphene composite more precisely, as shown in **Figure 2**. The fringe spacing is 0.31 nm, corresponds to (111) plane of Si.

Figure 3 confirms the inter-atomic distance in graphene (0.207 nm) and Si (0.316 nm). **Figure 4** shows the SEAD pattern of graphene composite SiNPs for different L values (i.e., L=205 mm and 150 mm, respectively). Energy dispersive spectroscopic (EDS) mapping verifies the composition of the sample, as shown in **Figure 5**, which, shows the presence of Si, graphene (i.e., C), and graphene composites through elemental mapping.

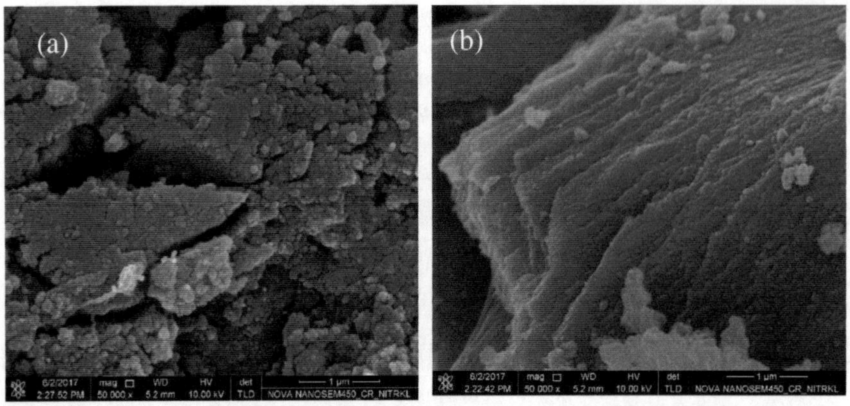

Figure 1. FESEM image of SiNPs-graphene composite (a) curly-flower structure (b) multi layered structure

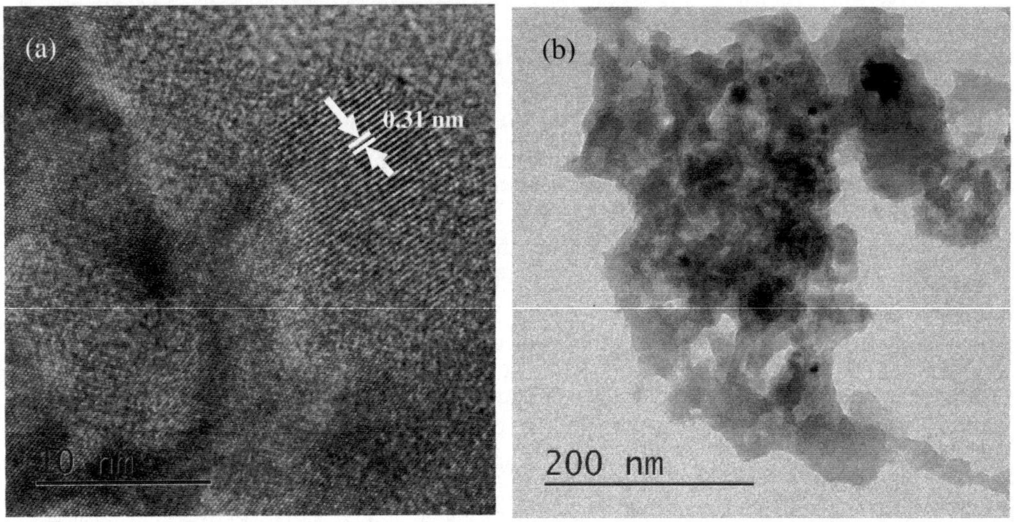

Figure 2. Graphene-wrapped SiNPs TEM image at (a) high resolution (b) low resolution

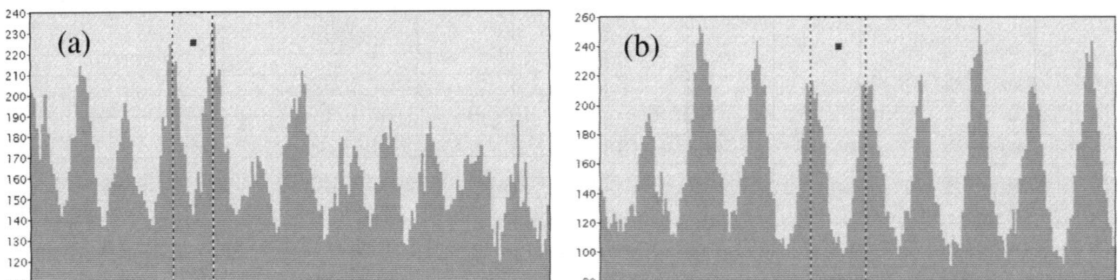

Figure 3. Interatomic distance profiles for (a) graphene 0.207 nm, and (b) Si 0.316 nm

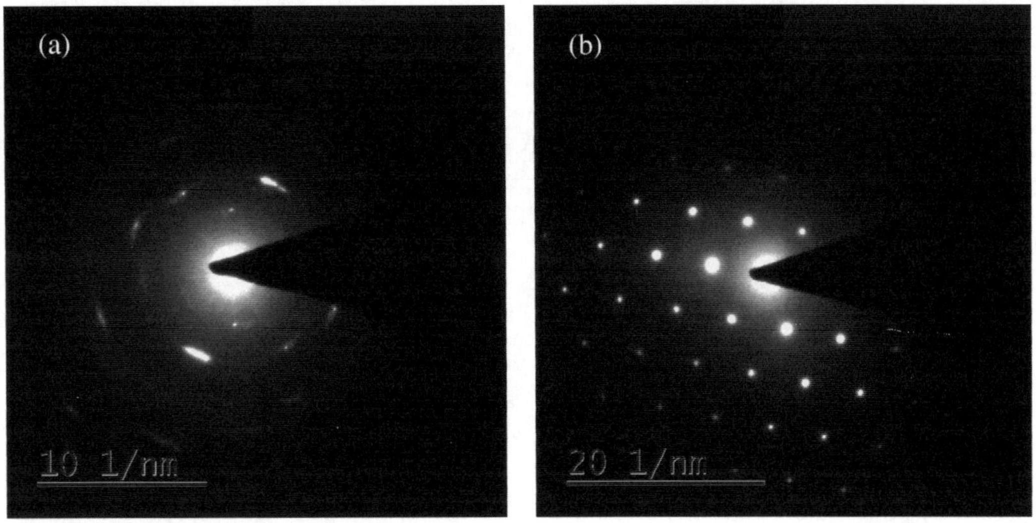

Figure 4. Diffraction pattern of SiNPs-graphene composite for (a) L=250 mm, and (b) L=150 mm

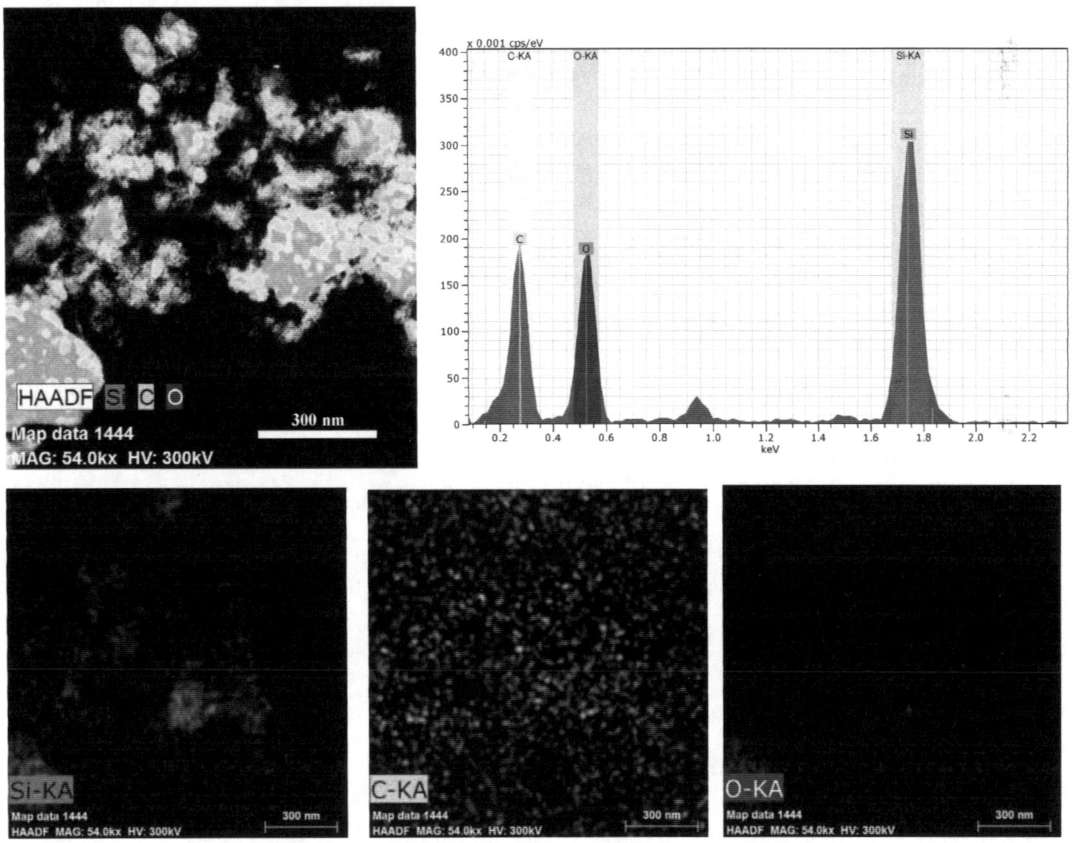

Figure 5. Elemental mapping of graphene-composite SiNPs

4. Conclusion

The PS, fabricated by electrochemical anodization, was coated with GO. The sample was characterized using FESEM, and HRTEM microscopy method. The characterization method confirms graphene and graphene composite SiNPs. The inter-atomic distance and crystal orientation of Si is 0.31 nm, (111) structure, respectively. The graphene-wrapped SiNPs can be used as the anode in LIB to overcome the lithiation problem.

References

[1] L. Oakes, A. Westover, J. W. Mares, S. Chatterjee, W. R. Erwin, R. Bardhan, S. M. Weiss and C. L. Pint, "Surface engineered porous silicon for stable, high performance electrochemical supercapacitors," *Scientific reports,* vol. 3, pp. 1-7, 2013.

[2] B. Su, X. Chen and E. Halvorsen, "Graphene synthesized on porous silicon for active electrode material of supercapacitors," in *Journal of Physics: Conference Series 773 (2016) 012057*, 2016.

[3] V. Starkov, D. Sedlovets, M. Knyazev and A. Red'kin, "Composite electrodes for current sources based on graphene-like films in porous silicon," *Protection of Metals and Physical Chemistry of Surfaces,* vol. 53, no. 1, pp. 85-87, 2017.

[4] M. Ko, C. S. and J. Cho, "Challenges in Accommodating Volume Change of Si Anodes for Li-Ion Batteries," *ChemElectroChem,* vol. 2, no. 11, pp. 1645-1651, 2015.

[5] C. Zhang, T. H. Kang and J. S. Yu, "Three-dimensional spongy nanographene-functionalized silicon anodes for lithium ion batteries with superior cycling stability," *Nano Research,* pp. 1-13.

[6] H. Tang, J. Zhang, Y. Zhang, Q. Xiong, Y. Tong, Y. Li, X. Wang, C. Gu and J. P. Tu, "Porous reduced graphene oxide sheet wrapped silicon composite fabricated by steam etching for lithium-ion battery application," *Journal of Power Sources,* vol. 286, pp. 431-437, 2015.

[7] Y. Honarpazhouh, F. R. Astaraei, H. R. Naderi and O. Tavakoli, "Electrochemical hydrogen storage in Pd-coated porous silicon/graphene oxide," *International Journal of Hydrogen Energy,* vol. 41, no. 28, pp. 12175-12182, 2016.

[8] H. G. Shiraz, "Efficient room temperature hydrogen gas sensing based on graphene oxide and decorated porous silicon," *International Journal of Hydrogen Energy,* vol. 42, pp. 15966-15972, 2017.

[9] P. Kale, P. Sharma and C. Solanki, "Synthesis and Characterization of Si Nanoparticles Obtained On Sonication of Porous Silicon Multilayer Films," *Journal of Nano Research,* vol. 17, pp. 13-25, 2012.

[10] P. Kale and C. Solanki, "Synthesis of si nanoparticles from freestanding porous silicon (PS) film using ultrasonication," in *Conference Record of the IEEE Photovoltaic Specialists Conference*, Honolulu, 2010.

IOP Conf. Series: Materials Science and Engineering **338** (2018) 012035 doi:10.1088/1757-899X/338/1/012035

Buckling analysis of SMA bonded sandwich structure – using FEM

Pankaj V Katariya*, Arijit Das, Subrata K Panda

Department of Mechanical Engineering, NIT Rourkela, Rourkela-769008, Odisha, India.

E-mail: pk.pankajkatariya@gmail.com

Abstract. Thermal buckling strength of smart sandwich composite structure (bonded with shape memory alloy; SMA) examined numerically via a higher-order finite element model in association with marching technique. The excess geometrical distortion of the structure under the elevated environment modeled through Green's strain function whereas the material nonlinearity counted with the help of marching method. The system responses are computed numerically by solving the generalized eigenvalue equations via a customized MATLAB code. The comprehensive behaviour of the current finite element solutions (minimum buckling load parameter) is established by solving the adequate number of numerical examples including the given input parameter. The current numerical model is extended further to check the influence of various structural parameter of the sandwich panel on the buckling temperature including the SMA effect and reported in details.

1. Introduction

The demand for the lighter and stronger materials has been increasing day by day in many aerospace and automotive industries. The structure as well as structural components of such industries experience the combined types of loadings (aerostatic, dynamic and aerodynamic heating) during their life cycle and are prone to buckle and can lose their stability because of the large geometrical distortion. Hence, the structural instability becomes a foremost concern in safe and reliable designs. The sandwich material strengthens via incorporating the Shape Memory Alloy (SMA) fibre, is one of the potential materials (ability to recover extremely large strains) which can handle such difficulties and perform the necessary function satisfactorily.

The buckling behaviour of the laminated as well as sandwich layered structures are studied by various researchers using different theories and few essential contributions in this field are presented here to establish the necessity of the current analysis. The finite element (FE) technique is adopted in the past [1-3] to investigate the structural responses (bending, vibration, buckling, post-buckling and acoustic) of SMA embedded composite structure of various geometries under the elevated thermal environment with the help of von-Karman strain-kinematics. Similarly, the static and the dynamic responses of SMA bonded structure are computed numerically as well as the experimental techniques [4] and compared with previously available 1-D models (Tanaka, Rogers, Liang and Brinson). The bifurcation type of buckling responses of the sandwich structure evaluated by Shiau and Kuo [5] using the higher-order triangular element for the sake of high-accuracy. Subsequently, the FE solutions of the nonlinear frequency values of SMA bonded layer composite structure obtained [6] via the first-order shear deformation theory (FSDT) and von-Karman nonlinear strain. In the recent past, the finite element method (FEM) is adopted to explore the buckling behaviour of the SMA fibre embedded composite structure [7-8] using the polynomial based mid-plane theories (FSDT and 3-D layer-wise theory; LWT). The first time, the higher-order shear deformation theory (HSDT) and Green-Lagrange strain-displacement relations are adopted by Panda and Singh [9] to compute the nonlinear fundamental frequencies of the SMA embedded layer composite spherical panel numerically using the FEM in association with the direct iterative method. Further, the sandwich plate structure bonded with SMA fibre modeled via Carrera's Unified Formulation (CUF) by Dehkordia and Khalili [10] for the frequency analysis using FEM technique and Kirchhoff's plate theory. Recently, the FEM technique in conjunction with the HSDT is adopted by Katariya *et al.* [11] to examine the buckling responses of SMA embedded sandwich shell panel structures including the geometrical nonlinearity via Green-Lagrange strain kinematics.

The review indicates that an extensive amount of study has already been reported on the buckling strength of the layered composite structure bonded with or without SMA fibres. Moreover, the reported studies are mainly utilized the FSDT kinematic model and von-Karman strain. Based on the authors' knowledge no study reported yet on the thermal buckling responses of SMA embedded sandwich structure via HSDT kinematics and Green's nonlinear strain-displacement relations. Hence, the current research aims to compute thermal buckling load parameter of SMA embedded sandwich flat panel structure numerically via the HSDT kinematics including the Green-Lagrange nonlinear strain and through thickness stretching effect. In order to compute the responses, a customized computer code (MATLAB environment) derived using the presently developed higher-order nonlinear FE model. The corresponding eigenvalue equation is derived by minimizing the total potential energy functional. After the necessary checking of the current numerical solution, it is engaged to compute the thermal load parameter of the layered sandwich structure bonded with and without SMA fibre and the final parametric (length-to-thickness ratio, aspect ratio and core-to-face thickness ratio) inferences discussed in details.

2. Mathematical modelling

The sandwich composite plate embedded with SMA fibre is considered here for the analysis purpose and Figure 1 represents the structural geometry. The geometrical dimension of the sandwich plate length, width and total thickness are represented as 'a', 'b' and 'h', respectively. In addition, the total thickness is the sum of two components i.e., core thickness 'h_c' and face sheet thickness 'h_f'.

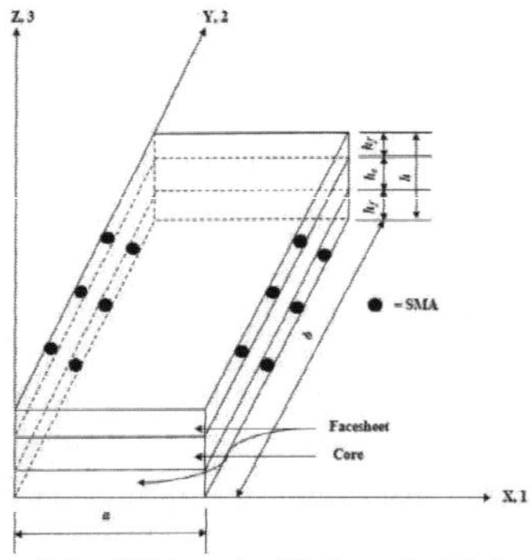

Figure 1. Configuration of the SMA embedded sandwich laminated composite plate

The higher-order displacement field kinematics [11] is adopted for the current modeling of layered sandwich panel embedded with SMA fibre and each node has ten degrees-of-freedoms (DOFs) per node. The panel model expressed mathematically in the following form as in Katariya *et al.* [11]:

$$\left.\begin{aligned} \bar{u} &= u_0 + z\theta_1 + z^2\phi_1 + z^3\zeta_1 \\ \bar{v} &= v_0 + z\theta_2 + z^2\phi_2 + z^3\zeta_2 \\ \bar{w} &= w_0 + z\theta_3 \end{aligned}\right\} \tag{1}$$

where, $(\bar{u},\bar{v},\bar{w})$ represents the displacement and (u_0, v_0, w_0) represents the mid-plane displacement of any point within the plate component along the principal material direction i.e. 1, 2 and 3, respectively. The rotation of normal to the mid-plane to the alternate axes represented as θ_1 and θ_2, respectively whereas θ_3 is the stretching through the thickness. Similarly, a few extra mathematical functions (Taylor's series expansion) i.e. $\phi_1, \phi_2, \zeta_1,$ and ζ_2 required to maintain the necessary stress profile for the sandwich panel.

The generalised strain-displacement relations utilised for the present SMA embedded layered sandwich plate structure as in Katariya *et al.* [11]:

$$\begin{aligned}\{\varepsilon_{Total}\} &= \{\varepsilon_{11}\ \ \varepsilon_{22}\ \ \varepsilon_{33}\ \ \gamma_{23}\ \ \gamma_{13}\ \ \gamma_{12}\} \\ &= \left\{\frac{\partial u}{\partial x}+\frac{w}{R_1}\ \ \frac{\partial v}{\partial y}+\frac{w}{R_2}\ \ \frac{\partial w}{\partial z}\ \ \frac{\partial v}{\partial z}+\frac{\partial w}{\partial y}+2\frac{w}{R_{12}}\ \ \frac{\partial u}{\partial z}+\frac{\partial w}{\partial x}-\frac{u}{R_1}\ \ \frac{\partial u}{\partial y}+\frac{\partial v}{\partial x}-\frac{v}{R_2}\right\}\end{aligned} \tag{2}$$

The required elastic constants associated with the SMA fibre and the composite are defined in Figure 1, along with the corresponding directions of the plate structure. Additionally, a uniform distribution of SMA including the close bonding with composite lamina is assumed throughout the analysis. The necessary elastic constants of SMA fibre within the plate structure are defined as same as in Panda and Singh [9]:

$$\left.\begin{aligned} &E_1 = E_{1m}V_m + E_sV_s, E_2 = \left(1\Big/\frac{V_s}{E_s}+\frac{V_m}{E_{2m}}\right) \\ &G_{12} = \left(1\Big/\frac{V_s}{G_s}+\frac{V_m}{G_{12m}}\right), G_{23} = G_{23m}V_m + G_sV_s \\ &\upsilon_{12} = \upsilon_{12m}V_m + \upsilon_sV_s, \alpha_2 = \alpha_{2m}V_m + \alpha_sV_s \\ &\alpha_1 = \left(\left(1\Big/\left(\frac{1}{\alpha_{1m}}+\frac{E_sV_s}{E_{1m}\alpha_{1m}V_m}\right)\right)+\left(1\Big/\left(\frac{E_{1m}V_m}{E_s\alpha_sV_s}+\frac{1}{\alpha_s}\right)\right)\right)\end{aligned}\right\} \tag{3}$$

Here, the subscripts 's' and 'm' indicate the SMA fibre and the composite matrix, respectively. Similarly, 'E' (elastic constant), 'G' (rigidity modulus), 'υ' (Poisson's ratio) and 'α' (linear thermal expansion co-efficient) are the necessary input material parameter assumed according to the analysis. Additionally, the corresponding volume fractions of the matrix material and the SMA fibre shown in Eq. (3) defined as 'V_m' and 'V_s', respectively.

Further, the thermoelastic constitutive relations of SMA embedded sandwich structure for any k^{th} lamina is derived from the source [9] by incorporating the blocking stress phenomena and conceded in the following form:

$$\{\sigma_{Total}\}^k = \left[\bar{Q}\right]^k \{\varepsilon_{Total}\}^k + \{\sigma_{recovery}\}^k V_s^k - \left(\left[\bar{Q}\right]_m \{\alpha\}_m V_m\right)^k \Delta T \tag{4}$$

where, $\{\sigma_{recovery}\}$ the recovery stress generated in SMA fibre due to the temperature increment (ΔT) and the details regarding the mathematical constituents (volume fractions, reduced transformed stiffness component and strain tensor etc.) in Eq. (4) can be seen in the reference [9]. Additionally, ΔT is the change in temperature increment due to the uniform temperature environment.

Now, the total energy (U) expression including the work done (W) due to the resultant in-plane thermal forces of the composite sandwich plate including the recovery forces are presented in the following lines by substituting the corresponding stress and strain values.

$$U = \frac{1}{2}\int_v \{\varepsilon_{Total}\}^T \{\sigma\} dV \tag{5}$$

$$W = \frac{1}{2}\int_A \{\varepsilon_G\}^T [D_G]\{\varepsilon_G\} dA \tag{6}$$

where, $\{\varepsilon_G\}$ and $[D_G]$ are the corresponding geometrical and the material property matrix.

The FEM has been adopted in the present analysis due to known advantages i.e. easy adoptability and capable of solving the complex structural problem (geometry and material related). In this regard, an isoparametric nine-noded quadrilateral element with ten DOFs per node is implementated for the domain discretisation and the modified displacement field variable $\{\delta\}$ is expressed as:

$$\{\delta\} = \sum_{i=1}^{9} N_i \{\delta_i\} \tag{7}$$

where, $\{\delta_i\} = [u_{0i}\ \ v_{0i}\ \ w_{0i}\ \ \theta_{1i}\ \ \theta_{2i}\ \ \theta_{3i}\ \ \phi_{1i}\ \ \phi_{2i}\ \ \zeta_{1i}\ \ \zeta_{2i}]^T$ is the nodal displacement filed vector and N_i is the interpolating functions associated with the 'i^{th} 'node.

Now, the modified form of the strain vector is rewritten after implementing nodal shape functions and represented as:

$$\{\bar{\varepsilon}\} = [B]\{\delta\} \tag{8}$$

where, $[B]$ is the generic form of the strain displacement relations.

Similarly, the energy and the work done expressions are modified further by substituting the nodal shape function variable (Eq. 7) in Eqs. (5) and (6) and the elemental matrices provided in the following lines.

$$U^e = \frac{1}{2}\int_A \left(\{\delta\}_i^T [K]^e \{\delta\}_i \right) dA + \{F_{\Delta r}\}_i - \{F_{\Delta T}\}_i \tag{9}$$

$$W^e = \frac{1}{2}\int_A \left(\{\delta\}_i^T [K_G]^e \{\delta\}_i \right) dA \tag{10}$$

where, $[K]^e = \int_{-1}^{+1}\int_{-1}^{1} [B]_i^T [D][B]_i |J| d\xi d\eta$, $\{F_{\Delta r}\}^e = \int_A [B]_i^T \{N_{\Delta r}\} dA$, $\{F_{\Delta T}\}^e = \int_A [B]_i^T \{N_{\Delta T}\} dA$,

$[K_G]^e = \int_{-1}^{+1}\int_{-1}^{+1} [B_G]_i^T [D_G][B_G]_i |J| d\xi d\eta$. $[K_G]^e$ represents the elemental geometrical stiffness matrix and $[B_G]$ is similar to the strain-displacement matrix obtained due to the geometrical distortion under the elevated thermal loading.

Finally, the governing equation of the buckled laminated composite sandwich plate bonded with SMA fibre is derived by minimizing the total potential energy (TPE) functional and expressed as:

$$\delta\Pi = 0 \tag{11}$$

where, $\Pi = U^e + W^e$

Now, the final form of the eigenvalue equation of the buckled smart laminated sandwich composite plate is expressed in the following line by substituting Eqs. (9) and (10) in Eq. (11).

$$\left([K_S] - T_{cr}[K_G]\right)\{\delta\} = 0 \tag{12}$$

where, $[K_S] = \left([K] + [K_{recovery}]\right)$ and $[K_G]$ are the global stiffness matrix and global geometrical stiffness whereas $[K_{recovery}]$ is the stiffness due to recovery stress behaviour of SMA. Additionally, T_{cr} represents the critical buckling temperature load.

3. Results and Discussion

The desired FE solutions for the thermal buckling load parameter of SMA bonded sandwich composite structure are obtained computationally via the suitable MATLAB code in association with higher-order model. In order to verify the stability and the degree of accuracy of the current numerical solution, the model is engaged to evaluate the buckling temperature for different mesh densities. Also, the values are compared with the published data. In this analysis, three different sets of end support conditions are utilized and provided in the following lines.

All edges clamped

$u_0 = v_0 = w_0 = \theta_1 = \theta_2 = \theta_3 = \phi_1 = \phi_2 = \zeta_1 = \zeta_2 = 0$ at $x = 0, a$ and $y = 0, b$

All edges simply-supported (SS)

$v_0 = w_0 = \theta_2 = \theta_3 = \phi_2 = \zeta_2 = 0$ at $x = 0, a$

$u_0 = w_0 = \theta_1 = \theta_3 = \phi_1 = \zeta_1 = 0$ at $y = 0, b$

Two opposite edges clamped and simply-supported

$u_0 = v_0 = w_0 = \theta_1 = \theta_2 = \theta_3 = \phi_1 = \phi_2 = \zeta_1 = \zeta_2 = 0$ at $x = 0, a$

$u_0 = w_0 = \theta_1 = \theta_3 = \phi_1 = \zeta_1 = 0$ at $y = 0, b$

For the computation purpose, the following material parameters of SMA laminated composites Kumar and Singh [8] are utilised:

Composite matrix: (graphite/epoxy)

$E_1 = 155\times10^9 N/mm^2; E_2 = 8.07\times10^9 N/mm^2; G_{12} = G_{13} = 4.55\times10^9 N/mm^2; G_{23} = 3.25\times10^9 N/mm^2;$

$\upsilon_{12} = \upsilon_{13} = 0.22;\ \upsilon_{23} = 0.3488; \alpha_0 = 10^{-6}; \alpha_1 = -0.07\times\alpha_0; \alpha_2 = \alpha_3 = 30\times\alpha_0$

SMA fibres - Nitinol: E is obtained from Kumar and Singh [8] and Park *et al.* [6] depending on the temperature increment and the recovery stress is obtained from Park *et al.* [6] depending on the prestrain of the SMA fibres $\left(\varepsilon_r = 1\%\right)$ and the temperature increment.

$E_0 = 0.1\times10^9 N/mm^2; G = 24.86\times10^9 N/mm^2; \upsilon_{12} = 0.33; \alpha_1 = 10.26\times10^{-6}/^\circ C$

3.1 Convergence and validation study

In order to check the required convergence behaviour and the validity, a square simply-supported (45°/S/S/45°) laminated composite plate bonded with SMA fibre example is solved including the material properties as defined earlier. For the computation purpose, the volume fraction of SMA (V_S) fibre 0.1, prestrain (ε_r) 1% and two length-to-thickness ratios (a/h = 50 and 100) are taken. In addition, the nondimensional critical buckling temperature is obtained using $T_{cr} = 10^3 \lambda \alpha_0 \Delta T$. Now, to check and demonstrate the convergence behaviour of the presently developed numerical model, the responses are obtained for a different number of elements and presented in Figure 2. From the figure, it is evident that the presently obtained results are showing very good rate of convergence with the mesh refinement. In this study, a (5×5) mesh size has been utilised for the computation of new results as well as the comparison purpose. The comparison study indicate an acceptable range of differences (≤10%) between the present and reference values.

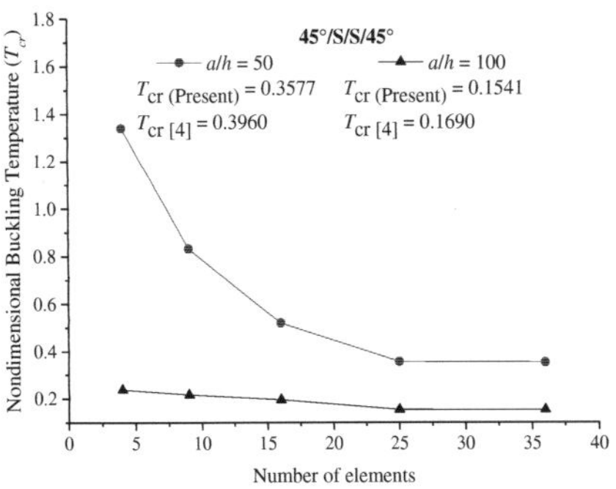

Figure 2. Convergence and comparison study of buckling temperature parameter

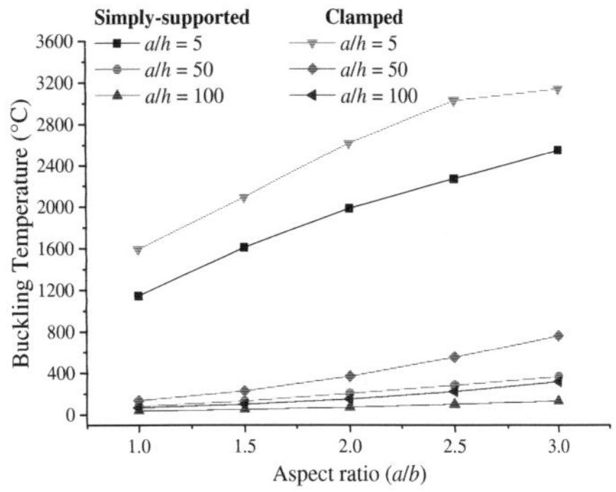

Figure 3. Effect of aspect ratios (a/b) on buckling temperature of sandwich plate embedded with SMA fibre (h_c/h_f = 25)

3.2 Numerical examples

The buckling responses are obtained for sandwich plate embedded with SMA fibre using different geometrical parameters and discussed in the detail in upcoming subsections. For the computation purpose, the SMA volume fraction (V_S) is considered as 0.1 and prestrain (ε_r) 1% until unless mentioned elsewhere. In the present research, the responses are obtained for (*S*/45°/-45°/*C*/-45°/45°/*S*) lamination scheme. It is important to mention that SMA properties for the present analysis are similar to the properties as mentioned in [8] whereas the material properties of the sandwich are taken same as [10]. The detailed properties of the face sheet and core layer of the sandwich construction are:

Facesheet: $E_1 = 275\times10^9 N/mm^2; E_2 = E_3 = 6.9\times10^9 N/mm^2; G_{12} = G_{23} = G_{13} = 6.9\times10^9 N/mm^2;$

$\upsilon_{12} = \upsilon_{13} = 0.25; \upsilon_{23} = 0.3$

Core: $E_1 = E_2 = E_3 = 0.5776\times10^9 N/mm^2; G_{12} = G_{13} = 0.1079\times10^9 N/mm^2; G_{23} = 0.22215\times10^9 N/mm^2;$

$\upsilon_{12} = \upsilon_{23} = \upsilon_{13} = 0.0025$

3.2.1 Buckling load variation due to change in aspect ratio (a/b)

The buckling temperature of SMA embedded sandwich plate is obtained for two different support conditions (SS and clamped) by varying five a/b = 1, 1.5, 2, 2.5 and 3 and three a/h = 5, 50 and 100 values. The temperature values are computed by taking the core-to-face thickness ratio as h_c/h_f = 25 and presented in Figure 3. The temperature parameter of the smart sandwich structure is increasing while the aspect ratios (a/b) increase whereas follow a reverse trend with a/h values. The responses are following an expected kind of behaviour i.e. the buckling temperature increase or decrease while the global structural stiffness values altered according to the related geometrical parameters (length-to-thickness ratio and aspect ratio) since it is directly proportional to the stiffness.

3.2.2 Variation of thermal buckling load due to different length-to-thickness ratio (a/h)

In this example, the buckling temperature of SMA embedded square sandwich plate is obtained to show the influence of the length-to-thickness ratios (a/h) and presented in Figure 4. For the computation purpose, two different support conditions (all edges simply supported and clamped) are considered with three core-to-face thickness ratios (h_c/h_f = 5, 15 and 25) and seven length-to-thickness ratios (a/h = 2, 5, 10, 15, 20, 25 and 50). From the figure, it is understood that the buckling temperature values are following the increasing trend for both the length-to-thickness ratios (a/h) as well as the core-to-face thickness ratios (h_c/h_f). The reason behind this is already discussed earlier and it is well known that the *a*/*h* increases, the structure becomes thin and it can be buckle easily.

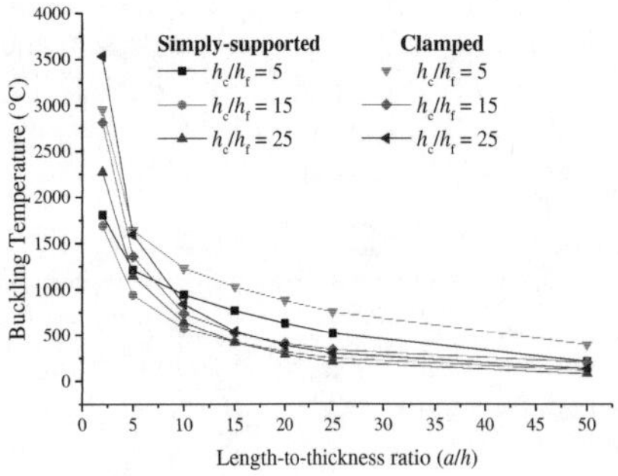

Figure 4. Variation of thermal buckling load parameter of SMA bonded sandwich plate due to different length-to-thickness ratios (*a*/*h*)

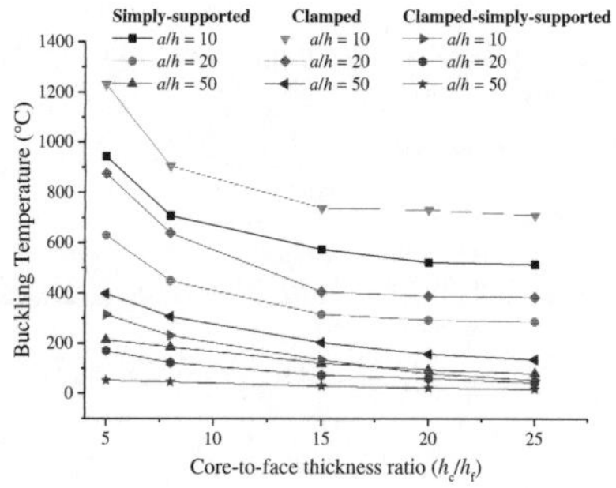

Figure 5. Core-to-face thickness ratio (h_c/h_f) influence on thermal buckling temperature of SMA bonded sandwich plate (*a*/*b* = 1)

3.2.3 Core-to-face thickness ratio (h_c/h_f) influence on thermal buckling

Now, one more example of square sandwich plate problem is solved for five, h_c/h_f = 5, 10, 15, 20 and 25 and presented in Figure 5. For the computational purpose, the necessary geometrical parameters are taken as: three length-to-thickness ratios (*a*/*h* = 10, 20 and 50) and three support conditions (SS, clamped and mixed). The buckling temperature curve is declining while the h_c/h_f value increases.

4. Conclusions

The buckling temperature of SMA embedded sandwich composite plate is obtained numerically via a suitable computer code in conjunction with the current higher-order kinematic model. The numerical model is developed by taking both the geometrical and material nonlinearity through the Green-Lagrange strain and marching technique. The final governing eigenvalue equations are solved using the isoparametric FEM steps and the correctness of the current numerical solutions verified by solving the adequate number of examples. After the necessary verification, the model is extended suitably to evaluate the influence of either one or more parameter on the buckling temperature parameter of SMA bonded sandwich structure including the recovery effect. The computed values indicate that the clamped sandwich structure show higher values of temperature load irrespective of geometrical parameters and SMA bonding.

References

[1] Rogers C A, Liang C and Jia J 1991 *Comput Struc* **38** 569–580
[2] Duan B, Tawfik M, Goek S N, Ro J J and Mei C 2000 *Proceedings of SPIE* **3991** 358–362
[3] Tawfik M, Ro J J and Mei C 2002 *Smart Mater Struct* **11** 297–307
[4] Zak A J, Cartmell M P, Ostachowicz W M and Wiercigroch M 2003 *Smart Mater Struct* **12** 338–346
[5] Shiau L C and Kuo S Y 2004 *Mech Based Des Struct Mach* **32** 57–72
[6] Park J S, Kim J H and Moon S H 2004 *Compos Struct* **63** 179–188
[7] Kuo S Y, Shiau L C and Chen K H 2009 *Compos Struct* **90** 188–195
[8] Kumar S K and Singh B N 2009 *J Aerosp Eng* **22** 342–353
[9] Panda S K and Singh B N 2013 *Aerosp Sci Technol* **29** 47–57
[10] Dehkordia M B and Khalili S M R 2015 *Compos Struct* **123**, 408-419
[11] Katariya P V, Panda S K, Hirwani C K, Mehar K and Thakare O 2017 *Smart Struct Syst* **20** 595–605

NANO-INDENTATION OF ALUMINIUM REINFORCED METALLIC GLASS COMPOSITES: A MOLECULAR DYNAMICS STUDY

D Yadav[*1], P Gupta[2], and N Yedla[2]
[1]Computational Materials Engineering Group Metallurgical and Materials Engineering Department, National Institute of Technology, Rourkela, India-769008

Email: deepakyadavkaos@gmail.com

Abstract
Molecular dynamics (MD) simulations are performed for nanoindentation on metal (Al)-metallic glass ($Cu_{50}Zr_{50}$) reinforced composites to investigate the mechanical properties and the effects of volume percentage on behavior of the load-displacement curves. The interaction among Al-Cu-Zr is modelled using a EAM (Embedded Atom Method) potential. Simulation box size of 100 Å (x) × 100 Å (y) × 100 Å (z) is modelled for investigating the properties of the sintered models by altering the volume percentage on the scale of 5%–20%. Nanoindentation is done along y-direction with a spherical diamond indenter at temperature of 300 K with constant indentation speed of 100 m/s. NVT ensemble is used with a timestep of 0.002 ps. Investigations on the effect of volume percentage show that as volume percentage of Metallic Glass (MG) increases, the corresponding Load required to penetrate inside the sample also increases. As a result of this Hardness also increase as volume percentage varies from 5% to 20%.

Keywords: Molecular Dynamics, Sintering, Nano- Indentation, Metal Matrix composites (MMC's), Volume percentage.

1. Introduction

The exceptional mechanical properties of aluminium based MMC's (metal matrix composites) are of great interest due to reduced density, high strength, enhanced elastic modulus as well as better wear resistance, and these exceptional mechanical properties make these MMC's captivating for utilization in automotive, defense and aerospace [1]. At ambient as well as elevated temperatures, ceramic materials like Al_2O_3 or SiC shows an outstanding combination of both stiffness and specific strength, which make these as commonly used reinforcements [2,3]. Ceramic reinforcement materials have porous interface due to their low wettability, which is a major problem because of increased corrosion sensitivity and ebbed mechanical properties of the composite [4]. Cu-Zr metallic glass (MG) as a reinforcement material in Al-alloy matrix is being used for the first time [5]. In recent times, Al alloy matrix is being reinforced with various types like Ni, Fe, Al, Cu and Zr based of MGs particles. They reports all in all enhancement in properties of Al alloy [5–7]. As a result, MG reinforcements as a potential substitute for ceramic reinforcements could be considered in MMC's.

Consolidation of powders with the help of thermal energy, which usually happens at about 2/3 of the melting temperature, is called sintering. In present work sintering is preferred over other manufacturing process because of the lower transition temperature of MG then the melting point temperature of the metal (Al). The other main advantages of this process over other manufacturing methods are less energy consumption, high ability to get complex shapes, dimensional precision in final product, excellent surface finish, repeatability and high reliability. Thus, sintering is one of the most economical manufacturing processes, especially in mass production [8]. In the present work Al based MMC's reinforced with Cu based MG ($Cu_{50}Zr_{50}$) nano-particles will be developed by sintering route through Molecular Dynamics (MD) simulations with volume percentage of MG to be 5% and 20% and termed as Model-I and Model-II respectively. Nano-indentation tests are done on this models to investigate the deformation behavior and response on the mechanical properties of the composite such as maximum load, hardness and elastic modulus through load-displacement plots [9].

2. Simulation method

2.1. Interatomic potentials

In MD simulations the interactions between atoms are calculated through the interatomic potential functions. Since in this work sintering of aluminium reinforced MG composites is investigated, there is a need for proper potential functions for the interactions between Copper and Zirconium atoms in Metallic glass and aluminium with Metallic glass. In order to obtain the forces between atoms, EAM potential is employed [10].
In present study we have done MD simulation using LAMMPS (Large Scale Atomic/Molecular Massively Parallel Simulator) an open source simulator developed by Sandia Laboratory is being employed [11]. The interaction among Al-Cu-Zr is modelled using a Zhou et al. EAM potential. For force calculations a cut-off distance of 7.1 Å is used. The potential is valid and other details of parameters can be obtained from Ref. [12]. Moreover, the potential is used by several researchers. Phase transitions in nanoparticles like FeCo and FeNi have been studied by Meng et al. [13]. Zhou et al. potential is used to study oxidation of Al layers on $Ni_{65}Co_{20}Fe_{15}$ surface [14]. Moreover, effect of temperature and strain rate on metal- metallic glass interface has been modelled using the above potential [15].

2.2. Modelling

The atomic position snapshot of Al (metal)-$Cu_{50}Zr_{50}$ (MG) model before and after sintering is shown in Fig. 1(a) & 1(b), with dimension as 10 nm (width) × 10 nm (height) × 10 nm (thick). To acquire Al (metal)-$Cu_{50}Zr_{50}$ (MG) sintered model we start with construction of Al and $Cu_{50}Zr_{50}$ nano clusters. The centre is filled with $Cu_{50}Zr_{50}$ alloy clusters, which is achieved through randomly substituting 50% of Cu atoms by Zr atoms and rest with Al clusters. The separation distance is set more than the equilibrium bond lengths, i.e., 3.5 Å. $Cu_{50}Zr_{50}$ MG is obtained by rapid cooling of $Cu_{50}Zr_{50}$ alloy from the temperature of 2300 K to 300 K at a cooling rate of 8.5×10^{11} K/s (timestep = 0.002 ps) with constant number of atoms (N), constant pressure and constant temperature (T) i.e NPT ensemble using the periodic boundary conditions. The sintered model then is equilibrated for 500 ps for relieving stresses generated during rapid cooling [3,5].

Indentation process was investigated using a spherical indenter of radius 15 Å. Fig 1(c) & 1(d) shows the model before and after indentations. The indenter deformability and thermal oscillations of indenter atoms were neglected. The hardness of Aluminium reinforced metallic glass composites is considerably lower than a diamond and therefore the rigid body model for the indenter is fully applicable in this model [16].

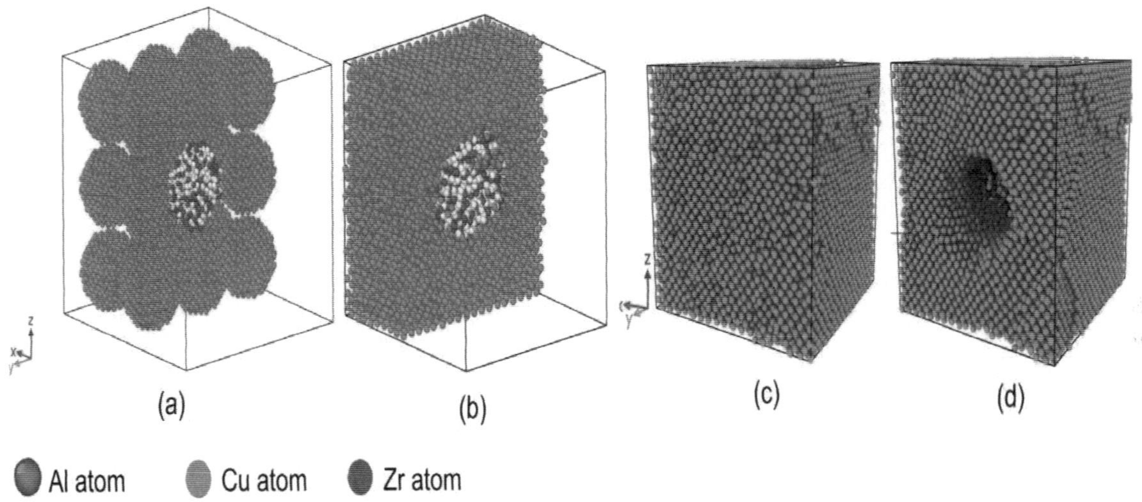

Figure 1: (a) initial model before sintering, (b) model after sintering, (c) model before indentation, and (d) model after indentation.

3. Results and discussion

Pressure sintering is used to model the two different models

1. *Model-I*: Volume percentage of MG in the model is taken as 5 %.
2. *Model*-II: Volume percentage of MG in the model is taken as 20 %.

Fig 1(a) & 1(b) shows the model-I with volume percentage of metallic glass to be 5%, before and after sintering. Fig 2(a) & 2(b) shows the model-I with volume percentage of metallic glass to be 20%, before and after sintering.

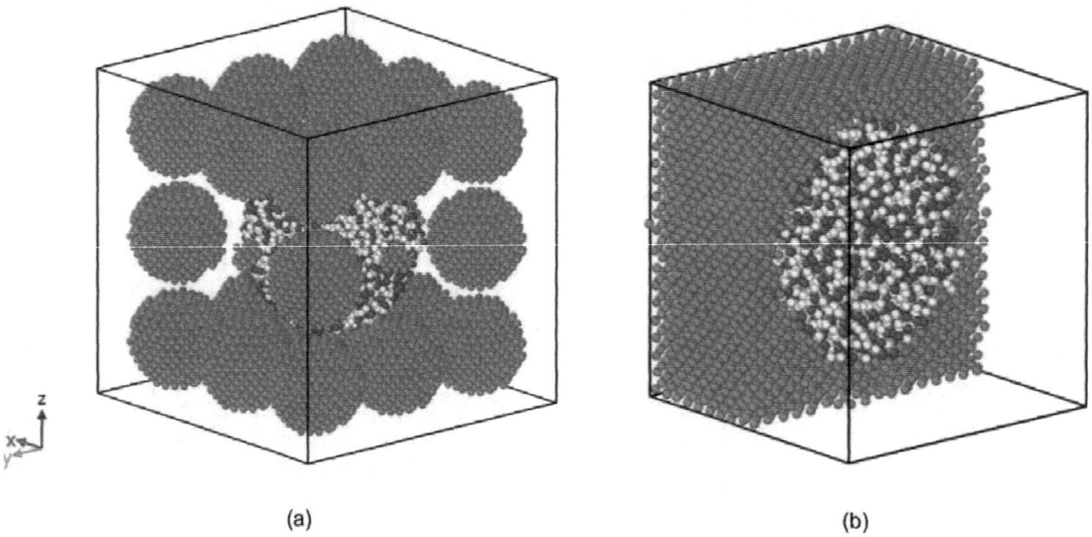

Figure 2: (a) initial model before sintering, (b) model after sintering, (c) model before indentation, and (d) model after indentation.

Nanoindentation process comprises of two stages in MD simulations, one being the equilibrium stage and other being the indentation stage. In equilibrium stage, the crystal surface and the indenter were positioned ~10 Å apart from each other and the system was simulated at 300 K for ~120 ps. In the indentation stage of the simulations, indenter is moved approximately 35 Å downward as per the modelled process at a perpetual speed of 100 m/s, and then retracted upwards at a perpetual speed of 50 m/s.

During displacement controlled indentations (depth-sensing approach), indenter is pressed against the surface up to a specified depth in the sample. The data of the load applied to the indenter and the displacement of the indenter is represented in *Fig 3(a),* which shows that, higher loads are required to penetrate inside the sample for higher MG vol%.

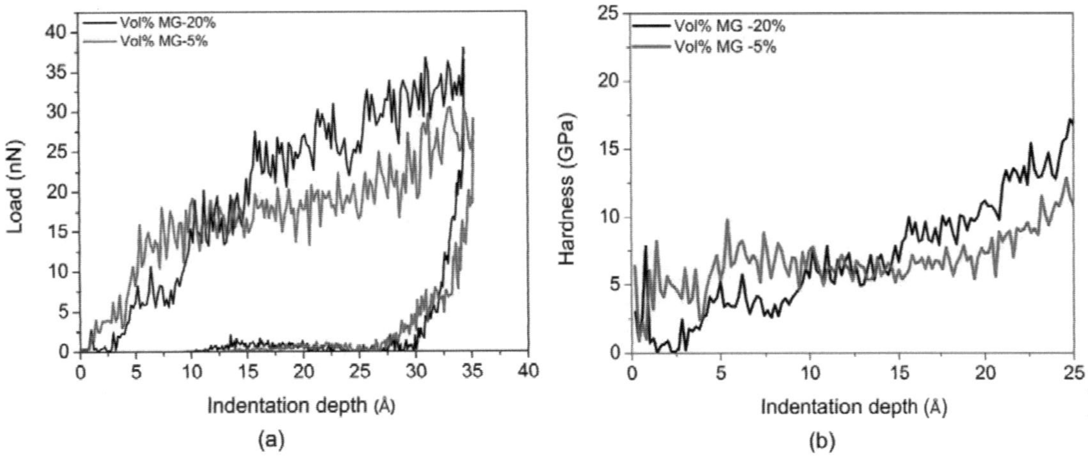

Figure 3: (a) Load-indentation depth plot (b) Hardness – indentation plot

Depending upon the Load vs indentation depth plot, Hardness has be calculated using the expression, H=P/A (where H is hardness, P is load and A is projected area) and plotted in *Fig 3(b).* Hardness increases as load required to penetrate inside the sample increases. Thus, as MG Vol% is higher, load required is higher and so is the hardness [17].

4. Conclusion

- As volume percentage of MG increases, the corresponding Load required to penetrate inside the sample also increases.
- Hardness of the sample increases with increase in volume percentage of MG and vice versa.

5. References

[1] M. Rosso, Ceramic and metal matrix composites: Routes and properties, J. Mater. Process. Technol. **175** (2006) 364–375.

[2] L. Zhang, H. Xu, Z. Wang, Q. Li, J. Wu, Mechanical properties and corrosion behavior of Al/SiC composites, J. Alloys Compd. **678** (2016) 23–30.

[3] M. Rahimian, N. Parvin, N. Ehsani, Investigation of particle size and amount of alumina on microstructure and mechanical properties of Al matrix composite made by powder metallurgy, Mater. Sci. Eng. A. **527** (2010) 1031–1038.

[4] J. Rams, M. Campo, A. Ureña, Sol-gel coatings to improve processing of aluminium matrix SiC reinforced composite materials, J. Mater. Res. **19** (2004) 2109–2116.

[5] P. Gupta, S. Pal, N. Yedla, Molecular dynamics based cohesive zone modeling of Al (metal)–Cu 50 Zr 50 (metallic glass) interfacial mechanical behavior and investigation of dissipative mechanisms, Mater. Des. **105** (2016) 41–50.

[6] R. Zheng, H. Yang, T. Liu, K. Ameyama, C. Ma, Microstructure and mechanical properties of aluminum alloy matrix composites reinforced with Fe-based metallic glass particles, Mater. Des. **53** (2014) 512–518.

[7] D. V Dudina, K. Georgarakis, M. Aljerf, Y. Li, M. Braccini, A.R. Yavari, A. Inoue, Cu-based metallic glass particle additions to significantly improve overall compressive properties of an Al alloy, Compos. Part A Appl. Sci. Manuf. **41** (2010) 1551–1557.

[8] M. Tavakol, M. Mahnama, R. Naghdabadi, Shock wave sintering of Al/SiC metal matrix nano-composites: A molecular dynamics study, Comput. Mater. Sci. **125** (2016) 255–262.

[9] A. Kumar, P. Gupta, N. Yedla, Nanoindentation studies of Zr50Cu50 metallic glass thin film nanocomposites via molecular dynamics simulations, Metall. Res. Technol. **113** (2016) 602.

[10] S.J. Zhou, D.L. Preston, P.S. Lomdahl, D.M. Beazley, Large-scale molecular dynamics simulations of dislocation intersection in copper, Science (80-.). **279** (1998) 1525–1527.

[11] S. Plimpton, Fast parallel algorithms for short-range molecular dynamics, J. Comput. Phys. **117** (1995) 1–19.

[12] [X.W. Zhou, R.A. Johnson, H.N.G. Wadley, Misfit-energy-increasing dislocations in vapor-deposited CoFe/NiFe multilayers, Phys. Rev. B. **69** (2004) 144113.

[13] [L.J. Meng, X.Y. Peng, K.W. Zhang, C. Tang, J.X. Zhong, Structural phase transitions of FeCo and FeNi nanoparticles: A molecular dynamics study, J. Appl. Phys. **111** (2012) 24303.

[14] B. Jeon, S.K.R.S. Sankaranarayanan, S. Ramanathan, Atomistic Modeling of Ultrathin Surface Oxide Growth on a Ternary Alloy: Oxidation of Al− Ni− Fe, J. Phys. Chem. C. **115** (2011) 6571–6580.

[15] N. Yedla, P. Gupta, Effect of Strain Rate and Temperature on Al (Metal)-Cu50Zr50 (Metallic Glass) Interface Strength: A Molecular Dynamics Study, (2016).

[16] G.B. Sushko, A. V Verkhovtsev, A. V Yakubovich, A. V Solov'yov, Molecular dynamics simulation of nanoindentation of nickel-titanium crystal, in: J. Phys. Conf. Ser., IOP Publishing, 2013: p. 12021.

[17] Y. Zhu, L. Zhou, Y.X. Yao, G.W. Kang, Molecular Dynamics Simulation for Single Crystal Aluminum Nano-Indentation Effect by Indenter Radius, in: Appl. Mech. Mater., Trans Tech Publ, 2008: pp. 401–405.

Fabrication and characterization of nano-Y_2O_3 and Al_2O_3 dispersed W-Ni alloys by mechanical alloying and pressureless conventional sintering

V. R. Talekar, A. Patra[1], S. K. Karak

Department of Metallurgical and Materials Engineering, National Institute of Technology, Rourkela-769008, Orissa, India

Corresponding author Email : patraa@nitrkl.ac.in/anspat.met@gmail.com

Abstract: Nano Y_2O_3 and Al_2O_3 dispersed W-Ni alloys with nominal composition of $W_{89}Ni_{10}(Y_2O_3)_1$ (alloy A), $W_{89}Ni_{10}(Al_2O_3)_1$ (alloy B) were mechanically alloyed for 10 h followed by compaction at 0.5 GPa pressure with 5 min of dwell time and conventional sintering at 1400°C with 2 h soaking time in Ar atmosphere with Ar flow rate of 100 ml/min. The microstructure of milled and sintered alloy was investigated using X-ray Diffraction (XRD), Scanning electron Microscopy (SEM), Energy dispersive spectroscopy (EDS) and Elemental mapping. Minimum crystallite size of 31.9 nm and maximum lattice strain, dislocation density of 0.23%, $9.12(10^{16}/m^2)$ respectively was found in alloy A at 10 h of milling. Uneven and coarse particles at 0 h of milling converted to elongated flake shape at 10 h of milling. Bimodal (fine and coarse) particle size distribution is revealed in both the alloys and minimum particle size of 0.69 μm is achieved in 10 h milled alloy A. Evidences of formation of intermetallic phases like Y_2WO_6, Y_6WO_{12} and $Y_{10}W_2O_{21}$ in sintered alloy A and $Al_2(WO_4)_3$, $NiAl_{10}O_{16}$, $NiAl_2O_4$ and $AlWO_4$ in sintered alloy B were revealed by XRD pattern and SEM micrograph. Minimum grain size of 1.50 μm was recorded in sintered alloy A. Both faceted and spherical W matrix is evident in both the alloys which suggests occurrence of both solid phase and liquid phase sintering. Maximum % relative sintered density and hardness of 85.29% and 5.13 GPa respectively was found in alloy A. Wear study at 20N force at 25 rpm for 15 min on ball on plate wear tester revealed that minimum wear depth (48.99 μm) and wear track width (272 μm) was found for alloy A as compared to alloy B.

1. Introduction

Tungsten is one of the desired choice as plasma facing material (PFM) in nuclear reactors and it also have several applications in other fields like electronics and electrical. This is due to its excellent properties like enhanced melting point (3410 °C), appreciable high temperature mechanical strength, maximum density (19.3 g/cm^3), 9.8 GPa hardness, elevated elastic modulus (411 GPa at 20 °C) in tensile mode and high thermal conductivity (155 W/m.K at 20 °C) [1-3]. Nanocrystalline materials display superior hardness, yield strength as compared to bulk materials [4, 5]. Improvement in mechanical properties of W after refinement of microstructure using mechanical alloying has been reported in literatures [6, 7]. It has been found that alloying of W helps to improve fabricability and also leads to improve utilization of tungsten. Addition of Ni to W imparts liquid phase sintering. Also

[1] patraa@nitrkl.ac.in/anspat.met@gmail.com (Prof. A. Patra)

Ni addition reduces Peierls stress by changing core structure of dislocation core which leads to improve ductility of W matrix [8, 9]. Y_2O_3 addition to W reduces the grain growth at high temperature and improves strength, hardness and wear resistance of W [10]. Nano-dispersed Y_2O_3 also increases creep resistance of W by pinning grain boundary at high temperature [11].

The present work implicates exploration of microstructure and mechanical properties (hardness, wear) of various oxides (Y_2O_3 and Al_2O_3) dispersed W based nanostructured alloys and investigate the superior dispersoid among them.

2. Materials and methods

W, Ni, Y_2O_3 (99.5% purity, Sigma Aldrich) with original particle size of <100 μm (W, Ni) and <50 nm (Y_2O_3 and Al_2O_3) has been mechanically alloyed by using high energy planetary ball mill (Fritsch, P5). During milling, powder sample at 0, 1, 5 and 10 h was taken out for characterization. The composition of the alloys is displayed in Table 1.

Table 1: Chemical composition selected for mechanical alloying.

Alloy	Composition (wt. %)			
	W	Ni	Y_2O_3	Al_2O_3
A	89	10	1	0
B	89	10	0	1

Compaction of 10 h milled powders (to get cylindrical pellets) has been carried out in a uniaxial press at 500 MPa for 5 min dwell time. The compacted pellets are sintered in a tunnel type sintering furnace (conventional sintering) at 1400°C for 2 h under continuous argon flow (100 ml/min). For particle size distribution of milled powder and grain size distribution of sintered powder 50 numbers of particles and grains has taken under consideration from microstructure SEM image of milled powder and sintered alloy respectively. For recording microhardness of sintered alloy, Vickers microhardness equipment (LECO, LM248AT) with 100-gf load and 10 s dwell time is used for hardness measurement. Wear study has been carried out at 20 N load and 25 rpm for 15 min on ball on plate wear tester (Make: DUCOM, Model: TR-208-M1) with initial theoretical track diameter of 4 mm. Methods used for various analysis were described in earlier reports [4].

3. Result and Discussion

3.1. Study of Phases in milled powder

XRD patterns study of milled alloy A and alloy B at various milling times (0, 1, 5 and 10) shows Decrease in intensity with increase (0.1968 at 0 h to 0.2952 at 10 h for alloy A, 0.4428 at 0 h to 0.4920 at 10 h for alloy B) in broadening (FWHM) of W (110) diffraction peaks with increase in milling time is obvious in both the alloys. This is anticipated to decrease in crystallite size of W and development of lattice strain [12]. Intensity of Ni (111) peak increases in both the alloys at 1 h of milling may reflect towards some localized cold welding of Ni particles and relaxation of lattice strain energy. Lowest crystallite size of 31.9 nm and highest lattice strain of 0.23% has been found for alloy A after 10 h milling. The rate of decrease of crystallite size of W and rate of increase of lattice strain is comparatively higher at 5 h of milling followed by decelerating the rate for both the alloys. This is because with advancement of milling time the refined particles are difficult for further size reduction owing to enhancement in required stress for dislocation formation or defect generation [6]. The Peierls Nabarro stress enhances with reduction in crystallite size and leads to decline in the crystallite size reduction rate with enhancement in dislocation density [4]. Dislocation density of alloy A ($9.12\times10^{16}/m^2$) is found higher than alloy B ($3.32\times10^{16}/m^2$) after 10 h of milling. This is because of maximum lattice strain achieved in alloy A against alloy B at 10 h of milling.

3.2. Microstructure of milled powder

Figure 1 represents morphology of milled powder at 10 h of milling of both the alloys. Uneven and coarse particle (round shape prior to milling) at zero h of milling are converted to both elongated and flake shape with the uniform size distribution of particles at 10 h of milling for both the alloys. Minimum particle size was evident in alloy A (0.69 μm) as compared to alloy B (1.69 μm) at 10 h milling. Particle Size distribution of fine (< 10 μm) and course (> 10 μm) particle was found to be 54 % and 46 % for alloy A, 24% and 76 % for alloy B respectively. Homogeneous distribution (almost equal percentage of fine and coarse particles) particles in alloy A causes the proper filling of interstitial voids between coarse particles which ultimately leads superior green and sintered density as compared to alloy B.

Figure 1. SEM images of powder morphology of (a) alloy A, (b) alloy B at 10 h of milling.

3.3. Phase analysis of sintered alloy

XRD pattern in Figure 2 (a) for alloy A and (b) for alloy B displays the phase formation in the alloys. Formation of intermetallic phases like Y_2WO_6, Y_6WO_{12} and $Y_{10}W_2O_{21}$ in sintered alloy A and $Al_2(WO_4)_3$, $NiAl_{10}O_{16}$, $NiAl_2O_4$ and $AlWO_4$ in sintered alloy B were revealed by XRD. For alloy A the W (110) and W (200) peak in milled powder (corresponding angle: 47.2302° and 68.966° respectively) are moved to lower diffraction angle (47.1132° and 68.8027° respectively) whereas for alloy B these peaks are shifted from (diffraction angle: 47.1611° and 68.9267°) in milled powder to (47.1007° and 68.8441°) in sintered alloy. This is attributed to W lattice distortion [13]. Peak broadening in sintered alloy is decreased as compared to milled powder at 10 h. For alloy A for (200) diffraction peak broadening of W in milled powder (FWHM: 0.5412) has decreased (FWHM: 0.3936) to lower value after sintering whereas for alloy B the W (200) peak broadening in milled powder (FWHM: 0.4920) has decreased (FWHM: 0.4428) to lower value after sintering. Coarsening of grain and lattice strain relaxation during sintering are the main reason for decreasing broadening of XRD peaks [4].

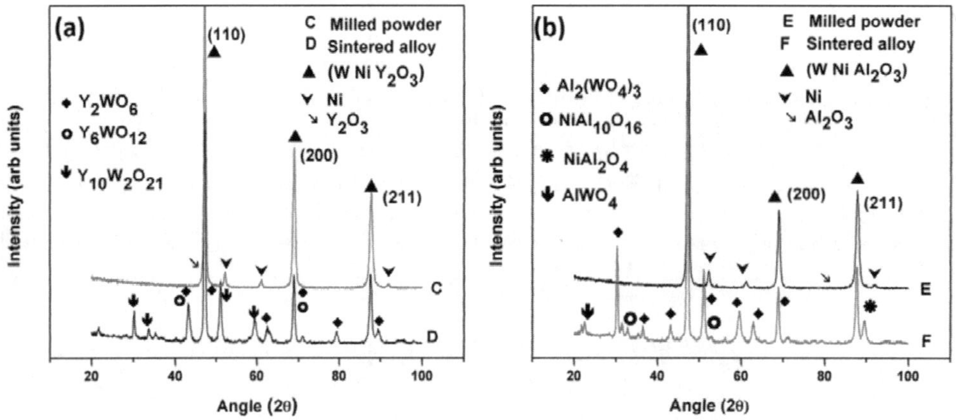

Figure 2. XRD pattern of sintered (a) alloy A, (b) alloy B at 1400 °C for 2 h.

3.4. Microstructure of sintered alloy

Microstructure of alloy A and alloy B sintered at 1400 °C are shown in Figure 3. It displays different phases like W (white phase) and Ni (gray phase) along with that formation of (Y-W) oxide (dark phase) intermetallic in alloy A (Figure 3(a)) and (Ni-Al) oxide and (Al-W) oxide intermetallic (dark phase) in alloy B (Figure 3(b)). Both faceted and spherical W matrix is evident in both the alloys which suggests occurrence of both solid phase and liquid phase sintering. Those phase formation was revealed by EDS, confirmed by elemental mapping and found good agreement with the phases found in XRD phase evolution study. The percentage of fine (<10 µm) W grains (matrix) in alloy A is found 42 % and it is only 24% in case of alloy B. Therefore Y_2O_3 is more effective in grain refinement as compared to Al_2O_3.

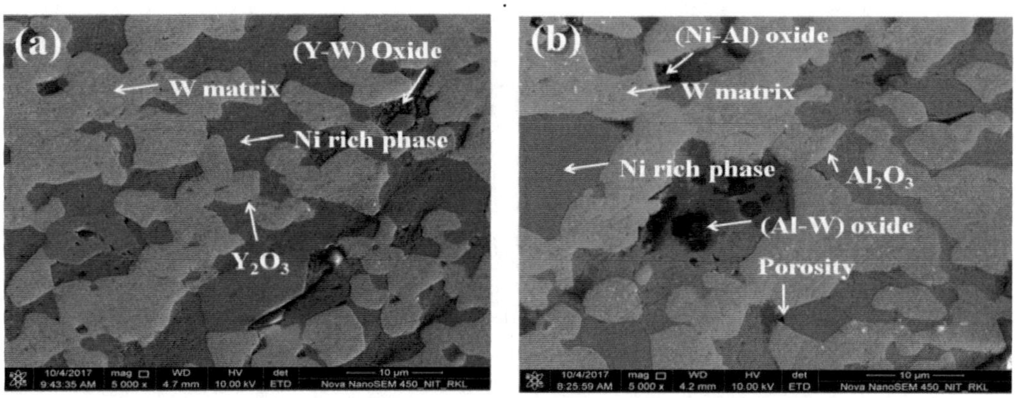

Figure 3. Microstructure of sintered (a) alloy A, (b) alloy B at 1400 °C for 2 h.

3.5. Densification study

The measured theoretical density for alloy A and alloy B is 16.853 gm/cm^3 and 16.675 gm/cm^3 Maximum % relative sintered density of 85.29 % found for alloy A which is superior than alloy B (83.95 %). Enriched densification of alloy A is due to lesser crystallite size after milling and more percentage of fine grains after sintering as compared to alloy B. Densification in both alloys is mainly due to the liquid phase formation(due to refinement of Ni, M.P. decreases and becomes less than sintering temperature used for current study) of Ni phase during sintering.

3.6. Evaluation of mechanical properties

Figure 4 (a) shows the variation of microhardness across cross section for both the alloys. Hardness is basically measured on tungsten matrix at 100 gf load from edge of the sample towards center, but still varying due to the compositional heterogeneity across the cross section. The load is selected so that the indentation is mostly confined within W matrix and effect of intermetallics at the interface on the hardness value is minimal. The average hardness for alloy A (5.13 GPa) was found superior as compared to alloy B (4.42 GPa). This is attributed to superior refinement microstructure of alloy A after milling as well as sintering as compared to alloy B.
Figure 4 (b) represent the alteration in wear depth with progressive increase in sliding distance travelled by diamond indenter with increase in time. Sliding distance of both alloys are also calculated according to previous literature [2]. Minimum wear depth and wear track width was found for alloy A as compared to alloy B (table 2). The superior wear properties of alloy A is attributed to higher hardness and densifications as compared to alloy B.

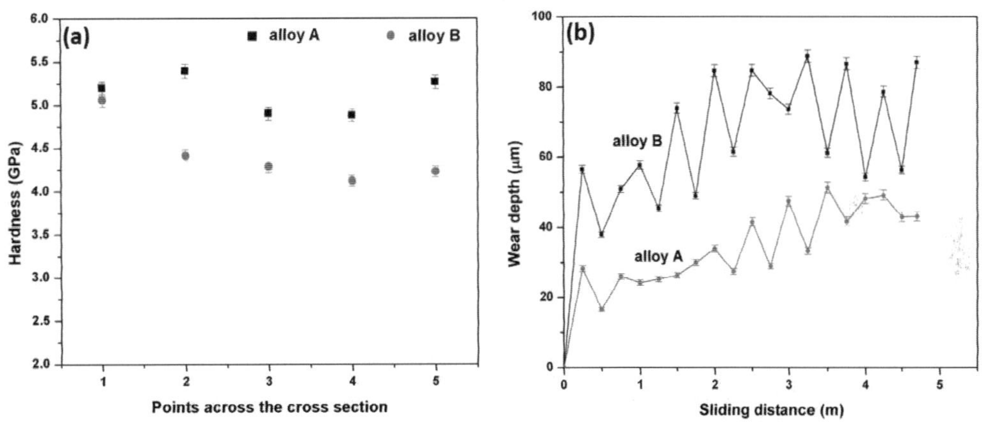

Figure 4. Variation of (a) Hardness of alloy A, alloy B across the cross section (b) Variation of wear depth of alloy A, alloy B with sliding distance sintered at 1400°C for 2 h.

Table 2. Wear study parameters of investigated alloys.

Alloy	Wear depth (μm)	Sliding Distance (m)	Mean Wear track width (μm)
A	35.42	4.7	272
B	89.23	4.7	302

4. Conclusion

After carrying out comparative study of microstructure and mechanical behavior of both the alloys following points were concluded.

- Minimum crystallite size, highest lattice strain and maximum dislocation density was found for alloy A as compared to alloy B.
- Uneven powder particle size distribution at 0 h of milling got converted to uniform size distribution at 10 h of milling, so mechanical alloying is the efficient method for synthesis of the oxide dispersed W-Ni alloys.
- Evolution of liquid phase in the course of sintering is the main densification mechanism which also leads to formation of intermetallic phases during cooling cycle of sintering.
- Higher hardness value of alloy A against alloy B was found due to the superior refinement of microstructure of alloy A after milling as well as sintering.
- Superior wear property of alloy A as compared to alloy B was found, which are mainly due to enhanced hardness of alloy A.

References

1. Zhou Z, Tan J, Qu D, Pintsuk G, Rodig M, Linke J, 2012 *J. Nucl. Mater* **431** 202–205.
2. Patra A, Saxena R and Karak S K 2016 *Int. J. Refract. Met. Hard Mater* **60** 131-146.
3. Smith W. F 1993 *Structure and Properties of Engineering Alloys*, second ed. McGraw-Hill, New York, 571–573.
4. Sahoo R. R, Patra A, Karak S K, 2017 *Proc. Nat. Conf. on Processing and Characterization of Materials (Rourkela)* vol. **178** (Bristol: UK/IOP Science) 012015.
5. Koch C. C 1997 *Nanostruct. Mater* **9** 13-22.
6. Suryanarayana C 2001 Mechanical alloying and milling, *Prog. Mater. Sci* **46** 1–184.
7. Suryanarayana C 2004 *Mechanical Alloying and Milling*, Vol. **13**, Marcel Dekker Inc, New York.
8. Wurster S, Baluc N, Battabyal M, Crosby T, Du J, García-Rosales C, Hasegawa A, Hoffmann A, Kimura A, Kurishita H, Kurtz R.J , Li H, Noh S, Reiser J, Riesch J, Rieth M, Setyawan W, Walter M, You J.H, Pippan R 2013 Recent progress in R&D on tungsten alloys for divertor structural and plasma facing materials *J. Nucl. Mater* **442** S181–S189.
9. Samolyuk G. D, Osetskiy Y. N, Stoller R. E 2011 First-principles investigation of the influence of alloying elements on the elastic and mechanical properties of tungsten, fusion reactor materials program, DOE/ER-0313/50, Vol. **50** pp. 124–128.
10. Patra A, Meraj Md, Pal S, Yedla N, Karak S K 2016 *Int. J. Refract. Met. Hard Mater* **58** 57–67.
11. Lassner E, Schubert W. D 1999 *Tungsten-Properties, Chemistry, Technology of the Element, Alloys, and Chemical Compounds* (Kluwer Academic/Plenum Publishers, New York) 255–268.
12. Nuthalapati M, Karak S K, Majumdar J D and Basu A 2014 *Metall. Mater. Trans. A* **45A** 3748- 3754.
13. Cullity B 1978 *Elements of X-ray diffraction* **2** (Addison Wesley: Philippines) 86-88.

Simulation of Texture Evolution during Uniaxial Deformation of Commercially Pure Titanium

B.Bishoyi[1], M.K.Debta[1], S.K.Yadav[1], R.K.Sabat[2], S.K.Sahoo[1]

[1]Dept. of Metallurgical & Materials Engineering, NIT Rourkela-769008, Odisha, India
[2]Dept. of Materials Engineering, IISc Bangalore, Bangalore-560012,Karnataka, India

Corresponding author Email ID : malayakumar.nm@gmail.com

Abstract: The evolution of texture in commercially pure (CP) titanium during uniaxial tension and compression through VPSC (Visco-plastic self-consistent) simulation is reported in the present study. CP-titanium was subjected to both uniaxial tension and compression upto 35% deformation. During uniaxial tension, tensile twin of {10$\bar{1}$2}<$\bar{1}$011> type and compressive twin of {11$\bar{2}$2}<11$\bar{2}\bar{3}$> type were observed in the samples. However, only tensile twin of {10$\bar{1}$2}<$\bar{1}$011> type was observed in the samples during uniaxial compression. Volume fractions of the twins were increased linearly as a function of percentage deformation during uniaxial tension. Whereas, during uniaxial compression the twinning volume fraction was increased up to 20% deformation and then decreased rapidly on further increasing the percentage deformation. During uniaxial tension, the general t-type textures were observed in the samples irrespective of the percentage deformation. The initial non-basal texture was oriented to split basal texture during uniaxial compression of the sample. VPSC formulation was used for simulating the texture development in the material. Different hardening parameters were estimated through correlating the simulated stress-strain curve with the experimental stress-strain data. It was observed that, prismatic slip {10$\bar{1}$0}<11$\bar{2}$0> operated as the primary deformation mode during uniaxial tension whereas basal slip {0001}<11$\bar{2}$0> acquired the leading role during deformation through uniaxial compression. It was also revealed that active deformation modes were fully depending on percentage deformation, loading direction, and orientation of grains.

Keywords: Tension, Compression, Texture, EBSD, XRD, VPSC

1. Introduction

Titanium and its alloys are most widely used for structural, chemical, automobiles, and medical application due to its large specific strength and excellent corrosion resistance [1-3]. Evolution of texture and microstructure in hexagonal material, thus, needs to be further exploited. Further VPSC simulation is successfully used to understand the texture developments in cubic as well as hexagonal materials [4-7]. In polycrystalline materials, the imposed plastic strain is accommodated by various active deformation mechanisms such as dislocation movement in a confined circumstance. Both slip and twin are the important deformation modes for dislocation movement in hcp material such as titanium [8, 9]. The activation of slip systems is generally depending upon the shear stress between two gliding planes, and angle between crystals and loading directions. In titanium, the most favorable slip systems being the prismatic slip with a-type Burgers vectors followed by basal and pyramidal slips with a-type Burgers vectors [10]. However according to Von Mice's, five independent slip systems are required for arbitrary shape change during the plastic deformation of polycrystalline titanium [11]. So pyramidal <a+c> slip or deformation twinning has to be activated during deformation in plastic range of hexagonal titanium [12]. The activities of slip and twin systems at room temperature deformation of pure titanium are observed to be mostly in the order of prismatic slip, basal slip, tensile twin, compressive twin and 1st order pyramidal slip

[13]. So pyramidal <a+c> slip or deformation twinning has to be activated during plastic deformation of hexagonal titanium [14].

Various authors have been exploited the texture evolution through different strain-paths. A dominant basal texture has been observed during uniaxial tension and compression of cp-titanium at room temperature [15, 16]. However, evolution of texture and microstructure during tensile and compression deformation along with modeling of texture has not been reported elsewhere. With this background study, in the present study the evolution of texture during tensile and compression deformation of cp-titanium has been investigated through experimentation and simulation.

2. Experimental details

(a)Material

CP-titanium (Grade-2) of 1mm thick sheet and 10mm diameter rod were used for uniaxial tension and compression respectively. The chemical composition (in wt %) of the material is shown in Table 1. Both tensile and compression specimens (**Figure 1.**) were prepared by using wire EDM (Electrical Discharge Machining) technique.

Table 1. Chemical composition of the material

Ti	Fe	C	N	H	O
0.820	0.034	0.004	0.004	0.004	0.134

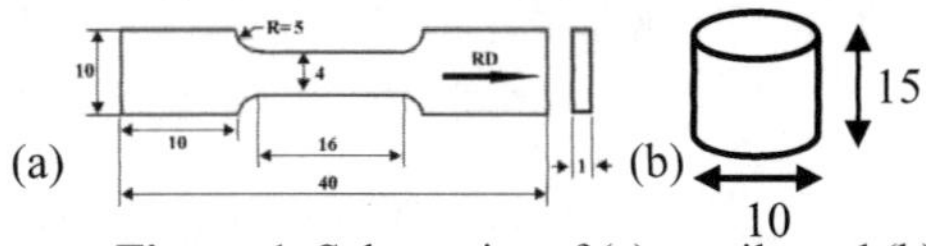

Figure 1. Schematics of (a) tensile and (b) compression specimen used (all are in mm)

(b) Uniaxial Tension

Tensile deformation was carried out along rolling direction of the specimen in an Instron 5567 screw driven UTM (universal testing machine) at room temperature. Subsequently, different specimens were subjected to tensile deformations of 5, 15, 25, and 35% deformation at the strain rate of $1\times10^{-3}s^{-1}$.

(c) Uniaxial Compression

Compression deformation was carried out similar to the extrusion direction (ED) of the specimen in an Instron 5567 screw driven UTM (universal testing machine) at room temperature. Different specimens were subjected to compression of 3, 10, 20, and 30% deformation at the strain rate of $1\times10^{-3}s^{-1}$.

(d) Sample Preparation

Deformed region of the samples were polished metallographically followed by electro-polished prior to the textural and microstructural characterizations. Metallographic polishing was accomplished in a Struers Labopol using different grit of SiC grinding papers, while electro-polishing was accomplished in a Struers Lectopol-5 using an electrolyte of methanol and perchloric acid (80:20) at -20 °C and 25V for 12sec.

(e) X-Ray Diffraction (XRD)

A Bruker D8 Advance XRD setup with Co Kα radiation was used to measure the bulk texture of the samples. (0002) pole figure was measured and was calculated using Labotex V 3.0 software.

(f) Electron Backscattered Diffraction (EBSD)

For the EBSD measurement of the samples, a FEI Quanta-3D-FEG scanning electron microscope was used. An approximately 1mm × 1mm area from two different locations of the samples was scanned using a step size of 0.5 μm. Both video and beam conditions were kept constant between the scans. A TSL-OIM Version 6.2 EBSD package was used for the analysis of the EBSD scans.

3. VPSC (Viscoplastic Self Consistent) Simulation

To pursue the effect of deformation on texture evolution during the deformation of cp-titanium in plastic range, visco-plastic self-consistent simulation was accomplished using the VPSC-7 software package. Deformation of the material was based on the mechanisms of crystal plasticity, where systems of slip and twin were activated by their respective resolved shear stress. The response of the single crystal is expressed with the help of rate sensitive constitute law.

$$\epsilon_{ij}(\bar{X})=\sum_s m_{ij}^s \gamma^s(\bar{X}) = \gamma_0 \sum_s m_{ij}^s \left(\frac{m_{kl}^s \sigma_{kl}(\bar{x})}{\tau_0^s}\right)^n \qquad (1)$$

Where 'ε' denotes strain rate in the grain, 'm^s' denotes the geometric of Schmid tensor, 's' denotes a label for slip systems, 'τ^s' denotes threshold stress to control the activation of a slip system and 'n' implies the rate sensitivity. The activation of slip system leads to self or latent hardening of it due to the activation of same slip system or any other slip system respectively. This is outlined by an empirical Voce law given as:

$$\hat{\tau}^s = \tau_0^s + (\tau_1^s + \theta_1^s \Gamma)(1\exp(-\Gamma \left|\frac{\theta_0^s}{\tau_1^s}\right|)) \qquad (2)$$

Where $\Gamma = \sum_s \Delta\gamma^s$ is the accumulated shear in the grain;τ_0,θ_0,θ_1,(τ_0+τ_1) are the initial CRSS value, the initial hardening rate, the asymptotic hardening rate and the back-extrapolated CRSS value respectively. While θ_0 and $\tau_1$are typically positive, their absolute values are used in (2) in order to entertain few special cases.

In this investigation, VPSC code with the intermediate n^{eff}= 10 approach was taken for modelling the deformation texture. The effect of strain path was modelled using a process IVGAR=0; where the code will impose the same velocity gradient in each step. In order to simulate the texture development during tensile and compression deformation, nearly 4000 random orientations were taken. The simulation was carried out till ε = 0.35 for tensile deformation using velocity gradient L_{11}= 1, L_{22}= -0.5, L_{33} = -0.5 and L_{12}= L_{13}= L_{21}= L_{23}= L_{31}= L_{32}=0 and till ε = 0.3 for compression deformation using velocity gradient L_{11}= 0.5, L_{22}= 0.5, L_{33} = -1 and L_{12}= L_{13}= L_{21}= L_{23}= L_{31}= L_{32}=0. Five deformation modes such as prismatic slip $\{10\bar{1}0\}<11\bar{2}0>$, basal slip$\{0001\}<11\bar{2}0>$, pyramidal slip $\{10\bar{1}1\}<11\bar{2}3>$, tensile twin $\{10\bar{1}2\}$ $<\bar{1}011>$ and compression twin $\{11\bar{2}2\}<11\bar{2}\bar{3}>$ were activated to match the stress strain curve obtained by simulation with experimental curve. The values of the hardening parameters for the best fit of the deformation through compression and tension are shown in table.2.

Table 2. Hardening parameter of deformation modes used for the simulation

Slip/Twin	τ_0	τ_1	θ_0	θ_1
Prismatic	110	20	600	120
Basal	170	60	1200	90
Pyramidal<a+c>	220	100	600	75
Tensile twin	120	120	500	150
Compression twin	180	110	1000	250

4. Results

4.1 Evolution of Microstructure and Texture

4.1.1 Uniaxial Tension

The microstructures (map) of the deformed samples as a function of % deformation are shown in **Figure 2.** The starting material had an equiaxed grain structure of average grain size of ~ 75 µm. Both 85° $<\bar{1}2\bar{1}0>$ type of tensile twins (red in colour) and 64.3°$<\bar{1}100>$ type of compression twins (blue in colour) were significantly observed during the deformation of the samples.

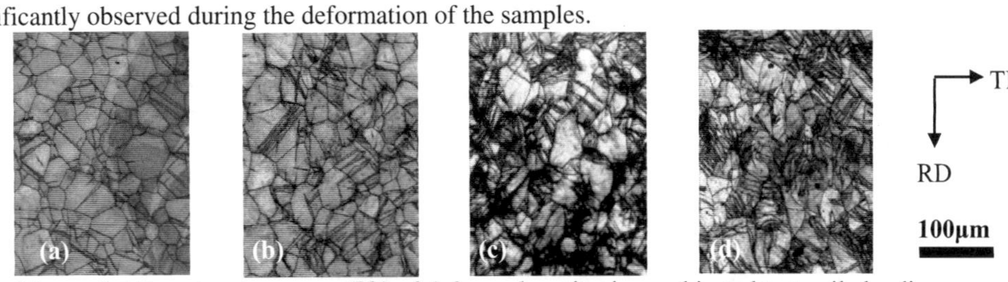

Figure 2. Microstructure maps (IQ) of deformed cp-titanium subjected to tensile loading of (a) 5%, (b) 15%, (c) 25%, and (d) 35% deformation

In **Figure 3(a),** the twin boundary length in the deformed materials is shown as a function of percentage deformation of the samples. It was observed that, twin boundary increased with increasing the % deformation. However the rate of activation of tensile twin was more with respect to compression twin. The average grain size as a function of % deformation during tensile deformation is shown in **Figure 3(b).** A progressive decrease in grain size with increasing % deformation was observed. The misorientation angle development within the grain of the deformed material during deformation is shown in **Figure 3(c).** It represents the average misorientation angle was increased with progressive deformation of the samples.

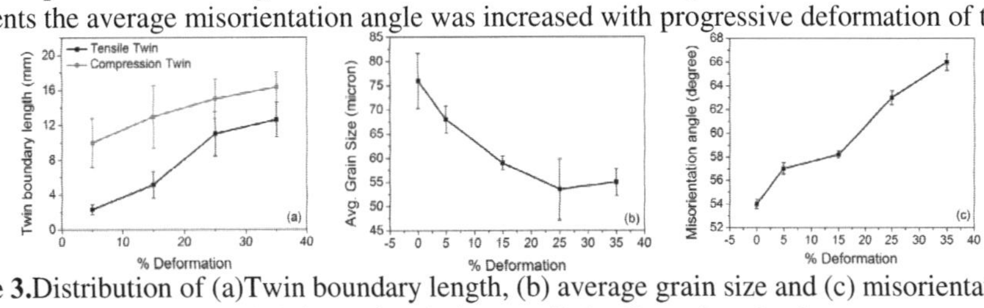

Figure 3. Distribution of (a)Twin boundary length, (b) average grain size and (c) misorientation angle w.r.t % deformation during tensile loading of cp-titanium

The texture evolutions in the deformed specimen of cp-titanium were represented by (0002) pole figure as shown in **Figure 4.** The initial sample had typical t-type texture, which also known as split basal texture inclined at 35° away from normal direction (ND) towards transverse direction (TD). The texture got strengthened with progressive deformation however position of maximum level of intensity of split basal texture remained at the early position.

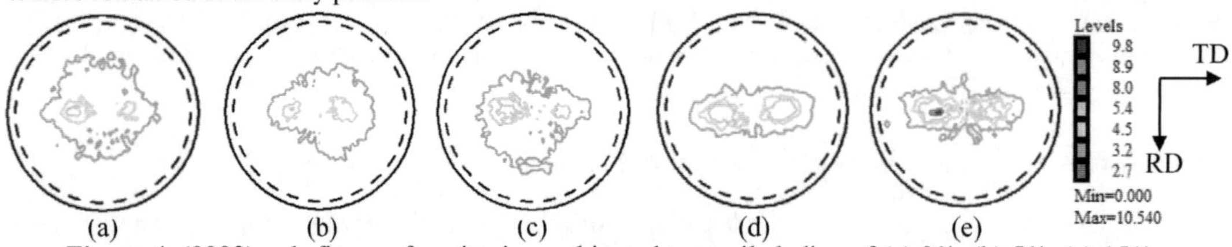

Figure 4. (0002) pole figure of cp-titanium subjected to tensile lading of (a) 0%, (b) 5%, (c) 15%, (d) 25%, and (e) 35% deformation

4.1.2 Uniaxial Compression

Figure 5 represents the microstructure of deformed samples during compression in the form of map. Un-deformed material with most of the grains oriented in the non-basal orientation had an equiaxed grain of ~ 14μm average size. It was observed that grain with non-basal orientations were prone to twinning. 85° <$\bar{1}2\bar{1}0$> type of tensile twin (red in color) was observed in significant quantity up to 20% deformation and then decreased as shown in **Figure 6(a)**. Average grain sizes, and misorientation angle development of deformed material with respect to percentage deformation during compression were shown in **Figure 6(b)** and **Figure 6(c)** respectively. It was observed that, average grain size of the grain decreased linearly with increase in deformation percentages. The misorientation development inside the specimen during deformation was increased up to 20% deformation and then decreased on further deformation.

Figure 5. Microstructure maps (IQ) of deformed cp-titanium subjected to compression of (a) 3%,(b) 10%, (c) 20%, and (d) 30% deformation.

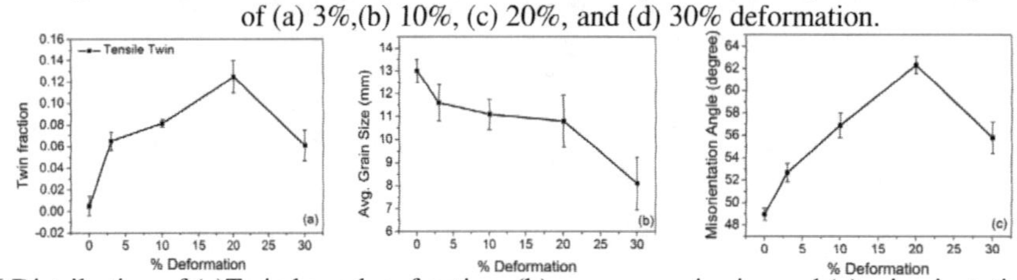

Figure 6. Distribution of (a)Twin boundary fraction, (b) average grain size and (c) misorientation angle w.r.t % deformation during compressive loading of cp-titanium

The texture evolution in deformed material during compression was represented by (0002) pole figure as shown in **Figure 7.**

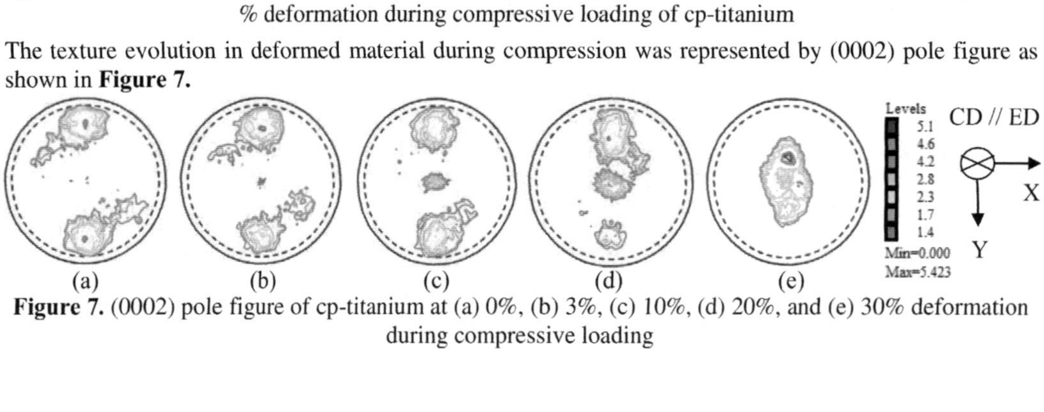

Figure 7. (0002) pole figure of cp-titanium at (a) 0%, (b) 3%, (c) 10%, (d) 20%, and (e) 30% deformation during compressive loading

The initial non basal texture of the un-deformed material was oriented towards basal texture during the compression. On further deformation, it ended with split basal texture.

4.2. VPSC Simulation

Hardening parameters (Table 2) were obtained from the best fit experiment and simulated true stress-strain curve of both tensile and compression Figure 8(a) & 9(a). The activity curve for the various deformation modes were plotted as shown in Figure 8(b) & 9(b). The (0002) pole figures of the deformed material obtained by simulation were shown in Figure 10 & 11. A considerable match is observed between simulated and experimental (0002) pole figure of both tensile and compression deformed material.

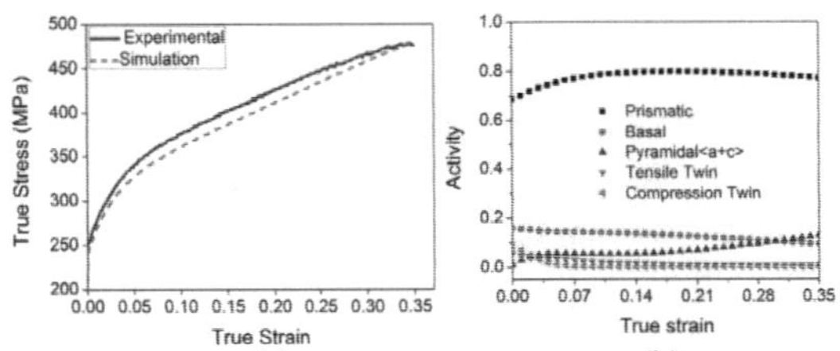

(a) (b)

Figure 8. (a) Comparison between experimental and simulated true stress-strain curve, (b) Activity of various deformation modes of cp-titanium material subjected to 35% deformation

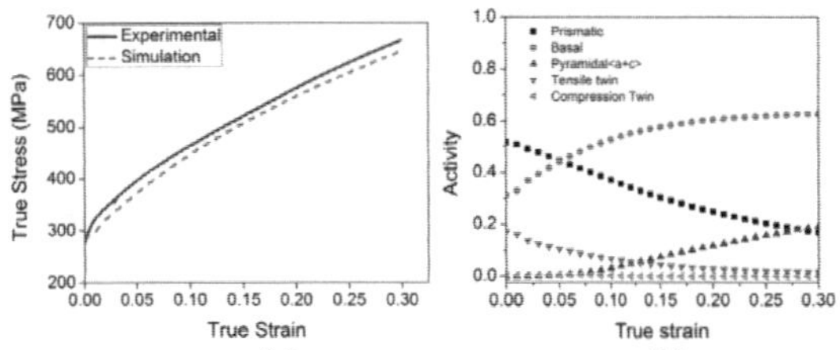

(a) (b)

Figure 9. (a) Comparison between experimental and simulated true stress-strain curve, (b) Activity of various deformation modes of cp-titanium material subjected to 30% compression loading

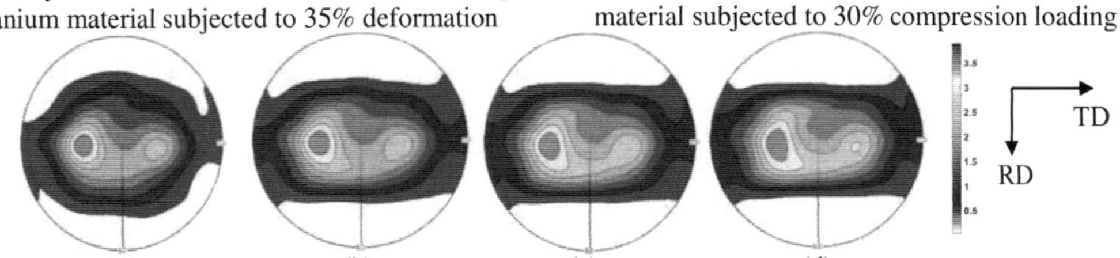

(a) (b) (c) (d)

Figure 10. Simulated (0002) pole figure of cp-titanium at (a) 5%, (b) 15%, (c) 25%, and (d) 35% deformation during tensile loading

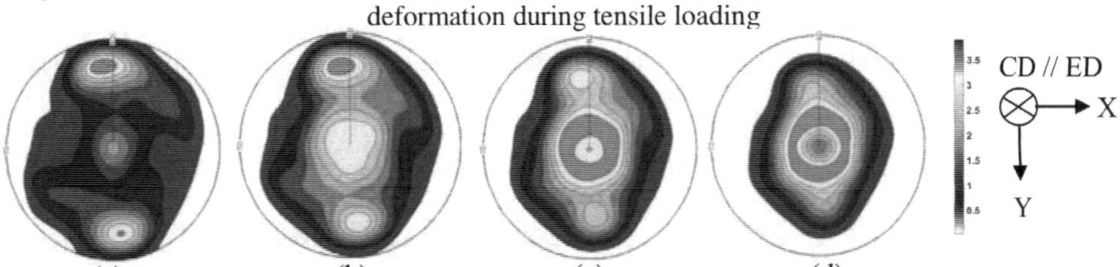

(a) (b) (c) (d)

Figure 11. Simulated (0002) pole figure of cp-titanium at (a) 3%, (b) 10%, (c) 20%, and (d) 30% deformation during compressive loading

In tensile deformation, prismatic slip system was activated as primary deformation mode and its contribution to overall deformation was higher. However the contributions of both twins were significant at the low strain for accommodating strain and diminished at higher strain. The slip activity obtained from the simulation indicated that, higher number of deformation modes was activated throughout the compression as compared to the tensile deformation. In compression, basal slip acted as preliminary deformation mode. At lower strain both prismatic slip and tensile twin played important role for plastic deformation. However at the high strain, basal and pyramidal slip systems were activated for accommodating the imposed strain.

5. Discussion

The difference in texture and microstructure during the deformation of both tensile and compression material was observed from both bulk and EBSD texture data. In tensile deformation, grains were fragmented due to the initial contribution of both tensile and compression twin. Grains with high schmid factor were prone to twining during the deformation. With increase in further deformation, the average misorientation angle was increased due to the continuous presence of the twin boundary which comprises both 85° and 64°misorientation in between the two adjacent boundary. Similarly the texture development was remained more or less same throughout the deformation of cp-titanium. This is due to the less participation of twin during the deformation and their sharing towards the evolution of preferred orientation was very less as compared to parent grain. From the VPSC, it has been anticipated that most of the grains in the material oriented along their prismatic slip plane along <$11\bar{2}0$> direction, and leads to forming split basal texture throughout the deformation. One can conclude that extension and contraction twinning were responsible for the grain refinement only and not for the texture evolution during the tensile deformation of cp-titanium material.

During compression deformation, the evolution of texture and microstructure were observed due to the presence of tensile twin of 85° type<$\bar{1}2\bar{1}0$>. The average grain size of the deformed grain was decreased with increase in deformation percentage due to the activation of tensile twin. Non basal grains with high Taylor factor value were prone to twinning. The fraction of twin increased with deformation percentage up to 20% and then decreased this may be due to the twin inhalation of whole grain during the deformation. At the higher deformation, all twin boundaries were converted to normal grain boundary with lesser misorientation in between twin grain and neighbour grain. So decrease in twin volume fraction and average misorientation angle was observed at the high deformed material. Similarly average misorientation angle increased up to 20% and then decreased for the above explained reason. The said tensile twin was also responsible for the 85° rotation of the non-basal oriented grain into basal direction in the initial stage of compression. On further increasing the deformation percentage, contribution of basal slip was more and played a role for the formation of split basal texture.

6. Conclusion

From VPSC simulation, it has been observed that prismatic slip acts as primary deformation mode during the tensile deformation. Both the twins have larger contribution towards accommodating imposed strain in the low strain deformation however their contribution is low at higher strain. In compression deformation, both microstructural and textural evolution has taken place due to the dominant tensile twin. Similarly basal slip acts as preliminary deformation mode for the deformation. It can be concluded that microstructure and texture evolution during the deformation of the cp-titanium material is dependent on the initial orientation of grain, direction of loading and size of the grain.

Acknowledgement

The authors would like to thank Prof. I. Samajdar for conducting EBSD measurements at IIT Bombay; Prof. S. Suwas, IISc Bangalore for providing tensile and compression testing of the samples; Prof. Kanjarala A.K., IIT Madras for guiding us through VPSC simulation; Council of Scientific and Industrial Research (CSIR), India for financial support (Grant No: 22(0714)/16/EMR-II dated 24/05/2016).

References

[1] R.R. Boyer1996 *Mater. Sci. Eng.A* **213** 103.
[2] A. Choubey, B. Basu, R. Balasubramaniam2005 *Trends Biomater. Artif.Organs* **18** 64.
[3] Z. Abdallah, M.T. Whittaker, M.R. Bache2013 *Intermetallics* **38**55.
[4] R.A. Lebensohn and C.N. Tome 1993*Acta Metall. Mater* **41**2611.
[5] C.N. Tome,G.R.Cannova,U.F.Kocks,N.Christodoulou and J.J.Jonas1984 *Acta Metall. Mater* **32** 1637.
[6] R. A. Lebensohn and C. N. Tome1994*Mater.Sci. Eng.A* **175** 71.
[7] H. Wang, B. Raeisinia, P.D. Wua, S.R. Agnew, C.N. Tomé2010*Int.J.Sol.& Str.***47** 2905.
[8] A.A. Salem, S.R. Kalidindi, R.D. Doherty2003 *Acta Mater***51** 4225.
[9] Y.B. Chun, S.L. Yu, S.L. Semiatin, S.K. Hwang 2005*Mater. Sci. Eng.A***398** 209.
[10] N. Bozzolo, N. Dewobroto, H.R. Wenk, F. Wagner 2007*J. Mater. Sci* **42** 2405.
[11] R. Von Mises 1928*App. Math.Mech* **8**161.
[12] I. Kim, J. Kim, D.H. Shin, C.S. Lee, S.K. Hwang 2003*Mater. Sci. EngA* **342** 302.
[13] M.G. Glavicic, A.A. Salem, S.L. Semiatin 2004 *Acta Mater***52** 647.
[14] Y. Zhong, F. Yin, K. Nagai 2008*Mater. Res* **23** 2954.
[15] Y.Ren, Xiyan Zhang, Ting Xia, Qi Sun, Qing Liu 2017*Mat. & Des.***126** 123.
[16]S.Wronski,K.Wierzbanowski,M.Jdrychowski,Tarasiuk,M.Wronski,A.Baczmanski,B.Bacroix2016*Mater. Sci. Eng.A***656** 1.

Optimization of Dimensional accuracy in plasma arc cutting process employing parametric modelling approach

Deepak kumar Naik, K.P. Maity*

Department of Mechanical Engineering, National Institute of Technology Rourkela, Odisha, India

Email address: naikdeepak81@gmail.com, kpmaity@gmail.com *

* Corresponding author

Abstract

Plasma arc cutting (PAC) is a high temperature thermal cutting process employed for the cutting of extensively high strength material which are difficult to cut through any other manufacturing process. This process involves high energized plasma arc to cut any conducting material with better dimensional accuracy in lesser time. This research work presents the effect of process parameter on to the dimensional accuracy of PAC process. The input process parameters were selected as arc voltage, standoff distance and cutting speed. A rectangular plate of 304L stainless steel of 10 mm thickness was taken for the experiment as a workpiece. Stainless steel is very extensively used material in manufacturing industries. Linear dimension were measured following Taguchi's L_{16} orthogonal array design approach. Three levels were selected to conduct the experiment for each of the process parameter. In all experiments, clockwise cut direction was followed. The result obtained thorough measurement is further analyzed. Analysis of variance (ANOVA) and Analysis of means (ANOM) were performed to evaluate the effect of each process parameter. ANOVA analysis reveals the effect of input process parameter upon leaner dimension in X axis. The results of the work shows that the optimal setting of process parameter values for the leaner dimension on the X axis. The result of the investigations clearly show that the specific range of input process parameter achieved the improved machinability.

Keywords: Plasma arc cutting, Dimensional accuracy, Taguchi method, ANOVA.

1. Introduction

PAC is a thermal cutting manufacturing process which used to cut the high strength material, such as stainless steel, carbon steel, aluminum and non-ferrous materials. To achieve the higher accuracy, better quality and reduced time, PAC process was introduced in 1950.This is a faster cutting process as compared to other with greater flexibility. The limitation of this process is that it can cut only electrically conductive materials [1, 2].

Plasma is a forth state of matter and highly energized in nature. It behaves like a high temperature gas. An electric arc is generated between workpiece and the electrode. The workpiece acts as the anode and the electrode acts as the cathode in this process. The physics behind the PAC process is that the arc is constricted by a copper nozzle (fine bore). Due to this, the temperature increases up to $20{,}000^0$ C . The plasma jet velocity emanating from the nozzle. Thus, the plasma gas the flow is increased so that the deeply penetrating plasma jet cuts the material and blown away the molten material [3]. Fig. 1 shows the mechanism of PAC process.

PAC process encompasses greater number of process parameter. To accomplish smooth operation of this process, it needs optimization of process parameters. Many researchers attained the investigations in this direction [4]. Optimization of quality indicators, such as dimensional accuracy, kerf characteristics and quality of the cut surface has been investigated by them [5-8] . In PAC operation, the effect of input variable onto the dimension accuracy performance were studied and introduced an optimal cutting condition using ANOVA and ANOM [9]. The optimal parameter setting was introduced using RSM and GRA coupled with PCA analysis in PAC on AISI 316 stainless steel [10]. Taguchi's L_{27} orthogonal array employed to design the experimental run and optimize the process parameter of PAC on1017 steel [11].

The proposed investigation focuses on the effect of input variables onto the dimensional accuracy of 304L stainless steel. An optimization was carried out employing full factorial design of experiment method.

2. Materials and methods

In this research work experiments were performed on 304L stainless steel which is very extensively used material in manufacturing industries. Mainly aerospace and automotive industries employs this grade of stainless steel to produce finish goods. CNC plasma arc cutting machine of MESSER Company named BURNY 1250 was used to conduct the experiment. A rectangular plate having 10 mm thickness was taken for all the experiment.

Working gas selection in PAC operation plays a vital role to achieve a precise cutting. Argon and oxygen was used as inert and shielding gas respectively. The supply of oxygen was fixed at 20 Mpa, while argon was at 1.2 Mpa. Voltage kept constant at 400 volts. 2mm diameter of swirl nozzle of tungsten material was taken as electrode.

The input process parameters were selected as arc voltage, standoff distance (SOD) and cutting speed (CS). Two levels were selected to conduct the experiment for each of the process parameter. Taguchi's L_{16} orthogonal array was choosed for the design of experimental run. The range of variables are furnished in Table 1 .

Table 1: Values of input variables

Symbol	Input parameters	Units	Level 1	Level 2
A	Arc current	A	70	80
B	Voltage	V	140	150
C	SOD	mm	3	4
D	CS	Mm/min	2500	3000

A rectangular cut of 40×50 mm was achieved for all 16 experiments as specified in the orthogonal matrix. In all experiments, clockwise cut direction was followed. Fig. 2 shows the direction of measurements. The leaner x dimension of the quadrilateral base was measured on its top surface. A digital vernier caliper of 1 micron was used to measure taken along X axis direction. Average of four measurements as well as deviation was measured for each workpiece. Deviation is the difference between the minimum and maximum value of four measurement. The result obtained thorough measurement is furnished in Table 2.

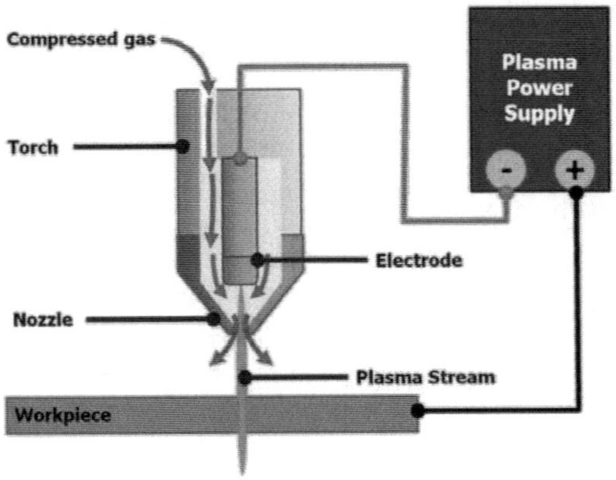

Fig. 1: PAC Mechanism

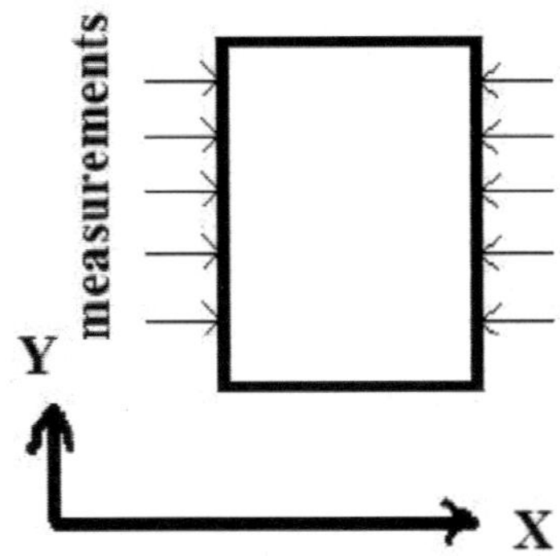

Fig. 2: Direction of measurement

Table 2: Experimental result

Run no	A	B	C	D	Average X (mm)	Deviation (mm)
1	70	140	3	2500	49.328	0.41
2	70	140	3	3000	49.300	0.39
3	70	140	4	2500	49.478	0.35
4	70	140	4	3000	49.460	0.33
5	70	150	3	2500	49.241	0.36
6	70	150	3	3000	49.223	0.34
7	70	150	4	2500	49.413	0.29
8	70	150	4	3000	49.393	0.28
9	80	140	3	2500	49.415	0.47
10	80	140	3	3000	49.360	0.46
11	80	140	4	2500	49.575	0.41
12	80	140	4	3000	49.554	0.43
13	80	150	3	2500	49.335	0.42
14	80	150	3	3000	49.320	0.41
15	80	150	4	2500	49.503	0.35
16	80	150	4	3000	49.499	0.36

IOP Conf. Series: Materials Science and Engineering **338** (2018) 012039 doi:10.1088/1757-899X/338/1/012039

3. Result and Discussion

3.1 Analysis of Means (ANOM)

ANOM (Analysis of means) was achieved onto the experimental result to find the rate of significant of process parameter. ANOM of leaner dimension and minimum deviation was performed and presented in Table 3 and Table 4 respectively. According to ANOM of linear dimension, SOD is the most significant process parameter followed by current, voltage and cutting speed. Also from the ANOM of minimum deviation, it is found that the current is the most significant process parameter followed by SOD, voltage and cutting speed. The ANOM graph was plotted for leaner dimension and minimum deviation and shown in Fig. *3* and Fig. 4 respectively. According to the ANOM diagram , the best optimal solution for optimizing the leaner X dimension is; current: 80 A, voltage: 140 V, SOD: 4 mm and cutting speed: 2500 mm/min. Minimization of the deviation is accomplished when; current: 70 A, voltage: 150 V, SOD: 4 mm and cutting speed: 3000 mm/min.

Table 3: ANOM (Leaner dimension)

Leaner dimension: X-Direction				
Means	L1	L2	Delta	Rank
A	49.35	49.45	0.09	2
B	49.43	49.37	0.07	3
C	49.32	49.48	0.17	1
D	49.41	49.39	0.02	4

Table 4: ANOM (Minimum deviation)

Deviation: (Maximum-Minimum)				
Means	L1	L 2	Delta	Rank
A	0.344	0.414	0.07	1
B	0.406	0.351	0.055	3
C	0.408	0.350	0.057	2
D	0.383	0.370	0.007	4

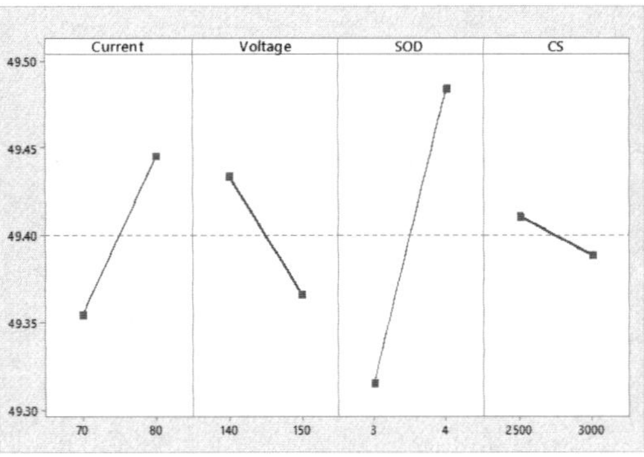

Fig. 3: ANOM Diagram (Leaner dimension)

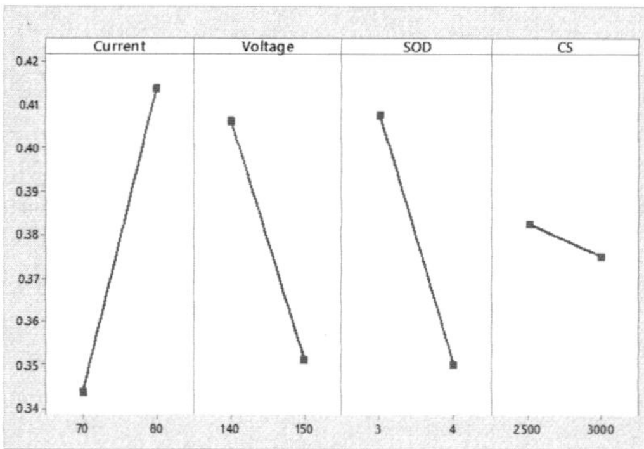

Fig. 4: ANOM Diagram (Deviation of measurement)

3.2 Analysis of Variance (ANOVA)

The impact of process parameter on leaner dimension and minimum deviation have been performed using ANOVA. This test reveals the significance of process parameter and performance indicator. Also needs for estimating the error variances for the parameters effects. The F-test has performed at 95% confidence level. The ratio of mean square deviations of each parameter and the mean square of corresponding parameter is the F-value. The factors are significant if P-value (probability of significance) is less than 0.05 at 95% confidence interval.

Table 5 and Table 6 portraits the result of ANOVA for leaner dimension and minimum deviation respectively. From Table 5, it reveals that all the cutting parameters are significant whose P-value is less than 0.05 at 95% confidence interval during the ANOVA test of leaner dimension. SOD is the most significant process parameter whereas it contributes 67.74 % followed by current 19.45 %, voltage 10.91 % and cutting speed 1.19 %. Also, Table 6 unveils that all the process parameter are significant except cutting speed. P-value of cutting speed is more than 0.05. Current is the most significant factor affecting deviation about 42.45 %. Also, SOD, current and cutting speed contributes 28.64 %, 26.21 % and 0.49 % respectively. This trend has also been observed from the ANOM graphs.

Table 5: ANOVA- Leaner dimension

Source	DF	Seq SS	Adj MS	F	P	%
Current	1	0.032852	0.032852	300.84	0	19.45
Voltage	1	0.018428	0.018428	168.76	0	10.91
SOD	1	0.114413	0.114413	1047.8	0	67.74
CS	1	0.002003	0.002003	18.34	0.001	1.19
Error	11	0.001201	0.000109			
Total	15	0.168896				

Table 6: ANOVA- Minimum deviation

Source	DF	Seq SS	Adj MS	F	P	%
Current	1	0.0196	0.0196	210.34	0	42.45
Voltage	1	0.0121	0.0121	129.85	0	26.21
SOD	1	0.013225	0.13225	141.93	0	28.64
CS	1	0.000225	0.000225	2.41	0.148	0.49
Error	11	0.001025	0.000093			
Total	15	0.046175				

4. Conclusions

In this research work the PAC process was experimentally investigated on 304L stainless steel. The influence of process parameters onto the X direction dimensional accuracy. Also, the deviation was established using ANOM and ANOVA. The following findings were concluded of this research experiment:

1. The optimal condition for the leaner dimensions on the X axis were; current: 80 A, voltage: 140 V, SOD: 4 mm and cutting speed: 2500 mm/min.
2. The optimum parameters values for the deviation were; current: 70 A, voltage: 150 V, SOD: 4 mm and cutting speed: 3000 mm/min.
3. ANOVA portraits that the X axis leaner dimension is contributed the most by the SOD (67.74 %) and the deviation is affected the most by the current (42.45 %).

References

1. Ilii, S.-M., M. Coteata, and A. Munteanu, *Experimental results concerning the variation of surface roughness parameter (Ra) at plasma arc cutting of a stainless steel workpiece.* International Journal of Modern Manufacturing Technologies, II/1, 2010: p. 31-36.
2. Salonitis, K. and S. Vatousianos, *Experimental investigation of the plasma arc cutting process.* Procedia CIRP, 2012. **3**: p. 287-292.
3. Rana, K., P. Kaushik, and S. Chaudhary, *Optimization of plasma arc cutting by applying Taguchi Method.* International Journal of Enhanced Research in Science Technology & Engineering, ISSN, 2013: p. 2319-7463.
4. Kumar, P., I. Ahuja, and R. Singh, *Application of fusion deposition modelling for rapid investment casting–a review.* International Journal of Materials Engineering Innovation, 2012. **3**(3-4): p. 204-227.
5. Kechagias, J., M. Billis, and S. Maropoulos, *A parameter design of CNC plasma-arc cutting of carbon steel plates using robust design.* International Journal of Experimental Design and Process Optimisation, 2010. **1**(4): p. 315-326.
6. Bini, R., et al., *Experimental study of the features of the kerf generated by a 200A high tolerance plasma arc cutting system.* Journal of materials processing technology, 2008. **196**(1): p. 345-355.
7. Chen, J.C., Y. Li, and R.A. Cox, *Taguchi-based Six Sigma approach to optimize plasma cutting process: an industrial case study.* The International Journal of Advanced Manufacturing Technology, 2009. **41**(7): p. 760-769.
8. Kechagias, J., et al., *On the multi–paramenter optimization of CNC plasma-arc cutting process quality indicators using Taguchi Design of Experiments.* Recent Advances in Electrical Engineering, 2014: p. 128-133.
9. Kechagias, J., et al. *Plasma Arc Cutting Dimensional Accuracy Optimization employing the Parameter Design approach.* in *ITM Web of Conferences*. 2017. EDP Sciences.
10. Maity, K. and D.K. Bagal, *Effect of process parameters on cut quality of stainless steel of plasma arc cutting using hybrid approach.* The International Journal of Advanced Manufacturing Technology, 2015. **78**(1-4): p. 161-175.
11. Bhuvenesh, R., M. Saifuldin, and N. MH, *Surface roughness and MRR effect on manual plasma arc cutting machining.* 2012.

Effect of fiber content on the thermal conductivity and dielectric constant of hair fiber reinforced epoxy composite

Bishnu Prasad Nanda and Alok Satapathy*

Department of Mechanical Engineering, National Institute of Technology, Rourkela, 769008, India

bishnuprasad.nanda726@gmail.com, alok@nitrkl.ac.in

Abstract. This paper reports on the dielectric and thermal properties of hair fibers reinforced epoxy composites. Hair is an important part of human body which also offers protection to the human body. It is also viewed as a biological waste which is responsible for creating environmental pollution due to its low decomposition rate. But at the same time it has unique microstructural, mechanical and thermal properties. In the present work, epoxy composites are made by solution casting method with different proportions of short hair fiber (SHF). Effects of fiber content on the thermal conductivity and dielectric constant of epoxy resin are studied. Thermal conductivities of the composites are obtained using a Unitherm^{TM} Model 2022 tester. An HIOKI-3532-50 Hi Tester Elsier Analyzer is used for measuring the capacitance of the epoxy-SHF composite, from which dielectric constant (D_k) of the composite are calculated. A reduction in thermal conductivity of the composite is noticed with the increase in wt. % of fiber. The dielectric constant value of the composites also found to be significantly affected by the fiber content.

Key words: Hair fiber, dielectric constant, thermal conductivity, biological waste

1. Introduction

Polymer matrix composites are commonly used in microelectronic application due to their easy fabrication, light weight, low cost and excellent insulation properties. The dielectric constant of a material plays an important role in functioning of the electronic equipment. Dielectric behaviour of polymer composite filled with different inorganic fillers has been reported by many researchers. Mishra et al. [1] studied the dielectric behaviour of epoxy composite with solid glass microsphere (SGM) as the filler material and observed that increase in filler content have significant effect on the dielectric constant of the composite. Agrawal et al. [2] investigated the mechanical, dielectric and thermal properties of polypropylene composite filled with AlN and SGM. Mishra et al. [3] have also studied the dielectric properties of hybrid composite filled with SGM and boron nitride.

Use of natural fibers as a reinforcing material in polymer matrix composite has been growing rapidly due to their good mechanical, tribological, dielectric and insulating properties. These fibers are also biodegradable in nature having low cost and low toxicity. In electrical and electronics application natural fiber reinforced polymer composite can also be a better substitute to synthetic fiber due to their better electrical insulating properties. Dielectric behaviour of polymer composite reinforced with different natural fiber has been researched by many researchers. Mishra et al. [4] investigated the dielectric properties of epoxy composite reinforced with chicken feathers. Joseph et al. [5] investigated the electrical properties of phenol formaldehyde composite reinforced with banana fiber with different fiber loading and chemical treatments. Elammaran et al. [6] investigated the dielectric properties of epoxy matrix composite reinforced with pineapple leaf fiber. Nayak et al. [7] reviewed the electrical insulating properties of polymer composite reinforced with natural fibers. A lot of work has been reported on dielectric behaviour of polymer composite reinforced with different natural fibers. But thermal and dielectric property of hair fiber reinforced epoxy composite has not been researched. So in the present work effect of fiber content on thermal conductivity and dielectric constant of hair fiber reinforced epoxy composite has been investigated.

2. Experimental Details

2.1. Composite Fabrication

Composite samples in the present study are fabricated by solution casting method. Raw hair fiber is collected from a saloon, washed with detergent and pure water. After washing, these fibers are kept in sunlight for few hours to remove the moisture content. After that with the help of a scissor these hair fibers are cut into 3-4mm length approximately. The epoxy resin (LY556) is mixed with its corresponding hardener (HY 951) in the ratio of 10:1 by weight and then short hair fiber is added into it with different weight proportions to prepare the composites. The mixture was mixed properly and then decanted into to a cylindrical mold of 1 cm diameter and 10 cm depth and also to a mold of size 5 cm diameter and 0.3 cm thickness. The casting is left as it is for 24 hours to cure at room temperature.

2.2. Thermal Conductivity Test

Unitherm^{TM} Model 2022 tester is used to compute the thermal conductivity of the epoxy-SHF composite. To conduct the test circular specimen of size 50mm diameter and 3mm thickness with both sides having flat surface are used. ASTM E-1530 standard is followed to perform the test.

2.3. Dielectric Test

The dielectric constant which is also known as the relative permittivity of a substance is the measure of permittivity of the substance relative to permittivity of vacuum. In the present research work dielectric constant of epoxy-SHF composite is measured in a HIOKI-3532-50 Hi Tester Elsier Analyzer. To conduct the test disc shaped composite specimen of 10mm diameter and 2-3 mm thickness are prepared. Both upper and lower surface of the test specimen are coated with silver paste. During the test 500mv AC voltages is applied and in a frequency range of 100Hz to 1 MHz dielectric constant of the composite samples are measured. The following formula can be used to compute the dielectric constant of the composite specimen

$$D_k = Ct/\varepsilon_0 A$$

where C represents the capacitance of test specimen, t represents the thickness of the discs, ε_0 is the dielectric constant of vacuum and A is the disc area.

3. Results and Discussion

3.1. Thermal Conductivity Test

The experimental thermal conductivity values of hair fiber reinforced composites with different proportions of SHF are shown in the Table.1. Theoretical values obtained by using rule of mixture [8] are also presented in this table for comparison. Fig.1. Compares the theoretically obtained effective thermal conductivity values of the composites with those obtained from experiments. It can be seen that the percentage error associated with each comparison is not very high. It is further noted that with the increase in fiber content, the effective thermal conductivity of the composite decreases which is attributed to the low intrinsic thermal conductivity of hair fiber.

Table 1. Thermal Conductivity of Epoxy-SHF Composites

SHF content (wt%)	Theoretical Thermal conductivity(W/mK)	Experimental Thermal Conductivity(W/mK)	Error (%)
0	0.363	0.363	0
2	0.359	0.342	4.7
4	0.355	0.335	5.6
6	0.351	0.327	6.8

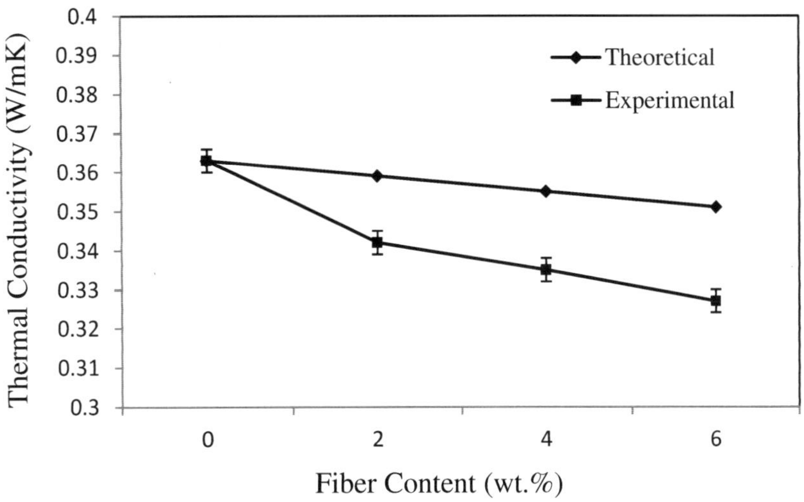

Fig 1. Thermal conductivity of composite at different wt. % of SHF

3.2. *Dielectric Test*

The dielectric constants of epoxy-SHF composites at room temperature with different wt. % of SHF content and at different frequencies (100Hz -10^6 Hz) are presented in the Table. 2. The variation of dielectric constant of the composite as a function of hair fiber content at different frequencies is shown in Fig. 2. It is observed that dielectric constant of the composite decreases with increasing fiber content and frequency. The reduction in dielectric constant with increasing frequency is the expected behaviour in most of the dielectric materials.

Table 2. Dielectric Constant of Epoxy-SHF Composites

Frequency(Hz)	Weight % 0	2	4	6
10^2	4.7	4.30	4.15	4.05
10^3	4.4	4.10	3.97	3.87
10^4	4.2	3.95	3.83	3.78
10^5	4.0	3.85	3.73	3.68
10^6	3.9	3.80	3.65	3.55

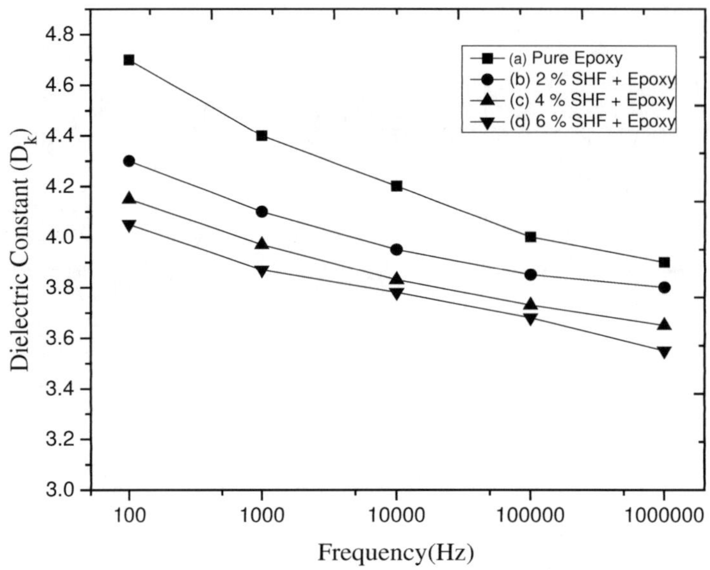

Fig 2. Dielectric constant of composite at different wt. % of SHF

4. Conclusions

This study reveals that the reinforcement of short hair fiber leads to substantial reduction in the effective thermal conductivity of epoxy composites and thereby improves their insulation capabilities. It is seen that addition of 6 wt. % of fiber results in 9.91 % reduction in the conductivity of epoxy. It also shows that the dielectric constant of the composites also varies with the fiber content and the operating frequency.

References

[1] Mishra D, Halder A, Panda A and Satapathy A 2012 APM.
[2] Agrawal A and Satapathy A 2017 CompComm **5** 36-39.
[3] Mishra D and Satapathy A 2012 APM.
[4] Mishra S C and Nayak N B 2010 J. Reinf. Plast. Compos. **29** 2691-2697.
[5] Joseph S and Thomas S 2008 J. Appl. Polym. Sci. **109** 256–263.
[6] Jayamani1 E, Heng1S K and Bakri M K B 2017 Key. Eng. Mater. **730** 42-47.
[7] Nayak N, Reddappa H N, Kalagi G and Bhat V 2017 IJERT **6** 165-169.
[8] Kumlutas D and Tavman I H 2006 J. Thermoplast. Compos. **19** 441-453.

Density Functional Theory Study of Structural and Electronic Properties of γ'-Ni_3Al and γ''-Ni_3Nb

D. K. Gorai*, T. K. Kundu
Department of Metallurgical and Materials Engineering,
Indian Institute of Technology Kharagpur, Kharagpur, WB, India-721302

*E-mail: deepakrai0134@gmail.com

Abstract: In the present paper, the structural and electronic properties of ordered face-centered cubic ($L1_2$) γ'-Ni_3Al and body-centered tetragonal ($D0_{22}$) γ''-Ni_3Nb are investigated by density functional theory calculation within projector augmented wave (PAW) potentials. The structures are optimized for both the phases and the results are found in good agreement with the experimental and theoretical data. In the geometrical optimization, all the forces on each atom are converged to less than 10^3 eV/Å, and the total energy less than 10^{-6} eV/atom. The Methfessel–Paxton smearing width of 0.2 eV is used here. The formation energy of γ'-Ni_3Al and γ''-Ni_3Nb phases are calculated and the values are found to be -40.67 kJ/mol and -27.55 kJ/mol respectively. The bonding character is analyzed based on the density of states (DOS).
Keywords: DFT, γ'- phase, γ''- phase, DOS

1. Introduction

Nickel-based superalloys are extensively used for high-temperature components such as gas turbine disks and aero engine[1]. ALLVAC 718+ is a nickel-based superalloy which can be used as a disc material for high-pressure turbine discs and compressor. This alloy has high-temperature stability and offers a full 55°C temperature advantage over alloy 718 [2].The principal strengthening phases in ALLVAC 718+ are γ'-Ni_3Al and γ''-Ni_3Nb. Hence, it is very important to calculate the structural, mechanical and electronic properties of γ'-Ni_3Al phase and γ''-Ni_3Nb phase.The first-principle density functional theory is extensively used to calculate the structural, mechanical and electronic properties of inter-metallic compounds. A number of studies, particularly on mechanical properties, is carried out on Ni_3Al[3][4]. Yasuda et al. [3] calculated the three elastic stiffness constants of different inter-metallic compounds such as Ni_3Al, Ni_3Mn, Ni_3Fe, Ni_3Si etc.. The structure, mechanical and magnetic properties of Ni_3Al, Ni_3In and Ni_3Ga was investigated by Boucetta et al. [4] using first principle and experimental studies. Yu et al. [5] studied the effects of pressure on elastic properties and thermal conductivity of γ''-Ni_3Nb phase. Cozar and Pineau [6] used X-ray diffraction to measure the lattice parameter of γ''-Ni_3Nb phase. However, there are few reports on the relationship between γ'-Ni_3Al phase and γ''-Ni_3Nb phase.

In this paper, we have used first-principle calculation within the framework of density functional theory (DFT) [7] to calculate the structural and electronic properties of γ'-Ni_3Al and γ''-Ni_3Nb phases. The bonding character is analyzed based on the total density of states (TDOSs) and partial density of states (PDOSs).

2. Materials and methods

In ALLVAC 718 plus, γ'-Ni_3Al and γ''-Ni_3Nb phases are the primary strengtheners by chemical and coherency strain hardening. The γ'-Ni_3Al belongs to ordered face-centered cubic (FCC) $L1_2$ crystal structure whose space group is Pm-3m (S.G. No 221). In the unit cell, there is one formula unit (Z=1) as shown in Figure 1(a). The γ''-Ni_3Nb belongs to ordered body-centered tetragonal (BCT) $D0_{22}$ crystal structure whose space group is 14/mmm (S.G. No 139). In the unit cell, there is two formula units (Z=2) as shown in Figure 1(b).

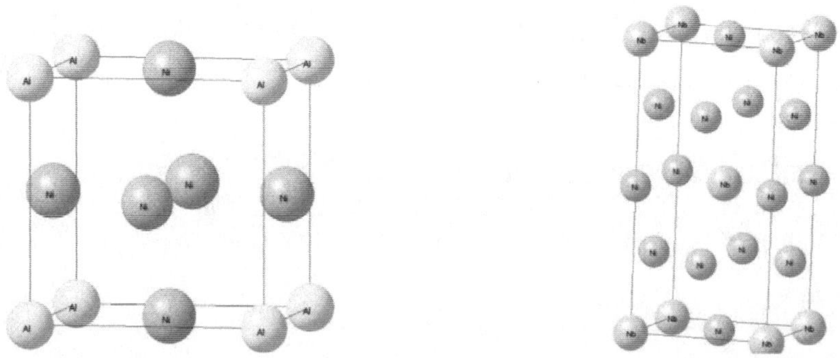

(a) (b)

Figure.1: The crystal structure of (a) γ'-Ni_3Al and (b) γ"-Ni_3Nb

In present work, the density functional theory implemented in the Vienna ab initio Simulation Package (VASP)[8] is used for the calculation. The projected augmented wave method (PAW) [9] is used to describe the ion-electron interactions. The structural calculations are performed independently using two separate approximation, (a) local density approximation (LDA) by Predew and Zunger[10] and (b) generalized gradient approximation of Perdew, Becke and Ernzerhof(GGA-PBE) [11] description of the exchange-correlation functional. Valence-electron configuration for pseudo-atoms are Ni-$3d^8 4s^2$, Al-$3s^2 3p^1$, and Nb-$4d^4 5s^1$.

In the calculation of γ'-Ni_3Al, initially the structure is optimized for different values of k-mesh as given in Table 1. From the Table 1, it is observed that the12x12x12 k-mesh optimized the structure with a minimum energy of -2111.55 kJ/mol. Therefore, the k-mesh 12x12x12 is used for the Brillouin Zone (BZ) sampling with a plane wave cut-off energy of 500 eV for both LDA and GGA method. Similarly, for γ"-Ni_3Nb phase the structure is optimized for different values of k-mesh as given in Table 2. From the Table 2, it is observed that the 10x10x5 k-mesh optimized the structure with a minimum energy of -2687.45 kJ/mol and therefore, the k-mesh 10x10x5 is used for BZ sampling with a plane wave cut-off energy of 500 eV for both LDA and GGA method. In the geometrical optimization of both the phases, all the forces on each atom are converged to less than 10^{-3} eV/Å, and the total energy less than 10^{-6} eV/atom. The atomic arrangements are relaxed using the Methfessel–Paxton method [12] with a smearing width of 0.2 eV. The Brillouin-zone integration is performed by the tetrahedron method with Blöchl corrections [13].

Table 1.. Optimized structural parameter of γ'-Ni_3Al for different values of k-mesh

k-mesh	**a=b=c(Å)**	**Volume (Å^3)**	**Density(Kg/m^3 x 10^3)**	**Energy (kJ/mol)**
16x16x16	3.57	45.38	7.43	-2111.55
12x12x12	3.57	45.37	7.43	-2111.55
11x11x11	3.57	45.37	7.43	-2111.55
10x10x10	3.57	45.38	7.43	-2111.47
8x8x8	3.57	45.38	7.43	-2111.33

Table 2. Optimized structural parameter of γ"-Ni_3Nb for different values of k-mesh

k-mesh	**a=b (Å)**	**c (Å)**	**Volume(Å^3)**	**Density(Kg/m^3 x 10^3)**	**Energy (kJ/mol)**
12x12x6	3.65	7.46	99.14	9.01	-2686.37
10x10x5	3.64	7.49	99.19	9.01	-2687.45
8x8x4	3.64	7.49	99.18	9.01	-2687.44
6x6x3	3.65	7.47	99.16	9.01	-2686.15

3. Result and Discussion

3.1 Structure Properties and stability

The total energy is calculated as a function of cell volume and fitted by the Brich-Murnaghan equation of states (EOS) [14] to find the equilibrium lattice constants. The optimized equilibrium cell volume and lattice parameters a_0 of Ni_3Al phase and a_0 and c_0 of Ni_3Nb phase is calculated. The theoretical and experimental results from the references[15][16][17][18][19][20], as well as the results calculated in the present work by both GGA and LDA methods, are given in Table 3 and Table 4 for Ni_3Al and Ni_3Nb respectively. It is found that for Ni_3Al, the value calculated by GGA method is in accordance with other calculations values [15][16] and as well as the experimental value [17][18]. The value calculated by LDA method is smaller than the experimental value. For the Ni_3Nb phase, the values of lattice constants a_0 and c_0 are respectively 0.55% and 0.78% higher than the experimental values, while the values of lattice constants a_0 and c_0 calculated by LDA method are respectively 1.74% and 0.78% smaller than the experimental values. Apart from this, the value of a_0/c_0 calculated by LDA method is not closer to the experimental value while the value of a_0/c_0 calculated by GGA method is closer to the experimental value. Hence, GGA method is used here for the subsequent calculation of stability of both Ni_3Al and Ni_3Nb phase. The LDA values are just used here as the reference for succeeding calculations.

Table 3: The calculated and experimental values of lattice constant (a_0), equilibrium cell volume (V_0) and bulk modulus (B_0) at 0 GPa of Ni_3Al phase

a_0(Å)	V_0(Å)^3	B_0(GPa)
3.57	45.37	-
3.48	42.17	-
3.57 [15]	45.65	182.0 [16]
3.57 [17]	45.58	179.3 [18]

Table 4: The calculated and experimental values of lattice constants (a_0,c_0), equilibrium cell volume (V_0) and bulk modulus (B_0) at 0 GPa of Ni_3Nb phase

a_0(Å)	c_0(Å)	c_0/a_0	V_0(Å)^3	B_0(GPa)	References
3.64	7.46	2.05	99.16	-	Present work (GGA-PBE)
3.56	7.75	2.06	92.57	-	Present work (LDA)
3.65	7.51	2.06	99.99	208.6	[19]
3.64	7.48	2.05	99.32	208	[20]

Generally, the formation enthalpy [21] is used as a measure of the stability of a phase. A negative formation enthalpy results in an exothermic process and a stable phase, while positive value implies in an unstable phase and endothermic process. In the present work, the following formula is used to calculate the formation enthalpy:

$$\Delta H(\ Ni_xAl_yNb_z) = \frac{E_{total}\ (Ni_xAl_yNb_z) - xE_{solid}\ (Ni) - yE_{solid}\ (Al) - zE_{solid}\ (Nb)}{(x+y+z)} \quad (1)$$

where, x, y, and z are the mole fraction of Ni, Al and Nb in the alloy. $E_{total}\left(Ni_xAl_yNb_z\right)$, $E_{solid}\,(Ni), E_{solid}\,(Al)\ and\ E_{solid}\,(Nb)$ are respectively the total energy of a $Ni_xAl_yNb_z$ alloy that relaxed to their equilibrium state and the total energies of elements Ni, Al and Nb (per atom), which are obtained from the first principle calculations. In case of Ni_3Al phase, the value of z is zero while, the value of y is zero in the case of Ni_3Nb phase. The total energies, symmetry group and the atomic radius of pure elements Ni, Al and Nb (per atom) in their ground state is given in Table 5. The values of total energies and formation enthalpies of Ni_3Al and Ni_3Nb phases are given in the Table 6. The results are in good harmony with the available theoretical result [20][23]. The negative value of formation enthalpy suggests an exothermic process and the thermodynamical stability of both the phases. It is noted, from the Table 6 that the value of formation enthalpy of Ni_3Al phase is -40.669 kJ/atom, which is smaller than the value of formation enthalpy -27.547 kJ/atom of Ni_3Nb phase. Hence it shows that the Ni_3Al phase with ordered face-centered cubic (FCC) $L1_2$ crystal structure is thermodynamically more stable than the Ni_3Nb phase with ordered body-centered tetragonal (BCT) $D0_{22}$ crystal structure. The less negative formation enthalpy of Ni_3Nb phase also indicates that it is thermodynamically not very strong. This is in good harmony with the experimental investigation which indicates that metastable γ"-Ni_3Nb phase is changed into more stable delta (δ- Ni_3Nb) phase below 900°C for a long time [24].

Table 5: Lattice constants, symmetry group, atomic radius, and total energies of pure elements per atom in their ground state.

Elements	a_0(Å)	Symmetry group	Atomic radius (Å)	Energy (kJ/mol)	Reference
Ni	3.52	FM3M	1.62	-529.33	This work
	3.52	-	-	-527.72	[22]
Al	4.04	FM3M	1.82	-360.96	This work
	4.05	-	-	-356.69	[22]
Nb	3.31	IM3M	2.08	-988.04	This work

Table 6: Calculated total energy (E_{tot}) and formation enthalpy (ΔH) of Ni_3Nb and Ni_3Al phases.

Ni_3Nb	E_{tot}(kJ/mol)	ΔH (kJ/mol)	Ni_3Al	E_{tot}(kJ/mol)	ΔH(kJ/mol)
GGA	-2687.19	-27.55	GGA	-2111.64	-40.67
LDA	-3115.30	-134.79	LDA	-4068.51	-135.88
Ref. [20]	-	-29.72	Ref.[23]	-	-43.98

3.2 Electronic Properties

Both GGA and LDA methods are used to calculate the total density of states (TDOSs) and partial density of states (PDOSs) of Ni_3Al phase and Ni_3Nb phase, at zero pressure to study underlying mechanism of mechanical properties and structural stability from the nature of the chemical bonding. It is found that the result calculated by both methods display the similar patterns. Hence, for the ease, we have just given the results calculated by the GGA method.

In Figure 2(a) and Figure 2(b), the total density of states and partial density of states of both the phases have been given. It is found that the value of the density of states for both the phases, at Fermi energy is non-zero, which reveals that both the phases have metallic characteristics. It is shown in Figure 2(a), that the Ni *3d* states and Al *3p* states control the contributions to the total density of states and also there exists a little difference between the spin-up and spin-down states, which indicates a weak magnetism of Ni_3Al[25]. It is found that there is no distinction between the spin-up and spin-down in Ni_3Nb phase (Figure 2(b)), hence there is no magnetism in this phase. Next, from the Figure 3(a) it is found that there is no overlapping in Al *3p* and Ni *4d*, so there is no covalent interaction in Ni_3Al, but from Figure 3(b), it is clear that there is a complete overlapping between Ni *3d* and Nb*4d* states, which indicates there exists covalent interaction in Ni_3Nb phase.

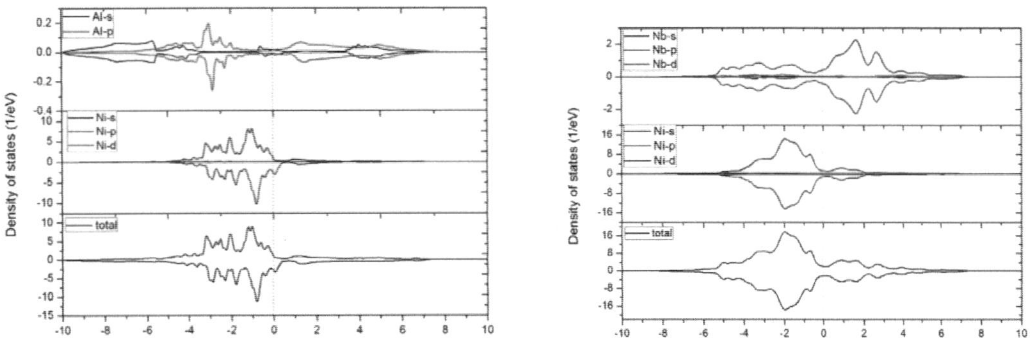

(a) (b)

Figure 2: Total and partial density of states (a) Ni_3Al phase and (b) Ni_3Nb phase

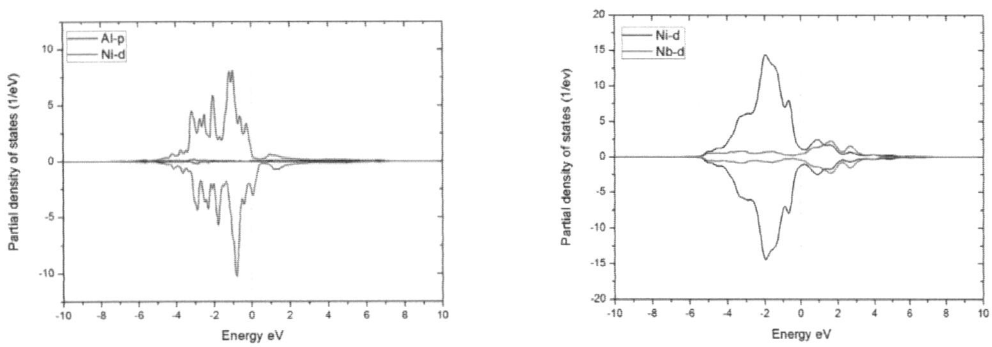

(a) (b)

Figure 3: Partial density of states (a) Ni_3Al phase and (b) Ni_3Nb phase

4. Conclusion

In the present work, we have calculated the structural and electronic properties of γ'-Ni_3Al and γ''-Ni_3Nb using first principle density functional theory calculation at 0 GPa pressure. The calculated lattice constants are in good harmony with the experimental and theoretical results. The negative value of formation energy points out that both phases are thermodynamically stable. The more negative value of formation energy in the case of γ'-Ni_3Al phase indicates that γ'-Ni_3Al is thermodynamically more strong than γ''-Ni_3Nb, which is in good accordance with the experimental investigation which suggests that metastable γ''-Ni_3Nb phase is changed into more stable delta (δ- Ni_3Nb) phase below 900°C for a long time. The density of states reveals that there exists a covalent interaction in γ''-Ni_3Nb.

References:

[1] Zhang H Y, Zhang S H, Cheng M, and Li Z X 2010 *Mater. Charact.* **61** pp. 49–53.
[2] Kennedy R L, Cao W D, Bayha T D and Jeniski R 2004 *Niobium for High Temperature Applications*, ed. by Young-Won Kim and Tadeu Carneiro TMS pp. 11-21.
[3] Yasuda H, Takasugi T, Koiwa M 1992 *ActaMetall Mater* **40** 381.
[4] Boucetta S, Chihi T, Ghebouli B and Fatmi M 2010 *Mater Sci Poland* **28** 347.
[5] Lin Y C, Luo S C, Chen M S and He D G 2016 *Journal of Alloys and Compounds* **688** pp

285- 293.
[6] Cozar R and Pineau A 1973 *Metall. Trans.* **4** pp 47-59.
[7] Kohn W and Sham L J 1965 *Phys. Rev. A* **140** 1133.
[8] Kresse G and Furthmuller J 1996 *Phys. Rev. B* **54** pp 11169-11186.
[9] Kresse G and Joubert D 1999 *Phys. Rev. B* **59** pp 1758-1775.
[10] Ceperley D M and Alder B J 1980 *Phys. Rev. Lett.* **45** 556.
[11] Perdew J P, Burke K and Ernzerhof W 1996 *Phys. Rev. Lett.* **77** 3865.
[12] Methfessel M and Paxton A T 1989 *Phys. Rev. B* **40** pp 3616-3621
[13] Blöchl P E, Jepsen O and Andersen O K 1994 *Phys. Rev. B* **49** pp 16223-16233.
[14] Murnaghan F D 1944 *Proc. Natl. Acad. Sci.* USA **30** 244.
[15] Guo G, Wang Y and Hsu LS 2002 *J MagnMagn Mater* **239** 91.
[16] Kim D, Shang S and Liu Z 2010 *Intermetallics* **18** 1163.
[17] Rao PM, Suryanarayana S, Murthy KS and Naidu SN 1989 *J Phys Condens Matter* 15357.
[18] Yasuda H, Takasugi T and Koiwa M 1992 *ActaMetall Mater* **40** 381.
[19] Lin Y C, Luo S, Chen M, He D and Zhao C 2016 *J. Alloys Compd.* **688** pp 285-293.
[20] Connetable D, Mathon M and Lacaze J 2011 *Coupling Phase Diagrams Thermochem* **35** pp 588-593.
[21] Duan Y H, Hu W C, Sun Y and Peng M J 2014 *J. Alloys Compd.* **614** pp 334-344.
[22] Kim D E, Shang S L and Liu Z K 2010 *Intermetallics* **18** pp 1163-1171.
[23] Wu Q and Li S 2012 *Comput. Mater. Sci.* **53** pp 436-443.
[24] Sundararaman M, Mukhopadhyay P and Banerjee P 1988 *Metall. Trans.* **19A (3)** 453.
[25] Hamid A S, Uedono A, Major Z, Haynes T D, Laverock J, Alam M A, Dugdale S B, and Fort D 2011 *Phys. Rev. B* **84** 235107.

Mechanical and microstructural characterization of W–Cu FGM fabricated by one-step sintering method through PM route

Rajat Gupta[1]*, Rohit Kumar[1], A.K. Chaubey[1], Shailesh Kanpara[2] and S.S. Khirwadkar[2]

[1]*CSIR- Institute of Minerals and Materials Technology (CSIR-IMMT), Bhubaneswar-751013, India*
[2]*Institute for Plasma Research, Gandhinagar-382428, India*

***Email:** rajat.msme@gmail.com

Abstract: Five layer W-Cu functionally graded material (FGM) for components in nuclear fusion application was fabricated by a one-step resistance sintering process, known as spark plasma sintering (SPS). In this study effect of sintering temperature (T_S) on physical, mechanical and surface property was investigated. Detailed microstructural study revealed that the graded structure of the composite layers with varying composition from 0 to 100 wt% W and Cu in opposite directions could be well densified after the SPS process. It also indicates that the fine microstructure within functionally graded layers can be maintained because of short sintering time. The sample sintered at 1050^{0}C shows more than 90% theoretical density, hardness greater than 239±5 Hv and excellent surface scratch resistance. The result demonstrates that SPS is promising and more suitable process for fabrication of W-Cu FGM.

Keywords: Mechanical properties; Functionally graded material; Plasma Facing Material; Powder metallurgy; Spark Plasma Sintering.

1. Introduction

To construct fusion reactors, developing plasma facing components (PFCs) for the divertor application is still an issue for material scientists and researchers to overcome because it will be exposed to high thermal loads (5-20 MW/m^2) and play an crucial role in the realization of a fusion reactor. According to the required material characteristics for PFCs in the International Thermonuclear Experimental Reactor (ITER), Tungsten and Copper alloys are the most suitable candidate for plasma facing and heat sink materials respectively in PFCs. Tungsten has high melting temperature, higher sputtering threshold, low tritium retention etc. which is suitable good thermal- mechanical properties [1-2]. Moreover, low coefficient of thermal expansion makes it a promising plasma facing material. On the other hand, copper is a well-known heat sink material option due to high thermal conductivity and high coefficient of thermal expansion. [3,4] However, the large difference in melting point and coefficient of thermal expansion (CTE) of tungsten and copper makes it difficult to fabricate by conventional methods (i.e. brazing, conventional sintering, liquid infiltration etc.) and resulted thermally induced stresses at the interface. FGM can be used as interlayer and could effectively implement the mitigation of thermally induced stresses effectively which subsequently improve the component's life.

FGM is an advanced composite material that shows continuous variation in material properties as the dimension varies. These are materials with a smooth gradient in one or more properties which is due to dependency of microstructure or chemical composition on position. The concept of property gradation can minimise thermal stresses due to difference in the CTE of W and Cu, and thus have achieved significant importance in high temperature environments applications where one component has to fulfil contrasting requirements in different positions and a joint is inappropriate. W-Cu FGM can additionally benefit due to an obvious strengthening effect, i.e. due to embedment of hard particle (W) in the soft matrix (Cu) [5]. It not only reduces the problems caused by the mismatch of the CTE and other physical properties of both metals, but also combines the outstanding qualities of W and Cu as described above. Hence, FGM is an appropriate and reliable option as interlayer between W and Cu for (PFC) to minimize these problems when used as interlayer.

Several techniques have been employed previously like chemical vapour deposition, plasma spraying slip casting etc. for the fabrication of PFCs, but they are expensive, complex and give porous microstructure to the FGM [6-11]. In recent decades, SPS process has gained heavy attention of the researchers and widely used for fabrication of FGM, inter metallic compounds, ceramics, and composites etc., because of ease in this process as compared to common sintering methods. SPS offers significant advantages with various kinds of new materials and consistently produces a highly dense compact due to rapid consolidation technique than conventional methods. Due to higher rate of heating and less sintering time grain growth is obstructed which will consequently keep the microstructure fine and ultimately provide better irradiation resistance [12-17]. Therefore, in this presented study W-Cu FGM was fabricated through SPS process to achieve higher density and microstructural homogeneity in the compacts for plasma facing applications.

2. Materials and Methods

W-Cu FGM was prepared from the elements of high purity with average particle size 4.3 μm and 35.78 μm of tungsten and copper respectively. Five-layered FGM samples were fabricated at three different temperatures according to the design as shown in Fig.1 with varying wt% of W and Cu (100W, 75W-25Cu, 50W-50Cu, 25W-75Cu, 100Cu) in each layer, respectively. Thickness of layers in FGM sample was kept constant is about 1 mm.

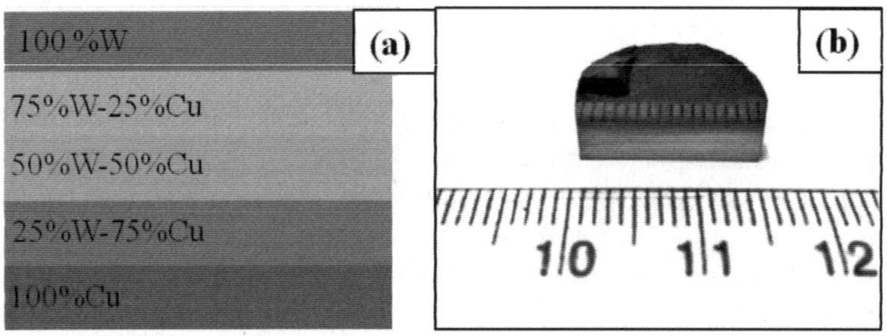

Fig.1 (a) Schematic of five layer stacking of W-Cu powder and **(b)** Fabricated FGM sample by SPS.

Initially W and Cu as received powders with calculated weight ratios were mixed in a planetary ball mill (Retsch PM 400) for 10h. X'Pert PRO PANalytical's Diffractometer (λ=1.54184 Å) was used for phase identification of the milled powders. Particle size of the powder were measured by particle size analyzer (MASTERSIZER) having water as dispersant and Scanning Electron Microscopy (SEM, Zeiss Supra-55) was used for particle morphology study.

In order to prepare bulk composite sample as shown in Fig.1 (b) with desired compositions powders were stacked layer wise (~1mm each layer) in a mould of 15 mm dia. made of graphite having height about 50mm. The sintering of the samples were done in a spark plasma sintering system (SPS Syntex, 725) at 950°C, 1000°C & 1050°C at vacuum < 10 Pa for 3 min with heating rate of 100 °C/min at a load of 60 MPa. Scratch testing (micro scratch tester, UMT-CETR, Bruker) has been performed to analyze the interlayer response toward load and ultimately the bond strength of FGM layers. The test was conducted under the progressive load from 5N to 50N on a single path with speed 0.016mm/s by diamond hard tip indenter with radius 1.5 μm.

3. Results & Discussions

3.1. Powder Analysis:

The SEM image of the milled powder and initial particle size distribution Fig.2 is clearly shown in figures below. Fig.2 (a) shows two types of particles big agglomerated equiaxed particles and small spherical particles. The distribution plot shows that the particle distribution of two separate range almost negligible overlapping area. The distribution plot of W shows that particle size ranges between 0.4μm and 10μm with mean particle size of about 3μm. Similarly Cu particles also distributed in a range of 1μm to 400μm with average 40μm.

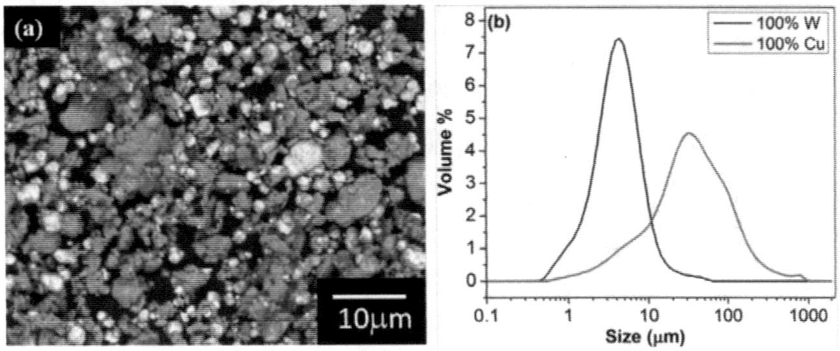

Fig.2. (a) FESEM micrograph of W-Cu powder milled for 10h in planetary ball mill and used for FGM synthesis (b) Graphs of showing initial size range of W-Cu powders particles before milling.

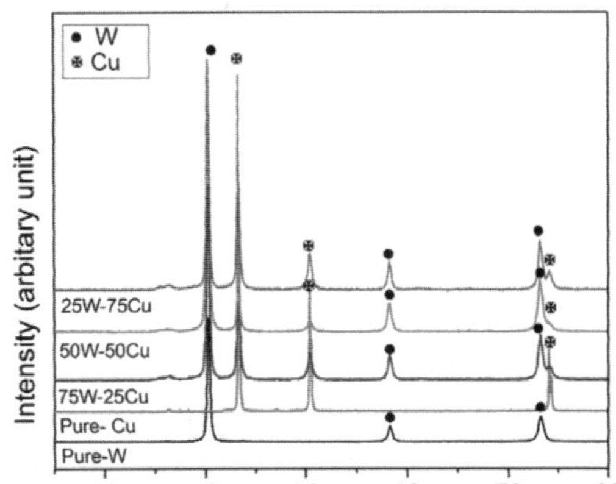

Fig. 3: XRD pattern of different functionally graded layers of FGM sample sintered at 1000^{0}C by spark plasma sintering.

Patterns observed through XRD for graded layers are shown in the Fig.3 in different colours to distinguish the pattern of each layer. The pattern shows only W and Cu peaks, indicates that there is no new phase formation during milling and sintering process. The only difference in the intensity is observed, as expected due to difference in the composition of the layers but all the patterns looks identical. The patterns also shows little broadening of the peaks due to fine size of the particles.

3.2. Microstructural study

SEM image in Fig.4 (a) clearly shows the different layers with distinct contrast. It also shows strong interfacial bonding between different layers of bulk FGM sample. The layers show uniformity throughout the sample and gradation of Cu and W is clearly visible and differentiable through elemental mapping Fig.4 (b). With increase in W % in the layers the microstructure is changing in a continuous manner, as it is visible in the figures Cu is uniform as a matrix and W particles are well surrounded by the Cu.

Fig.4. SEM micrograph (a) and corresponding elemental area mapping (b) showing clear gradation pattern of Cu (Red) & W (White).

3.3 Physical and Mechanical properties:

Sintered samples were taken for density measurement and density was measured with the help of Archimedes principle and values were accounted by average from repeated experiments. The black bars in Fig.5 exhibits the % change in relative density of W/Cu FGM samples at different sintering temperatures. It increases with increase in sintering temperature. The increase of density with increase of sintering temperature is due to better covering and filling the space between the W particles by the molten Cu-and also for the reason that at higher sintering temperature the pores filled-up by molten metal resulting increase of densification [18]. The maximum value of densification reaches about 90.5 % at the sintering temperature of 1050 °C.

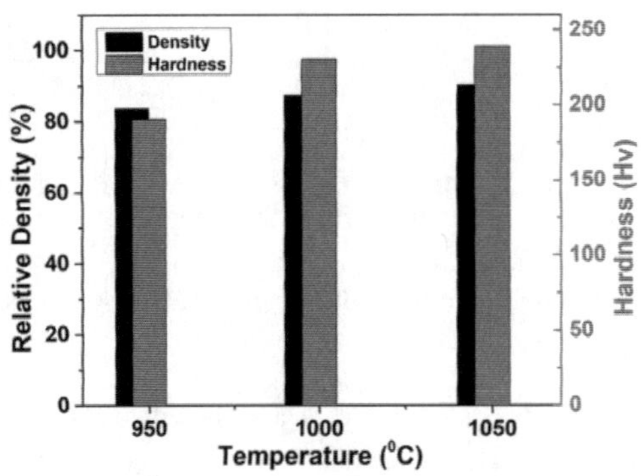

Fig.5. Relative density variation and average Vickers hardness plots of sintered FGM samples at different temperatures (T_S).

Fig.5 also shows the Vickers hardness of tungsten richest part in W/Cu FGM samples sintered for 3 minutes at different temperatures. The plot reveals as usual results that the hardness value increases with increase of sintering temperature. This is due to shrinkage of the open pores and the grain boundary diffusion at higher sintering temperature resulting increase of density and hardness [19]. The hardness values are highest at 1050 °C compared to other sintering temperature.

3.4 Scratch testing:

Scratch test was conducted on the polished surface of the W-Cu FGM bulk sintered samples and results are shown in Fig.6(a&b). Scratch mark on the surface of the sample shows almost uniform as the load increase from one end to other end because the sample hardness also increase from one end to other end. Optical image of the scratch profile shows more detailed insight of the surface behaviour of FGM samples, the scratch width along the length from one end to other end was almost same. Stylus holder attached with acoustic emission (AE) sensor records the high frequency signal generated during scratching. When the material undergoes some damage, deformation, fracture triggering a sudden release of elastic energy in the form of acoustic radiations and thease radiation signals are recorded as an indication of material failure.The material hardness depends upon the generated signal of acoustic emmission.The scratch hardness of material at different temperature has been observed by generated signal of acoustic emmission. Uniform signals of Acoustic emmission throught out the samples shows that uniform deformation during scratch test without interfacial hinderance and layers are intact to each other. Similar behavior was also found for the coefficient of friction (COF) curve shows that no significant change in friction behaviour, only sudden hump at the interfaces of each layer interfaces due to change in compositional gradient. Moreover, the AE plot is almost homogeneous throughout the test which also supports the bond strength factor of our fabricated FGM samples.

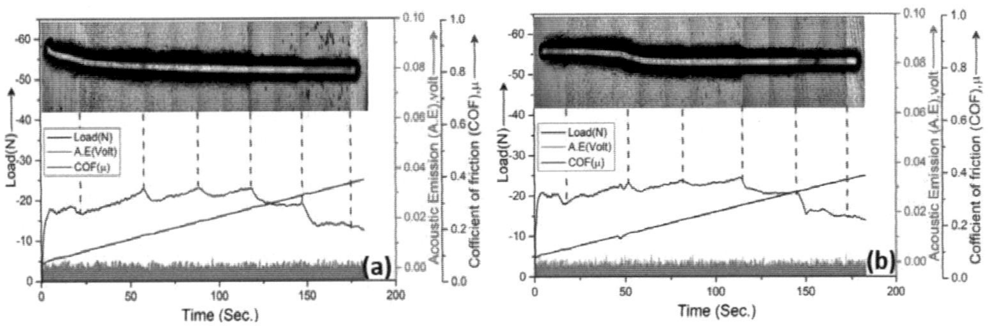

Fig.6. Scratch test data plots for FGM samples sintered at 1000 °C (a) and 1050 °C (b) respectively.

4. Conclusions

Five layered W /Cu FGM (100W, 75W-25Cu, 50W-50Cu, 25W-75Cu, 100Cu (wt %) samples were successfully fabricated through SPS method. Each layer (~1mm thickness) is clearly visible in the FSEM micrographs. The image contrast at interface due to compositional changes is clear and well bonded at the interface. The density and hardness values of each FGM samples shows different values and it increase with increase in sintering temperature. The bond strength study through scratch test revealed good intact between interlayers throughout the sample as supported by COF and AE plots. The optimization study shows that 1050 °C sintering temperature and pressing force 60 MPa is best possible parameters to fabricate W-Cu FGM using SPS technique for use as plasma facing component application.

5. References

[1] Diegele E, Krüssmann R, Malang S, Norajitra P and Rizzi G 2003 Modular He-cooled divertor for power plant application *Fusion Eng. Des.* **66–68** 383–7

[2] Cottrell G A 2006 A survey of plasma facing materials for fusion power plants *Mater. Sci. Technol.* **22** **(8)** 869–880

[3] Ivanov A D, Nikolaev A K, Kalinin G M and Rodin M E 2002 Effect of heat treatments on the properties of CuCrZr alloys *J. Nucl. Mater.* **307–311** 673–6

[4] Merola M, Orsini A, Visca E, Libera S, Moreschi L F, Storai S, Panella B, Campagnoli E., Ruscica G and Bosco C 2002 Influence of the manufacturing heat cycles on the CuCrZr properties *J. Nucl. Mater.* **307–311** 677–680

[5] You J H, Brendel A, Nawka S, Schubert T and Kieback B 2013 Thermal and mechanical properties of infiltrated W/CuCrZr composite materials for functionally graded heat sink application *J. Nucl. Mater.* **438** 1–6

[6] Kieback B, Neubrand A and Riedel H 2003 Processing techniques for functionally graded materials *Mater. Sci.and Eng. A* **362** 81–105

[7] Liu R, Hao T, Wang K, Zhang T, Wang X P, Liu C S and Fang Q F 2012 Microwave sintering of W/Cu Functionally graded mateial *J. Nucl. Mater.* **431** 196–201

[8] Zhou Z J, Du J, Song S X, Zhong Z H and Ge C C 2007 Microstructural characterization of W/Cu functionally graded materials produced by a one-step resistance sintering method *J. Alloys Compd.* **428** 146–50

[9] Zhou Z J, Song S X, Du J, Zhong Z H and Ge C C 2007 Performance of W/Cu FGM based plasma facing components under high heat load test *J. Nucl. Mater.* **363–365** 1309–14

[10] Pintsuk G, Brunings S E, Doring J E, Linke J, Smid I and Xue L 2003 Development of W/Cu-functionally graded materials *Fusion Eng. Des.* **66-68** 237– 240

[11] Miyamoto Y, Kaysser W A, Rabin B H, Kawasaki and Ford R G 1999 *Functionally Graded Materials* (US: Springer)

[12] Ding L, Xiang D P, Li Y Y, Li C and Li J B 2012 Effects of sintering temperature on fine-grained tungsten heavy alloy produced by high-energy ball milling assisted spark plasma sintering *Int. J. Refract. Met. Hard Mater.* **33** 65–9

[13] Munir Z A, Anselmi-Tamburini U and Ohyanagi M 2006 The effect of electric field and pressure on the synthesis and consolidation of materials: a review of the spark plasma sintering method *J. Mater. Sci.* **41** 763–777

[14] Feng H, Meng Q, Zhou Y and Jia D 2005 Spark plasma sintering of functionally graded material in the Ti–TiB2–B system *Mater. Sci. Eng. A* **397** 92–7

[15] Orrù R, Licheri R, Locci A M, Cincotti A and Cao G 2009 Consolidation/synthesis of materials by electric current activated/assisted sintering *Mater. Sci. Eng. R* **63** 127–287

[16] Qiao Z, Räthel J, Berger L M and Herrmann M 2013 Investigation of binder less WC–TiC–Cr_3C_2 hard materials prepared by spark plasma sintering (SPS) *Int. J. Refract. Met. Hard Mater.* **38** 7–14

[17] Anselmi-Tamburini U, Gennari S, Garay J E and Munir ZA 2005 Fundamental investigations on the spark plasma sintering/synthesis process II. Modeling of current and temperature distributions *Mater. Sci. Eng. A* **394** 139–148

[18] Ardestani M, Rezaie H R, Arabi H and Razavizadeh H 2009 The effect of sintering temperature on densification of nanoscale dispersed W–20–40%wt Cu composite powders *Int. J. Refract. Met. Hard Mater* **27** 862–7

[19] Chu H L, Wang C L, Lee H E, Sie Y Y, Chen R S, Hwang W S and Wang M C 2013 Effect of sintering process parameters on the properties of 3Y-PSZ ceramics *IOP Conf. Ser. Mater. Sci. Eng.* **47** **(1)** (2005)

IOP Conf. Series: Materials Science and Engineering **338** (2018) 012043 doi:10.1088/1757-899X/338/1/012043

Effects of Mn addition on microstructure and hardness of Al-12.6Si alloy

Prosanta Biswas, Surajit Patra and Manas Kumar Mondal*

Department of Metallurgical and Materials Engineering, National Institute of Technology, Durgapur-713209, West Bengal, India

Email :- manas.nitdgp@gmail.com

Abstract. In this work, eutectic Al-12.6Si alloy with and without manganese (Mn) have been developed through gravity casting route. The effect of Mn concentration (0.0 wt.%, 1 wt%, 2 wt% and 3 wt%) on microstructural morphology and hardness property of the alloy has been investigated. The eutectic Al-12.6 Si alloy exhibits the presence of combine plate, needle and rod-like eutectic silicon phase with very sharp corners and coarser primary silicon particles within the α-Al phase. In addition of 1wt.% of Mn in the eutectic Al-12.6Si alloy, sharp corners of the primary Si and needle-like eutectic Si are became blunt and particles size is reduced. Further, increase in Mn concentration (2.0 wt.%) in the Al-12.6Si alloy, irregular plate shape $Al_6(Mn,Fe)$ intermetallics are formed inside the α-Al phase, but the primary and eutectic phase morphology is similar to the eutectic Al-12.6Si alloy. The volume fraction of $Al_6(Mn,Fe)$ increases and $Al_6(Mn,Fe)$ particles appear as like chain structure in the alloy with 3 wt.% Mn. An increase in Mn concentration in the Al-12.6Si alloys result in the increase in bulk hardness of the alloy as an effects of microstructure modification as well as the presence of harder $Al_6(Mn,Fe)$ phase in the developed alloy.

1. Introduction

Al-Si alloys are extensively used in various engineering industries such as aerospace, automobile, structural and packaging and industries for their excellent properties like high wear and seizure resistance, improved specific strength and stiffness, low thermal expansion coefficient, improved damping capacity and low weight to strength ratio [1-2]. However, the Al-Si alloys mechanical properties are not adequate due to presence of coarse needle shape and irregular eutectic and primary silicon particles [3] and those sharp corners of Si particles will act as a suitable crack initiation sites during loading [4]. Therefore, modification of Al-Si alloy is essential to enhance the mechanical properties. Several elements such as Na [5], Ba [6], Ca [6], Y [6], Yb [6-7], Eu [7], Sb [8] and Sr [8-9] are generally used to modify the Al-Si alloy. Manganese can be used to modify the Al-Si alloy. The Mn has significant effect to improve the mechanical properties of cast Al and Al-Si alloys without harming the chemical resistance [10-11]. It has role to adjust the iron containing phase in aluminium based alloy [12].

However, limited number of investigations on the effect of Mn addition have been carried out on microstructures modification and mechanical properties enhancement of Al-Si based alloys. A narrow range of Mn concentration in Al-Si alloy has been studied. In this investigation, Al-12.6Si alloy with and without Mn addition has been developed through gravity casting process. The effect of Mn concentration (0.0 wt.%, 1 wt.%, 2 wt.% and 3 wt.%) on microstructure and hardness property of eutectic Al-12.6 alloy has been investigated. Further, a correlation between microstructure morphology and hardness has been established.

2. Experimental procedure

The eutectic Al-12.6Si alloy with (1wt.%, 2wt. % and 3 wt.%) and without Mn addition has been developed through gravity casting process. Commercially pure Al (99.7%), Si (99.8%) and Al-80Mn master alloy were used as starting materials. Initially, small blocks of commercially pure Al were melted at 760°C in a clay graphite crucible of 3kg capacity in an electrical resistance furnace. Then, the pure Si granules were added to the molten Al and hold for 45 to 60 minute for complete dissolution of Si. After that, aluminium foil wrapped small species of Al-80Mn master alloy were added in the melt as per requirement and again hold for 5-10 minute. Afterward, degassing was performed using 0.1 wt.% hexachloroethane and immediately slag was removed and the molten metal was poured into a preheated (160°C for 1hrs) permanent cast iron mould. The metallographic and hardness test samples were collected from the middle portion of the cast billets. All the samples were prepared with standard procedure for aluminium specimen preparation and Keller's reagent (1% HF, 1.5% HCl, 2.5% HNO_3 and distilled water) was used to etch the metallography sample after polishing. The optical microscope (Leica DM 2500M) and scanning electron microscope (JEOL, JSM-5800) equipped with energy dispersive X-ray spectroscopy (EDS) detector (Oxford INCA EDS) were used for metallography and phase identification. The bulk hardness and micro hardness test were carried out using a Vickers hardness testing machine (BV 250 (S)) and a micro-hardness testing machine (MMT-X7B). The different phases present in the modified and unmodified Al-Si alloys were identified by X-ray diffraction (XRD) analysis. The microstructural features such as volume percentage (serologically equated with area percentage) and the Si particle size were measured using the ImajeJ image analysis software.

3. Results and Discussion

3.1 Microstructural Evaluation

The optical microstructure (Fig. 1(a)) of the unmodified Al-12.6Si alloy exhibits the presence of plate, needle and rod like grey colour irregular small eutectic particles with very sharp corners and coarser grey colour primary particles within the α-Al grains. Further, the EDS spot analysis (Fig. 2) on the grey color particles, identify the particles as primary silicon particles and eutectic Si particles.

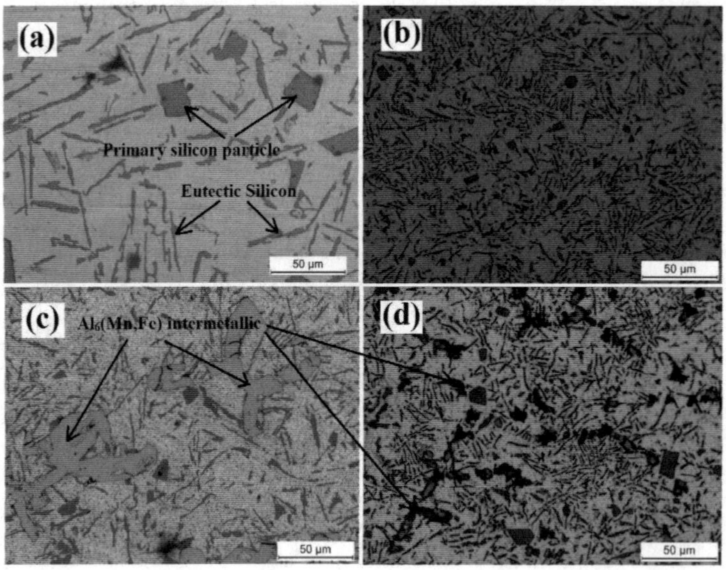

Figure 1. Optical microstructures of the gravity cast (a) Al-12.6Si-0 wt.% Mn (b) Al-12.6Si-1.0 wt.% Mn (c) Al-12.6Si-2.0 wt.% Mn and (d) Al-12.6Si-3.0 wt.% Mn alloy.

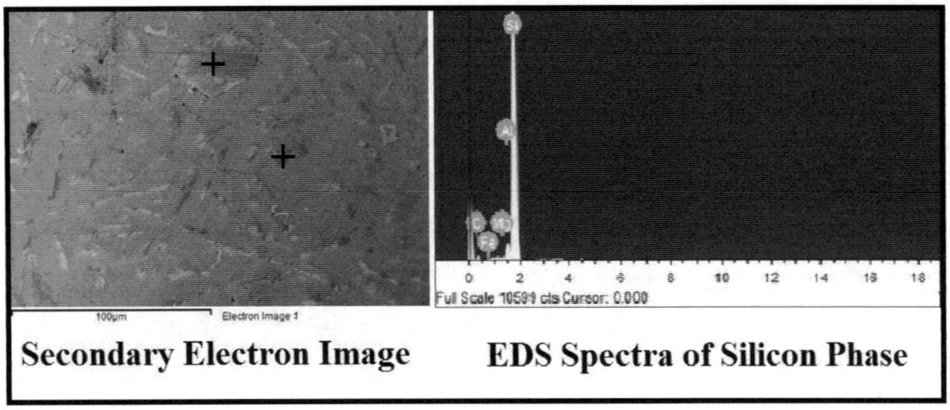

Figure 2. Identification of primary Si particles by SEM based EDS spot analysis in Al-12.6Si-0.0Mn alloy.

Manganese addition in Al-12.6Si alloy effectively modify the primary Si particles as well as eutectic Si morphology and distribution. In addition of 1wt.% of Mn in Al-12.6Si alloy, sharp corners of the primary Si particles and needle like morphology of eutectic Si become blunt and the primary Si particles as well as eutectic Si particles become finer (Fig. 1(b)). It is also clear from Fig. 1(b) that the eutectic phase volume fraction is increased and the distribution of both the silicon phases becomes uniform and primary silicon particles become compact with addition of 1 wt.% Mn with respect to unmodified alloy. The roundness/sphericity of the primary silicon particles is also increased. It is interesting to report that with the addition of 2.0 wt. % of Mn in the base alloy some irregular intermetallic is formed, but the eutectic silicon phase morphology is like the base Al-12.6Si alloy and the thickness of the eutectic Si phase is found to be reduced (Fig. 1(c)). The primary silicon size is decreased but the sharp corner of the primary silicon particles is remain unchanged. Further, the X-ray diffratrogram of the Al-12.6Si and Al-12.6Si-2Mn alloy (Fig. 3) exhibits the presence of Si phase and $Al_6(Mn,Fe)$ intermetallic phase apart from the α-Al and Si phase and those irregular intermetallics are $Al_6(Mn,Fe)$ (Fig. 1(c) and (d)).

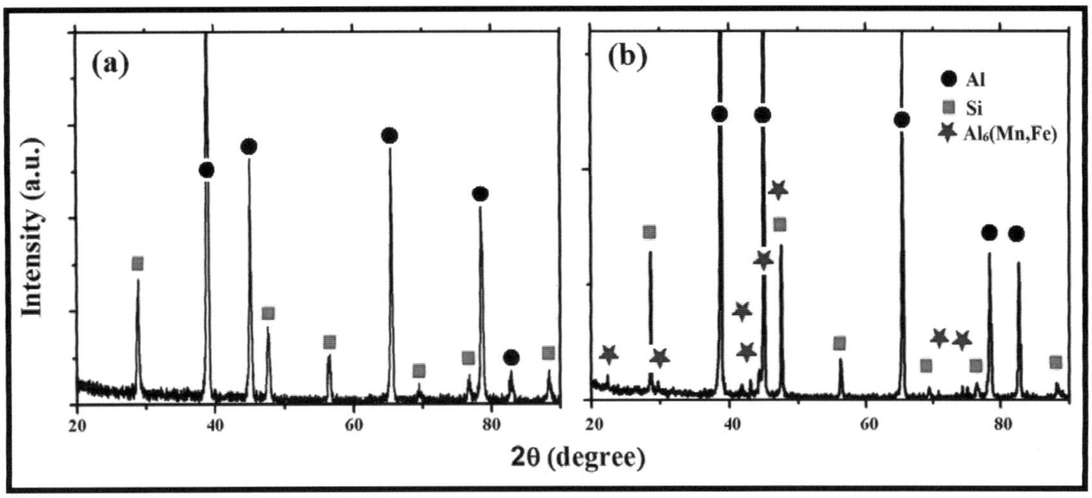

Figure 3. XRD patterns of the (a) Al-12.6Si and (b) Al-12.6Si-2 wt.% Mn alloy.

It is also clear from Fig. 1(c) that the volume fraction of eutectic silicon is decreased. It is notable that when intermetallic of Al and Mn formed in the alloy, there is not sufficient amount of Mn present to

IOP Conf. Series: Materials Science and Engineering **338** (2018) 012043 doi:10.1088/1757-899X/338/1/012043

modify the eutectic and primary silicon morphology. Further increase in Mn concentration (3 wt.%) in the alloy increases the volume fraction of $Al_6(Mn,Fe)$ intermettalics and $Al_6(Mn,Fe)$ intermettalics are appear as like a chain, but the shape is changed into compact and small size quadrilateral shape particles (Fig. 1(d)). Further, the EDS spot analysis of this quadrilateral shape phase (Fig. 4) again conform that those are the $Al_6(Mn,Fe)$ intermettalics and it also verify that the Fe is present in the intermettalics ($Al_6(Mn,Fe)$ as a supplementary of Mn atom.

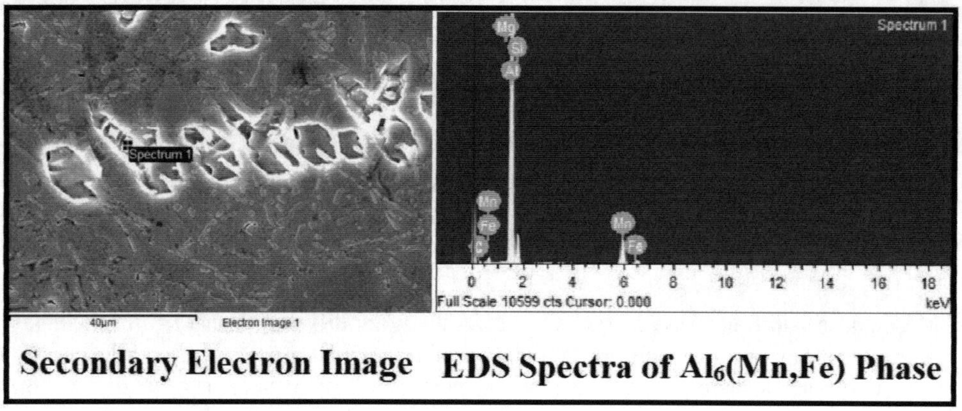

Figure 4. Identification of $Al_6(Mn,Fe)$ intermetallic particle by SEM based EDS spot analysis in the Al-12.6Si-3.0 wt.% Mn alloy.

3.2 Hardness

The alloying elements and reinforcing phase concentration as well as morphology and volume fraction of phases present in an alloy have strong impact on the bulk hardness of the Al based alloy and composites [13]. Figure 5 shows the bulk hardness of the Al-12.6 Si alloy as a function of Mn concentration. It is found that the hardness is significantly increased with increase in Mn concentration in the alloy.

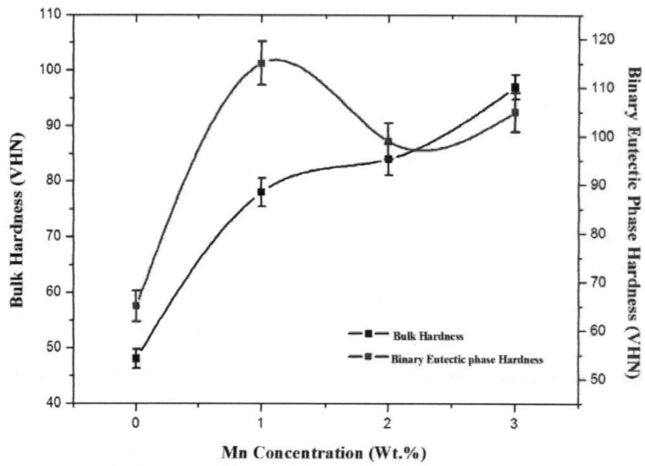

Figure 5. Graphical representation of change in the bulk hardness and the binary eutectic phase hardness with increasing Mn concentration.

The hardness of the unmodified alloy is 48±1.7 VHN, but 1 wt.% Mn modified alloy hardness (78±2.6 VHN) is much higher than the base alloy due to combine modification of primary and eutectic Si particles and uniform distribution of both the phases as well as reduction in sharp corners and increase in compactness of the primary Si particles. Fine and compact primary Si particles without sharp corners are more efficient to protect the plastic deformation during hardness test as the stress concentrated points are reduced due to modification effects of Mn. Furthermore, when the Mn concentration again increased to 2 wt.% and 3 wt.%, again hardness of the alloy increased to 84±2.9 and 97±2.2 respectively because of the presence of harder $Al_6(Mn,Fe)$ intermetallics phase. As the Mn concentration increased, the volume fraction of the $Al_6(Mn,Fe)$ intermetallics phase increase because of this the alloy with 3 wt.% of Mn has high hardness than the alloy with 2 wt.% of Mn.

The micro hardness of constituent phases is shown in Fig. 5 and Fig.6. It is found that that the micro hardness of Al-Si binary eutectic phase is high in Mn modified alloy. In addition of 1 wt.% of Mn, the Al-Si binary eutectic phase hardness is increased significantly as the eutectic Si become uniform and finer with a higher volume fraction.

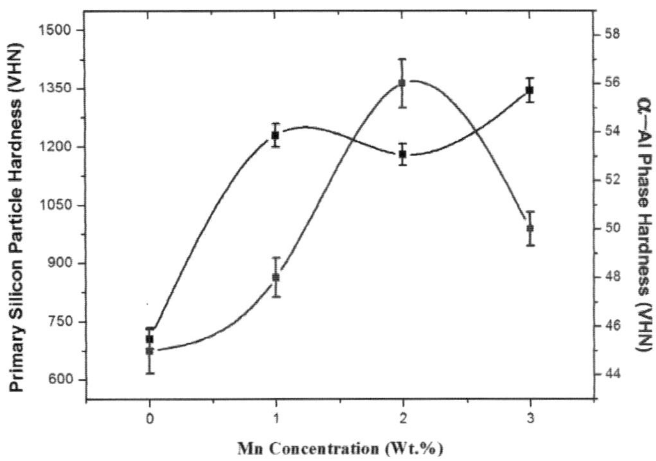

Figure 6. Graphical representation of change in the primary Al phase hardness and primary Si particles hardness with increasing Mn concentration.

But, 2 wt.% and 3 wt.% Mn modified alloy binary eutectic phase hardness is lower than the 1 wt. % Mn modified alloy because the eutectic phase is not modified significantly and the eutectic Si phase volume fraction is less. The micro hardness of the primary Si particles is increased in modified alloy, because the compactness and roundness are higher in the modified alloy. The α-Al phase hardness is not significantly changed after Mn addition.

4.0 Conclusion

The microstructure and hardness properties of gravity cast Al-12.6Si alloy with and without Mn addition have been studied and some conclusions are summarised:

- The Al-12.6Si alloy has coarse irregular primary and eutectic Si particles with very sharp corners.
- The Mn addition in the alloy effectively modifies the morphology of primary and eutectic Si phase.
- Addition of Mn on and above 2 wt.% result in formation of $Al_6(Mn,Fe)$ intermetallic phase.

- Increase in Mn concentration in the alloy, the volume fraction of $Al_6(Mn,Fe)$ intermetallic phase is increased.
- The bulk hardness and micro harnesses of the alloy increased after Mn addition.

Acknowledgement:

Authors are very thankful to Mr. S C. Das and Mr M. K. Das of National Institute of Technology Durgapur, India, for their support in several phases of this investigation. Authors are also like to thank NIT, Durgapur RIG # 2 project for financial support and the Director of National Institute of Technology Durgapur, India for his continuous encouragement.

References:

[1] A. Mazahery and M. O Shabani, 2014, JOM, 66, 726–738.

[2] W. S Miller, L. Zhuang, J. Bottema, A. J Wittebrood, P. De Smet, A Haszler and A. Vieregge, 2000, Materials Science and Engineering: A, 280, 37-49.

[3] S. A Kori, B. S Murty, M. Chakraborty, 2000, Materials Science and Engineering: A, 280, 58-61.

[4] J. R Davis, 1993, Aluminum and aluminum alloys, J. R. Davis & Associates, ASM International, 3-8.

[5] W.X Shi, B. Gao, G.F. Tu and S.W. Li. 2010, Journal of Alloys and Compounds, 508, 480–485.

[6] A. Knuutinen, K. Nogita, S.D. McDonald and A.K. Dahle, 2001, Journal of Light Metals, 1, 229–240.

[7] K. Nogita, H. Yasuda, M. Yoshiya, S.D. McDonald, K. Uesugi, A. Takeuchi and Y. Suzuki, 2010, Journal of Alloys and Compounds, 489, 415–420.

[8] A.K. Dahle, K. Nogita, S.D. McDonald, C. Dinnisa and L. Luc, 2005, Materials Science and Engineering A, 413–414, 243–248.

[9] H. Liao, Y. Sun and G. Sun, 2002, Materials Science and Engineering A, 335, 62–66.

[10] L. Li, Z. Zhao, Y. Zuo, Q. Zhu and J. Cui, 2013, Journal of Materials Research, 28, 1567-1573.

[11] S. G Shabestari, 2004, Materials Science and Engineering A, 383, 289–298.

[12] L.F. Mondolfo, 1976, Aluminum Alloys: Structure and Properties, Butterworth, London, p. 324.

[13] P. Biswas, M. K Mondal, H. Roy and D. Mandal, 2017, Canadian Metallurgical Quarterly, 56, 340-348.

IOP Conf. Series: Materials Science and Engineering **338** (2018) 012044 doi:10.1088/1757-899X/338/1/012044

Evaluation of the optical characteristics of *c*-axis oriented zinc oxide thin films grown by sol gel spin coating technique

K. Baisakh[1], S.Behera[1] and S.Pati[2]*

[1]Department of physics, School of Applied Science, CUTM, Bhubaneswar, 752050, India.

[2]Department of Physics, N. C. (Auto.) College, Jajpur, 755001, India

Abstract. In this work we have systematically studied the optical characteristics of synthesized wurzite zinc oxide thin films exhibiting (002) orientation. Using sol gel spin coating technique zinc oxide thin films are grown on pre cleaned fused quartz substrates. Structural properties of the films are studied using X-ray diffraction analysis. Micro structural analysis and thickness of the grown samples are analyzed using field emission scanning electron microscopy. With an aim to investigate the optical characteristics of the grown zinc oxide thin films the transmission and reflection spectra are evaluated in the ultraviolet-visible (UV-VIS) range. Using envelope method, the refractive index, extinction coefficient, absorption coefficient, band gap energy and the thickness of the synthesized films are estimated from the recorded UV-VIS spectra. An attempt has also been made to study the influence of crystallographic orientation on the optical characteristics of the grown films.

Keywords: Crystal texture; Grain growth; Optical property; Envelope method

*Corresponding author: Sumatipati.11@gmail.com, Phone-+919438855121

1. Introduction

The optical characteristics of metal oxide films are greatly influenced by their crystallographic orientation. The influence of preferential orientation on the optical characteristics of the thin films has scarcely been reported in the literature. As explained by F. Paraguay-Delgado, et al. texture has strong influence on the optical properties of the tin oxide thin films [1]. Various techniques have been used to grow such oriented thin films. This includes CVD (chemical vapor deposition), radio frequency sputtering, ion-beam evaporation, MBE (molecular beam epitaxy), spray pyrolysis, and so on [2-3]. Usually these techniques are expensive and also need special approach for the growth. In these methods careful regulation of various parameters during growth facilitates the growth of films with desired orientations. However, in the literature there are hardly any reports on the synthesis of thin films by simple and cost effective sol gel spin coating technique [4]. In this method the growth along specific directions depends on several factors, such as the concentration and viscosity of the sol, speed of the spin coater unit, drying and annealing schedule, etc. Hence using this method, precise control of all these parameters and acquiring a metal oxide thin film with desired orientation is really a challenging issue. Among metal oxides, zinc oxide is considered to be an efficient candidate in optoelectronic applications [5]. In this work, *c*-axis oriented zinc oxide thin films have been synthesized by the cost effective sol gel spin coating technique. The phase formation behaviour and the microstructure evolution of the samples are investigated by X-ray diffraction (XRD) pattern and field emission scanning electron microscopy (FESEM) analyses, respectively. The optical characteristics are analyzed from the measured transmission and reflection spectra using UV-Vis spectroscopy. From these spectra, the values of the refractive index (n), extinction coefficient (k), absorption coefficient (α), band gap energy (E_g) and the thickness (d) of the films are estimated. An attempt has also been made to study the influence of crystallographic orientation of the grown films with their optical properties.

2. Material synthesis and characterization

With an aim to synthesize zinc oxide thin films with specific (002) orientation precursor sol was prepared by adding 2-methoxyethanol with Zn $(CH_3CO_2)_2{\cdot}2H_2O$. Mono-ethanolamine was used as stabilizer. The solution was stirred for 2 h at 60 ^{0}C and is used as precursor sol. In a spin coater unit the precursor sol was put onto pre cleaned quartz substrates and required heat treatment is provided. The structural properties of the synthesized films are studied by X-ray diffraction (XRD) analyses using Cu Kα radiation. The micro structural properties of the films were analyzed by field emission scanning electron microscopy (FESEM) analysis. The optical characteristics of the films were evaluated from the recorded transmission and reflection spectra using UV-VIS absorption spectrometer (Lambda 750, Perkin Elmer, USA).

3. Results with discussion

3.1 Structural and micro structural properties

X-ray diffraction (XRD) pattern of the synthesized zinc oxide thin films is shown in Fig. 1. As observed from the figure it shows only one peak along (002) plane. This textured behavior of the film indicates the *c*-axis orientation of the grains i.e., along the normal to the substrate. As reflected in the literature, since in zinc oxide the surface energy is minimum along (002) plane hence there is a preferential growth along this direction [6].

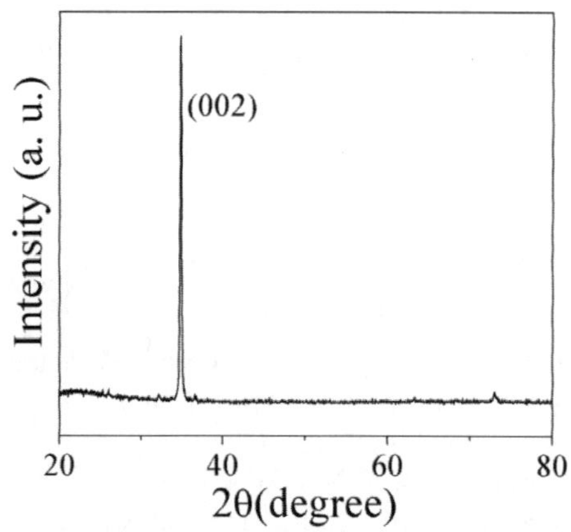

Fig.1. X-ray diffraction pattern of zinc oxide thin films synthesized by sol gel spin coating method, on fused quartz substrate.

From the studied X-ray diffraction pattern the average crystallite size (D) is estimated to be 46.5 nm using the well known Debye-Scherer formula [6], which is $D = 0.9\lambda / \beta\cos\theta$, where $\lambda = 0.154$ nm, is the wavelength of the used X-ray radiation, 2θ is the diffraction angle of the (002) peak and β is the full width at half maximum (FWHM) of the diffraction peak at 2θ measured in radian.

Fig. 2 shows a typical micro structure of the synthesized zinc oxide thin films. From this figure the continuous and uniform nature of the film is observed, thus indicating the good quality of the film.

IOP Conf. Series: Materials Science and Engineering **338** (2018) 012044 doi:10.1088/1757-899X/338/1/012044

The cross section of the same sample is shown in the inset of fig. 2. From the cross section the thickness of the film is estimated to be ~418nm. In addition, the orientation of the grains perpendicular to the substrate is clearly observed from the figure, which could be the effect of annealing. Thus both the structural and micro structural properties indicate the *c*-axis orientation of the thin films.

Fig.2. FESEM image of zinc oxide thin films synthesized by sol gel spin coating method, on fused quartz substrate. The inset shows the cross sectional view indicating the thickness of the film.

3.2 Optical characterization

The optical transmission and reflection spectra of the synthesized zinc oxide thin films, in the wavelength (λ) range 200 – 800 nm, is shown in Fig. 3. As the maxima in the transmission spectra matches well with the minima of the reflection spectra and vice versa, so "envelope method" is suitable for evaluation of its optical constants. This method was originally developed by Manifacier et al. [7] and latter modified by Swanepoel [8].

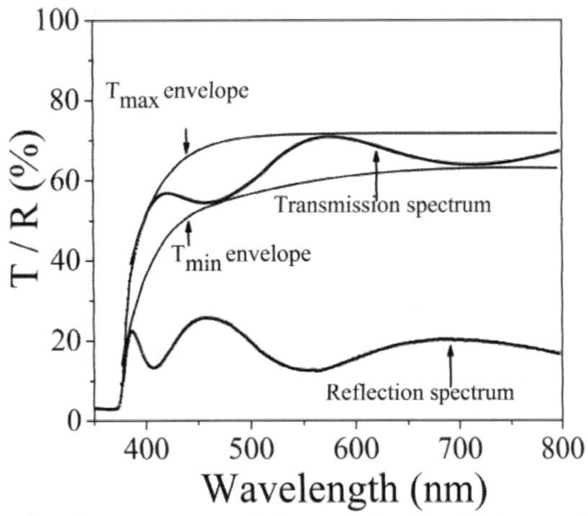

Fig.3. The transmission and reflection spectra of the synthesized zinc oxide thin films and the envelope drawn on the transmission spectra

Using this we have calculated the optical constants like refractive index(n), extinction coefficient(k), absorption coefficient(α), band gap energy (E_g), and thickness (d) of the grown films from the transmission spectra. Using the model proposed by Manifacier et al. [7] α of the film is given by

$$\alpha = -lnx/d \tag{1}$$

and is related to k and λ by $k = \frac{\alpha\lambda}{4\pi}$ (2)

and n is expressed as $n = [N + (N^2 - n_s^2)^{0.5}]^{0.5}$ (3)

where $N = 0.5(1 + n_s^2) + 2n_s \frac{(T_{max} - T_{min})}{(T_{max} T_{min})}$ (4)

Using the expression given below the value of refractive index of fused quartz (n_s) as a function of λ in the range 200 – 6700 nm can be estimated

$$n_s = 1 + \left[\frac{0.6961663\lambda^2}{\lambda^2 - (0.684043)^2}\right] + \left[\frac{0.4079426\lambda^2}{\lambda^2 - (0.1162414)^2}\right] + \left[\frac{0.8974794\lambda^2}{\lambda^2 - (9.896161)^2}\right] \tag{5}$$

Where λ is in micron [9]. Knowing the value of n, x can be found using the following equation

$$x = \frac{C_1}{C_2}\frac{[(T_{max} - T_{min})^{1/2} - 1]}{[(T_{max} - T_{min})^{1/2} + 1]} = \frac{(n+1)(n_s+1)[(T_{max}/T_{min})^{1/2} - 1]}{(n-1)(n_s-1)[(T_{max}/T_{min})^{1/2} + 1]} \tag{6}$$

3.2.1 Determination of thickness (d) of the film

The value of d of the film can be calculated using the following relation [8]:

$$d = \frac{\lambda_1\lambda_2}{2[n(\lambda_1)\lambda_2 - n(\lambda_2)\lambda_1]} \tag{7}$$

Where $n(\lambda_1)$ and $n(\lambda_2)$ are the refractive indices at two adjacent maxima (or minima) in the transmittance spectrum at λ_1 and λ_2, respectively.

Table 1. Determination of thickness *(d)* of the ZnO thin film using "envelope method"

Composition	λ (nm)	n	d' (nm)	d'_{av} (nm)	m'	m	d (nm)	d_{av} (nm)	d (FESEM) (nm)
ZnO	417 (p)	2.35	482	454	5.12	5	443	437	418
	571 (p)	2.03			3.24	3	420		
	440 (v)	2.26	426		4.66	4.5	438		
	695 (v)	1.94			2.53	2.5	447		

p = peak position, v = valley position

Using Eq. (7) different values of d are obtained for different positions of interference maxima and minima in the transmittance spectra and their average (d_{av}) is calculated. Then from the formula for interference fringes

$$2nd = m'\lambda \tag{8}$$

m' values are obtained using the value of d_{av} in place of d. The obtained values of m' are rounded off to the nearest integer for maxima and half integer for minima and the values thus obtained is denoted by m. Now again using Eq. (8), different values of d are obtained using the values of m (in place of m'), λ and n and their average is taken to get the value of d of the film.

3.2.2 Evaluation of α and k of the film

Using the value of x as obtained from Eq. (6) and d value, α and k of the film can be calculated from Eq. (1) and (2), respectively. The value of band gap energy (E_g) is estimated from Fig. 4 (c) by extrapolating the linear portion of the graph using the expression given below [6]:

$$\alpha h\nu^2 = B(h\nu - E_g)^{n} \tag{9}$$

where $h\nu$ is the energy of photon and B is a constant. In case of direct band gap materials, such as zinc oxide, $n = ½$. The value of E_g can be estimated from the intercept of this plot on the energy axis.

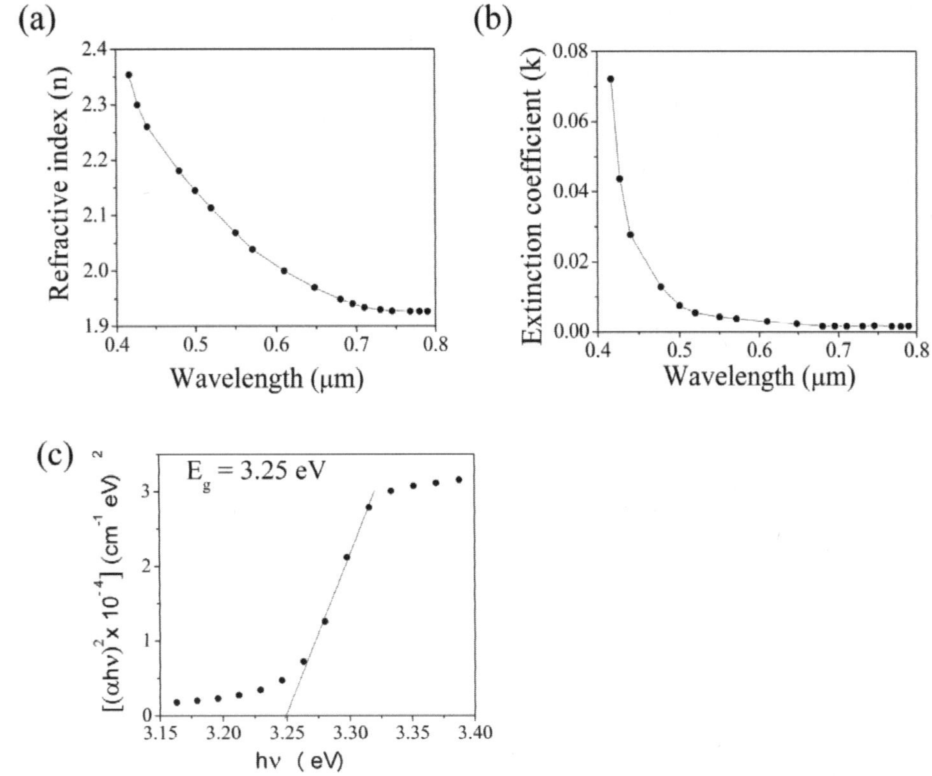

Fig. 4. Optical characteristics of the synthesized zinc oxide thin films, (a) refractive index (n) verses wavelength (λ), (b) Extinction coefficient (k) verses wavelength (λ), and (c) $(\alpha h\nu)^2$ *vs* $h\nu$ plot (Tauc plot).

As shown in Fig. 3 two envelopes T_{max} and T_{min} are drawn taking the maxima and minima points of the transmittance spectra. Then the value of n of the film as a function of λ is calculated using Eq. (3) and is plotted in Fig. 4(a). As observed from the figure the value of n decreases with increasing λ. Table 1 summarizes the procedure for determination of thickness (d) of the film using the obtained values of n. This value of d is used to calculate the value of k of the film with the variation of λ using Eq. (1), (2) and (6) and is shown in Fig.4 (b). Higher value of n and lower value of k than the reported value [10] signifies the good optical quality of the film. Fig.4 (c) shows the plot of $(\alpha h\nu)^2$ versus $h\nu$. Then using Eq. (9) the value of E_g is calculated to be 3.25 eV, and is near to the band gap of zinc oxide.

The values of optical parameters thus obtained are compared with the reported values. For example, Gaspera et al. reported the value of n of ZnO thin film at 600 nm as 1.65 [11] and Suvorova et al. reported the n value of 2 at 632 nm which is same as that of single crystal [12]. In our study we obtained the value of n as 1.95 at the wavelength 632 nm. The higher value of n than that reported by Gaspera et al. [11] could be due to the textured behavior of the thin film, as the value of n depends on the direction of propagation of light, which in turn changes with the crystallographic orientation. Moreover, it can be seen that our value of n almost equal to that of Ref. 12, which is same as that of single crystal and may be attributed to the c-axis orientation of the grains in the thin film.

4. Conclusion

In the present work, c-axis orientated zinc oxide thin films are grown on fused quartz substrates using sol gel spin coating method. The surface morphology, micro structural analysis and optical properties of the grown films are investigated. The films are textured along (002) planes with an average grain size of 46.5 nm. Optical constants such as refractive index, extinction coefficient and film thickness are evaluated from transmission and reflection spectra using the envelope method. The lower value of extinction coefficient and higher value of refractive index than the reported value suggest good surface quality of the films and is attributed to the *c*-axis orientation of the films. The thickness of the films estimated from the optical characteristics (437 nm) is in consistent with that obtained from the FESEM image (418 nm). The film is found to have optical band gap energy of 3.25 eV which is near to the band gap energy of zinc oxide. Finally, the optical properties are correlated with the crystallographic orientation of the zinc oxide thin films.

Acknowledgments
Authors gratefully acknowledge Indian Institute of Technology, Kharagpur for providing various experimental facilities. Also, in this work the authors sincerely thank school of applied science, CUTM, Bhubaneswar.

References

[1] Paraguay-Delgado F, Miki-Yoshida M, Antunez W, González-Hernández J, Vorobiev Y V and Prokhorov E 2008 *Thin Solid Films* **516** 1104.
[2] Shewale P S, Agawane G L, Shin S W, Moholkar A V, Lee J Y, Kim J H and Uplane M D 2013 *Sens Actuators B* **177** 695.
[3] Kim D, Yun I and Kim H 2010 *Curr Appl Phys* **10** S459.
[4] Znaidi L, SolerIllia G J A A, Benyahia S, Sanchez C and Kanaev A V 2003 *Thin Solid Films* **428** 257.
[5] Chawla A K, Kaur D and Chandra R 2007 *Opt Maters* **29** 995.
[6]Pati S 2017 *J Mater Sci Mater El* **28** **(2)** 1756.
[7] Manifacier J C, Gasiot J and Fillard J P 1976 *J. Phys. E: Sci. Instrum.* **9** 1002.
[8] Swanepoel R and Phps J 1983 *E Sci Instrum* **16** 1214.
[9] MalitsonI H 1965 *J Opt Soc Am* **55** 1205.
[10] Khoshman J M and Kordesch M E 2007 *Thin Solid Films* **515** 7393.
[11] Gasperaa E D, Guglielmia M, Martuccia A, Giancaterinib L and Cantalinib C 2012 *Sens Actuators B* **164** 54.
[12] Suvorova N A, UsovI O, Stan L, DePaula R F, Dattelbaum A M, Jia Q X and Suvorova A A 2008 *Appl Phys Lett* **92** 141911.

Investigation of consolidation kinetics and microstructure evolution of Al alloys in direct metal laser sintering using phase field simulation

Bubloom Bimal Satpathy, Jyotirmoy Nandy, Seshadev Sahoo*
Department of Mechanical Engineering, Institute of Technical Education and Research, Siksha 'O' Anusandhan, Odisha, Bhubaneswar-751030

Email: seshadevsahoo@soa.ac.in

Abstract: Direct metal laser sintering is one of the very efficient processes which comes under the field of additive manufacturing and is capable of producing products of good mechanical and physical properties. The process parameters affect the physical and mechanical properties of the final products. Rapid solidification plays an important role in the consolidation kinetics as the powdered material sinters and forms a polycrystalline structure. In the recent times, the enormous use of computational modeling has helped in examining the utility of final products in a wide range of applications. In this study, a phase field model has been implemented to foresee the consolidation kinetics during the liquid state sintering. Temperature profiles have been used to study the densification behavior and neck growth which is caused by the surface diffusion of particles at initial stage. Later, importance of grain boundary and the volume diffusion during densification process is analyzed. It is also found that with rise in temperature, neck growth also increases rapidly due to the interaction of adjacent grains through grain boundary diffusion and stabilization of grain growth.

1. Introduction

Direct Metal Laser Sintering (DMLS) is a powder bed type of additive manufacturing technique which uses heat from a high-powered laser source. In this process powders of metal/alloys are used in the powder bed as inputs for the formation of a defined solid layer. They are partially melted or fully melted in the particular layer level by an energy source and solidified after cooling into a solid shape. Energy sources can be from a single laser, electron beam or an infrared panel heater [1, 2]. The key features of DMLS process are production of complex geometries, act of rapid prototyping to reduce development time, production of strong and durable components, minimum tooling requirement, high accuracy detailing and excellent platform for testing of prototypes [3]. A schematic diagram of the process and its components are shown in fig. 1. Microstructures are structural inhomogeneities, created throughout the procedure of manufacturing. The study of microstructure evolution is considered very important due to the reduction of total free energies. This occurs due to the external effects of thermodynamic equilibriums which are present in different states in the whole system [4]. Hence, during the process of additive manufacturing and its capability of producing products using the layer by layer build technique, many intricate characteristics develop during the evolution of microstructure.

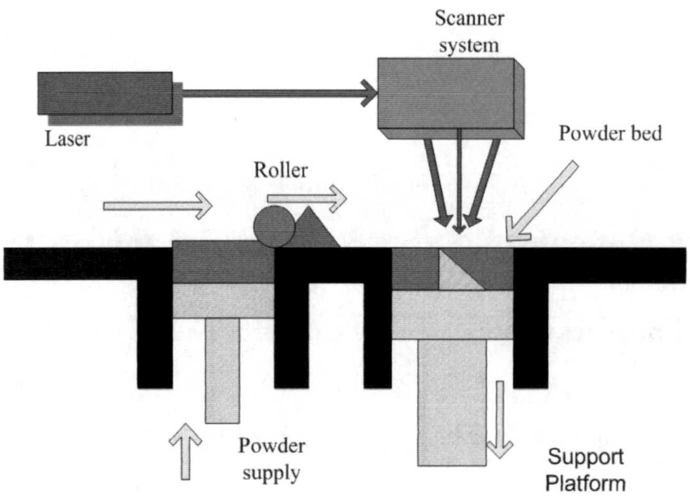

Figure 1. Schematic diagram of Direct Metal Laser Sintering

In one hand, the multiple phases of the commercially used alloys give rise to various microstructures. And on the other hand, these alloy systems are enhanced with added properties to improve the final microstructure patterns. Also, the muti-phase characteristic gives rise to the various stable as well as metastable states. The presence of brilliant properties such as mechanical, biomedical and corrosion resistance in Al alloys has widened the scope for research work in DMLS using this alloy [5, 6]. Also, the good castability and weldability of specifically Al-Si alloys have provided depth in studying the microstructural properties under the DMLS process.

In a couple of past decades many works have been carried out in the simulation and modeling of additive manufacturing. Wang [7] developed diffuse interface model to study the solid-state sintering. It was used for studying various physical processes. Ahmed et al. [8] studied phase field model for examining the grain growth in polycrystalline ceramics containing porosity. Moelans et al. [9] studied the diffuse-interface approach, which allowed for the evolution of arbitrary complex grain morphologies without any presumption on their shape or mutual distribution. Biswas et al. [10] developed a model and observed that at the initial stage of sintering interactions between powder particles are initiated due to surface diffusion and at the later stage, densification is caused by volume and grain boundary diffusion. Boettinger et al. [11] formulated a phase-field variable and a corresponding governing equation to describe the state (solid or liquid) in a material as a function of position and time. Yana et al. [12] developed a model that evaluates the manufacturability and performance of AlSi10Mg periodic cellular lattice structures. Sercombe et al. [13] developed a model that describes the development of an infiltrated Al system that can be used to make complex components directly from computer models. Nandy et al. [14] carried out simulations using the diffuse interface model to examine the microstructure evolution of AlSi10Mg using a set of temperature gradients. The study showed the increase in dendritic tips with the increase in temperature gradients. Sahoo and Chou [15] developed a phase field model for microstructure evolution in electron beam additive manufacturing by using FORTRAN programming language. Till now no such study has been carried out for studying the consolidation kinetics of laser sintered Al alloys using phase field modelling. The objective of this paper is to study the consolidation kinetics and the sintering behavior of an Al alloy in DMLS using phase field modelling.

2. Phase field model formulation

A phase field model is formulated numerically to simulate the consolidation kinetics and microstructure evolution of Al alloys in DMLS process. The model is formulated using two field variables such as the conserved density field ρ, which acquires the value of 1 at the solid state and 0 at the pores, with rapid

change at the solid–pore interface. The second variable consists of the non-conserved order parameter, η_i, which discriminates the different particles in the microstructure.
The microstructure field variables are represented by free energy function of a system that is as given below

$$F = \int_V \left[f(\rho, \eta_{1....p}) + \frac{K_\rho}{2}(\nabla_\rho)^2 + \sum_i \frac{K_\eta}{2}(\nabla \eta_i)^2 \right] dv \quad (1)$$

κ_ρ=Gradient energy coefficients of concentration, κ_η=Grain boundary energies

The evolution equation for ρ follows the Cahn–Hilliard equation [16] that can be written as below

$$\frac{\partial \rho}{\partial t} = \nabla.\left(D\nabla \frac{\partial F}{\partial \rho} \right) = \nabla.D\nabla\left(\frac{\partial f}{\partial \rho} - \kappa_\rho \nabla^2 \rho \right) \quad (2)$$

For evolution of the non-conserved order parameters η_i that represents particles and a grain boundary follows Allen Cahn equation [11], which is given below

$$\frac{\partial \eta_i}{\partial t} = -L\frac{\delta F}{\partial \eta_i} = -L\left(\frac{\partial f}{\partial \eta_i} - \kappa_\eta \nabla^2 \eta_i \right) \quad (3)$$

L is the grain boundary mobility.
D=microstructure-dependent diffusivity or mobility coefficient and D is taken as [8]

$$D = D_{vol}\varphi(\rho) + D_{vap}[1-\varphi(\rho)] + D_{surf}\rho(1-\rho) + D_{GB}\sum_i \Sigma \eta_i \eta_m \quad (4)$$

D_{vol}=bulk diffusivity in the lattice, D_{vap}=diffusivity of the vapour phase

D_{surf}=surface diffusivity, D_{GB}=grain boundary coefficient

Interpolation function $\varphi(\rho)$ is taken as

$$\varphi = \rho^3(10 - 15\rho + 6\rho^2) \quad (5)$$

The above equation ensures that the bulk diffusivity is zero at pores and maximum at the solid regions. The equation (4) takes into account different diffusion a mechanism that explicitly tracks surface and grain boundaries.

3. Numerical approach and simulation tool for the phase field simulation

The simulations are carried out by extracting the temperature from a thermal model and then integrating it with the phase field model using mobility function. Finite difference technique is used by taking five-point stencil in 2D spaces to solve the governing equations. The Laplace operator and the dimensional derivatives are approximated and used in finite difference method. The time assimilation is carried by using explicit Euler time marching format. MATLAB software was used to solve the phase field equations. The modelling is further carried out by considering the top layer of powder bed in the build geometry which is in X and Y directions. The computational domain is carried out for a square domain taken with 200μm × 200μm in two dimensions. The domain was divided into 100 × 100 number of grids. The grid spacing between two grid points taken as dx=dy=0.5μm. The simulations were started with 2 circular seeds with a diameter of 20μm at the centre of the computational domain for equal sized particles. For unequal sized particles, a diameter of 20μm was used for the larger particle and the smaller particle has a diameter of 12μm.
The simulations were carried out in MATLAB using the following procedure:

I. Define the initial parameters and compute the phase field variable in the defined domain.
II. Prepare the material specific parameters field and microstructure modelling by enabling the order parameters in simulation cell.

III. Build the microstructure evolution and derive the free energy functional of the concentration field at every grid point in the simulation cell and storing it in a function and stabilize the boundary conditions.
IV. Build the mobility field values and calculate at all grid points with time integration of concentration field.
V. Solve the phase field equations using finite difference method.
VI. Input boundary conditions of the temperature in the diffusion coefficients.
VII. Update the parameters and time steps with the help of new values and repeat steps (III) and (IV) for next iteration until the field reaches the boundary.

4. Results and Discussion

The simulation was done with accordance to different process parameters as used in direct metal laser sintering. It is seen that temperature as well as time plays an important role in sintering process. The sintering process of two equal sized particles with respect to time has been shown in the figure 2. The simulation is done to show the mechanism of neck and grain boundary development between two particles of equal size. From figure 2(a), it can be seen that there is almost negligible point of contact between the particles at 0.01ms. The neck growth starts at 0.5ms where the necking is in its initial stage and as seen in figure 2(c) the intermediate necking stage has already been reached where the neck width increases dominantly. Figure 2(d) shows the final sintered structure which is achieved at 5ms.

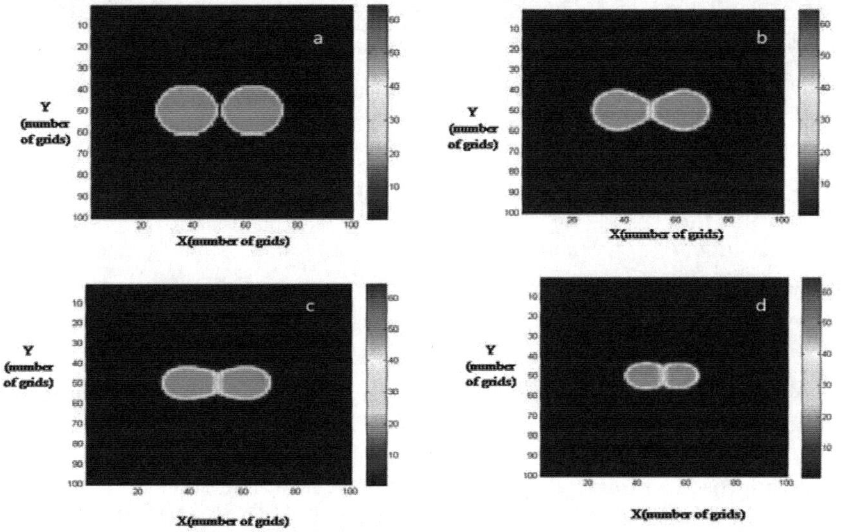

Figure 2. Phase profiles of two equal sized particles at different time (a) t=0.01ms (b) t=0.5ms(c) t=2ms (d) t=5ms.

4.1. Temperature effect on equal sized particles

Figure 3 shows the effect of increase in temperature on the phase profiles of two equal sized particles. It can be seen that the initial point of contact is almost made at 0.01ms. As the temperature maintained is high, the mobility also rises and the grain boundary diffusion takes place at a higher rate. From figure 3(b) and (c) it can be clearly seen that the final sintering condition is achieved within 2ms. As compared to figure 2, the total sintering time taken is low during this case. With the rise in temperature the sintering time decreases and sintering rate increases. Also, it can be observed that the equal size particles maintain the solid-state sintering condition till 2ms.

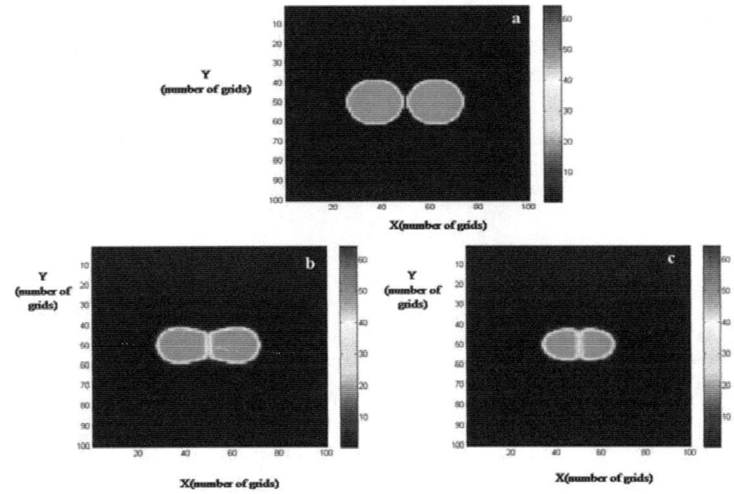

Figure 3. Phase profiles of two equal sized particles at higher temperature and different time (a) t=0.01ms (b) t=0.5ms (c) t=2ms.

4.2. Temperature effect on unequal sized particles

Figure 4 shows the phase profiles of unequal sized particles at a similar temperature range as that in the previous section. The point of contact and the initial stage of neck formation can be seen to occur in figure 4(a). In figure. 4(b) and 4(c), neck width increases with the increase in time as grain boundary diffusion takes place. The phenomenon of surface diffusion occurs at 3ms as seen in figure 4(d). The size of the smaller particle keeps reducing as the volume diffusion starts. From figure 4(e), it can be observed that at 50ms the particles experience the liquid state sintering. The grain boundary diffusion and volume diffusion has completely merged the particles.

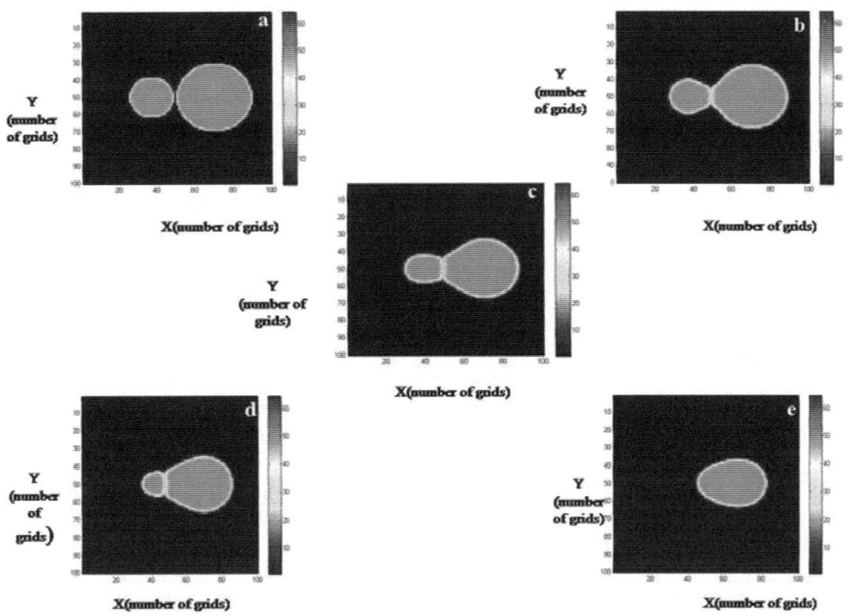

Figure 4. Phase profiles of two unequal sized particles at high temperature and different time (a) t=0.01ms (b) t=0.4ms (c) t=10ms (d) t=30ms (e) t=50ms.

5. Conclusion
Using the phase field modelling, the consolidation kinetics and sintering behaviour of DMLS produced Al alloys was studied. Using MATLAB, the set of conserved and non-conserved field parameters were calculated in both cases (equal and unequal size particles). Input parameters were extracted from the thermal profiling and then were integrated with the mobility parameter in the phase field model. From the results, it was observed that the temperature plays a vital role in the microstructure evolution. Also, it was observed that the unequal size particles sinter faster than the equal sized particles as liquid state sintering takes place.

References
[1] Gebhardt A, Schmidt FM, Hötter JS, Sokalla W and Sokalla P. 2010 *Physics Procedia* **5** 543.
[2] Simchi A. 2006 *Materials Science and Engineering: A* **428** 148.
[3] Chen LQ. 2002 *Annual review of materials research* **32(1)** 113.
[4] Rosenthal I, Stern A and Frage N. 2017 *Materials Science and Engineering: A* **682** 517.
[5] Wang LZ, Chen T and Wang S. 2017 *Optik-International Journal for Light and Electron Optics* **143** 173.
[6] Vilaro T, Abed S and Knapp W. 2008 In *Proc. of 12th European Forum on Rapid Prototyping.*
[7] Ahmed K, Yablinsky CA, Schulte A, Allen T and El-Azab A. 2013 *Modelling and Simulation in Materials Science and Engineering* **21(6)** 065005.
[8] Moelans N, Blanpain B and Wollants P. 2008 *Calphad* **32(2)** 268.
[9] Biswas S, Schwen D, Singh J and Tomar V. 2016 *Extreme Mechanics Letters* **7** 78.
[10] Boettinger WJ, Warren JA, Beckermann C and Karma A. 2002 *Annual review of materials research* **32(1)** 163.
[11] Yan C, Hao L, Hussein A, Bubb SL, Young P and Raymont D. 2014 *Journal of Materials Processing Technology* **214(4)** 856.
[12] Sercombe TB and Schaffer GB. 2003 *Science* **301** 1225.
[13] Kumar V, Fang ZZ and Fife PC. 2010 *Materials Science and Engineering: A* **528(1)** 254.
[14] Nandy J, Sarangi H and Sahoo S. 2017 Materials Science and Engineering **178** 1.
[15] Sahoo S and Chou K. 2016 *Additive Manufacturing* **9** 14.

Microstructure and Current-Voltage Characteristics of Erbium Oxide Doped Multicomponent Zinc Oxide Varistors

Samarpita Roy[1]*, Tapatee Kundu Roy[2], Debdulal Das[1]

[1]Department of Metallurgy and Materials Engineering, Indian Institute of Engineering Science and Technology, Shibpur, Howrah - 711 103, West Bengal, India

[2]Variable Energy Cyclotron Centre, 1/AF Bidhan Nagar, Kolkata - 700064, West Bengal, India

*Corresponding author Email : samarpitaroy89@gmail.com

Abstract: The present work emphasizes the influence of Er_2O_3 addition on the microstructure and nonlinear current-voltage characteristics of ZnO based varistors prepared by mixing in a high energy ball mill followed by compaction and sintering at a temperature of 1100 °C for duration ranging from 0.5 to 8 h. Increasing sintering time is found to enhance the size of ZnO grains of the sintered pellets and thereby, degrades the electrical properties. However, Er_2O_3 addition retards the grain growth of ZnO due to the generation of secondary spinel phases ($ErVO_4$ and Er-rich) at grain boundaries and triple points that restrict the grain boundary migration. Er_2O_3 modified ZnO varistor sintered at 1100 °C for 0.5 h exhibits considerably improved electrical property with nonlinear exponent and breakdown field of 27 and 3880 V cm^{-1}, respectively.

1. Introduction

Varistors or variable resistors are voltage dependent switching devices in which their resistance varies (from 10^{12} to 1 Ωcm) instantaneously (almost in picoseconds) in response to the change in voltages appearing across it [1, 2]. Above a certain threshold voltage, known as the breakdown voltage (V_b), their resistance reduces substantially which allows the current to flow through the varistor rather than the main electrical circuit. As a result, varistors are extensively used as an important element of lightning arrester in electrical power systems and as surge absorber in electronic circuits due to their excellent nonlinear properties, quick response in transient voltage and high surge energy absorption capability ranging from a few joules to thousands of joules [3]. Microstructurally varistors consist of semiconducting *n*-type ZnO grains with highly resistive grain boundaries [2]. The nonlinear properties are mainly attributed to the double Schottky electrical potential barrier at the grain boundaries due to the segregation of large ionic forming oxides [4,5]. Multicomponent varistors are generally prepared by sintering of ZnO powder combined with a number of other minor additives such as Bi_2O_3, Sb_2O_3, V_2O_5, Pr_6O_{11} *etc* [6]. Microstructure of ZnO-based varistor is important since it controls electrical properties. Specifically, the average grain size of the varistor ceramics is crucial, because it is well established that the breakdown voltage of a varistor is inversely related with its grain size. In other words, the grain growth should be minimized during consolidation process to achieve a better varistor property. It is reported that rare earth (Er, Y, Dy, La and Tb) oxides are grain growth inhibitor [7,8]. These oxides restict the mobility of grain boundaries and hence, limit the grain coarsening during sintering. In recent times, continued efforts have been directed to develop high performance varistors specifically by improving the nonlinear property with proper dopant addition. Nahm [9] has proposed that of rare earth oxides doping not only restricts the grain boundary migration but also improves nonlinear propertiesand stability against DC accelerated ageing stress. Beneficial effects of Erbium oxide (Er_2O_3) on microstructural characteristics and nonlinear electrical property of Zinc oxide based varistors have been highlighted in some recent investigations [10,11]. However, little efforts have been directed till date to understand the influence of isothermal sintering duration on structure and property of Er_2O_3 modified ZnO-based varistor ceramics.

Considering the present state of understanding, this work aims to investigate the role of sintering duration of ball milled varistor powder mixtures. The emphasis has been directed to understand the influence of Er_2O_3 addition on microstructural characteristicsand nonlinear current-voltage properties of multicomponent ZnO-based varistors with reference to Er_2O_3 free samples.

2. Materials and Methods

In the present study, varistors have been developed by using the high purity (≥ 99.9%) oxide powder mixtures available commercially with a nominal composition of (97.4 – x) mol.% ZnO, 0.5 mol.% V_2O_5, 2.0 mol.% MnO_2, 0.1 mol.% Nb_2O_5 and x mol.% Er_2O_3 (x = 0.0 and 0.5). The process of preparing the varistor is comprised of mixing all the powders for a predetermined time of 35 h in a high energy ball milling with a ball to powder ratio of 16:1. The ground powders have been compacted into disc shaped pellets with a diameter of 10 mm and thickness around 3 to 4 mm on a hydraulic press at a pressure of 500 MPa. All the pellets have been subjected to sintering at a temperature of 1100 °C for various times (0.5 to 8 h). The uniformity of heating and cooling the specimen has been maintained throughout the sintering process at a rate of 5 °C min^{-1}. The sintered pellets have been measured for their bulk density based on Archimedes principle in the microbalance using deionized water as liquid medium. The formation of different phases in the varistor pellets has been determined by using X-ray diffraction (XRD) technique. The surface microstructures of the sintered pellets have been examined by a field emission scanning electron microscope (FESEM). Reduction of adverse effects due to charging and improving the resolution of the recorded microstructure has been done by coating of thin Pt alloy layer on the sintered pellets. Spectrometric analysis has been performed at a particular area of interest for selected samples using energy dispersive spectrometer (EDS) attached with the FESEM unit. Linear intercept method has been used to measure the average grain size *(d)* of the recorded microstructures. A minimum of ten randomly taken FESEM images have been used to find the average grain size based on the following relation [7]:

$$d = \frac{1.56 \times L}{MN} \tag{1}$$

where, L is the random line length on the micrograph, M is the magnification of the micrograph, and N is the number of the grain boundaries intercepted by the line length L. The nonlinear electric field-current density (E-J) characteristics were measured using a high voltage source-measurement unit. The breakdown field (E_{1mA}) has been determined at a current density of 1.0 mA cm^{-2} and the nonlinear exponent (α) has been estimated from the following expression [12]:

$$\alpha = \frac{(\log J_2 - \log J_1)}{(\log E_2 - \log E_1)} \tag{2}$$

where, E_1 is the electric field at J_1= 1.0 and E_2 are the electric fields at J_2= 10 mA cm^{-2}.

3. Results and Discussion

A set of FESEM microstructures in Fig. 1 illustrates the effect of sintering time and Er_2O_3 addition on densification and mean grain size of the sintered ZnO pellets. All the sintered pellets reveal a well-defined grain structure (Fig. 1). Microstructure of the sintered pellets consists of uniformly distributed homogeneous ZnO grains as a primary phase. The secondary intergranular phases are found to present at the grain boundaries, triple points and occasionally within the grains. With Er_2O_3 addition, the amount of secondary phases increases. The bulk density of all the specimens varies within a narrow range of 97-99% theoretical density; this indicates no significant effect of sintering time on the consolidation process. In contrast, the grain size is found to increase monotonically with increasing sintering time due to obvious grain coarsening. However, Er_2O_3 addition is found to be considerably reduced the grain size at any particular time duration. For example, the mean grain values of Er_2O_3

free and 0.5 mol.% Er_2O_3 added samples are 32.6 and 10.8 μm, respectively, when both pellets were sintered for 8 h. EDS microanalyses results (Fig. 2) corresponding to the intergranular regions of the pellets sintered for 4 h confirm the presence of V, Mn, Nb and O along with Zn in Er_2O_3 free sample as compared to relatively high concentration of Er (33.66 wt.%) in addition to V, Mn, Nb, O and Zn in the Er_2O_3 added ones. These observations indicate that Er_2O_3 mainly presents at the grain boundaries and triple points.

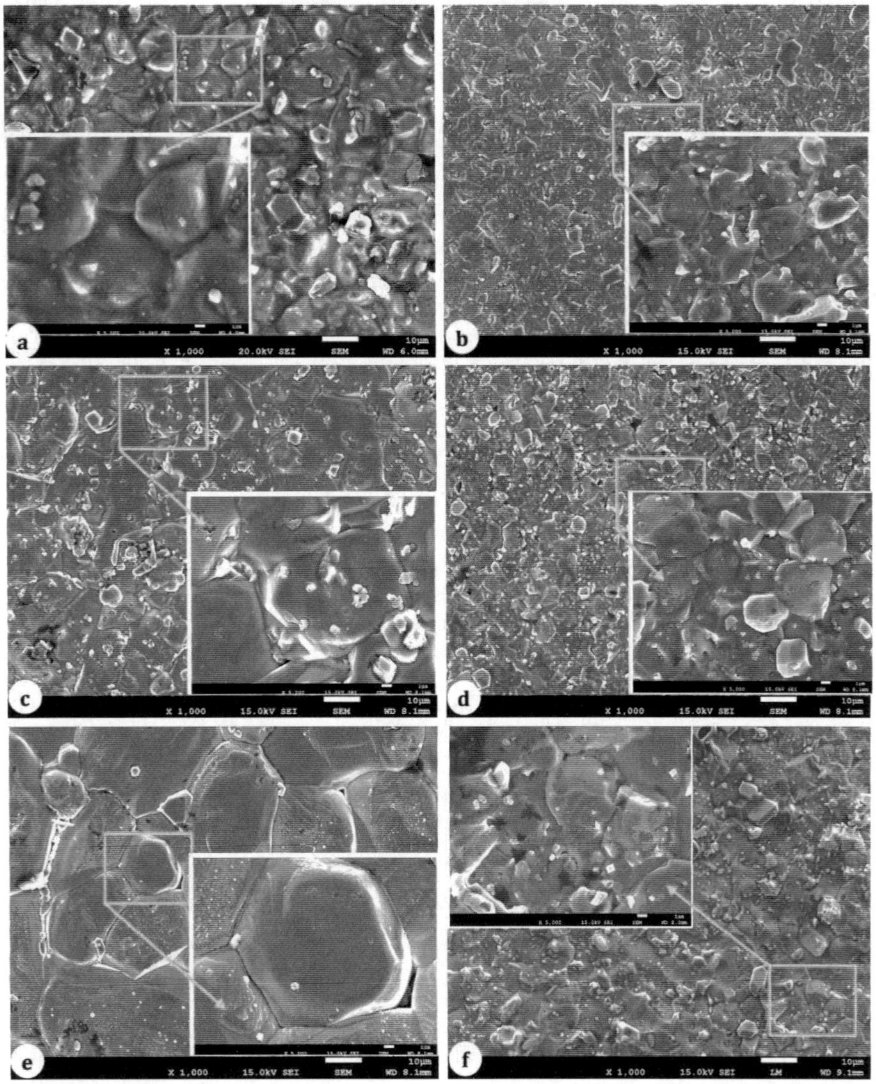

Figure 1. FESEM micrographs of Er_2O_3 free varistor pellets sintered for (a) 0.5 h, (c) 2 h and (e) 8 h, and 0.5 mol.% Er_2O_3 added pellets sintered for (b) 0.5 h, (d) 2 h and (f) 8 h. Inset images at 5 kX exemplify the effects of sintering time and Er_2O_3 addition on grain size.

All the developed phases in the varistor pellets are characterized by XRD analysis. Fig. 3 represents the typical XRD patterns of some selected sintered pellets. The secondary phases like $Zn_3(VO_4)_2$, $Zn_4V_2O_9$,V_2O_5, Zn_2MnO_4 and Mn-rich in addition to primary phase of major hexagonal ZnO has been identified in the Er_2O_3 free sample. The Er_2O_3 added samples additionally reveal the presence of Er related secondary phases like $ErVO_4$ and Er-rich phasesas secondary phases. The Er-rich spinel phases are found to be mainly segregated in the intergranular region of grain boundary area and triple points (Fig. 2). It is worthy to mention here that the intensities of the secondary phases appear to be higher with increasing sintering time and/or Er_2O_3 addition indicating higher degree of formation of secondary phases.

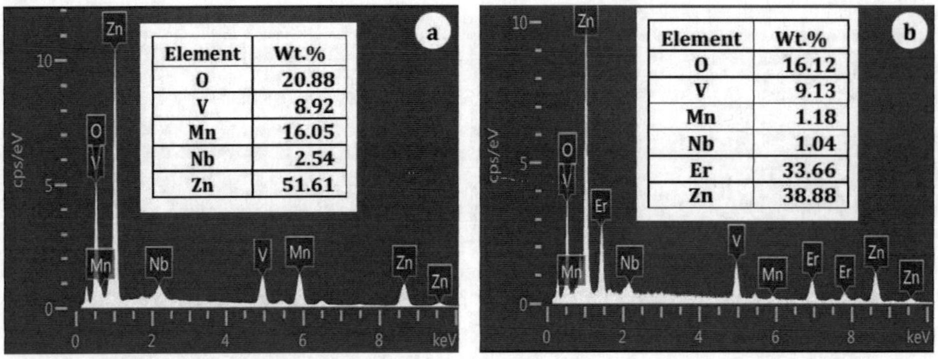

Figure 2. EDS results related to the intergranular layer in the (a) Er_2O_3 free and (b) 0.5 mol.% Er_2O_3 added varistor samples.

Fig. 4 depicts the variations of mean grain size as calculated by image analysis of FESEM micrographs as function of sintering time. It can be observed that with increasing the time of sintering from 0.5 to 8 h, the mean size of ZnO grain increases; however, the rate of grain coarsening is significantly higher for the Er_2O_3 free samples than that of the Er_2O_3 added samples. For the Er_2O_3 free sample, the grain size increases from 10.7 to 32.6 μm, whereas, the grain size increases from 6.3 to 10.8 μm for Er_2O_3 added samples within the investigated sintering time duration. The obtained results assist to conclude that Er_2O_3 segregates at the intergranular layers and forms Er-rich secondary spinel phases. These spinel phases restrict the grain boundary movement [2], and thereby diminishes the degree of grain coarsening during sintering. Therefore, the ZnO grain size is found to be considerably finer for Er_2O_3 added specimens as compared to that of Er_2O_3 free samples at a given sintering duration (Fig. 4). Fig. 5 depicts the grain size distribution of Er_2O_3 free and 0.5 mol.% Er_2O_3 added varistor pellets sintered at a temperature of 1100 °C for 4 h. It can be seen that Er_2O_3 addition results a narrower distribution as compared to Er_2O_3 free sample which indicates the restriction of grain growth during sintering. For example, the grain size distribution range of Er_2O_3 free sample varies from

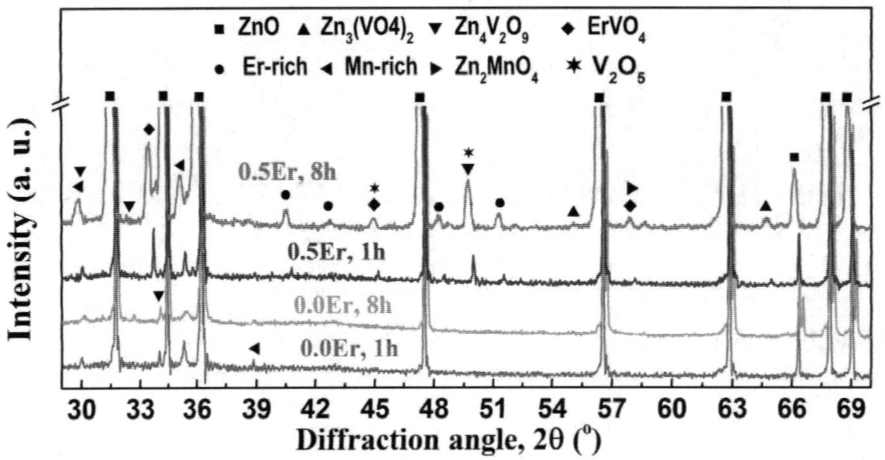

Figure 3.XRD patterns of sintered pellets.

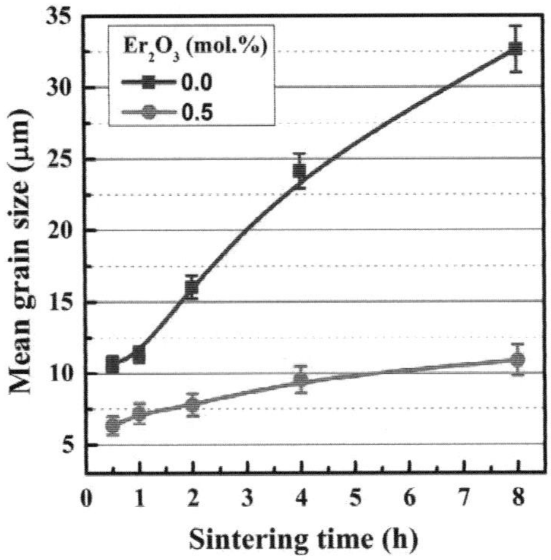

Figure 4. Effect of sintering time with mean grain size of varistor specimens.

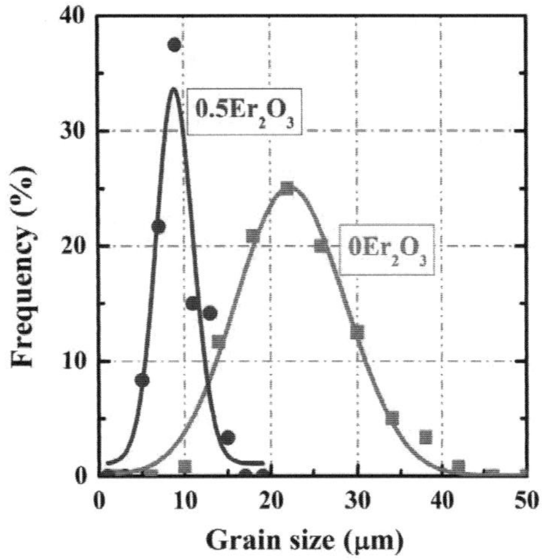

Figure 5. Grain size distribution of the varistor pellets sintered for 4 h.

1.8 to 49.5 μm; whereas, it is limited in between 0.84 to 19 μm for Er_2O_3 added sample. Nahm [12] has reported that incorporation of Er_2O_3 reduces the average grain size of Er_2O_3 doped ZnO varistors.

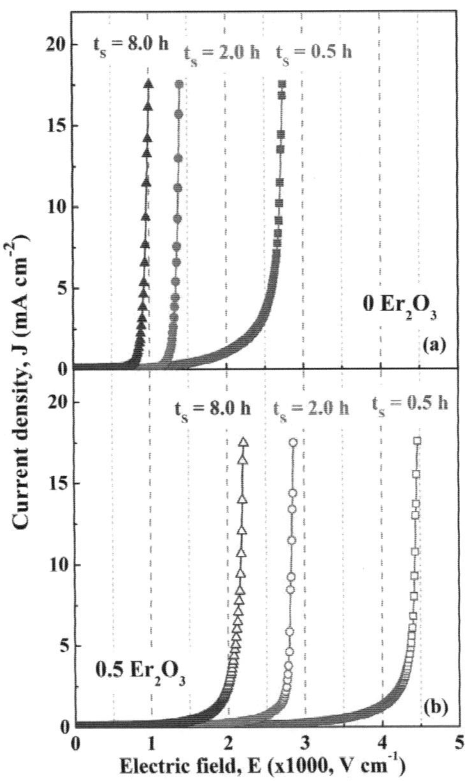

Figure 6. E-J characteristics of (a) Er_2O_3 free and (b) 0.5 mol.% Er_2O_3 added varistor specimens sintered for different time duration.

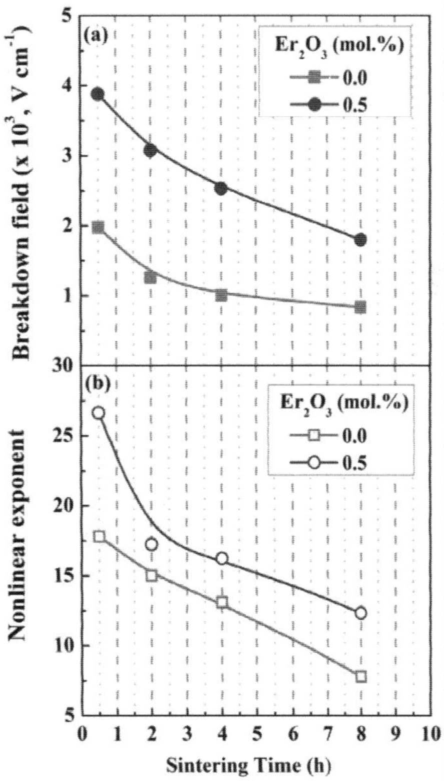

Figure 7. Variations of (a) breakdown field and (b) nonlinear exponent with sintering time.

The nonlinear electrical property of some selected samples is estimated by their electric field–current density (*E–J*) response and the obtained results are shown in Fig. 6. The *E–J* curves are noticeably divided into non-conduction and conduction regions [12].The transition point between these two regions is known as breakdown field (E_{1mA}) [1]. The results in Fig. 6 reveal that the nonlinear property of the selected varistor ceramics enhances due to incorporation of Er_2O_3, however, the same reduces with increasing sintering duration. The magnitudes of the breakdown field (E_{1mA}) andnonlinear exponent (α) are estimated from the *E-J* curves. Fig. 7 graphically represents the variations of the nonlinear parameters (E_{1mA} and α) with isothermal sintering time. Both the obtained breakdown field and nonlinear exponent values are found to reduce continuously with increasing sintering time. The Er_2O_3 addition is found to enhance the electrical property of the varistor pellets. The value of breakdown field and nonlinear exponent increases from 1976 to 3880 V cm^{-1} and 17 to 27 with Er_2O_3 addition for the samples sintered for 0.5 h. Considerable improvement in breakdown field is associated to the reduction of average grain size occurred due to the incorporation of Er_2O_3 (Fig. 4). These observations can be easily explained considering well established relationship [7]:

$$E_{1mA} = V_{gb}/d \tag{3}$$

where, V_{gb} is the voltage per grain boundary and d is the average grain size. The breakdown field is expected to decrease with increasing average grain size. Therefore, the variation of the nonlinear properties with sintering time (Fig. 6) is quite natural if one considers the effect of sintering time on the size of ZnO grain (Fig. 4). In the present investigation, the Er_2O_3 added varistor sample sintered for 0.5 h exhibits the best nonlinear property witha breakdown field of 3880 V cm^{-1}andnonlinear exponentof 27.

4. Conclusions

The effects of Er_2O_3 addition and isothermal sintering duration (0.5-8 h) on microstructure and current-voltage behaviour of ball milled (35 h) $ZnO–V_2O_5–MnO_2–Nb_2O_5$ semiconducting varistor powder mixtures have been investigated at sintering temperature of 1100 °C.The microstructure of the sintered pellet characterized by FESEM examinations as well as EDS and XRD analyses assist to conclude that varistor samples consists primarily of ZnO grains with secondary spinel phases along the intergranular layers. With increasing sintering time, nonlinearelectrical properties of the investigated varistor ceramics diminish due to coarsening of ZnO grains. Whereas, Er_2O_3 addition enhances the nonlinear electrical properties by reducing the ZnO grain size due to the development of mainly $ErVO_4$ and Er-rich secondary spinel phases which acts like a grain growth inhibitor. ZnO-based varistor sintered at 1100 °C for 0.5 hwith 0.5 mol.% Er_2O_3 addition exhibits the best nonlinear electrical properties withhighest electrical field and nonlinear exponent valueof 3880 V cm^{-1} and 27, respectively.

Acknowledgement

Authors would like to acknowledge the financial assistantship received from the COE, TEQIP-II to carry out a part of this work.

References

[1] Gupta T.K.1990 *J. Am. Ceram.Soc.***73** 1817.
[2] Eda K. 1989 *IEEE Electr. Insul.Mag*. **5** 28.
[3] Levinson L.M., Philipp H.R. 1977 *IEEE Trans. Parts, Hybrids, Packag.***13**338.
[4] Mantas P.Q., Baptista J.L. 1995 *J. Eur. Ceram. Soc.* **15** 605.
[5] Ivon A.I., Glot A.B., Lavrov R.I., Bubel T.A. 2016*J. Alloys Compd.***656** 740.
[6] Viswanath N. S. M. Pandey S., Kumar D. 2016 *J. Mater. Sci. Mater. Electron.* **27** 3748.
[7] Chen G.H., Li J.L., Chen X., Kang X.L., Yuan C.L. 2015 *J. Mater. Sci.* **26** 2389.
[8] Bernik S., Daneu N., Recnik A. 2004 *J. Eur. Ceram. Soc.* **24** 3703.
[9] Nahm C.W. 2012 *Ceram. Int.* **38** 6651.
[10] Roy S., Das D., Roy T.K. 2017 *J. Mater. Sci. Mater. Electron.***28** 14906.
[11] Roy S., Roy T.K., Das D. 2017 *Mater. Sci. Forum.***880** 105.
[12] Nahm C.W. 2015 *J. Rare Earth* **33** 403.

IOP Conf. Series: Materials Science and Engineering **338** (2018) 012047 doi:10.1088/1757-899X/338/1/012047

Thermal shock effect on Mechanical and Physical properties of pre-moisture treated GRE composite

A.P. Chakraverty,[*,1] A. B. Panda,[1] U. K. Mohanty,[2,3] S. C. Mishra[2] and B. B. Biswal[4]

[1]Department of Physics,
C. V. Raman College of Engineering, Bhubaneswar-752054, Odisha, India
[2]Department of Metallurgical and Materials Engineering,
National Institute of Technology, Rourkela-769008, Odisha, India
[3]Department of Mechanical Engineering
Institute of Technical and Educational Research (ITER),
SOA University, Bhubaneswar-751030, Odisha, India
[4]Director, National Institute of Technology, Maghalaya-793003, Meghalaya, India
*Corresponding Author, Email: *anantac8@gmail.com*

Abstract: Many practical situations may be encountered under which a GFRP (Glass fibre reinforced polymer) composite, during its service life, is exposed to the severities of sudden temperature fluctuations. Moisture absorption of GRE (Glass fibre reinforced epoxy) composites followed by various gradients of temperature fluctuations may cause thermo-mechanical degradation. It is on this context, the hand layed GRE composite samples are exposed to up-thermal shock (-40°C to +50°C) and down-thermal shock (+50°C to -40°C) for various time interval after several periods of moisture (hydrothermal/hygrothermal) conditioning. The thermally shocked GRE specimens are put to 3-point bend test to divulge inter laminar shear strength (ILSS). Least ILSS values are recorded for the samples with maximum period of moisture treatments under with both up-thermal and down-thermal shock conditions. Lower glass transition temperature (T_g) values, as revealed through the low temperature DSC test, are exhibited at maximum durations of both up-thermal and down-thermal shock for the samples with higher periods of hygrothermal/hydrothermal treatments. SEM fractographs of representative GRE specimens after optimum period of moisture treatments and thermal shock show the various modes of failures.

Key Words: GRE composite, thermal shock, ILSS, glass transition temperature (T_g)

1. Introduction

GFRP (Glass Fibre reinforced polymer) composites have established their performance for various engineering applications [1-3]. Now-a-days, aerospace structures require polymeric composites for exhibiting excellent specific strength and high corrosion resistance [4, 5]. Such materials during their use in concerned environment are exposed to more or less temperature fluctuation combined with/without moist environment [6, 7]. Thus, durability of FRP composites subject to their application in such stringent environment is a great challenge with respect to lower damage tolerance [8]. Moisture is a prime deteriorating agent for long term sustainable performance of GFRP composites. Moisture induced degradation like plasticization [9] and swelling [10] can reduce the performance of polymer matrix as well as decrease the fibre/matrix interfacial bond strength. In addition, temperature variation can change the moisture absorption rate, which in turn can result local swelling in some location of the composite body by non-equilibrium moisture absorption/desorption kinetics [11]. Thermal shock due to sudden change in thermal gradient can cause unequal thermal expansion of fibre and matrix [12]. This may result strain misfit [13] between the components of the FRP composite system and generation of residual stress [14]. A particular example may be found in the aviation industry where an aircraft descends to the ground level from a height, the temperature suddenly changing from the cryogenic level to the atmospheric temperature. It is only desirable to evaluate the composite under such severe ambient conditions to predict its performance and utility. In addition to it, if the composite has prior exposure to moisture coupled with certain temperature above ambience, the ingressed moisture plays vital role in influencing the performance of the composite under the exposure to thermal shock conditions.

Catastrophic failure of polymer composites may be resulted due to creation of thermal stresses around the defects across the fibre/matrix interface [15]. The environment pertaining to instant variation of temperature is a questionable issue, where temperature of moisture environment affects the relative rates of moisture absorption/desorption process. Post-thermal shock to pre-moisture treated composite may enhance the increasing/decreasing rate of desorption of moisture (which absorbed during moisture conditioning), which may responsible for the fluctuation of inter laminar adhesion strength as well as glass transition temperature (T_g) [16, 17]. This may cause diminution of ILSS as well as glass transition temperature (T_g) in some cases [18, 19]. Frozen moisture during exposure of the composite samples to sub-ambient temperature may increase the mechanical locking between fibre and polymer matrix. This may cause enhancement of mechanical properties in some cases [20]. The choice of polymer and fibre is motivated by the literature [21, 22]

The present investigation is aimed at monitoring the mechanical and physical properties of pre-moisture treated E-glass fibre/epoxy composite exposed to thermal shock environment (both up-thermal shock and down-thermal shock) for various time periods. The property of pre-moisture treated glass fibre reinforced epoxy (GRE) composite is thoroughly investigated and documented [17]. In the present work, the mechanism of fluctuation of mechanical properties are interpreted with respect to moisture absorption/desorption kinetics during transition of sub-ambient to elevated temperature and vice-versa of thermal shock exposure.

2. Materials and methods

Composite material comprising matrix phase as epoxy resin (LAPOX L-12) and reinforcing agent as E-glass fibre was fabricated by hand laying method. The properties of the constituting components were reported in our previous paper [9]. Diamine (LAPOX K-6) was added to epoxy resin in the proportion of 1:10 by weight for curing purpose. The composite was cured for 48 hours at room temperature. Then, short beam shear (SBS) specimens were prepared adopting ASTM D 2344-84 (1984) reference standard for 3-point bending test.

Two batches of such samples were exposed to moisture treatment (95% of R.H. with 65° C as hygrothermal conditioning for 3 months and immersion of samples in distilled water as hydrothermal conditioining for 4 months). The details of the mechanical and physical properties of GRE composites subject to above mentioned moisture treatments were rigorously investigated in our previous paper [17]. In the present investigation, the moisture treated GRE composite specimens were exposed to thermal shock exposure. For up-thermal shock, different batches of pre-moisture treated SBS specimens of GRE composite were first kept in -40° C in a cryogenic chamber for various time of exposure (1, 2, 3, 4 and 5 hours) and then suddenly transferred to +50° C in electric oven for 2 hours of fixed duration. Similarly, for down-thermal shock treatment, different batches of pre-moisture treated SBS specimens of GRE composite were first kept in +50° C in electric oven for various time periods (1, 2, 3, 4 and 5 hours) and then instantly exposed to -40° C in cryogenic chamber for 2 hours of fixed time.

The thermally shock treated samples were put to 3-point bend test in INSTRON-1195 to determine load at peak. ILSS (Inter laminar shear strength) thermally shocked composite samples were also documented by the formula [4]:

$$\text{ILSS} = 0.75\ {}^{P_b}\!/_{bt} \qquad (1)$$

where, p_b as breaking load (load at rupture in kN), b as width of the specimen and t as thickness of the specimen.

Low temperature differential scanning calorimetry (DSC) test was performed for the representative samples with extreme periods of moisture conditioning and thermal shock treatment. DSC test was carried out by ADSC module of Mettler-Toledo-821 interfaced with STAR software for thermal analysis purpose. Temperature scanning was considered in the range 30-150° C at 10° C/min scan rate. Glass transition temperature (T_g) for the test specimens are determined by indicating the first change in DSC thermogram. The fracture surface of the GRE composite samples after 3-point bend test are also imaged through scanning electron microscope (SEM: ZEISS EVO 60) to reveal the various failure mechanism for the sample subject to maximum duration of moisture conditioning and both types of thermal shock treatments of optimum periods.

3. Results and Discussion

3.1 Inter laminar shear strength (ILSS)

Results pertaining to ILSS for all pre-moisture treated (hygrothermal and hydrothermal conditioning) samples exhibit a zig-zag trend for shorter durations of both up-thermal and down-thermal shock followed by a general lowering trend as the duration of thermal shock increases (Figure 1 and Figure 2). Least ILSS values are exhibitted for the samples exposed to maximum period of both hygrothermal/hydrothermal moisture conditioning with maximum durations of up-thermal and down-thermal shocks.

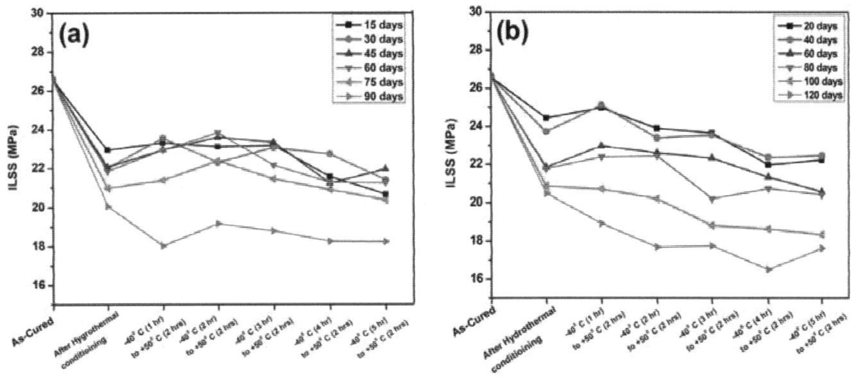

Figure 1. Variation of ILSS for GRE composite after up-thermal shock treatment (a) pre-hygrothermally treated (b) pre-hydrothermally treated

The more or less zig-zag trend of variation in ILSS may be attributed to complex deformation behaviour due to local inhomogeneity owing to local swelling of some locations of matrix and/or fibre/matrix interfacial region. These phenomena are more prominent for lower periods of moisture exposures when the time of exposure is not sufficient to show for average stimuli by the composite body to the above mentioned cause. The samples, however, when exposed to longer periods develop compressive residual stresses resulting in reduction of mechanical properties.

The reason for lowering of ILSS with time of up-thermal shock is many fold i.e., it is the combined effect of many factors. The moisture desorption associated with the thermal shock may form fine whiskers or inter-connected pores when the composite after an initial cryogenic exposure is subjected to +50° C under up-thermal shock conditions [11]. These outlets as a result of moisture desorption can generate residual stresses in the matrix or at the interface, which finally bring in a decreasing trend in ILSS values. The elevated temperature (+50° C) further may generate thermal stresses due to mismatch of thermal expansion coefficients of matrix and fibre. These phenomena may result in decrease in interfacial bond strength giving rise to strain mismatch along the interface. Thus, the ILSS decreases indicating the deterioration of the mechanical property of the composite.

Figure 2 exhibit the variation of ILSS of composite samples exposed to down-thermal shock after pre exposure to hygrothermal/hydrothermal conditioning. It is interesting to see that for lower periods of exposure to down-thermal shock, the ILSS initially increases. This may be due to matrix hardening [23, 24], the duration at -40°C being almost equal to that at +50°C. However, with increase of down-thermal shock duration the material is exposed to +50° C for a longer time (for all durations of down-thermal shock exposure to -40°C is for 2 hours only). This may result in desorption of moisture causing formation of fine whiskers or inter-connected pores. This is why the ILSS decreases with longer durations of exposure to down-thermal shock. Similar interpretations can be made in the case of hydrothermally treated samples also. As reported out by Loken *et. al.* [24] and Salin & Sefaris [25], these damage phenomena in composites exposed to cryogenic conditioning consist of pot-holing, delamination and micro-cracking. All such failures are associated with the thermo-mechanical degradation [26] process due to thermal shock exposures.

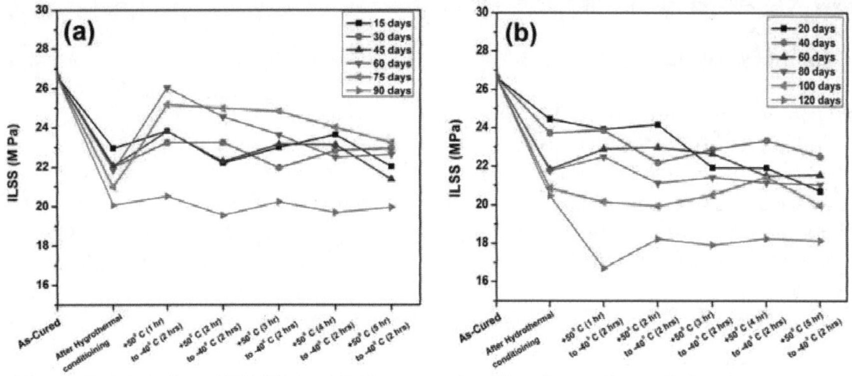

Figure 2. Variation of ILSS for GRE composite after down-thermal shock treatment (a) pre-hygrothermally treated (b) pre-hydrothermally treated

3.2 Glass transition temperature (T_g)

The DSC thermograms interpreting the T_g values of representative selected samples are shown in Figure 3 and Figure 4. All the T_g values of representative samples (Figure 5) are compared with that of the value of as-cured sample (107.04° C) [17]. The small variations in T_g values for the cases observed is due to minute change of free volume due to hygrothermal/hydrothermal conditioning followed by up and down-thermal shock. Glass transition temperature (T_g) is a function of cross-linking density of polymer, affecting free volume. T_g got depressed for all moisture treated samples exposed to higher durations of both types of thermal shock treatments. This result may be observed due to creation of swelling stress during +50° C exposure of final stage of up-thermal shock and breaking of the brittle polymer due to action of excess cryogenic stress during -40° C of final stage of down-thermal shock treatment [4, 16].

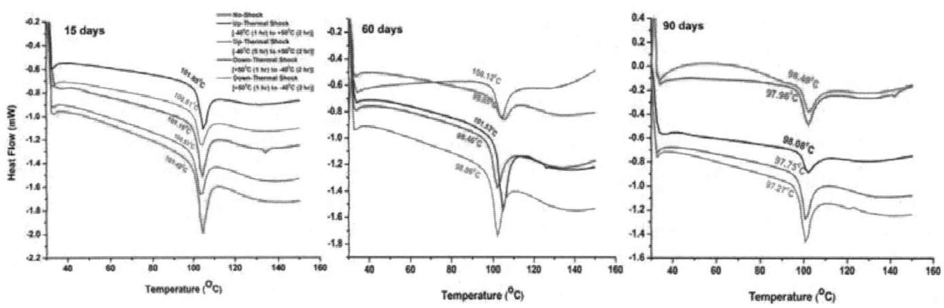

Figure 3. DSC thermograms for hygrothermally conditioned samples subject to optimum duration of thermal shock treatments

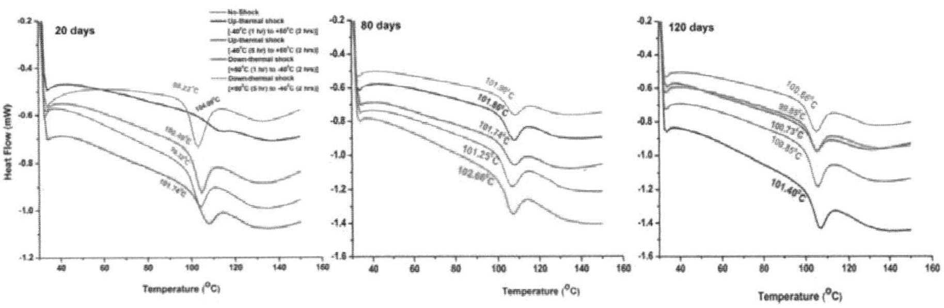

Figure 4. DSC thermograms for hydrothermally conditioned samples subject to optimum duration of thermal shock treatments

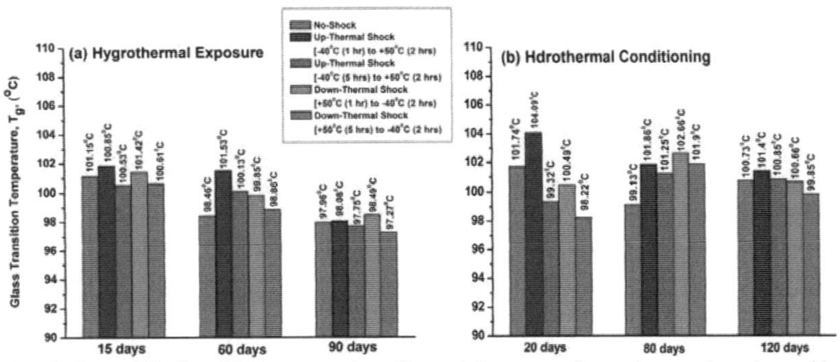

Figure 5. Variation of T_g for GRE composite after minimum and maximum duration of both up and down-thermal shock treatment
(a) pre-hygrothermally treated (b) pre-hydrothermally treated

Though cryogenic compressive stress is responsible for increase of T_g of the polymer [15], the excess amount of same can lower the T_g value up to very small extent by volumetric increase of frozen moisture. The frozen moisture may sometimes tend to get accommodated in the polymer matrix by creating voids or sometimes by breaking up of the epoxy chain structure.

3.3 SEM fractographs

SEM fractographs of pre-moisture conditioned (with maximum periods) composite specimens with concurrent post-thermal shock (minimum and maximum duration of thermal shock treatments) reveal that the mode of failure may be fibre breakage, debonding between fibre and matrix, fibre pull-out, crazing of matrix (cusp formation), matrix cracking etc. (Figure 6, Figure 7). The causes of failure are observed due to anyone of the mentioned modes or any combinations.

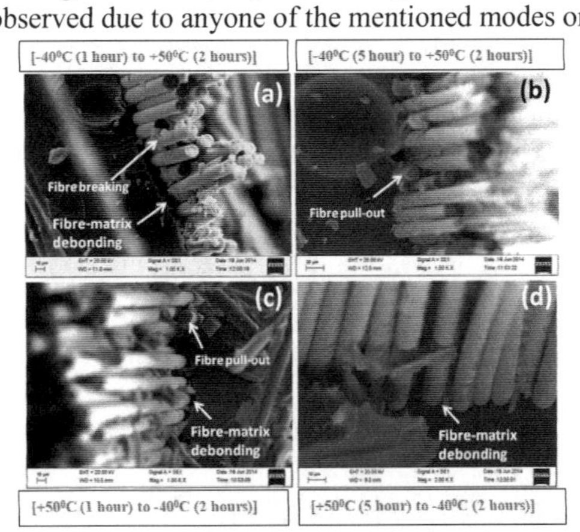

Figure 6. SEM fractographs for of hygrothermally treated samples (90 days)

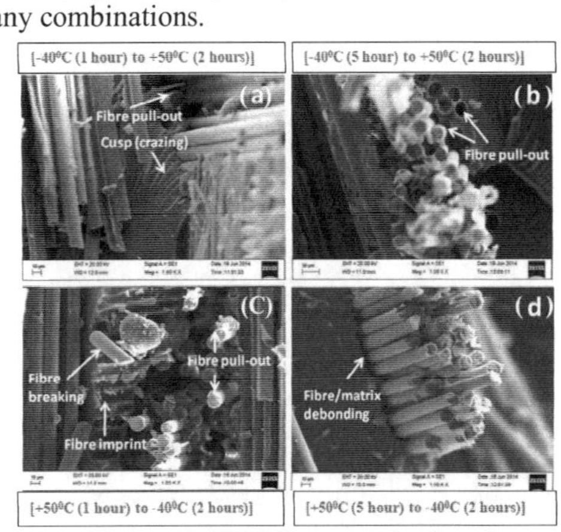

Figure 7. SEM fractographs for of hydrothermally immersed (120 days)

4. Conclusions

Mechanical properties of GRE composite sample show initial fluctuating trend due to non-equilibrium kinetics of moisture absorption/desorption process. ILSS of pre-moisture treated GRE composite samples decrease with increasing duration of up-thermal shock. Modulus values for all moisture treated GRE samples are lowered at higher exposing periods of both up-thermal and down-thermal shock treatments, despite of exhibiting initial non-equilibrium increasing trend. GRE composite samples exposed to maximum period of moisture treatment (both hygrothermal and hydrothermal), exhibited early glass transition phenomenon for higher durations of up-thermal and down-thermal shock treatments. The failure modes of thermally shocked GRE composite with pre-moisture conditioning are mostly due to debonding process between fibre and matrix as a consequence of fibre pull-out owing to strain misfit phenomenon.

Acknowledgement
The authors sincerely thank National Institute of Technology, Rourkela and C. V. Raman College of Engineering, Bhubaneswar for supporting various experimental facilities.

References

[1] Xin H, Liu Y, Mosallam A S, He Jun and Du A 2017 *Compos Struct* **182** 283.
[2] Ansari M M, Chakrabarti A and Iqbal M A 2017 *Composites: Part B* **125** 211.
[3] Pinto M, Chalivendra V B, Kim Y K and Lewis A F 2016 *Compos Struct* **154** 333-337.
[4] Chakraverty A P, Mohanty U K and Biswal B B 2012 *Emer Mat Res* **1** 263.
[5] Fitriah S N, Abdul Majid M S, Ridzuan M J M, Daud R, Gibson A G and Assaleh T A 2017 *Compos Struct* **159** 350.
[6] Berketis K S, Tzetzis D and Hogg P J 2008 *Mater Des* **29** 1300.
[7] Fitriah S N, Abdul Majid M S, Ridzuan M J M, Daud R, Gibson A G and Assaleh T A 2017 *Compos Struct* **159** 350.
[8] Hawa A, Abdul Majid M S, Afendi M, Marzuki H F A, Amin N A M, Mat F and Gibson A G 2016 *Mater Des* **89** 455.
[9] Chakraverty A P, Mohanty U K, Mishra S C and Satapathy A 2015 *IOP Conference Series: Materials Science and Engineering* **75** 012029.
[10] Ellyin F and Rohrbacher C 2000 *J Reinf Plast Comp* **19** 1405.
[11] Mishra G, Mohapatra S R, Behera P R, Dash B, Mohanty U K and Ray B C 2010 *Aircraft Engineering and Aerospace Technology: An International Journal* **82** 258.
[11] Nishizaki I and Meiarashi S 2002 *J Compos Constr* **6** 21.
[12] Berketis K S, Tzetzis D and Hogg P J 2008 *Mater Des* **29** 1300.
[13] Papanicolaou G C, Xepapadaki A G and Tagaris G D 2009 *Compos Struct* **88** 436.
[14] Bhuyan M K, Bhuyan M S, Rodriguez-Devora J I and Yanez M 2012 *J Compos Mater* **47** 3421.
[15] Chakraverty A P, Mohanty U K and Biswal B B 2012 *Int J Curr Res* **4** 67.
[15] Loos A C and Springer G S 1979 *J Compos Mater* **13** 131.
[16] Chakraverty, A P, Mohanty U K, Mishra S C and Biswal B B 2016 *IOP Conference Series: Materials Science and Engineering* **115** 012004.
[17] Chakraverty, A P, Mohanty U K, Mishra S C and Biswal B B 2017 *IOP Conference Series: Materials Science and Engineering* **178** 012013.
[18] Xu L R, Krishnan A, Ning H and Vaidya U 2012 *Composites: Part B* **43** 2480.
[19] Jang B Z 1994 *ASM International*, Materials Park, OH, USA.
[20] Hancox N L 1998 *Mater Des* **19** 93.
[21] Wang S, Zhang W, Zhou H and Li Q 2002 *Polym Test* **21** 195.
[22] Incropera F P, De Witt D P 1990, Fundamentals of Heat and Mass Transfer (3rd ed.), John Wiley & Sons, ISBN 0-471-51729-1.
[23] Kim H Y, Park Y H, You Y J and Moon C K 2006 *KSCE J Civil Engg* **10** 291.
[24] Loken H Y, Nollen D, Wardle D and Zahr G 1988 DuPont Corporation, Wilmington, NC.
[25] Salin I and Sefaris J C 1996 *Polym Compos* **17** 430.
[26] Bai Y, Keller T and Vallée T 2008 *Compos Sci Technol* **68** 3099.

Effect of alumina on grain refinement of Al-Si hypereutectic alloys

J Majhi*, S K Sahoo, S C Patnaik, B Sarangi and N K Sachan
Metallurgical and Materials Engineering, IGIT, Sarang, Odisha, 759146, India

****Author for Correspondence*** E-mail: jogia7924@gmail.com.

Abstract. The size, volume fraction and distribution of primary as well as eutectic silicon affect the mechanical properties of the Al-Si hypereutectic alloys. It is very difficult for the simultaneous refinement and modification of primary and secondary Si particles in hypereutectic Al-Si alloys through traditional processes. This paper explores the role of γ-Al_2O_3 nanoparticles on Si particles in the course of solidification in hypereutectic Al-Si alloys at particular pouring temperature. The present study involves incorporation of varying contents dispersed γ-Al_2O_3 nanoparticles into a molten base metal during stir casting and followed by solidification. It has been reported that the synthesized composites having good interfacial bonding (wetting) between the dispersed phase and the liquid matrix was achieved in order to provide improved mechanical properties of the composite. The cast product of hypereutectic Al-16Si alloy with the reinforcement of nanoparticles, illustrated a significant improvement in both wear behaviour and hardness. The dry sliding wear test has been performed on a group of specimens with varying parameters (different loads and sliding velocities) in a pin on disc wear testing machine. Moreover, the wear rate and specific wear rate also affected in different load and different sliding velocities. However in XRD analysis of the samples, the enhancement of wear resistance as well as hardness was due to the formation of brittle phases like SiO_2, Al_2O_3 and Al-rich intermetallic compounds. The hardness value of the materials increases nearly 6% in addition to increase in the density of only 0.8%. As per literature, the large plate eutectic Si particles were modified in to the fine core particles and it acquires enough potential for various applications.

Key words:- *γ-Al_2O_3, Al-rich intermetallic compound.*

1. Introduction

Al-Si alloys are familiar as the potential candidate in automobile engines in place of the conventional use of cast iron so as to enhance savings in fuel consumption & reduce vehicle emission. The mechanical behaviour and wear resistance are two factors which determine the correlation between synthesis and performance of engineering materials. In current scenario the modelling of materials has provided a renewed stimulus to develop a good understanding of mechanism and factors involved. Aluminium based materials are one example of which to optimise the microstructure as well as the mechanical properties through different processing or by addition of the grain refiner. The demand for the development of light weight metal matrix composites has been increasing in automobile and aerospace industries to enhance the fuel economy and diminish green house gas emissions [1]. The most quickly rising field of composites is that of aluminium alloy based metal matrix composites because of their enhanced physical, mechanical and tribological properties. In MMCs as engineering materials, Al-Si alloys are extensively used as matrix phase as a result of their unusual combination of properties like good castability, high specific strength, and improved wear resistance with reasonable cost [2-4]. Conventional casting of hyper eutectic Al-Si binary alloys gives rise to various morphology of eutectic and primary silicon and the shape, size along with distribution of the constituents play vital role in deciding the mechanical and tribological properties.

Instead of conventional star, polygonal, feathery or plate like primary Si with coarse needle or flake like eutectic Si, achievement of fine spherical & uniformly distributed phases is able to improve mechanical as well as wear properties of hyper-eutectic Al-Si alloys [5-7]. Incorporation of reinforcement particles like TiB_2, WC, SiC or Al_2O_3 to the matrix of aluminium alloys causes remarkable enhancement in hardness, strength along with wear properties at the cost of slight increase in density [8]. AMCs ensure achieving good interfacial bonding and wettability with matrix and there by resulting in excellent properties and extensive application of the composite [9].

In this paper to explore the synthesis and characterization of Al-Si hyper-eutectic alloys with different wt. % Al_2O_3 through stir casting method.

2. Materials and experimental methods

The materials used for Al-16Si-xAl_2O_3 (x = 0,1 and 2wt. %) composite samples are 99.97% purity aluminium ingot, 99.9% purity alumina (gamma-Al_2O_3) and Al-50%Si master alloy. Al-16Si-xAl_2O_3 have been manufactured by stir casting methods and the specimens have been inspected for their physical and mechanical properties.

2.1 Synthesis of Al- 16Si- xAl_2O_3 composite

Cp aluminium and Al-50 wt% Si master alloy was melted in a clay-graphite crucible in a pit type melting f/c at 720°C and added the different wt%Al_2O_3 particles to prepare Al-16Si-xAl_2O_3 (x = 0,1 and 2wt%) alloy. The Al-16Si-xAl_2O_3 composites were manufactured by conventional stir casting method. Dross was removed and melt was degassed with the help of C_2Cl_6 before casting, shown in figure 2.1.

Figure 2.1 Casting of AMC Al-16Si-2%Aluminium by stir casting method.

2.2 Characterization

The samples of size 10mm x 10 mm x 2mm was taken for XRD analysis which was carried out with Cu-Kα target to recognize different phases present in the sample by matching of obtained peaks with JCPDS data files.

2.3 *Hardness and density*

The hardness values of the specimens have been obtained using Microvicker's hardness tester with square base diamond pyramid indenter and applied load of 1kgf for fifteen seconds. Care was taken make both the horizontal faces of the test sample parallel by polishing. Vickers Hardness Number of the prepared specimens were determined from the diagonals of the impressions made at five selected spots of single test sample. The density tests were carried out with the use of a Mettler density tester following standard procedure.

2.4 Wear test

To investigate the dry sliding wear behaviour of the composite sample with 10mm diameter and 30mm height were tested in a pin on disc wear testing machine. The load was varied from 40 to 60N with rotation speed of 300, 400 and 500rpm (having 40mm radius of track) for 5 minutes at room temperature without use of any lubricating stuff. The apparatus used is a microprocessor controlled machine and provides simultaneous data for height loss (in micron). After every test the mass loss due to wear of each specimen was determined. The role of applied load and alumina content on wear behaviour of the prepared composites was studied.

3. Results and discussions

3.1. XRD analysis of composites

The major peaks obtained from the XRD test describe the presence of mainly of aluminium and silicon where as the minor peaks signify the existence SiO_2 and Al_2O_3. As the % of Al_2O_3 increases the intensity of peaks of the Al-rich intermetallic compound ($Al_{3.21}Si_{0.47}$) was increased significantly as shown in figure 3.1(a-c).

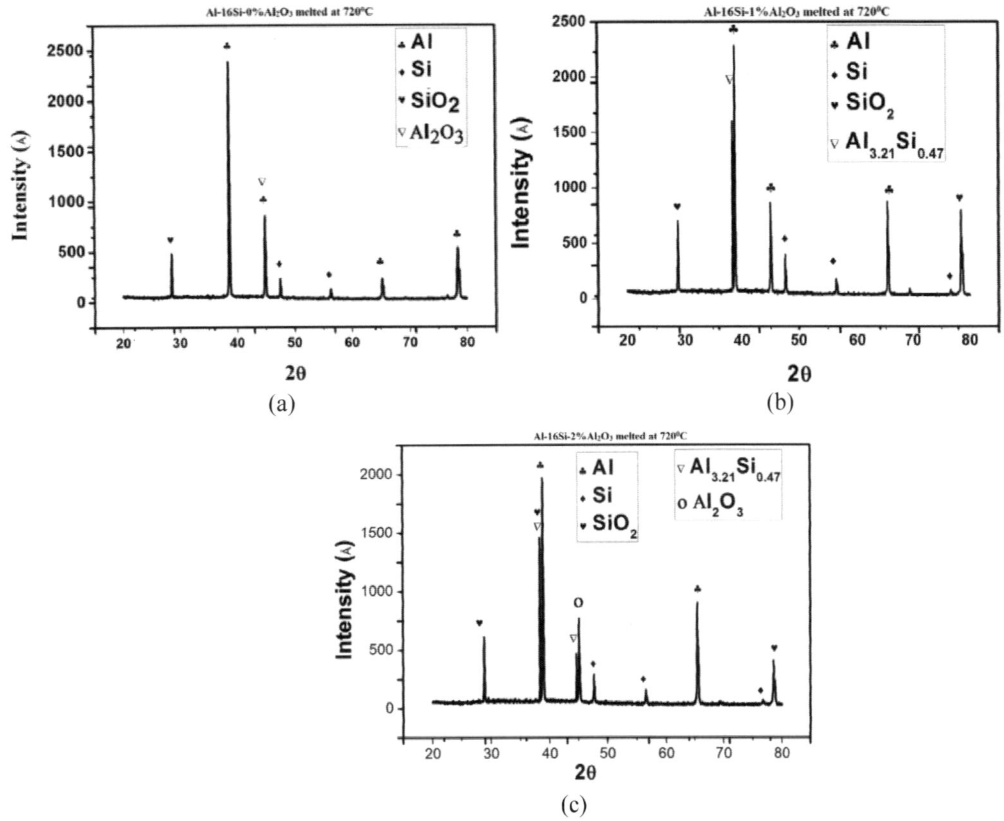

Figure 3.1 X-ray diffraction patterns of (a) Al-16Si-0%Al_2O_3 (b) Al-16Si-1%Al_2O_3 and (c)Al-16Si-2%Al_2O_3 melted at 720^{0}C.

3.2. Hardness and density

Table 1. The Density and Hardness of the different compositional samples.

Sample	Density(g/cc)	Hardness(VHN)
Al-16Si	**2.6213**	**52.565**
Al-16Si-1%Al_2O_3	**2.6388**	**54.29**
Al-16Si-2%Al_2O_3	**2.644**	**55.69**

From the table 1, it is found that the density and hardness increases with increasing of alumina. The density of Al_2O_3 is about 3.95g/cc. The increase in formation of Al-rich intermetallic harder phase causes for increasing hardness.

3.3. Wear test

The wear increases with load and sliding speed as shown in figure 3.2 and 3.3 respectively. The figure 3.4 shows that the wear resistance was significantly increased with increase in alumina content. This may be attributed to the role of γ-Al_2O_3 in grain refinement of the base alloy along with modification of Si phases.

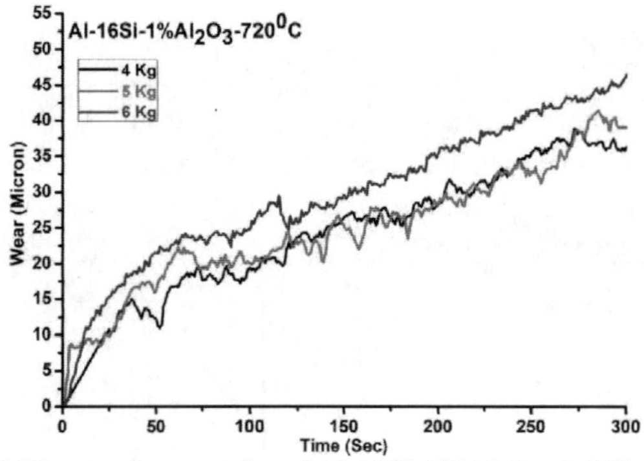

Figure 3.2 Wear vs time graphs of Al-16Si-1%Al_2O_3 at different load.

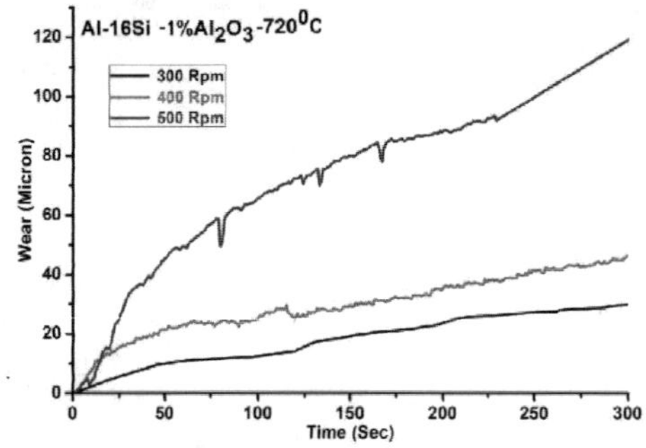

Figure 3.3 Wear vs time graphs of Al-16Si-1%Al_2O_3 at different speed.

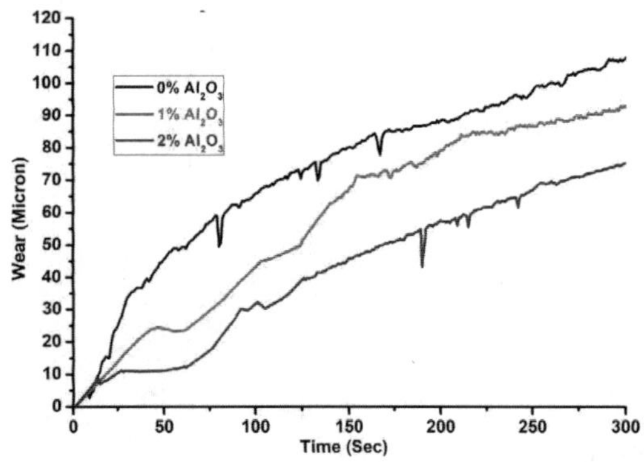

Figure 3.4 Wear vs time graphs of 6kg load having different composition.

4. Conclusions

- The XRD analysis shows the major peaks of aluminium and silicon and minor peaks of SiO_2 and Al_2O_3. As the increasing alumina content the peaks of Al-rich intermetallic compound also shown into the picture.

- The Hardness and density of the MMC increases with alumina content because of formation of harder intermetallic compound and higher density of the alumina respectively.
- The wear of the alloys was increased with increasing load and rotation speed where as less wear was detected in Al-16Si-2%Al_2O_3 sample.

References

[1] Zhou J, Duszczyk J and Korevaar B M 1991 *Journal of materials science* **26**(11) 3041-50.
[2] Gupta M and Lavernia E J 1995 *Journal of materials processing technology* **54**(1-4) 261-70.
[3] Zhou W and Xu Z M 1997 *Journal of Materials Processing Technology* **63**(1-3), 358-63.
[4] Tee K, Lu L and Lai M O 1999 *Journal of Materials Processing Technology* **89** 513-19.
[5] Tjong S C, Wu S Q and Zhu H G 1999 *Composites Science and Technology* **59**(9) 1341-47.
[6] Ipina J P, Yawny A and Oliver C G 2003 *Journal of composite materials* **37**(22) 2001-10.
[7] Ge L L, Liu R P, Li G, Ma M Z and Wang W K 2004 *Materials Science and Engineering A* **385**(1-2) 128-32.
[8] El-Labban H F, Abdelaziz M and Mahmoud E R 2016 *Journal of King Saud University-Engineering Sciences* **28**(2) 230-39.
[9] Wasly H S 2016 *International Research Journal of Engineering and Technology* **03** (11) 1350-55.

IOP Conf. Series: Materials Science and Engineering **338** (2018) 012049 doi:10.1088/1757-899X/338/1/012049

INDUSTRIAL APPLICATION OF VALUABLE MATERIALS GENERATED FROM PLK ROCK-A BAUXITE MINING WASTE

Ranjita Swain[1], Sunita Routray[2], Abhisek Mohapatra[3] and Biswa Ranjan Patra[4]

[1,2]Assistant Professor, C.V. Raman College of Engg., Bhubaneswar 752054, India
[3]JRF, C.V. Raman College of Engineering, Bhubaneswar 752054, India
[4]Chemical Engineering. Department, C.V. R. C. E, Bhubaneswar 752054, India
E-mail: ranjitaswain79@gmail.com

Abstract.

PLK rock classified in to two products after a selective grinding to a particular size fraction. PLK rocks ground to below 45-micron size which is followed by a classifier i.e. hydrocyclone. The ground product classified in to different sizes of apex and vortex finder. The pressure gauge was attached for the measurement of the pressure. The production of fines is also increasing with increase in the vortex finder diameter. In order to increase in the feed capacity of the hydrocyclone, the vortex finder 11.1 mm diameter and the spigot diameter 8.0 mm has been considered as the best optimum condition for recovery of fines from PLK rock sample. The overflow sample contains 5.39% iron oxide (Fe_2O_3) with 0.97% of TiO_2 and underflow sample contains 1.87% Fe_2O_3 with 2.39% of TiO_2. The cut point or separation size of overflow sample is 25 μm. The efficiency of separation, or the so-called imperfection I, is at 6 μm size. In this study, the iron oxide content in underflow sample is less than 2% which is suitable for making of refractory application. The overflow sample is very fine which can also be a raw material for ceramic industry as well as a cosmetic product.

Key words: *Bauxite, PLK rock, Hydro cyclone, Overflow, Underflow, Cut point*

1. INTRODUCTION

Bauxite mining belt covers the east coast region of Odisha and Andhra Pradesh. NALCO's Bauxite Mines at Panchpatmali, Koraput, Odisha. During bauxite mining of 63.4 lakh tonnes per annum, an equal amount of Partially Laterized Khondalite (PLK) rocks associated with kaolinised khondalite rocks are also being generated and dumped as waste material at mining area or sometimes used for back filling of abandoned mines. Fig.1 shows the mine site at NALCO, Damanjodi. This can be a valuable product after suitable beneficiations.

The hydro cyclone is a one of the classifier which separates the coarse and fine materials in liquid medium by using the fluid pressure to generate centrifugal force.

Prasad Rao et al. (Prasad Rao et al., 1995) studied on Shimoga kaolin clay. The beneficiation studies involving hydrocyclone and sluicing to improve the clay content from 25% to 95% which is suitable for industrial use. Qian et al. (2007) studied the effects of different inlet particle concentrations on the separation efficiency.

Gao et al. (Gao et al., 2008) studied on low-grade diasporic bauxite by using a hydrocyclone of small cone-angle. He studied about the separation indices with different process parameters like apex diameter, feed pressure and feed concentration. Recovery of Al_2O_3 increases in underflow with increase in apex diameter. The separation indexes were improved by increase in feed pressure, but were worst with increase in feed concentration.

IOP Conf. Series: Materials Science and Engineering **338** (2018) 012049 doi:10.1088/1757-899X/338/1/012049

Kuang et al. (2012) studied the effects of feed solid concentrations on the hydro cyclone separation performance. In their experiments, the solid concentration varied from 1% to 25% by volume, and the separation efficiency decreased as the feed solid concentration increased.

Liu et al. (2014) studied the circulation flow in a hydro cyclone, and they determined that higher separation efficiency results in a narrower circulation flow distribution area and a lower flow rate. If the feed flow rate is excessively large, the circulation flow will negatively affect the separation efficiency. Zhu and Liow (2014) found that the entrainment of small particles within the wake of a large particle is the major cause of the fishhook effect. As the particle size (DP) increased, the separation efficiency was higher because of a more dominant centrifugal effect.

Abdollahzadeh et al. (Abdollahzadeh et al., 2015) observed that the separation performance of the mini-hydro cyclones increased as the inlet flow rate increased, and the separation efficiency decreased when the feed flow rate continuously increased. The feed flow rate ($P < 0.001$) significantly affected the separation efficiency.

Ghodrat et. al. (Ghodrat et. al., 2014) studied on the multiphase flow and performance of hydrocyclone. Separation efficiency was much more significant with vortex finder than shape and diameter of hydrocyclone. It was also showed opposite trend at low and high feed solid concentration.

In this paper, the sample (PLK rocks) prepared and classified into two products. Here the both product characterisation are projected, based on which the material applies for industrial applications.

2. MATERIALS AND METHODS

PLK rocks crushed and ground to a particular size below 100 μm size. Sample was classified using hydrocyclone. Fig. 2 shows the ROM of PLK rocks which is to be crushed for a particular size.

Fig.1 Bauxite mining area of NALCO, Damanjodi

Fig.2 PLK rock sample

The Mozley C 155 one-inch hydrocyclone was used for the classification study. The different sizes of apex and vortex finder were used for the classification of the sample. The pressure gauge was attached for the measurement of the pressure. The dimensions of different parts of the hydrocyclone were given below:

Cyclone size:	mm	50
Vortex finder diameter:	mm	14.3, 11.1, 8.1
Spigot diameter:	mm	8.0, 6.3, 4.7
Pressure:	psi	5

The particle size analysis of overflow and underflow were determined by using standard sieves and as well as by using particle size analyzer (Cilas 1064). The iron content of overflow and underflow samples was determined by wet chemical method. The tromp curve was plotted with respect to the average size vs. percentage of underflow responded to the feed. This was referred as d_{50} size. The efficiency of separation, or also called as imperfection I, has been given by:

$$I = (d_{75} - d_{25}) / 2 * d_{50} \quad \text{(Eq. 1)}$$

At first instance an attempt has been made to recover superfine particles from the optimum mesh of PLK rock ground sample using hydrocyclone at different operating conditions. The main aim of the cyclone study is to only recovery of fines for leaching studies. Hence, the details of cyclone design and other theoretical aspects are not given as priority.

3. RESULT AND DISCUSSION

The results of these studies are given in Table 1; the data indicate that with increase in the spigot diameter for a constant vortex finder diameter, the production of the fines is increasing. This observation is same for other vortex finder diameters. The production of fines is also increasing with increase in the vortex finder diameter. In order to increase in the feed capacity of the hydrocyclone, the vortex finder 11.1 mm diameter and the spigot diameter 8.0 mm has been considered as the best optimum condition for recovery of fines from PLK rock sample. The physical and chemical analyses of hydrocyclone overflow and underflow samples are given in Table 2.

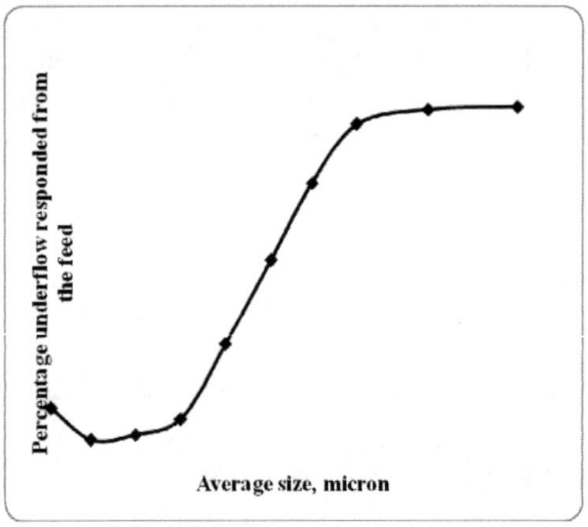

Fig. 3. Particle size analysis of hydrocyclone overflow sample

Table 1.Hydrocyclone studies on PLK sample at different conditions

Exp No.	Vortex finder, mm	Spigot dia., mm	Pressure, psi	Products	Wt., %	Fe_2O_3, %
1	14.3	8.0	5	Overflow	59.7	5.81
				Underflow	40.3	2.10
				Total	100	4.31
2	14.3	6.3	5	Overflow	58.6	5.30
				Underflow	41.4	2.91
				Total	100	4.31
3	14.3	4.7	5	Overflow	42.1	5.91
				Underflow	57.9	3.14
				Total	100	4.30
4	11.1	8.0	5	Overflow	69	5.39
				Underflow	31	1.91
				Total	100	4.31
5	11.1	6.3	5	Overflow	56.7	5.61
				Underflow	43.3	2.62

				Total	100	4.31
6	11.1	4.7	5	Overflow	40.8	6.31
				Underflow	59.2	2.93
				Total	100	4.30
7	8.0	8.0	5	Overflow	33.3	6.82
				Underflow	66.7	3.11
				Total	100	4.34
8	8.0	6.3	5	Overflow	31.3	6.91
				Underflow	68.7	3.23
				Total	100	4.38
9	8.0	4.7	5	Overflow	25	7.11
				Underflow	75	3.41
				Total	100	4.33

Table 2 Physical and chemical characteristics of the hydrocyclone overflow and underflow samples

Details	Particle size analysis			Chemical analysis				
	d_{10}, μm	d_{50}, μm	d_{90}, μm	Al_2O_3,%	SiO_2, %	Fe_2O_3,%	TiO_2,%	LOI, %
Overflow	0.86	4.11	13.09	37.09	40.92	5.39	0.97	15.6
Underflow	4.49	31.45	73.55	37.71	41.89	1.87	2.39	16.1

The data indicate that the overflow sample contains 5.39% Fe_2O_3 whereas underflow sample contains 1.87% Fe_2O_3. The overflow sample having 0.97% of TiO_2 whereas underflow sample contains 2.39% of TiO_2. The performance or partition curve for the hydrocyclone is shown in Fig. 3. This helps in measuring the efficiency of hydrocyclone.

This gives the relation to the weight fraction of each particle in feed which reports to apex or undersize particle. The is also found that the cut point or separation size of overflow sample is 25 μm. As per the definition of cut size, the size is determined at 50% of the particles passing through that particular sieve. The efficiency of separation, or the so-called imperfection I, is at 6 μm size.

$$I = (d_{75} - d_{25})/d_{50} \quad \text{(Eq. 2)}$$

$$I = (31-19)/2 = 6$$

The complete chemical analysis of hydrocyclone overflow and treated as well as preconcentrated samples are given in Table 2. The data indicate that the overflow sample contains 5.39% Fe_2O_3 and underflow sample contain 1.87% of Fe_2O_3. Particle size analysis data of the sample are also given in Table 2. The data for the classified sample indicate that that d_{10} passing size is 0.86 μm, d_{50} is 4.11 μm and d_{90} is 13.09 μm. The data for sample 2 indicate that that d_{10} passing size is 1.12 μm, d_{50} is 5.47 μm and d_{90} is 20.33 μm.

The XRD of the hydrocyclone overflow sample is shown in Fig. 4. It is observed from this data that it contains major minerals such as kaolinite, gibbsite and goethite followed by minor minerals such as orthoclase, quartz, hematite and magnetite. The SEM of the overflow sample is also shown in Fig. 5. It is clearly found that the iron patches are present within the clay. The XRD of the hydrocyclone underflow sample is shown in Fig. 6. The major minerals are kaolinite and quartz whereas the minor minerals are orthoclase, hematite and goethite. The SEM of the underflow sample is also shown in Fig. 7. A small stone like feature found in this figure, which represents the presence of quartz with other impurities.

The characterisation data indicates that the underflow sample is one of sample can be utilised in refractory and ceramics as it contains less than 2% iron oxide. The fine samples i.e. overflow sample contains high amount of iron oxide and it can be utilised as a high-grade filler material after a suitable beneficiation technique.

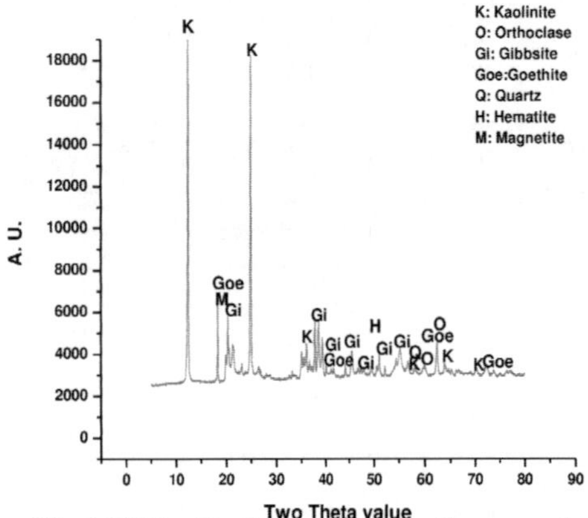

Fig.4 XRD of hydrocyclone overflow sample

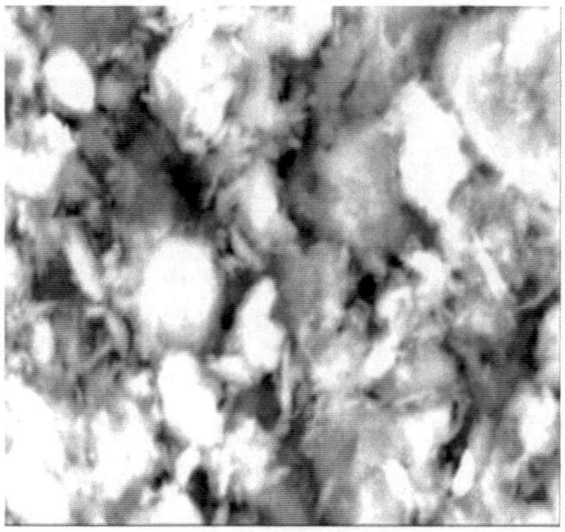

Fig. 5 SEM of hydrocyclone overflow sample

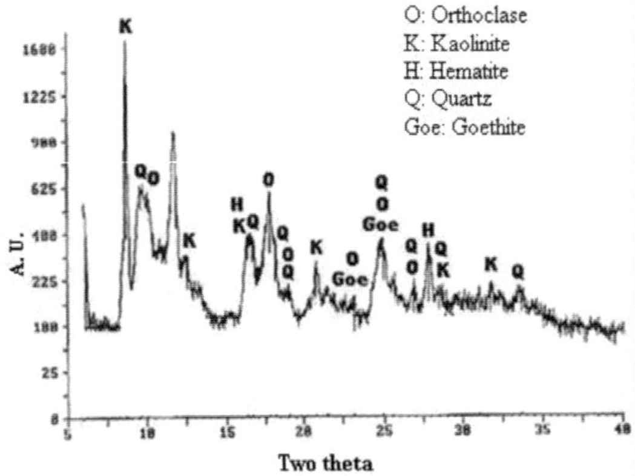

Fig. 6 XRD of hydrocyclone underflow sample

Fig.7 SEM of hydrocyclone underflow sample

4. CONCLUSIONS

This bauxite mining waste can be a potential mineral resource, which can fulfil the demand of refractory, ceramic, filler and other industries. The hydrocyclone overflow sample contains 5.39% iron oxide (Fe_2O_3) and underflow sample contains 1.87% Fe_2O_3. Underflow sample is suitable for refractory brick preparation as it contains less iron content. Overflow sample also can be filler material after suitable beneficiation techniques.

5. ACKNOWLEDGEMENT

Authors are thankful to The Chairman of C. V. Raman Group of Institutions, Bhubaneswar, Odisha, India for providing support to carry out the research facilities. One of the authors Dr. Ranjita Swain is also thankful to NALCO for funding and JNARDDC for its support to carry out the research successfully. One of the authors, Mr. Abhisek Mohapatra is thankful to NALCO for granting JRF.

6. REFERENCE

[1] Gao, Shu-ling Li, Xia-an Wei, De-Zhou Fang and Ping Jia, Chun-yun, Liu, Wen-gang and Han, Cong. 2008 Beneficiation of low-grade diasporic bauxite with hydrocyclone *Transactions of Nonferrous Metals Society of China* **18**(2) 444-48.

[2] Prasad Rao, Rao, R Bhima and Sastri, S R S 1995, Preliminary studies on beneficiation of Shimoga kaolin clay, Report of Regional Research Laboratory, Bhubaneswar, India.

[3] Qian F, Huang Z, Chen G and Zhang M, 2007 Numerical study of the separation characteristics in a cyclone of different inlet particle concentrations. *Comput. Chem. Eng.* **31** 1111–22.

[4] Kuang S B, Chu KW, Yu A B and Vince A 2012 Numerical study of liquid-gas-solid flow in classifying hydro-cyclones: effect of feed solids concentration. *Miner. Eng.* **31**, 17–31.

[5] Liu Y, Yang Q, Qian P and Wang H 2014 Experimental study of circulation flow in a light dispersion hydro cyclone. *Sep. Purif. Technol.* **137**, 66–73.

[6] Zhu G and Liow J 2014 Experimental study of particle separation and the fishhook effect in a mini hydro-cyclone *Chem. Eng. Sci.* **111** 94–105.

[7] Abdollahzadeh L, Habibian M, Etezazian R and Naseri S 2015 Study of particles shape factor, inlet velocity and feed concentration on mini hydro cyclone classification and fishhook effect *Powder Technol.* **283** 294–301.

[8] Ghodrat M, Kuang S B, Yu A B, Vince Andrew, Barnett G D and Barnett P J 2014 Numerical analysis of hydrocyclones with different vortex finder configurations *Minerals Engineering*. **63** 126-138.

IOP Conf. Series: Materials Science and Engineering **338** (2018) 012050 doi:10.1088/1757-899X/338/1/012050

Comparison of wear behaviour and mechanical properties of as-cast Al6082 and Al6082-T6 using statistical analysis

Sandhya Rani Rana, A.B. Pattnaik and S.C. Patnaik

Department of Metallurgical and Materials Engineering

Indira Gandhi Institute of Technology, Sarang, Odisha, India

Email: amulyapattnaik@igitsarang.ac.in , amulyapattnaik@gmail.com

Abstract. In the present work the wear behavior and mechanical properties of as cast Al6082 and Al6086-T6 were compared and analyzed using statistical analysis. The as cast Al6082 alloy was solutionized at 550°C, quenched and artificially aged at 170°C for 8hrs. Metallographic examination and XRD analysis revealed the presence of intermetallic compounds Al_6Mn.Hardness of heat treated Al6082 was found to be more than as cast sample. Wear tests were carried out using Pin on Disc wear testing machine according to Taguchi L9 orthogonal array. Experiments were conducted under normal load 10-30N, sliding speed 1-3m/s, sliding distance 400,800,1200m respectively. Sliding speed was found to be the dominant factor for wear in both as cast and aged Al 6082 alloy. Sliding distance increases the wear rate up to 800m and then after it decreases.

Keywords: Metallographic examination, Pin-on-disc wear test, Taguchi, L9 orthogonal array, sliding speed

1. Introduction

Recent applications require materials of high strength, light weight, and less expensive. With increased demand of fuel economy and improved engine performance in automobiles it is necessary to develop materials with high strength to weight ratio [1].

The 6xxx aluminium alloys possess high specific strength and good corrosion resistance which makes them worthy to replace steels for automotive applications. These alloys are heat treatable which again enhances the mechanical properties [2].G. Mrówka-Nowotnik et.al [3] studied the intermetallic phase particles in 6082 aluminium alloy and found seven phases in between the aluminium dendrites of as-cast Al6082. G.Mrowka-Nowotnik et.al [4] during their investigation on 6005 and 6082 aluminium alloys studied the influence of solution heat treatment on kinetics of ageing. They observed that during cooling after homogenization, β-Mg_2Si phase readily precipitates. The generally accepted precipitation sequence in 6xxx alloys [5-7] is.

Super saturated solid solution ⟶ Mg,Si clusters ⟶ Guiner Preston Zones ⟶ β^{II} ⟶ β^{I} ⟶ β

Jadhav et.al [8] investigated the effect of heat treatment on mechanical properties and microstructure of Al6082 and observed that the type and distribution of the phases significantly affect the mechanical properties. Recently industries are using statistical analysis to find out the parameter having significant influence on the response. Number of statistical analysis techniques(Response surface methodology, Factorial design, random design etc) can be applied to determine

IOP Conf. Series: Materials Science and Engineering **338** (2018) 012050 doi:10.1088/1757-899X/338/1/012050

the significant controlling parameter during wear test, design of experiment approach using Taguchi method not only decreases the number of experiments and time but also doesn't affect the accuracy and quality of results [9].

2. Materials and Methods

2.1. Alloy preparation and heat treatment

In the present investigation, Al6082 alloy was used which contains 1% Si, 0.8%Mn, 1%Mg and balance aluminium was prepared by stir casting process. The process involves (i) melting of pure aluminium in a graphite crucible in an electrical resistance furnace (ii) continuous addition of manganese and Al-Si master alloy followed by mechanical stirring (iii) addition of magnesium turnings (iv)casting and solidification of the melt in cast iron mould. The alloy Al6082 was solutionized at 550^{0}C for 2hrs in an electric resistance furnace and then subjected to water quenching to room temperature. Then the homogenised alloy was aged for 8hrs at 170^{0}C in the furnace.

2.2. Microstructure

The microstructural characterization was done using CARL ZEISS optical microscope. The Al-6082 and Al-6082- T6 samples were polished using emery paper and diamond paste and etched using Keller's reagent.

2.3.Hardness

Hardness of both Al-6082 and Al-6082- T6 samples was measured using a micro Vickers's hardness testing machine Modd FMV-1, FSA.

2.4.X-Ray diffraction studies

The X-Ray diffraction studies were performed on both as-cast and heat treated samples to identify the phases present. The X-Ray diffraction plot with various phases present is shown in Fig.1.

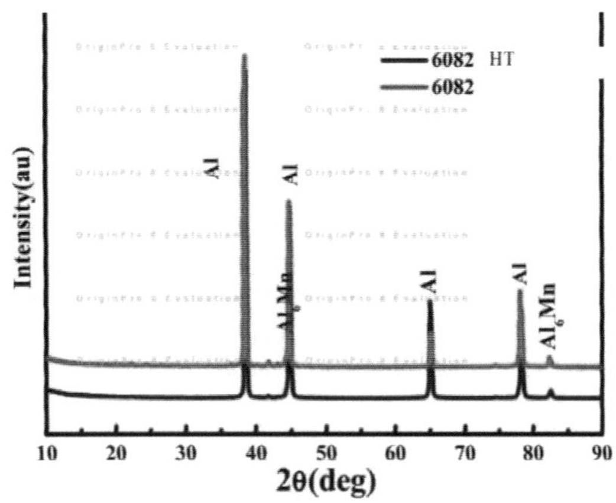

Figure1. XRD plot of as-cast and heat treated (T6) Al6082 alloy

2.5. Wear test

Wear test of both as-cast and heat treated samples were carried out using a Pin-on-disk wear testing machine as per ASTM G99. The specimens were prepared of dimensions 10mm diameter and 30mm height. The weights of the samples were measured using an electronic balance.

2.6. Statistical analysis using Taguchi method

Taguchi method is an important tool for determining the effect of design parameters on the response (output). The statistical analysis using Taguchi method is carried out in various stages (i) identification of response or output, (ii) determination of design parameters influencing the output, (iii) selection of orthogonal array according to the design parameters, (iv) conducting experiments in accordance with the orthogonal array, (v) calculation and tabulation of signal to noise ratios and (vi) the range is determined from the S/N ratios and higher rank is assigned to higher range. In this work, the influence of three factors, i.e. load, sliding speed, sliding distance on wear response were studied. The factors and levels used in the experiment are shown in Table 1.

Table.1: Levels of Factors

Control Factors	Levels			Units
Load (A)	10	20	30	N
Sliding speed (B)	1	2	3	m/sec
Sliding distance (C)	400	800	1200	m

The L_9 orthogonal array design with S/N ratio for both as-cast and heat treated conditions are given in Table 2 and 3 respectively. The experimental observation was converted into a logarithmic function called signal-to-noise ratio(S/N). This can be determined as logarithmic transformation of a loss function as: The smaller is the better characteristics: $\frac{S}{N} = -10\log\frac{1}{n}(\sum y^2)$ .... (1)

Where y is the observed data and n is the number of observations.

Table 2—L9 orthogonal array with response and S/N ratio for as-cast Al 6082 alloy

Sl.No	Load (A), N	Sliding speed (B), m/sec	Sliding distance (C), m	Mass Loss (gm)	S/N Ratio, (dB)
1	10	1	400	0.0022	53.1515
2	10	2	800	0.0084	41.5144
3	10	3	1200	0.0176	35.0897
4	20	1	800	0.0258	31.7676
5	20	2	1200	0.1026	19.7771
6	20	3	400	0.0976	20.2110
7	30	1	1200	0.0165	35.6503
8	30	2	400	0.0177	35.0405
9	30	3	800	0.1032	19.7264

Table 3—L9 orthogonal array with response and S/N ratio for Al 6082-T6 alloy					
Sl.No	Load (A), N	Sliding speed (B), m/sec	Sliding distance (C), m	Mass Loss (gm)	S/N Ratio, (dB)
1	10	1	400	0.0840	21.5144
2	10	2	800	0.0062	44.1522
3	10	3	1200	0.0150	36.4782
4	20	1	800	0.0068	43.3498
5	20	2	1200	0.0978	20.1932
6	20	3	400	0.0071	42.9748
7	30	1	1200	0.0115	38.7860
8	30	2	400	0.0377	28.4732
9	30	3	800	0.4383	7.1646

3. Results and Discussion

3.1. Microstructural characterization

Figures 2(a) and 2 (b) shows the optical micrographs of Al 6082 and Al 6082T6 alloy. The microstructural analysis reveals the presence of intermetallic compounds such as Al_6Mn that is confirmed by X-ray diffraction analysis shown in Fig.1. It can also be seen in the micrograph that grain growth has taken place in the heat treated sample.

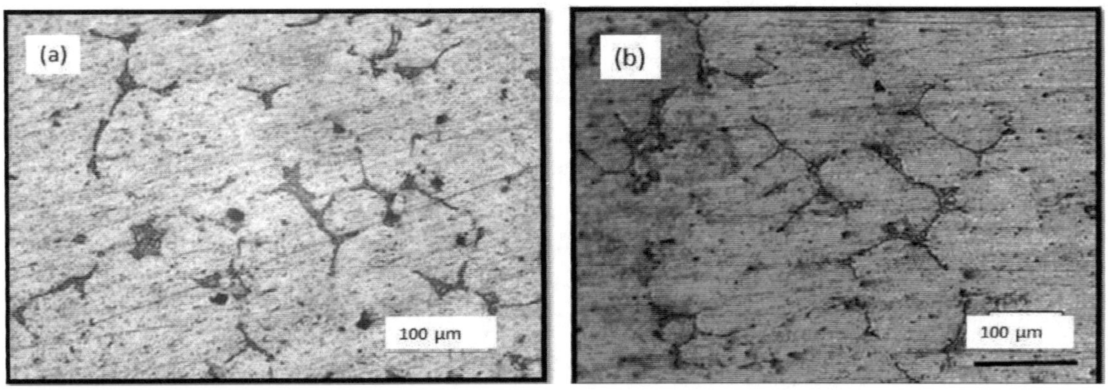

Figure2. Optical micrographs of (a) as-cast Al6082 (b) Al6082-T6

3.2. Hardness

Hardness of both Al-6082 and Al-6082- T6 samples were measured using a micro hardness tester and are reported in Table.4. It was found that the Al6082-T6 alloy was having higher hardness than the as-cast alloy that is Al6082-T6 alloy had 88.28 VHN and as-cast Al6082 it was 55.67 VHN. The increase in hardness can be due to the precipitation strengthening after ageing.

Table 4. Hardness values of as-cast Al6082 and Al6082-T6 alloy	
Microstructural condition	Hardness values (VHN)
As-cast Al6082	55.67
Al6082-T6	88.28

3.3. Analysis of wear behaviour using Taguchi Methodology

The Taguchi orthogonal arrays design of Al-6082 and Al6082-T6 alloys with all experimental data's and S/N ratios are shown in Table 2 and 3 respectively. MINITAB 16 software was used to determine the values of S/N ratios for each experiment. The response table of the S/N ratios for both Al6082 and Al6082-T6 are shown in Tables 5 and 6 respectively [9].

Table 5—Response table of S/N ratio for as-cast Al6082 (smaller is the better characteristics)

Level	A	B	C
1	43.25	40.19	36.13
2	23.92	32.11	31.10
3	30.14	25.01	30.17
Delta(Δ)	19.33	15.18	5.96
Rank	1	2	3

Table 6—Response table of S/N ratio for Al6082-T6 (smaller is the better characteristics)

Level	A	B	C
1	34.05	34.55	30.99
2	35.51	30.94	31.56
3	24.81	28.87	31.82
Delta(Δ)	10.70	5.68	0.83
Rank	1	2	3

The rank and the delta values in the response table were generated basing on S/N ratios. From Table 5 it is clearly observed that, load is the prominent design parameter influencing the wear behaviour followed by sliding speed and sliding distance. In case of Al6082-T6 alloy, again load was found to be the major contributor to wear loss as can be seen from Table 6. This is mainly attributed to the soft nature of aluminium alloy. This also shows that there is no change in the significant controlling factor affecting wear behaviour with change in microstructure due to heat treatment. The main effect plot for S/N ratios of as-cast Al6082 alloy is shown in Figure 3.

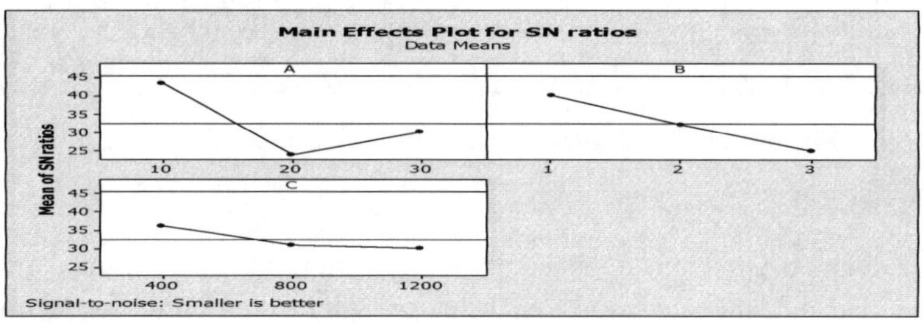

Figure 3. Main effects plot for S/N ratios for as-cast Al6082 alloy

The above plot shows that the combination of A_2, B_3, C_3 factors would result in less wear in as-cast Al6082 alloy. The main effect plot of S/N ratios of Al6082-T6 is shown in Figure 4. The plot shows that lesser wear will result with the combination of A_3, B_3, and C_1 factors.

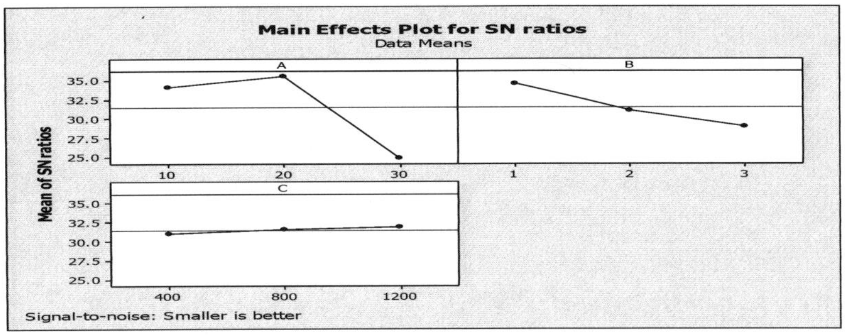

Figure 4. Main effects plot for S/N ratios for Al6082-T6 alloy

4. Conclusions

The conclusions are as follows.

I. The comparison of wear behaviour of as-cast Al6082 and Al6082-T6 was studied using Taguchi experimental design approach.

II. Microstructural characterization and XRD results showed the presence of intermetallic compounds.

III. The hardness of the Al6086-T6 was found to be 88.28VHN and that of Al6082 as-cast was found to be 55.67VHN.

IV. From the response table of Taguchi orthogonal array, load was found to be the prominent factor affecting wear loss in both as-cast Al6082 and Al6082-T6 samples.

5. References

[1] Bodunrin MO, Alaneme KK and Chown LH 2015 *Journal of Material Research Technology*, V**4**, 434-445.

[2] Hosseinifar M and Malakhov DV 2011 *Metallurgical and Materials Transactions A*, Vol**42**, 825-833.

[3] Nowotnik GM, Sieniawski J, Wierzbinska M 2007 *Archives of Material Science and Engineering*, Vol**28**, 69-76.

[4]Nowotnik GM and Sieniawski J 2005 *Journal of Material Processing Technology*, Vol**4**, 367-372.

[5] Edwardsa GA, Stillerb K, Dunlopa GL and Couperc MJ 1998 *ActaMaterialia*,Vol**46**, 3893-3904.

[6] Marioara CD, Andersen SJ, Jansen J and Zandbergen HW 2001*ActaMaterialia*, Vol **49**, 321-328.

[7] Muryama M, Honoa K, Sagab M, Kikuchib M 1998 *Materials Science and Engineering: A*, Vol **250**, 127-132.

[8] Jadhav S, Singh R, Pawar V, Mane S 2017 *Mechanical and Aerospace Engineering, IEEE.*

[9] Pattnaik AB 2014 *Indian Journal of Engineering and Material Sciences*, Vol **21**, 379-386.

Sea water Corrosion of Nickel based Plasma Spray Coating

M Parida[1]*, S P Nanda[1], S K Bhuyan[2] and S C Mishra[2]
[1] Dept. Of Chemistry, CUTM, Odisha
[2] Department of M&ME, N.I.T, Rourkela.

*email: rinaa26@yahoo.com

Abstract
Different types of erosion resistant coatings are applied/deposited on aero components, depending on the operating/working temperatures. Nickel based coating are applied on the air craft (compressor) components, which can sustain up to working temperature of 650°C. In the present investigation, to improve the compatibility between substrate (i.e. the machine component) and the top coat, application of bond coat is there. The application of Nickel based coating by thermal plasma spray technique has proven to be a satisfactory means of producing acceptable sealing surface with excellent abradability. Before the corrosion study, coated sample is subjected to hardness, thickness and porosity testing. Hence the result is being evaluated. The corrosion behavior of coating was studied by sea water immersion with a time period of 16 weeks. It is observed that, up to 9 weeks increase in weight of coating occurs in a sharp trend and then takes a decreasing trend. The weight gain of the samples has varied from 37.23% (with one week immersion in sea water) to a maximum of about 64.36% for six weeks immersion. Coating morphology and composition analysis of the coatings are studied using SEM and EDS. This behavior shows adsorption/deposition of the foreign particles with polygonal shape on the coating surface by sea water interaction. Foreign particles with polygonal shape deposited on the coating and with increase in immersion/treatment time, washing out of the deposited materials starts, which reflects the decreasing trend of weight gain of the specimen.

Key words: thermal plasma, erosion resistant coating, nickel base coating, plasma spraying, sea water corrosion.

1. Introduction

Thermal plasma spray coating is one of the effectively developed modern technology to resist the surface from degradation. Coating made with plasma technique exhibit excellent wear, corrosion and thermal resistance. Nickel based abradable coating is used with thermal plasma spray techniques in air craft (Compressor components), which can sustain up to working temperature of 650°C, by improving engine efficiency. Characteristics of coating varies due to substrate surface preparation and critical plasma spray parameters [1-4]. Spraying condition like CPSP affects the final coating microstructure, hardness, porosity and adhesion strength etc. Hence selected CPSP is taken for better coating result. Plasma spray coating is composed of layers that are formed when molten liquid droplets flatten and solidify on the surface [5-7]. Till date, no effort has been taken on the characterization and evaluation of sea water corrosion behavior of different plasma spray coating used in aero engines. Present piece of work is intended to study corrosion behavior of Nickel base plasma spray coating, when treated with sea water at different time length.

2. Experimental

2.1. Substrate Preparation

Nimonic plates were used for substrate surface (as this is been used in aero-components). Substrate surface was roughened using Al_2O_3 powder of grit size 60 to produce a surface roughness of 4-6 micron Ra.[8]. The roughened surface was protected from contaminations up to spraying to avoid disintegration of coating.

2.2. Preparation of coating powder

Powders of Nickel, Copper Oxide, Cadmium Oxide and Hexagonal Boron Nitride powder are sieved, after baking at 130^0 -$150\ ^0$ C for 2-3 hr. Nickel powder 54- 68 %, Copper Oxide 10 -12%, Cadmium Oxide 3-5%, Boron Nitride 16-19%, Graphite 1-4 % and Aluminium Powder were mixed in a ball mill along with liquid glass. The mixture is rubbed over sieve and baked at 150^0C and finally sieved to get the final before use..

2.3. Spray Deposition

Plasma spraying condition i.e. CPSP parameters were selected carefully for coating, where Argon (53 lpm) is used as Primary plasma Gas & Hydrogen (4 lpm) is used as Secondary Plasma- gas with Current(460 Amp).After coating the samples were heat treated at 650^0C for half an hour.

2.4. Specimen preparation for corrosion test

With the help of EDM wire cut the coating specimens were cut into slices of 3-4 mm thickness and 3-6 gm weight. The cut samples were treated with sea water (Bay of Bengal) to study the corrosion behavior at different time length. The chemical composition was analyzed before corrosion test, is tabled below.

Table 1. Composition of Bay of Bengal sea water.

Type of ion	Element	Concentration (Mg/ltr)	% by wt.
Cation	K^{+1}	380	1.10
	Na^{+1}	10500	30.42
	Sr^{+2}	8	0.02
	Mg^{+2}	1350	3.91
	Ca^{+2}	400	1.16
Anion	F^{-1}	1	0.003
	Cl^{-1}	19000	55.04
	Sio_3^{-2}	8	0.02
	So_4^{-2}	2655	7.04
	Bo_3^{-3}	20	0.06
	$C0_3^{-2}$	140	0.41
	Br^{-1}	65	0.19

3. Results and discussion

3.1. Hardness test

Coated sample surfaces were polished using emery paper of distinctive grit size for hardness estimation. Brinell hardness test was performed at room temperature on 10 different areas and to get the average hardness value, and were observed to be in the range of 7-13 HB.

3.2. Corrosion Test

Corrosion behavior of coating was studied by sea water immersion with a time period of 16 weeks. Weight of specimens were taken before and after treatment at different time interval and the results are tabled below.

Table 2. % change in weight with time.

Sl No.	Time in weeks	Initial Weight (gm)	Final Weight (gm)	Change in weight (gm)	% change
1	16	5.01	6.88	1.87	37.33
2	15	4.04	5.88	1.84	45.54
3	14	5.48	7.52	2.04	37.23
4	12	4.02	5.95	1.93	48.01
5	11	3.04	4.86	1.82	59.87
6	9	3.03	4.98	1.95	64.36
7	8	4.7	6.68	1.98	42.13
8	7	4.62	6.72	2.1	45.45

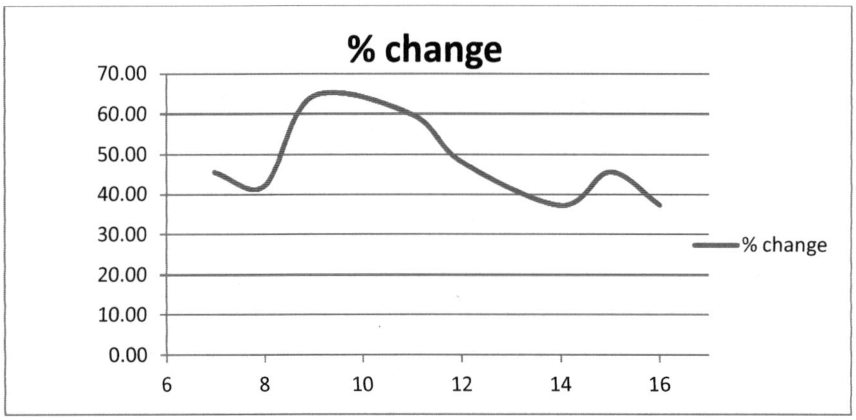

Figure 1. Time Vs wt. change

The graph shown in Fig. 1 shows that deposition of the foreign elements taking place on coating surface occurs during sea water interaction and with increase in immersion/treatment time, washing out of the deposited materials starts, which reflects the decreasing trend of weight gain of the specimen.

IOP Conf. Series: Materials Science and Engineering **338** (2018) 012051 doi:10.1088/1757-899X/338/1/012051

3.3. Coating Morphology

SEM micrograph and EDS analysis of raw & corroded samples, in Fig. 2 and Fig. 3.

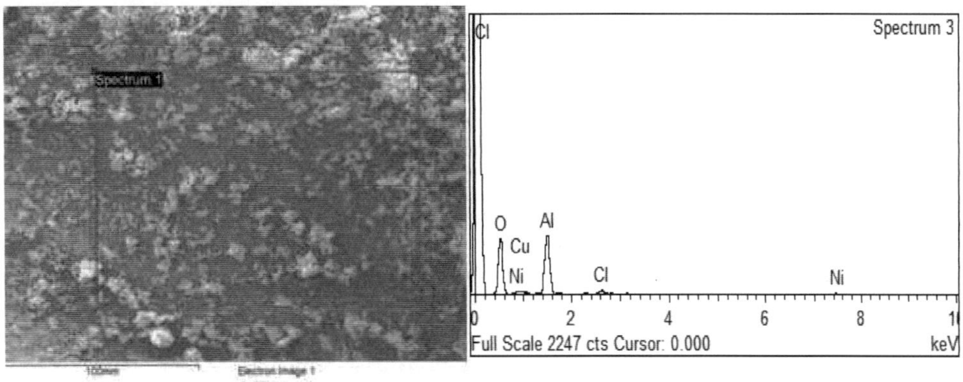

Figure 2. Untreated specimen EDS and SEM spectra.

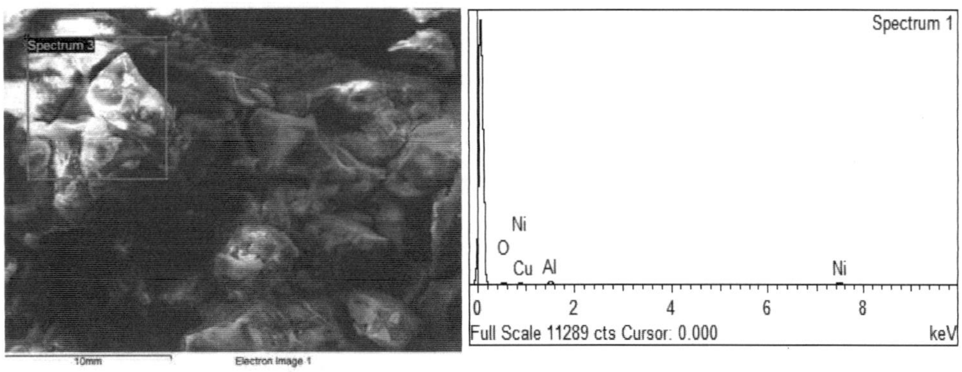

Figure 3. Treated specimen EDS and SEM spectra.

SEM micrograph of coating deposited at CPSP by plasma spray in Fig. 2 and Fig. 3, show uniform distribution of molten /semi molten particles .Coating deposited are smooth, more homogenous, having least porosity with the selected CPSP parameters. This might be the reason to get a suitable dense coating. SEM morphology of corroded sample in Fig. 3 shows small foreign particles with polygonal shape deposited on the coatings. EDS analysis data of corroded sample from Table 3 and Table 4 showed the increase in oxygen in large amount where as very less increase in Cl is there. Which implies, some kind of ionic reactions are taking place. Further detailed examination is in progress. The polygonal shaped particles are deposited on coating surface and EDS analysis exhibit the increase/presence of Cl, which is also another clue to confirm the formation of some chlorine compound; as the cation (Cl) concentration is more in sea water than the anion concentration.

Table 3. Coating elemental analysis before treated by EDS .

Element	Weight%	Atomic%
Al K	15.98	16.66
Si K	2.20	2.20
Ti K	4.80	2.82
Ni K	35.86	17.18
O K	32.62	57.35
Cu K	8.53	3.78
Totals	100.00	

Table 4. Elemental analysis of the coating after dipped in sea water for 10 weeks by EDS .

Element	Weight%	Atomic%
Al K	26.41	18.84
Cl K	2.38	1.29
O K	64.71	77.84
Ni K	2.76	0.91
Cu K	3.74	1.13
Totals	100.00	

4. Conclusions

From this investigation it is found that, the Ni base abardable coatings used in aero component could be plasma sprayed with thermal plasma spray technique. The coating hardness is found to be 7-13HB. There is a weight gain of the coatings with initial time period of deeping in sea water but after few weeks there is a decrease in weight; may be due to washing out of compounds formed with interaction of the coating with sea water with prolonged time of treatment; implies weakening of the coating materials. From the SEM and EDS observations, the formation of polygonal shaped particles on the coating surface and increase in Cl component implies and confirms the formation of other compounds with sea water.

References

[1] Mishra S 2012 *Thermal Plasma Application in Metallurgy* Germany Lap Lambert Academic Publishing ISBN: 978-3-659-23522-1.

[2] Das S 2007 *Study of Plasma Spray Alumina-Aluminide Composite Coating on Metals* (Doctoral dissertation).

[3] Parthasarathi N L and Duraiselvam M 2010 Effect of sliding velocity and plasma spraying parameters on wear resistance of NiCrBSiCFe plasma sprayed on austenitic steel at high temperature *Computer Modelling and New Technologies* **14** 17–29.

[4] Kulkarni A, Wang Z, Nakamura T, Sampath S, Goland A, Herman H, Allen J, Ilavsky J, Long G, Frahm J and Steinbrech R W 2003 Comprehensive microstructural characterization and predictive property modeling of plasma-sprayed zirconia coatings *ActaMaterialia* **51** 2457–75.

[5] Yang C Y, Lee T M, Lu Y Z, Yang C W, Lui T S, Kuo A and Huang B W 2010 The influence of plasma-spraying parameters on the characteristics of fluorapatite coatings *Journal of Medical and Biological Engineering* **30** 91-8.

[6] Lin Y B, Shen Y Z, Chen T F and Taoa J 2015 Optimization of Deposition Parameters for α-Al2O3 Coatings by Double Glow Plasma Technique *Kemija u industriji-journal of chemists and chemical engineers* **64** 457-66.

[7] Miller R A 1987 Current status of thermal barrier coatings-An overview, Surface Coatings and Technology **30** 1-11.

[8] Thirumalaikumarasamy D, Shanmugam K, Balasubramanian V 2012 Influences of atmospheric plasma spraying parameters on the porosity level of alumina coating on AZ31B magnesium alloy using response surface methodology *Progress in Natural Science: Materials International* **22** 468-79.

IOP Conf. Series: Materials Science and Engineering **338** (2018) 012052 doi:10.1088/1757-899X/338/1/012052

Continuum simulation of heat transfer and solidification behavior of AlSi10Mg in Direct Metal Laser Sintering Process

Akash Ojha, Mihir Samantaray, Dhirendra Nath Thatoi and Seshadev Sahoo*
Department of Mechanical Engineering, Institute of Technical Education and Research, Siksha 'O' Anusandhan, Bhubaneswar, OD, India-751030

Email: seshadevsahoo@soa.ac.in

Abstract: Direct Metal Laser Sintering (DMLS) process is a laser based additive manufacturing process, which built complex structures from powder materials. Using high intensity laser beam, the process melts and fuse the powder particles makes dense structures. In this process, the laser beam in terms of heat flux strikes the powder bed and instantaneously melts and joins the powder particles. The partial solidification and temperature distribution on the powder bed endows a high cooling rate and rapid solidification which affects the microstructure of the build part. During the interaction of the laser beam with the powder bed, multiple modes of heat transfer takes place in this process, that make the process very complex. In the present research, a comprehensive heat transfer and solidification model of AlSi10Mg in direct metal laser sintering process has been developed on ANSYS 17.1.0 platform. The model helps to understand the flow phenomena, temperature distribution and densification mechanism on the powder bed. The numerical model takes into account the flow, heat transfer and solidification phenomena. Simulations were carried out for sintering of AlSi10Mg powders in the powder bed having dimension 3 mm × 1 mm × 0.08 mm. The solidification phenomena are incorporated by using enthalpy-porosity approach. The simulation results give the fundamental understanding of the densification of powder particles in DMLS process.

1. Introduction

Aluminum alloys and composites have been extensively used in various sectors, such as automotive, aerospace and domestic industries due to the combination of excellent weldability, high heat conductivity, sufficient hardenability and good corrosion resistance [1]. Among diferent types of aluminum alloys and composites, some of them are complex structure which are difficult to manufacture using conventional casting processes. Furthermore, it produces the coarsened grain structure and large degrees of segregation due to the slow cooling rates during the conventional casting processes [2]. Therefore, novel processing methods are highly demanded to satisfy the need for obtaining complex components with fine and uniform microstructures. Among different types of rapid prototyping technology, Direct Metal Laser Sintering (DMLS) offers great advantages for the processing of metallic powders and composites. In DMLS process, the metallic powders interact with a high intensity laser energy, which helps to melt and solidify the powders in a laminated manner. In this rapid manufacturing process high rate of heating and cooling takes place, which directly affects the microstructure and mechanical properties of the built parts. The advantages of such layer-based manufacturing technology are high degree of freedom in design, minimal material wastage, potential elimination of tooling, light-weight design, elimination of production steps [3, 4]. This process is accompanied by fluid flow, heat transfer and solidification phenomena which makes it very complex. The schematic diagram of the direct metal laser sintering process is shown in Figure 1.

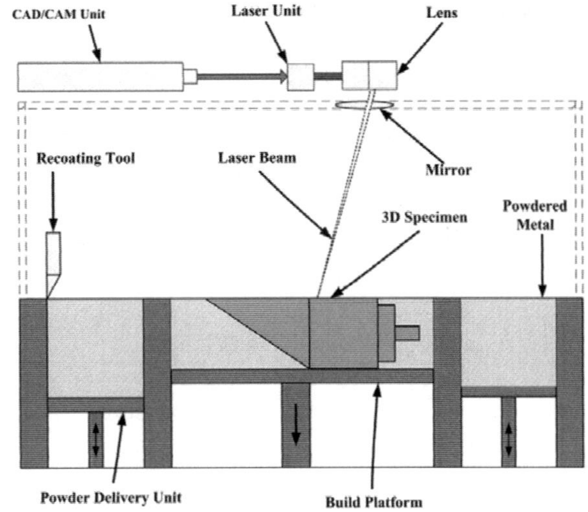

Figure 1: Schematic diagram of DMLS process

Over the past decades, several researchers tried to understand the fundamentals of sintering process during liquid phase sintering. A comprehensive thermal model of selective laser sintering has been developed by Zeng et al. [5]. The thermal behavior in the powder bed was investigated by Tang et al. [6] and studied the quality and dimensional accuracy of the solid structure based on processing condition. The authors considered the laser beam as a moving heat source. Dong et al. [7] proposed a transient three-dimensional finite element model to simulate the phase transformation of powders during the laser sintering process. In their model, they consider the thermal and sintering phenomena. Yuan et. al. [8] developed a numerical model for selective laser melting of Aluminum nano-composites using finite volume method. Also, they had investigated the molten pool behavior and temperature distribution with respect to process parameters. Xiao et al. [9] have examined a three-dimensional numerical model to understand the physical mechanism occurring in the laser sintering process. Sahoo et al. [10] have developed a thermal and microstructure model for Ti alloys in EBAM process by considering moving heat source. A thermal model was developed to predict the temperature distribution on a single powder layer, when laser beam irradiated with powder bed by Patil and Yadava [11]. They also studied the effect of laser spot size, hatching space and laser power on thermal behavior.

In DMLS process, rapid heating and cooling occurs due to short interaction of laser beam and powder particles. This leads to a large temperature gradient and thermal stress in the powder bed, which may cause defects in the final products. However, it is difficult to investigate the rapid melting and solidification phenomena experimentally during the sintering process. Therefore, it is important to explore numerical method to investigate fluid flow, the heat transfer and solidification behavior in DMLS process. In the present study, a mathematical model has developed for direct metal laser sintering of AlSi10Mg alloy by considering fluid flow, heat transfer, and solidification characteristics in the liquid pool.

2. Mathematical model

A mathematical model for sintering of AlSi10Mg alloy powder in direct metal laser sintering process is developed. For simplification of the model, the following assumptions are made:

(i) The powder bed is homogeneous and uniformly distributed;
(ii) Temperature dependent thermo-physical properties of the powders are considered;
(iii) Conduction, convection and radiation mode of heat transfer is considered;
(iv) A constant heat transfer coefficient is considered between powder bed and surrounding environment.

(v) Laser beam is taken as the input heat source which follows Gaussian heat source distribution

Considering the above mentioned physical assumption, governing equations for mass, momentum and energy are as follows:

Continuity Equation:

$$\frac{\partial \rho}{\partial t} + \nabla.(\rho.\vec{V}) = M_s \quad (1)$$

Momentum equation:

$$\rho\left(\frac{\partial \vec{V}}{\partial t} + \vec{V}.\nabla\vec{V}\right) = \mu\nabla^2\vec{V} - \nabla p + M_s.\vec{V} + F \quad (2)$$

Energy equation:

$$\rho\left(\frac{\partial T}{\partial t} + \vec{V}.\nabla T\right) = \nabla.(\kappa\nabla T) + S_H \quad (3)$$

where ρ, κ, μ, p represent density, thermal conductivity, dynamic viscosity and pressure, respectively. $\vec{V}$ is melt velocity, M_s is source term, and F refers to body force, S_H is the source term of the energy equation, which is defined as

$$S_H = -\rho\left(\frac{\partial \Delta H}{\partial t} + \nabla.(\vec{V}\Delta H)\right) \quad (4)$$

where ΔH , is the latent heat of phase change in the medium.

In the current model, the input laser heat source follows a Gaussian distribution [12]. So, the heat flux is defined as

$$q(r) = \frac{2AP}{\Pi r_0^2} e^{-2r^2/r_0^2} \quad (5)$$

where,
A = laser absorptance of the powder system;
P = Laser power;
r_o= radius of the laser beam;
r = radial distance between the laser beam and the center of the spot generated on the top surface of the powder bed.

Initial & Boundary Conditions

In this process, the laser beam interacts with the powder bed and the heat transfer takes place on the powder bed surface.
Initially, the powder bed layer is assumed to be at room temperature at t=0,

$$[T(x, y, z, t)]_{t=0} = T_0 \quad (6)$$

where, T_0 is the ambient temperture
Also, the input laser energy from the moving laser beam at initial stage is assumed to be zero.

The heat will transfer in convection and radiation mode on the upper surface of the powder bed and other sides of substrate and powder bed and in conduction mode within the powder bed. The boundary condition of heat transfer are as follows:

The net amount of heat transfer from each surface except the bottom surface is given by Equation (7);

$$-k\left(\frac{\partial T}{\partial z}\right) = \varepsilon\sigma(T^4 - T_0^4) + h(T - T_0) \quad (7)$$

where, k= thermal conductivity, ε= emissivity, σ= lattice-Boltzmann constant, T= powder bed temperature and T_0= ambient temperature

Assume that there is no heat loss on the bottom of substrate, so it meets the Equation (8)

$$\left[k\left(\frac{\partial T}{\partial z}\right)\right] at(Z=0)=0 \tag{8}$$

Solution method

The numerical modelling of fluid flow, heat transfer and solidification are carried out by using ANSYS FLUENT 17.1.0 platform. With suitable boundary conditions, the mass, momentum and energy equations are discretized using control volume approach and then solved using Semi-Implicit Method for Pressure Linked Equation Consistent (SIMPLEC) algorithm. Characteristically, the time step is restricted to be no more than 10^{-5} s to satisfy the convergence criteria necessitated by the governing equations. The thermos-physical properties of AlSi10Mg and the processing parameters of DMLS process are given in Table 1 and Table 2.

Table-1: Thermo-physical properties of AlSi10Mg alloy [13]

Properties	Values
Thermal conductivity(k)	183 W/mK
Specific heat capacity (cp)	940J/kgK
Heat transfer coefficient (h)	80 W/m2K
Density (ρ)	2680 g/cc
Emissivity (ε)	0.36
Latent heat	5321018 J/kg
Solidus temperature	830 K
Liquidus temperature	869 K

Table-2: Process parameters of DMLS process

Parameters	Values
Laser Power	100 Watt
Scanning speed	2 m/s
Laser spot size	0.2 mm
Thickness of layer	0.08 mm
Laser absorptivity	0.95

3. Results and Discussion

A comprehensive numerical model for direct metal laser sintering of AlSi10Mg alloy is developed in the present investigation and the effect of laser power and scan speed on heat transfer and solidification behaviour are investigated. The model takes into account the coupled fluid flow, heat transfer and solidification phenomena in the powder bed. Simulations are carried out for sintering of AlSi10Mg in a powder bed having 3 mm × 1 mm × 0.8 mm. Figure 2 shows the computational domain used for simulation. The simulation results show the temperature distribution and melt pool characteristics on the powder bed.

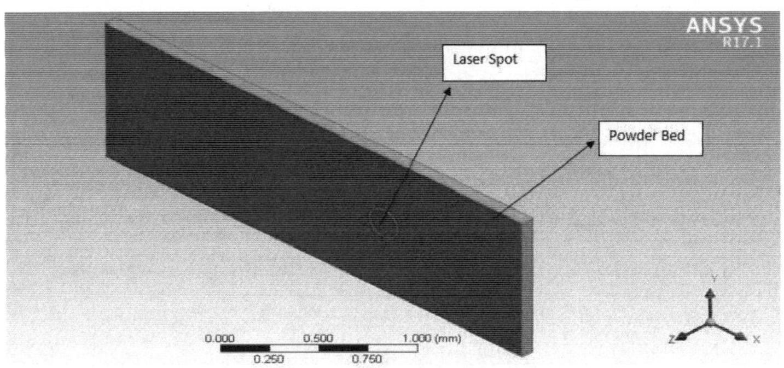

Figure 2: Computational domain used for simulation

Figure 3 shows the temperature distribution in the powder bed when the laser power is 100 Watt and the scan speed is 2 m/s. Color gradient represents the temperature distribution in the molten pool region. In the melt pool region, the solid-liquid region is differentiated by interface temperature i.e., 830 K. The temperature is higher in the melt pool region than in the other regions. The maximum temperature at the molten pool region which is around 906 K. From the temperature profile it clearly seen that, AlSi10Mg alloy is in liquid form below the laser spot.

Figure 3: Temperature profile of AlSi10Mg powder bed in DMLS process

The velocity vector plot in the melt pool region of the powder bed is shown in Figure 4. The flow direction is indicated by the arrow head, and the color gradient represents the quantitative value of the velocity. It shows that the velocity of the liquid metal below the laser spot is approximately equal to the scan velocity. This is because due to no slip condition of the laser beam.

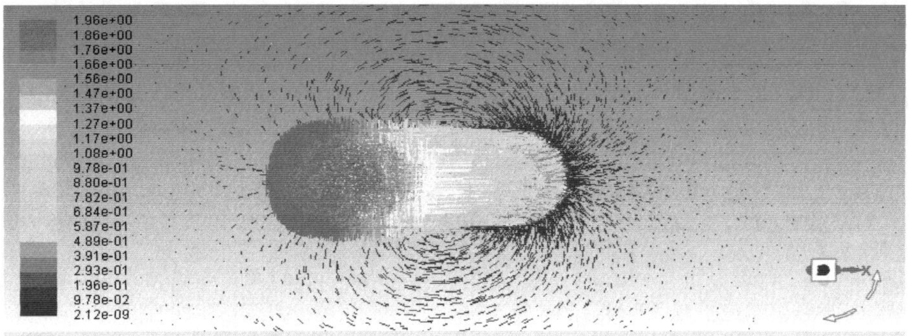

Figure 4: Velocity vector profile of AlSi10Mg powder bed in DMLS process

In this process, when the laser beam strikes at a particular location melting will occur and when the beam moves to the next location, solidification at the previous location takes place. Figure 5 represents the solidification profile of AlSi10Mg. It is represented by the variation of liquid fraction of AlSi10Mg, which varies from 0 to 1. From the solidification profile it is observed that, when the laser beam scans the powder bed, melting of the powder will occur due high rate of heat transfer. As the beam moves on the powder bed, the melting layer rapidly solidifies. So, the melting and solidification phenomena continues with the movement of laser beam.

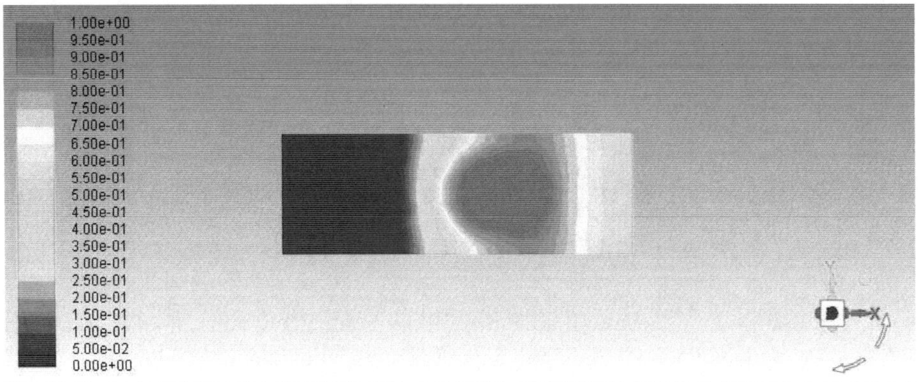

Figure 5: Solidification profile of AlSi10Mg powder bed in DMLS process

4. Conclusion

A three-dimensional numerical model was developed by using finite volume method (FVM) for DMLS process to investigate the temperature evolution, melt pool behavior and solidification phenomena. The simulation results provide a fundamental understanding of the heat transfer and solidification phenomena in this process. From the simulation results, it is quite evident that, the process parameters and material properties have a great influence on the quality and property of the sintered part.

Reference

[1] Olakanmi EO, Cochrane RF and Dalgarno KW. 2015, Progress *in Materials Science.* **74** 401.
[2] Nandy J, Sarangi H and Sahoo S. 2018, *Advances in Manufacturing*, DOI:10.1007/s40436-018-0213-1.
[3] Paul R and Anand S. 2012, *Journal of Manufacturing System.* **31** 429.
[4] Wong KV and Hernandez A. 2012, *ISRN Mechanical Engineering*, **2012** 1.
[5] Zeng K, Pal D and Stucker B. 2012, A review of thermal analysis methods in Laser Sintering and Selective Laser Melting. *23rd Annual International Solid Freeform Fabrication Symposium - An Additive Manufacturing Conference*, SFF 2012. 796.
[6] Tang Y, Loh HT, Wong YS, Fuh JYH, Lu L and Wang X. 2003, *Journal of Materials Processing Technology.* **140** 368.
[7] Dong L, Makradi A, Ahzi S and Remond Y. 2009, *Journal of Materials Processing Technology.* **209** 700.
[8] Yuan P and Gu D. 2015, *Journal of physics D: Applied Physics.* **48** 1.
[9] Xiao B and Zhang Y. 2009, *6th International Symposium on Multiphase Flow, Heat Mass Transfer and Energy Conversion*, Xi'an, China, Paper No. MN-18.
[10] Sahoo S and Chou K. 2016, *Additive Manufacturing.* **9** 14.
[11] Patil RB and Yadava V. 2007, *Int. J. Mach. Tools Manuf.* **47(7)** 1069.
[12] Roberts IA and Wang CJ. 2009, *International Journal of Machine Tools and Manufacture.* **49** 916.
[13] Nandy J, Sarangi H and Sahoo S. 2017, *IOP Conf. Series: Materials Science and Engineering.* **178** 1.

Effect of Bauxite addition on Adhesion Strength and Surface Roughness of Fly ash based Plasma Sprayed Coatings

S K Bhuyan[1*], S Samal[1], D Pattnaik[1], A Sahu[1], B Swain[1], T K Thiyagarajan[2] and S C Mishra[1]

[1]Metallurgical & Materials Engineering Department, N.I.T. Rourkela.

[2]Laser and Plasma Technology Division, BARC, Mumbai.

Email: subratbhuyan2014@gmail.com

Abstract

The environment is being contaminated with advancement of new technology, day by day. One of the primary sources for this contamination is the industrial waste. Industrialization is the prime reason behind the prosperity of any country to meet the materialistic demand. To run the industries, a huge amount of (electric) power is needed and hence need for thermal power plants to serve the purpose. In present scenario, coal fired thermal power plants are set up which generates a huge quantity of Fly ash. Consumption of industrial waste (Fly ash), continually a major concern for human race. In recent years, fly ash is being utilized for various purposes i.e. making bricks, mine reclamation, production of cements etc. The presence of Silica and Alumina in fly ash makes it useful for thermal barrier applications also. The plasma spray technology has the advantage of being able to process any types of metal/ceramic mineral, low-grade-ore minerals etc. to make value-added products and also to deposit ceramics, metals and a combination of these to deposit composite coatings with desired microstructure and required properties on a range of substrate materials. The present work focuses on utilization of fly ash mixing with bauxite (ore mineral) for a high valued application. Fly ash with 10 and 20% bauxite addition is used to deposit plasma spray overlay coatings at different power levels (10-20kW) on aluminum and mild steel substrates. Adhesion strength and surface roughness of the coatings are evaluated. Phase composition analysis of the coatings were done using X-ray diffraction analysis. Surface morphology of the coatings was studied using a scanning electron microscope (SEM). Maximum adhesion strength of 4.924 MPa is obtained for the composition fly ash and bauxite (10%), coated on mild steel at 16kW torch power level. The surface roughness (Ra) of the coatings is found to vary between 10.0102 to 17.2341 micron.

Keywords: Fly ash; Bauxite Ore; Plasma Spraying; Adhesion Strength.

1. Introduction

Coal is an indispensable part in electricity generation around the world. As per the World Coal Institute, coal fuelled power plants generates 41% of worldwide electricity. It is estimated that coal fuelling 44% of electricity generation worldwide by 2030 [1]. Huge amounts of fly ash is generated in coal-fired thermal power stations [2]. In the year 2016-17 it is expected that the production of fly-ash increases around 300-400 MT/year .Now a days, in manufacturing of concrete, bricks, cement and tiles fly ash is used. Also it finds a remarkable contribution in mine filling and road embankment [3]. In recent years, industrial wastes are used for making value-added products [4, 5]. Some research have been done to develop plasma spray coatings using these wastes [6, 7]. Also these wastes finds application in wear resistant coatings [8]. Plasma spray technology is one of the most widely used techniques of preparing overlay coating on any surface to improve properties and increase life span of parts [9]. The plasma spray technology can apply various minerals, metals and ceramics to obtain the desired microstructure of the near homogenous composite coatings. While considerable emphasis is being placed to process low-grade ore minerals through plasma spray process, fly ash was found to be a productive substitute for conventional extenders in high performance industrial coatings [10].

Bauxite ore contains alumina, silica, iron oxide and titania as the major oxides. The two very important properties of protective coatings are adhesion and thermal shock resistance. The adhesion strength of the coating strongly determines the performance and quality of protective coatings. Here, in the present work, an attempt has been made to utilize an industrial waste in addition of an ore mineral to examine its usefulness and functionality. The overlay coatings of 10% and 20% bauxite addition with fly-ash on metals are developed using plasma spray technique. The objective of the current work is to use an industrial waste to develop a ceramic coating on metal substrates.

2. Materials and Methods

Fly-ash (FA) and bauxite ore are the major raw materials used in the present research work. Sieve analysis method is used particle size analysis where both materials contains particles in the range of 20 to 100 micron. Raw materials of three different compositions are homogeneously mixed prior to plasma spraying and are; FA, FA + 10% bauxite and FA + 20% bauxite. The detailed chemical compositions of two raw materials as mentioned above are listed in table 1 given below.

Table 1 Chemical analysis of raw materials.

Wt. %

Raw material	Compounds (Oxides)					LOI
	SiO_2	Al_2O_3	Fe_2O_3	TiO_2	CaO	
FA	61.2	31.5	2.6	1.25	2.75	0.7
Bauxite Ore	4.0	43.0	23.0	3.0	3.0	24.0

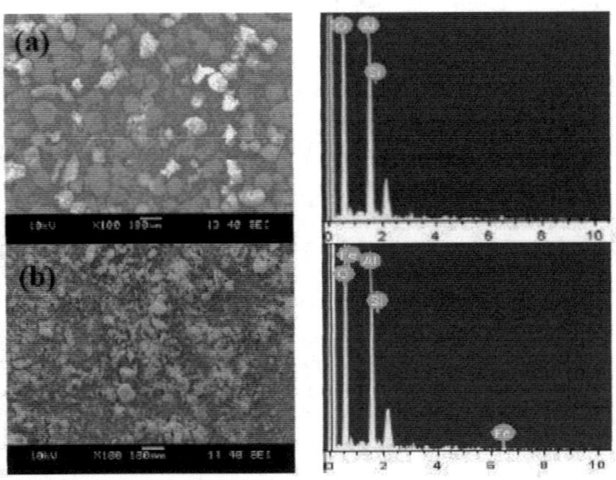

Element	Wt. %	At. %
O K	56.45	69.16
Al K	15.48	11.24
Si K	28.07	19.59

Element	Wt. %	At. %
O K	52.84	68.95
Al K	32.30	24.99
Si K	1.37	1.02
Fe L	13.50	5.05

Figure.1 Scanning Electron images of a) FA and b) Bauxite ore with their respective EDS spectrums.

SEM images and EDS analysis of raw fly ash and bauxite samples are given in figure 1. Both materials contain oxides of silicon and aluminium but significant amount of iron oxide is found to be present in bauxite.

Mild steel and aluminium plates of size 25.4mm x 25.4 x 3mm are taken as the substrate material. Sand blasting was adopted to obtain a minimum surface roughness value of ~3.5 micron Ra. 10, 12, 16 and 20 kW DC power level was maintained for plasma spraying with a non-transferred arc. Argon is chosen for both carrier as well as plasma forming gas where flow rates of 30 and 6 LPM are maintained for plasma forming and carrier gases respectively. Apart from this nitrogen gas of flow rate 2LPM are included as secondary carrier gas for enhancement of flame enthalpy.

The input feeding of powder was maintained at a rate of 10gm/min. A distance of 80mm is maintained from plasma torch to substrate. The surface morphology and phase identification studies were done by using JEOL (JSM-6480LV) scanning electron microscope and BRUKER D8 Advance type diffractometer with a Co-Ká radiation (1.79285 Å) respectively. A jig is fabricated to evaluate the adhesion strength of coating. Cylindrical dummy samples with 13 mm top diameter was prepared for the same purpose. The adhesive used for the test was Epoxy 900-C.Coating pull out method in INSTRON machine was adopted to evaluate the adhesion strength. The surface roughness was measured with Stylus Surface Profilometer.

3. Results and discussion

Further all coating surfaces are observed in scanning electron microscope and are shown in the figure 2 where figures are in increasing order of power level i.e. coated with application of 10 to 20 KW.

Figure 2 SEM images of FA + bauxite 10% coated samples (a-10kW, b-12kW, c-16kW, d-20kW).

With increase in power level the densification rate is increased which can be clearly seen from the above micrographs. Highest flattening of deposited coatings is observed at 16 kW. In figure 2^{d} the lesser flattening is observed as that of 16kW coated surface; may be due to the faster quenching rate as occurs in plasma spraying. The temperature difference in the axial zone of plasma jet and flame tip is the key reason behind the rapid cooling rate; which in turn resulted in deposition of small fragmented particles at some regions [4].

Further X-Ray analysis was performed for proper understanding of spraying kinetics and are the diffractograms thus obtained are shown in figures 3 and 4. The presence of Quartz, Mullite, Iron Oxide (Fe_2O_3) and Corundum phases are detected in the XRD pattern of raw FA. After spraying transformation of quartz has into tridymite (the high temperature polymer) and Iron Oxide (Fe_2O_3) to Iron Oxide (Fe_3O_4) is observed. Also, with increase in power level the intensities of tridymite, mullite and Iron Oxide (Fe_3O_4) phases are increased as evident from the XRD patterns shown in figure 3.

The XRD patterns of raw composition (fly ash 80 + bauxite 20%) as well as coatings at different power levels viz. 10 to 20 kW are shown in the figure 4. With bauxite addition to fly ash, along with Quartz,

Mullite, Iron Oxide (Fe2O3) and Corundum, presence of gibbsite phase is also found. From figure 4 it observed that, with 20% addition of bauxite to fly ash, the intensities of mullite and corundum phases increases with increase in power level.

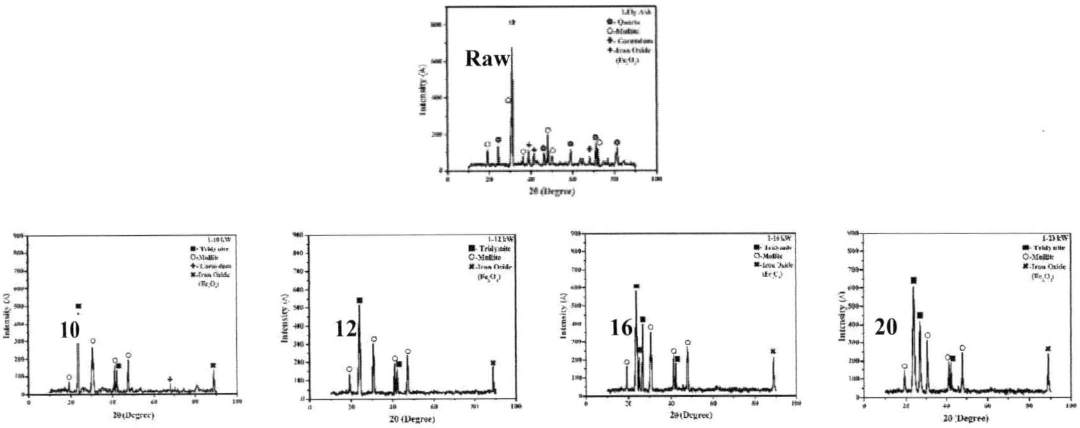

Figure.3 XRD patterns of Raw FA and FA coatings made 10, 12, 16 and 20 kW.

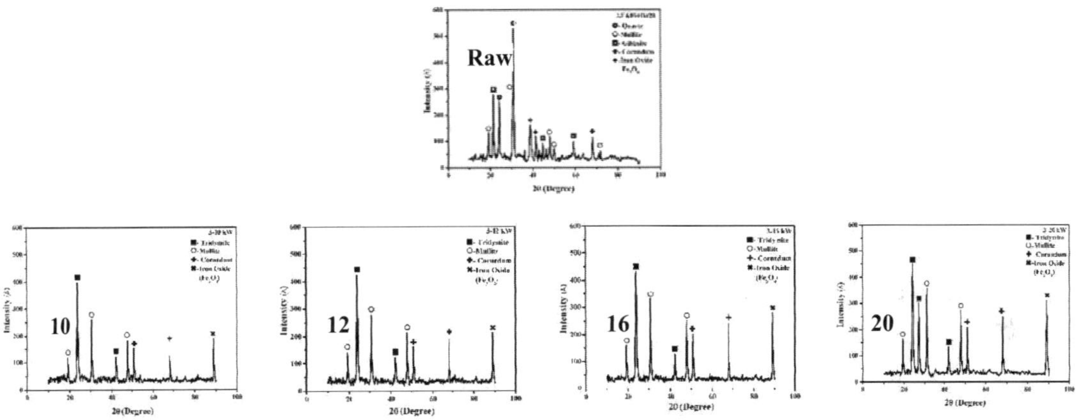

Figure. 4 XRD patterns of Raw (FA + bauxite 20%) and coatings 10, 12, 16 and 20kW.

The adhesion strength of the coatings were evaluated with implementation of coating pull out technique following ASTM C-633 standard. While performing the test, it is observed that at the interface of coating and substrate gets fractured in case of all coated samples. Figure 5 represents the plot showing the variation in coating adhesion strength w.r.t operating power level. In both the cases the adhesion strength shows an increasing trend up to 16kW power level and after that the bond strength decreases i.e. at 20kW. This may be due to the reason that the melting of raw/fused particles is favored by high temperature plasma flame, but at upper power input the plasma jet velocity is too high with the elevated temperature which causes/enhances particle fragmentation [9]. Maximum adhesion strength values of 4.924 MPa is achieved for the fly ash 90+ bauxite 10% coatings on Mild steel substrate at 16kW. This may be due to the reason that aluminium has higher thermal conductivity than mild steel substrate. For this reason in mild steel substrate the molten/semi molten particles get more heat at the interface of the coating and substrate resulting a better interlocking [10].

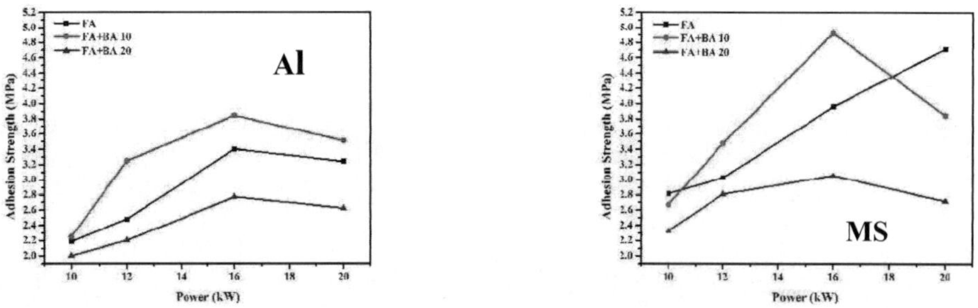

Figure. 5 Adhesion Strength Comparison between FA, FA+ Bauxite 10% and FA + Bauxite 20% coatings made at various power levels for Al and MS substrate.

The variation of surface roughness for various compositions w.r.t operating power levels is shown below in figure. 6.

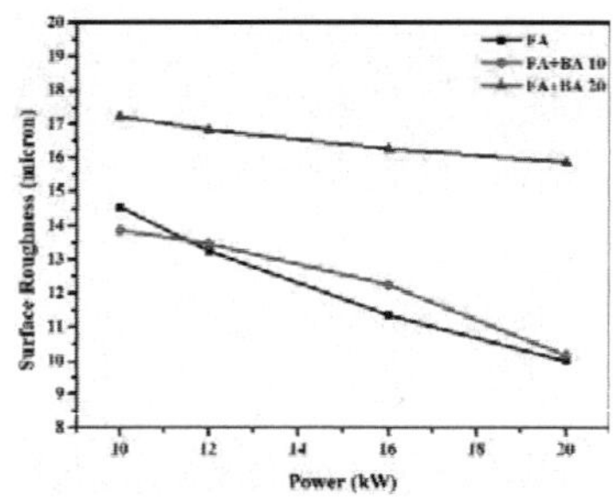

Figure. 6 Surface Roughness Comparison between FA, FA+ Bauxite 10% and FA + Bauxite 20% coatings at different operating power levels.

In figure 6, it is clearly observed that surface roughness of the each composition decrease with rise in torch power. Surface roughness value is minimum i.e. 10.01 micron (Ra) for fly ash coating at 20kW power level and maximum in fly ash 80 + bauxite 20% coatings i.e.17.23 micron (Ra) at 10kW. It may be because of; with increase in power level, high temperature-high enthalpy plasma flame favors the softening/melting of the particle, creates more homogenous distribution of sprayed particles from molten state.

Conclusions

1. Efficient melting of particles occurs at 16kW power level as seen from SEM images and proper bonding/melting etc. is restricted at higher power level.
2. The intensities of tridymite, mullite and iron oxide (Fe_3O_4) peaks increase with increased operating power level. When Bauxite is added to fly ash, it also favours the increase in formation of corundum phase (i.e. alpha alumina).
3. The adhesion strength shows an increasing trend up to 16kW power level and after that the bond strength decreases except for fly ash coating, on MS substrate. This may be due to uniform melting of charge composition (Fly ash) and increase in power level favoring the melting/bonding of particles with MS substrate.
4. Minimum surface roughness value of 10.01 micron (Ra) is obtained for fly ash coating (without bauxite) at 20kW.

References

[1] A. A. Cha´vez-Valdez , A. Arizmendi-Morquecho , G. Vargas , J.M. Almanza ,J. Alvarez-Quintana , 2011 *Ultra-low thermal conductivity thermal barrier coatings from recycled fly-ash cenospheres* **59** 2556-2562.
[2] Openshaw SC 1992 *Utilization of coal fly ash Ph.D. Thesis University of Florida.*
[3] Haque Md 2013*Waste Resources* **3** 22-25.
[4] Mishra S C, Mishra P C, Rout K Cand Anantapadmanabahn P V 1999 *Fly ash utilization for value added products NML. Jamshedpur* 131-135.
[5] Sathpathy A 2005 *Thermal Spray Coating of Red Mud on Metals PhD Thesis, NIT Rourkela, India.*
[6] Sathpathy A and Mishra S C 2003 *National Conference on Materials and Related Technology, TIET.*
[7] Mishra, S. C., Das, S., Satapathy, A., Sarkar, S., Ananthapadmanabhan, P. V., and Sreekumar, K. P. (2009), “Investigation on Composite Coating of Low Grade Minerals,” *Journal of Reinforced Plastics and Composites* **24** 3061-3067.
[8] Mishra, S.C., Rout, K.C., Padmanabhan, P.V.A. & Mills, B. 2000 Plasma spray coating of fly ash pre-mixed with aluminium powder deposited on metal substrates. *Journal of Materials Processing Technology* **102** 9–13.
[9] Satapathy A, Mishra SC, Mohanty U, Mishra TK, Anantha padmanabhan P V and Sreekumar K P 2004 *Proceedings of the International Conference on Industrial Tribology, Tribological Society of India, Mumbai.*
[10] Mishra S C, Praharaj S and Satapathy A 2009 *Journal of Manufacturing Engineering* **4** 241-246.

Effect of carbonization temperatures on biochar formation of bamboo leaves

D Pattnaik[*1]**, S Kumar**[1]**, S K Bhuyan**[1] **and S C Mishra**[1]
[1]Metallurgical & Materials Engineering Department, N.I.T. Rourkela.

Email: pattnaikdeepankar@gmail.com

Abstract
Bamboo is a typical plant native in Asia, been used in many sectors, which also produces a large volume of leaves which goes waste and not find its application for any useful purposes; is often considered as a bio-waste and normally incinerated or dumped; as its applications are not yet fully explored. However, some research work done on bamboo fibers for use as a reinforcement in making polymer matrix composite. In the present piece of research work, the influence of burning/carbonization of bamboo leaves (at different temperatures) have been studied and characterized. Proximate analysis gave the fixed carbon content (of ~nearly21%). X-Ray diffraction results revealed the presence of various phases viz. cristobalite (SiO_2), Calcite (Ca_2O_3) etc. accompanied with changes in crystal structures. Fourier transform infrared spectroscopy results showed various modes of vibrations viz. O-H stretching bending of other bonds; (for aromatic benzene derivatives) etc. Scanning Electron Microscopic observation (of morphology) showed irregular stacking arrangements between the randomly spaced lamellae structure, with variation in carbonizing temperature. Results revealed the advantages of pyrolysis process in biochar production/formation. It appears that, the bamboo biochar can have suitable properties for its use as an alternative energy source and also for agricultural applications. Its high porosity and carbon content suggest its application as activated carbon also; after physical or chemical treatments. The present research focuses on extending the frontiers of use of bamboo leaves from being an unutilized biowaste to its conversion into a value added product, which can be compassed in terms of sustainable applications.

Keywords: Bamboo leaf, Biochar, Carbonization, FTIR analysis, Morphology

1. Introduction

Biochar is a stable solid containing carbon and can last in soil for thousands of years. The carbon in biochar resists degradation and can hold carbon in soils for hundreds to thousands of years. Biochar is produced through gasification or pyrolysis — processes that heat biomass in the absence of oxygen. During pyrolysis process micro pores are developed, so biochar has a large microscopic surface area and can be used as a soil amendment, nutrients retention, adsorption of pollutant and ion exchange capacity **[1]**. Bamboo biomass energy can possibly be an option for non-renewable energy source. The thermal or biochemical conversion of bamboo biomass to produce energy products like syngas, charcoal and biofuels can be substitutions for existing fossil fuel products. Cellulose, hemicelluloses, lignin and extractives are the major chemical constituents of bamboo **[2]**. Bamboo fibers are used to make paper, textiles and board. Bamboo has many desirable fuel characteristic such as low alkali index and ash content. The moisture content in bamboo is relatively low in comparison to other type of plant. Thus bamboo leaf pyrolysis or carbonization can produce certain genuine products like solid-biochar, hydrocarbon gases, etc. which can be potentially used for either environmental advantages or technological applications depending upon suitability.

In the present study, emphasis has been given for analysis of bamboo leaves carbonized at different temperatures (ranging from 250°C to 900°C). Carbonization or pyrolysis of bamboo leaves was done to compare the properties of biochar obtained with the raw bamboo leaves. Proximate analysis was done to know the fixed carbon content. X-Ray diffraction (XRD), Fourier Transform Infrared Spectroscopy

(FTIR), Scanning Electron Microscopy (SEM) and Energy Dispersive Spectroscopy (EDS) was performed to know the various phases, functional groups as well as morphological details of the biochar. The distribution of phases as well as resulting end-products as an outcome of carbonization was important in order to determine the potential of the biochar to be used in various suitable applications.

2. Materials and Methods

2.1. Raw materials:

Bamboo leaves biowaste was collected and washed with water. The bamboo leaves obtained was then sun-dried for a week in order to get rid of the moisture. Using planetary ball mill at an rpm of 300 for 30 minutes the leaves was ground to fine powder. After sometime the fine powder from the ball mill was taken out and was kept in air tight pouch to prevent any contaminants.

2.2. Proximate analysis of raw material:

Proximate analysis of the sample was done to know the fixed carbon content. To determine the moisture present 2(g) of sample in a crucible was placed at 110°C in hot air oven for an hour. For volatile content same amount of bamboo leaves powder was taken in a closed crucible and was kept in a muffle furnace at 900°C for 10 minutes. To calculate the ash content, an open crucible containing 2(g) of sample was placed at 775 °C for an hour in the muffle furnace. Finally the fixed carbon content percent was evaluated to know about the correctness of the sample to be carbonized.

2.3. Carbonization experiment of raw material:

Pyrolysis of bamboo leaves sample at temperatures 250°C, 350°C, 450°C, 600°C and 900°C were done in a furnace with inert atmosphere. Once the desired pyrolysis temperature was attained in a furnace, 5g of the sample was placed isothermally for 2 hours followed by furnace cooling. Next day the sample was taken out of the furnace and weight loss was measured. All the carbonized samples were kept in air tight pouch to prevent oxidation and moisture contamination. Five carbonized and a raw sample were taken for further characterization.

2.4. Characterization:

The Scanning Electron Microscope with an Energy dispersive X-ray spectroscopy analysis was done to study the microstructure and elemental composition. Wide angle X-Ray diffraction (WAXRD) was done in the range of 20 – 100° with step size of 0.02° and a scan rate of 10 °/min. Fourier transform infrared spectra (FTIR) using Perkin Elmer Spectrum Two was done to determine the chemical functional groups by subjecting the disc shaped pellets formed to IR radiations.

3. Results and discussion

3.1. Proximate analysis of bamboo biomass:

Proximate analysis is done to know the percentage of volatile matter, moisture content, ash content and fixed carbon present in a particular organic material. *T. Rout et al. (2016)* **[3]** stated that favorable conditions for bio-oil production during pyrolysis are: high volatile matter, low content of moisture and ash. In the present study, the moisture and ash content was very low (given in Table 1). All of these values are included in the range of values compiled by Scurlock et-at. (2000) **[4]**.

Bamboo biomass	Wt.%
Moisture content	7
Volatile matter	57
Ash content	15
Fixed carbon	21

Table 1. Proximate analysis of the raw material

The volatile matter in bamboo biomass is more than 55% enabling a stable flame of the biomass. Biomass fuels contain more volatile components and are more reactive than coal. Due to high volatile matter the combustion of biomass is rapid and tough to control. However the fixed carbon content was found to be 21%.

3.2. Characterization and analysis of bamboo biomass and its biochars:

3.2.1. Micro-structural analysis

Scanning Electron Microscope analysis showed the morphology of the six samples prepared. The raw bamboo leaf exhibited a cross-linked structure as marked in fig 1. The cross linking reveals the presence of a network containing lignin, cellulose and hemicellulose. In raw sample few pores were seen when compared to its biochars. From fig.4 it was observed that the biochars above 450°C had rougher surface than the raw bamboo biomass. Porous structures were formed on the surface. The biochar was observed to contain more porous structure than the original bamboo biomass samples.

Fig. 1) Raw bamboo leaves.

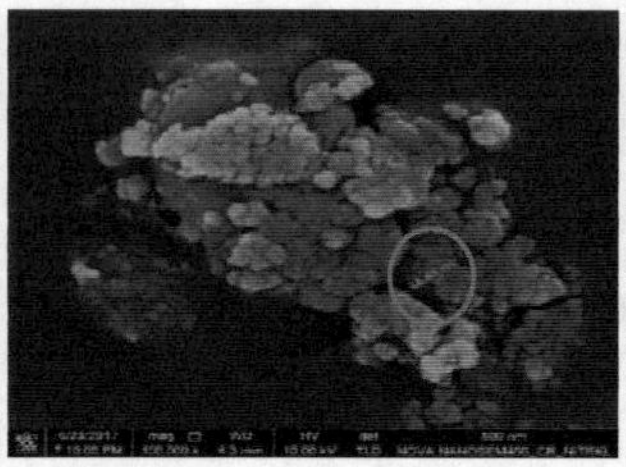

Fig. 2) Biochar pyrolysed at 250°C.

Fig. 3) Biochar pyrolysed at 350°C.

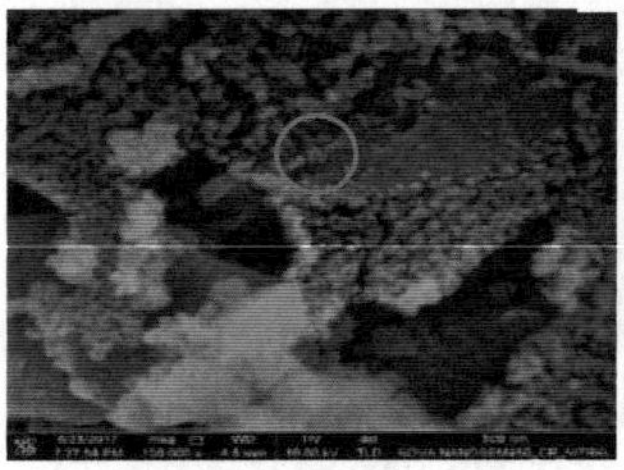

Fig. 4) Biochar pyrolysed at 450°C.

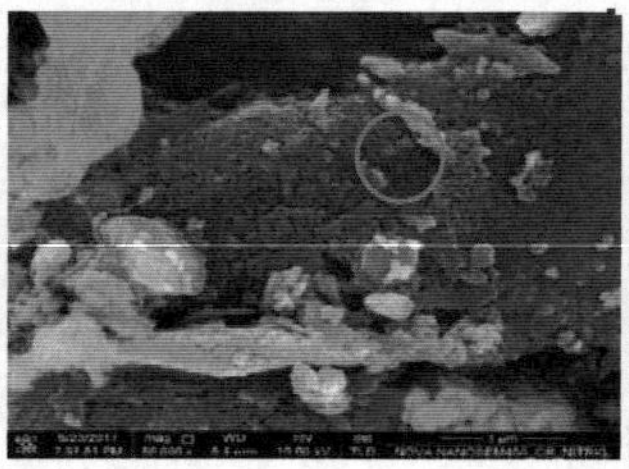

Fig. 5) Biochar pyrolysed at 600°C.

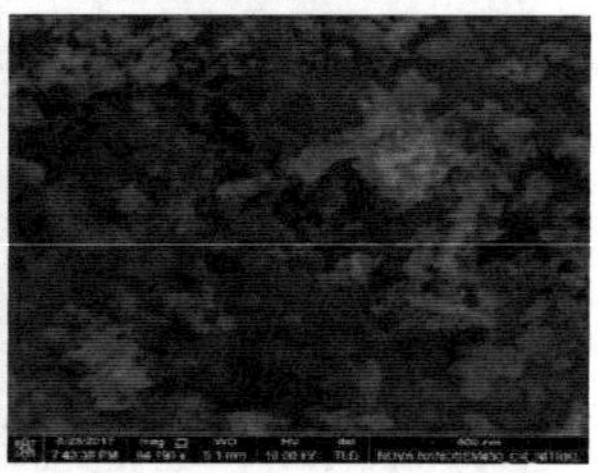

Fig. 6) Biochar pyrolysed at 900°C.

Carbonization at lower temperature produces large quantity of char but due to thermal breakdown of hemicellulose and cellulose at 450°C, ample pore structures are observed. Above figures when compared the bulk density of raw bamboo sample is more and particle density is less than that of its biochars. Large pores present in bamboo biochar can provide an environment for microorganisms thereby increasing the soil quality (Thies and Rilling, 2009) **[5]**. At 900°C the biomass burns completely as revealed by SEM image as shown in fig. 6.

3.2.2. Elemental analysis

The composition of elements present in bamboo biomass and its biochars was determined using EDS analysis. The Fig. 7 shows the variation amount of different elements with pyrolysing temperature for raw bamboo leaves and the biochar.

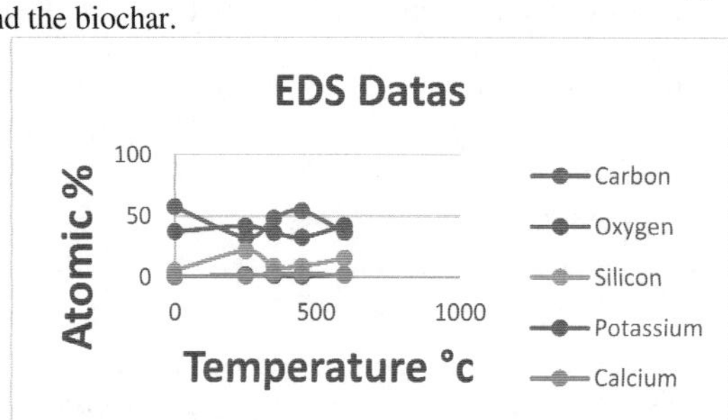

Fig. 7) Variation of amount of different element with paralyzing temperature for bamboo leaves.

Sharp peaks of carbon (C), oxygen (O), silicon (Si), calcium (Ca) and potassium (K) is seen in the carbonized bamboo leaves samples. Peaks of other elements like chlorine (Cl), nitrogen (N), magnesium (Mg) and aluminium (Al) is also seen in some carbonized samples.

3.2.3. X-Ray Diffraction of Bamboo leaves and its biochars

XRD analysis can potentially determine the phases and planes of components present within a material. A comparison has been made for raw bamboo leaves and all other biochars to know the different between the original bio residue and its derivatives.

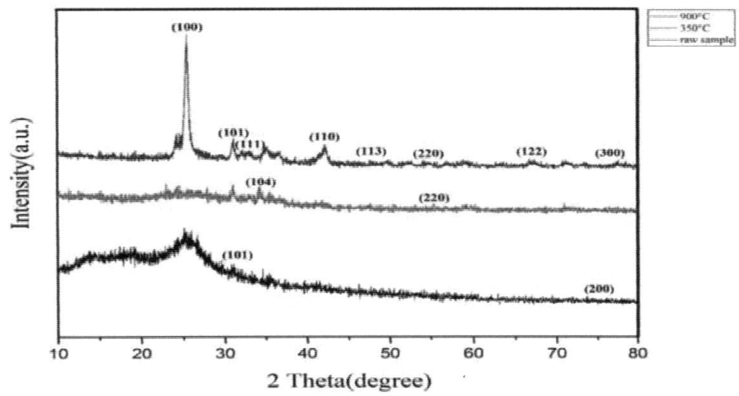

Fig. 8) XRD pattern for raw bamboo leaves, biochar at 350°C and biochar at 900°C.

Peak number	Peak position(°)	Mineral name	Formula	(hkl)	Structure
1.	25.447	Silicon oxide	SiO_2	100	Hexagonal
2.	31.040	Silicon oxide	SiO_2	101	Hexagonal
3.	33.051	Silicon	Si	111	Cubic
4.	42.281	Silicon oxide	SiO_2	110	Hexagonal
5.	45.642	Calcium Carbonate	Ca_2O_3	113	Rhombohedral
6.	54.654	Calcium Fluoride	CaF_2	220	Cubic
7.	67.157	Calcium Carbonate	Ca_2O_3	122	Rhombohedral
8.	75.957	Iron oxide	Fe2O3	300	Rhombohedral

Table 2. The peaks for biochar at 900°C are shown.

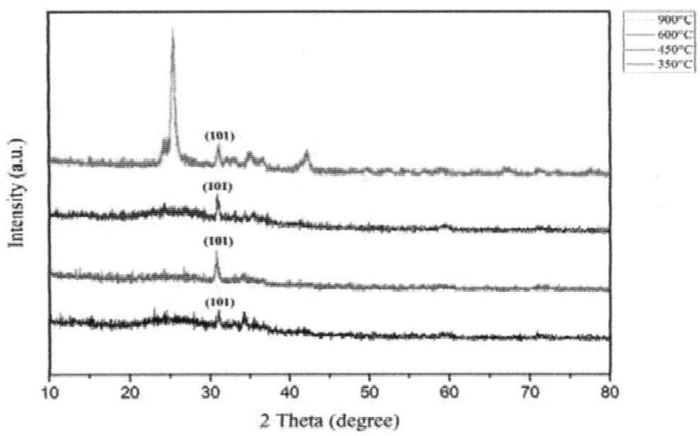

Fig. 9) XRD pattern for bamboo biochars.

As seen from the figure 8, absence of well-defined diffraction peaks in case of raw bamboo leaves, except for a weak peak at ~31°; it is worth mentioning it to be an amorphous or non-crystalline material as stated by Macedo et al. **[6]**. However a single weak peak at ~31° was detected and ascribed to the (101) plane of hexagonal silicon oxide. The amorphous nature of raw bamboo leaves arises due to the presence of hemicellulose, lignin and other similar components.

As shown in fig. 9 a peak at ~ 31° show the presence of silicon oxide (SiO_2) ascribed to the (101) plane for all the biochars. The results revealed the presence of various phases viz. cristobalite (SiO_2), Calcite (Ca_2O_3) etc. accompanied with changes in crystal structures.

3.2.4. FTIR analysis
The functional groups present in bamboo biomass and its biochar was determined by FITR analysis. Fig 10 is the FTIR plot for bamboo biomass before and after treatment, thereby revealing the function groups present. Table 3 shows the characteristic bands of raw bamboo leaves and their assignment to different functional groups.

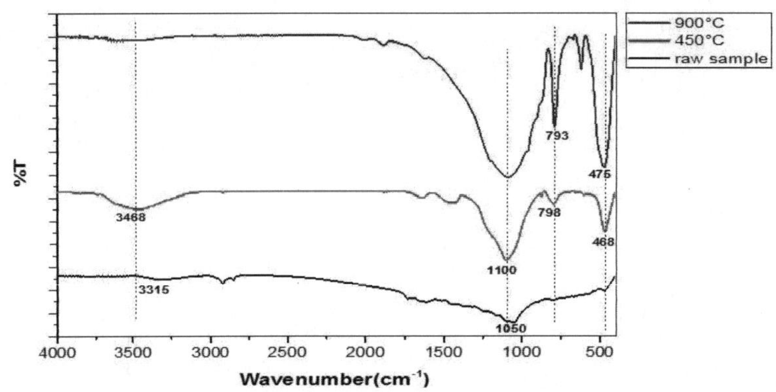

Fig. 10) FTIR pattern for raw bamboo leaves and biochars carbonized at 450°C and 900°C.

Stretching Frequency	Functional Group
3650-3200	Broad and strong, O-H stretch from hydroxyl group of alcohols
1700-1500	Broad, amine N-H / carbonyl group C=O
1100-1000	C-O-C stretch of the ethers present in lignin Or Si-O-Si stretch
800-700	C=O absorption/ Si-O-Si bond bending vibrations
700-600	Out-of-plane ring deformation or weak vibration $-CH_2$ rocking
500-400	rings in benzene derivatives / Si-O-Si out-of-plane rocking motion

Table 3. Band assignment to the peaks of raw bamboo leaves

In figure 10, a broad peak noticed between 3500 and 3300 cm^{-1} was attributed to the O-H stretch from the hydroxyl groups of the alcohols present. The extremely weak peaks between 3500 and 3800 cm^{-1} in biochar 900ºC could be assigned to OH stretching which may have occurred due to retention of some hydroxides as well as water in the sample, although mineral based Si-OH in this region could be expected as well. The decomposition of hemicellulose during carbonization was indicated by the disappearance of the carbonyl peak at 1635cm^{-1} with increased treatment temperature. *Chia et al. 2012* **[7]** stated that the main chemical processes which occur during the pyrolysis of lignocellulosic substances are depolymerisation and dehydration resulting in the formation of C=C bonds, carbonyl and carboxylic functional groups. In a fingerprint region a prominent peak was seen between 1100 – 1000 cm^{-1} can be attributed to C-O-C stretch of the ethers present in lignin **[8]** Or Si-O-Si stretch. C=O absorption or Si-O-Si bond bending vibrations around 798 cm^{-1} was observed in all biochars. The weak mode of vibration in between 700 and 600 cm^{-1} was –CH_2 rocking or out-of-plane ring deformation. Also with increase in temperature the aromaticity of biochars increases.

4. Conclusions

The present study dealing with raw bamboo leaves and the biochar obtained from it at different temperatures and conditions, revealed the effects of carbonization temperature leading to a conclusion that the type of carbon found in it was isotropic which does not fuse at higher temperatures. From the proximate analysis we found the fixed carbon content in the raw bamboo leaves to be 21%. High carbon content in biochar was revealed on characterisation. The carbon content is affected with increase in pyrolysis temperature. Variation was also noticed for elements like silicon, oxygen, magnesium, nitrogen etc. with carbonization temperature. Formation of new bond and change in bond structure with increase in temperature was observed. Also formation of new phases in biochars was revealed from the XRD observation. FTIR analysis verified the presence of Si-O-Si stretch and the Si-C peaks in biochars. Also with increase in temperature the aromaticity of biochars increases. The study showed that a biowaste can be made valuable thereby decreasing environmental pollution. Thus the present work can be helpful in the determination of some high-class biochar properties for further use in the near future.

References

[1] Laidy E. Hernandez-Mena , Arai A. B. Pécoraa and Antonio L. Beraldo 2014 Slow pyrolysis of bamboo biomass: Analysis of biochar properties **37**.

[2] Gates B. C., Huber G. W., Marshall C. L., Ross P. N., Siirola J., Wang Y. 2008 Catalysts for emerging energy applications. *MRS Bulletin* **33(4)** 429–435

[3] Rout, T., Pradhan, D., Singh, R. K., & Kumari, N. (2016). Exhaustive study of products obtained from coconut shell pyrolysis. *Journal of Environmental Chemical Engineering* **4** 3696-3705.

[4] Scurlock, J.M.O., Dayton, D.C. & Hames, B. 2000. Bamboo: an overlooked biomass resource? Biomass & Bioenergy **19** 229-44.

[5] Thies J. E., Rillig M. C., 2009, Characteristics of biochar: biological properties. In: J. Lehmann, S. Joseph (Eds.), Biochar for Environmental Management. Gateshead, UK, Earthscan, 85–105

[6] De Souza Macedo, J., da Costa Júnior, N. B., Almeida, L. E., da Silva Vieira, E. F., Cestari, A. R., de Fátima Gimenez, I., ... & Barreto, L. S. 2006 Kinetic and calorimetric study of the adsorption of dyes on mesoporous activated carbon prepared from coconut coir dust. *Journal of colloid and interface science*, **298(2)** 515-522.

[7] Chia, C. H., Gong, B., Joseph, S. D., Marjo, C. E., Munroe, P., & Rich, A. M. (2012). Imaging of mineral-enriched biochar by FTIR, Raman and SEM–EDX. *Vibrational Spectroscopy* **62** 248-257.

[8] Garcia-Perez, M., & Metcalf, J. 2008 The formation of polyaromatic hydrocarbons and dioxins during pyrolysis: A review of the literature with descriptions of biomass composition, fast pyrolysis technologies and thermochemical reactions.

IOP Conf. Series: Materials Science and Engineering **338** (2018) 012055 doi:10.1088/1757-899X/338/1/012055

Effect of reduced graphene oxide-carbon nanotubes hybrid nanofillers in mechanical properties of polymer nanocomposites

Kadambinee Sa, Prakash C. Mahakul, B.V.R.S. Subramanyam, Jagatpati Raiguru, Sonali Das, Injamul Alam, Pitamber Mahanandia*

Department of Physics & Astronomy,
National Institute of Technology, Rourkela, Odisha, India

Corresponding author Email : pitam@ nitrkl.ac.in

Abstract: Graphene and carbon nanotubes (CNTs) have tremendous interest as reinforcing fillers due to their excellent physical properties. However, their reinforcing effect in polymer matrix is limited due to agglomeration of graphene and CNTs within the polymer matrix. Mechanical properties by the admixture of reduced graphene oxide (rGO) and CNTs in Poly (methyl methacrylate) (PMMA) prepared by solution mixing method has been investigated. The prepared samples are characterized using X-ray diffraction (XRD), scanning electron microscope (SEM), transmission electron microscope (TEM) and Raman spectroscopy. The hybrid composite shows improvement in the mechanical properties compared to rGO/PMMA and MWCNTs/PMMA composites due to better interaction between rGO-MWCNTs and polymer matrix.

1. Introduction

Two dimensional (2D) graphene and one dimensional (1D) carbon nanotubes (CNTs) exhibit exceptional physical properties and are being used for different applications [1,2]. Such properties make them ideal nanofillers for polymer composites with enhanced electrical, mechanical and thermal properties for various potential applications [3,4]. However, poor dispersion of graphene and CNTs in polymer matrices hinders the potential applications of graphene and CNTs polymer composites. In order to fully realize the reinforcing effect of graphene and CNTs in polymer matrix, two critical issues have to be resolved (i) dispersion of graphene and CNTs, (ii) interfacial adhesion between fillers (graphene or CNTs) and polymer matrix [5]. Due to large surface area of graphene and CNTs, they possess large van der Waals forces which makes very difficult to be effectively dispersed in polymer matrix. Several efforts have been taken in order to obtain homogeneous dispersion of CNTs/graphene in the polymer matrix and better interfacial interaction between nanofillers and matrix [6-8]. Graphene-CNTs hybrid filler polymer composites are now developing as a new hybrid nanofiller material with better properties.

Polymer composites containing hybrid carbon nanofillers with enhanced electrical, mechanical and thermal properties have been reported by different groups. Chatterjee et al. [9] demonstrated that graphene nanoplatelets/CNTs epoxy composites shows better mechanical properties than single filler systems. Graphene oxide (GO)–CNT/ Polyvinyl alcohol (PVA) composites show superior mechanical properties compared to GO/PVA and CNTs/PVA composite due to the complete exfoliation of GO and CNT via good dispersion [10]. Enhancement in the electrical, flexural and fracture properties are observed by Kim et al. [11] in the multiwalled carbon nanotubes (MWCNTs)-graphene nanoplatelets/epoxy composite. Kumar et al. [12] reported that polyetherimide (PEI)/(graphitic

nanoplatelets (GNP)/functionalized MWCNTs) composite shows enhancement in electrical conductivity, thermal conductivity and dynamic mechanical properties than PEI/GNP and PEI/MWCNTs composites at same filler loadings . Yang et al. [5] demonstrated improvement of thermal conductivity and tensile strength of MWCNTs/multi-graphene platelets/epoxy composite. Zhang et al. [13] reported that increase in mechanical and thermal properties is observed in PVA/reduced graphene oxide (rGO)-treated CNT (t-CNT) composite as compared to PVA/t-CNT and PVA/rGO composites. Patole et al. [14] reported that admixture of two types of carbon fillers provide better improvement in thermal and mechanical properties compared to the neat polymer.
In this study effect of rGO-MWCNTs hybrid nanofiller on the mechanical properties of polymer composites have been investigated. The mechanical properties of rGO-MWCNTs/polymer hybrid composite has been compared with MWCNTs/polymer and rGO/polymer composite.

2. Materials and methods

Commercially procured MWCNTs and graphite powder (GFG 50), sodium nitrate ($NaNO_3$), sulphuric acid (H_2SO_4), potassium permanganate ($KMnO_4$), hydrogen peroxide (H_2O_2), hydrazine hydrate, N-N Dimethyl formamide (DMF), Poly(methyl methacrylate) (PMMA) have been used in this work.
Graphite oxide (GO) has been prepared by Modified Hummers method [15]. In order to prepare GO, 1.0 g of graphite powder and 0.5 g of $NaNO_3$ have been mixed together and then 25 mL of concentrated H_2SO_4 has been added to that mixture. The mixture have been placed in an ice bath in order to decrease its temperature. After stirring for 2 hr, 3 g of $KMnO_4$ has been added slowly and again stirred for another 1 hour continuously. After that the ice bath is removed and the mixture is cooled down to room temperature. The temperature of the mixture is increased after addition of 100 mL of distilled water. Again 300 mL of distilled water is added and again continuously stirred for another one and half hour. At this stage the colour of the mixture changes to mud brown. For the termination of the reaction 10 mL of H_2O_2 has been added to the mixture. For the removal of unwanted ions and acid the prepared GO has been washed using distilled water until it becomes neutral. The material has been recovered by filtering and the obtained material is dried in vacuum at 100 ^{0}C for 3 hours. For the reduction of GO hydrazine hydrate is used [16].
The mostly used methods for the preparation polymer composites are melt mixing [17,18], in-situ polymerization [19] and solution casting [20]. Each and every method has their own advantages and disadvantages. In the present case, we have adopted solution mixing method to prepare the desired rGO-MWCNTs/PMMA hybrid nanocomposite (HC). In this method for the preparation of HC, MWCNTs, rGO and PMMA have been separately dissolved in DMF. The resulting solutions are sonicated and stirring in order to get homogeneous dispersion. After that the three solutions are mixed together and again sonicated followed by stirring. In order to get the hybrid nanocomposites films, the solutions are put in a glass petri dish and dried in an oven. Several batches of HC are prepared by varying rGO wt% (0.05, 0.1, 0.3, 0.5, 1, 3 wt%) keeping MWCNTs wt% constant (0.3 wt%). Similar procedure is adopted for the preparation of MWCNTs/PMMA and rGO/PMMA composites, in order to compare the mechanical properties of HC with MWCNTs/PMMA and rGO/PMMA composites. The as-prepared samples are characterized using Scanning electron microscope (SEM), transmission electron microscope (TEM), X-ray diffraction (XRD), Raman spectroscopy and tensile test using ASTM D 3039.

3. Result and Discussion

The morphology of rGO and CNTs are investigated through SEM. Randomly oriented crumpled sheets of rGO are shown in Figure 1(a) which is due to good exfoliation of rGO sheets. Entangled CNTs are observed from SEM image of CNTs (Figure 1(b)). The HRTEM images of rGO and CNT are shown in Figure 1(c) and (d). From the HRTEM image of rGO, it is observed that the prepared rGO are mostly multilayer in nature. Multilayer nature of CNTs is observed from HRTEM image of CNT.

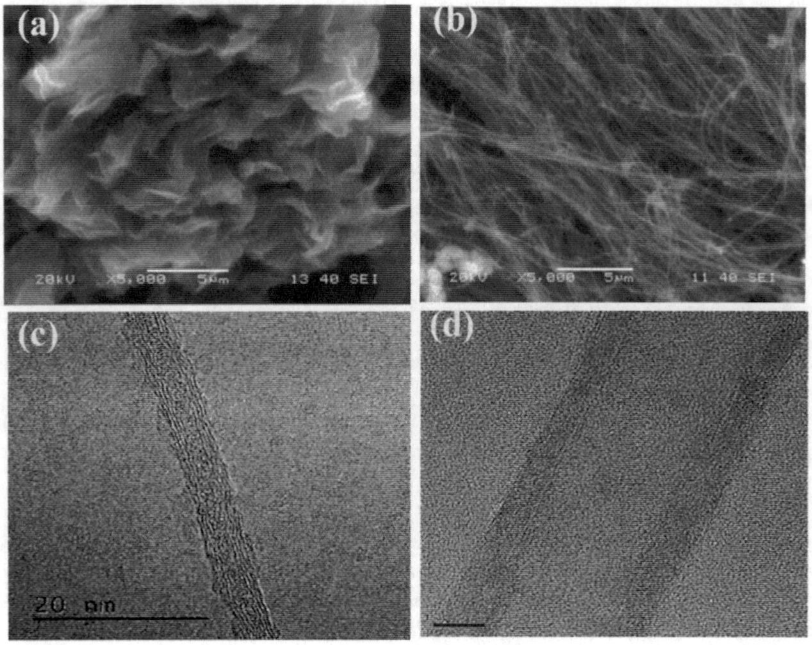

Figure 1 (a), (b) SEM images and (c), (d) HRTEM images of rGO and MWCNTs respectively.

XRD pattern of graphite powder, GO, rGO and MWCNTs are shown in Figure 2 (a). A characteristics peak of graphite is obtained at 26.4^0 with interlayer spacing of 3.34 A^0. In the XRD pattern of GO due to incorporation of oxygen functional groups in graphite the interlayer spacing increases and the peak is shifted to 11.5^0, but after reduction due to removal of oxygen functional groups the interlayer spacing again decreases and peak shifted to 24.7^0. The diffraction peak of MWCNTs is obtained at 26.1^0 which arises from graphitic carbon of MWCNT.

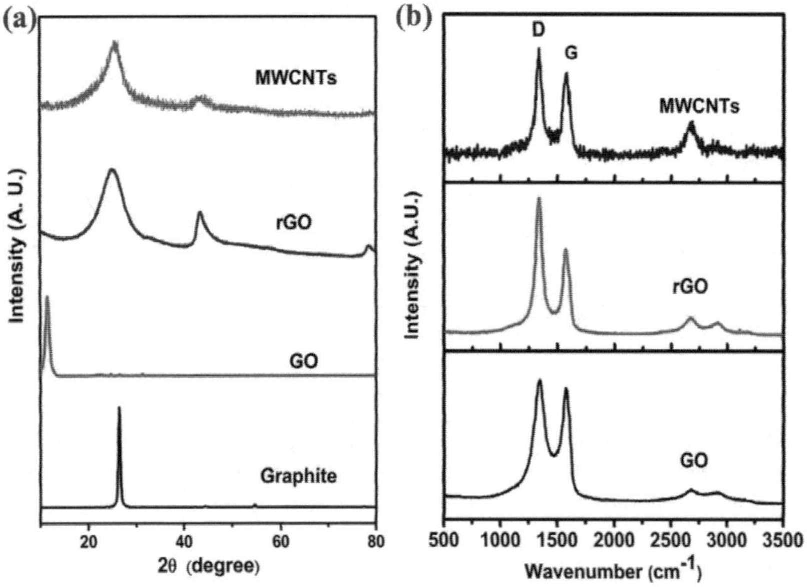

Figure 2 (a) XRD pattern of graphite, GO, rGO and MWCNTs, (b) Raman spectra of GO, rGO and MWCNTs.

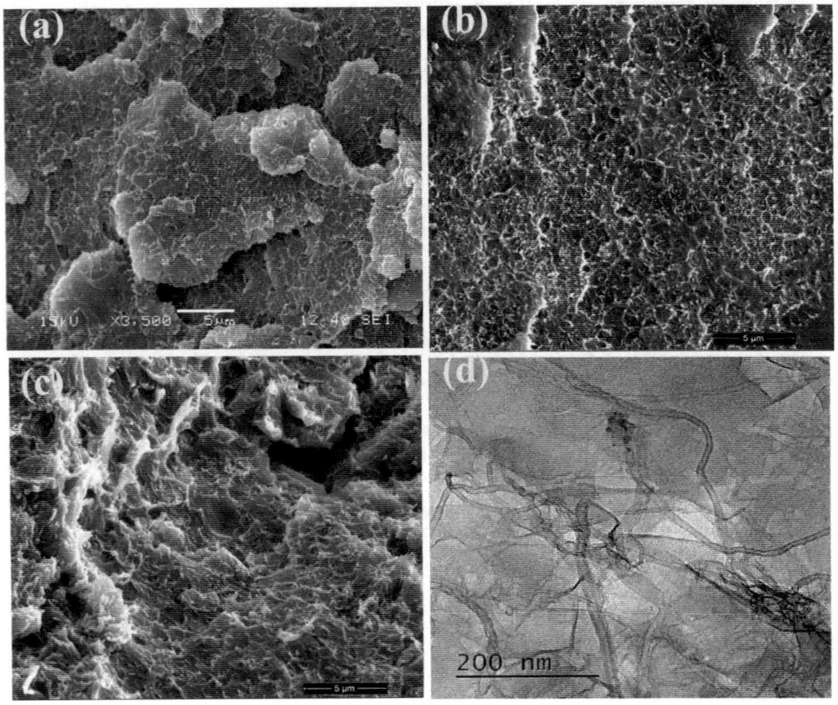

Figure 3 SEM images of as prepared HC with (a) 0.05, (b) 0.3, (c) 3 wt% rGO, (d) TEM image of HC with 0.3 wt% rGO

Raman spectroscopy is generally used to characterize crystal structure, disorder and defects in graphene-based materials. The Raman spectra of GO, rGO and MWCNTs are shown in Figure 2(b). For GO and rGO the D peak which is due to the defect is at 1347 and 1339 cm^{-1} respectively. Since there are no content of amorphous carbon in the GO and rGO, the D peaks are attributed to the defects. The G peaks of GO and rGO are present at 1572 and 1576 cm^{-1} respectively which arises due to crystalline nature of graphitic structure [16]. From the Raman spectrum of MWCNTs is seen that the D and G peaks are at 1339 and 1584 cm^{-1}.

From the SEM images of HC (Figure 3), it is observed that most of the 2D rGO and 1D MWCNTs are buried and exfoliated in the PMMA matrix. From the SEM images it is seen that most of the MWCNTs and rGO are covered by PMMA and some folded rGO sheets are visible. With increasing rGO wt% more folded rGO sheets are visible. Also from SEM and TEM images it is seen that MWCNTs and rGO form a network structure in the polymer matrix which may help for improving the mechanical properties of the resulting HC as compared to MWCNTs/PMMA composite and rGO/PMMA composite. Presence of 1D MWCNTs on the surface of rGO sheets in the polymer matrix are also seen that may reduce the restacking of rGO sheets.

The XRD pattern of PMMA, HC with 0.05, 0.1, 0.3, 0.5, 1 and 3 wt% of rGO are shown in Figure 4(a). It is seen that for PMMA the two broad reflection peaks are obtained at 13^0 and 30^0 which show amorphous character of PMMA. But in case of HC the reflection peaks are only due to PMMA, here XRD peaks due to MWCNT and rGO are not observed probably due to higher PMMA concentration and better exfoliation of MWCNTs and rGO in polymer matrix [21].

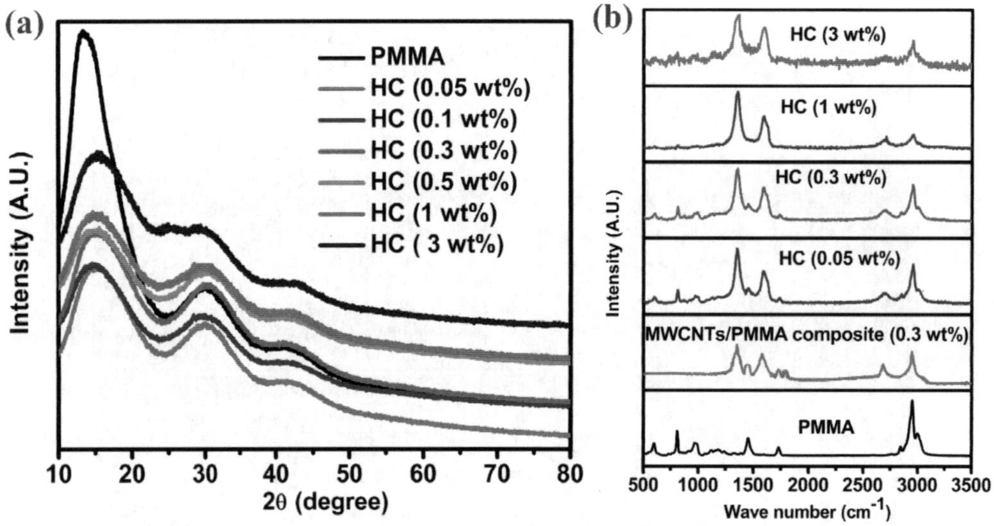

Figure 4 (a) XRD pattern of PMMA and HC with various wt% of rGO, (b) Raman spectra of PMMA, MWCNTs/PMMA composite (0.3 wt%), HC at different wt% rGO.

Raman spectroscopy is also used to study the interaction between polymer matrix and nanofiller. Raman spectra of PMMA, MWCNTs/PMMA composite with 0.3 wt% and HC at 0.05, 0.3, 1 and 3 wt% of rGO is shown in Figure 4(b). Decrease in the Raman peak intensity of PMMA is observed with addition of MWCNTs in PMMA. Again with the addition of rGO in MWCNTs/PMMA shifting in the D and G peaks occurs. Also in the HC, the intensity of the D and G peaks are less than that of rGO and MWCNT which is attributed may be due to the network structure of rGO-MWCNTs within the polymer matrix, grafting of the polymer on the surface of rGO sheets and MWCNTs provide higher defect level. Also with increase in rGO concentration the D and G peaks are shifted indicating increase in interfacial interaction between filler and matrix. Tensile measurements of the composites are done in order to study the mechanical properties. Figure 5(a) shows the tensile stress–strain graphs of the composites. The tensile strength and Young's modulus of PMMA are 21.19 MPa and 1.42 GPa respectively.

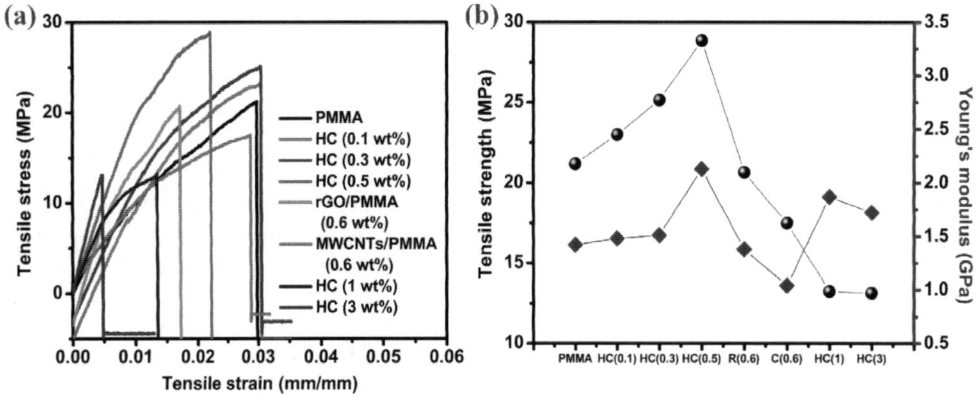

Figure 5 (a) Stress-Strain curves of HC at different wt% of rGO, rGO/PMMA composite with 0.6 wt% (R (0.6)) and MWCNTs/PMMA composite with 0.6 wt% (C (0.6)) (b) Tensile strength and Young's modulus of HC at different rGO wt%.

The tensile strength and Young's modulus of PMMA is increased to 23 MPa and 1.48 GPa with the addition of 0.1 wt% rGO into MWCNTs/PMMA composite with 0.3 wt%. More tensile strength and Young's modulus is observed in HC than MWCNTs/PMMA composite and rGO/PMMA composite at the same total filler wt%. Similarly Young's modulus of HC more compared to MWCNTs/PMMA composite but lower than that of rGO/PMMA composite. These increase in mechanical properties of HC is may be due to the formation of the network structure by rGO and MWCNTs which provide larger contact area for stress transfer at the polymer matrix and the nanofiller interface and better contact between polymer and rGO-MWCNTs hybrid nanofillers [22]. Further increase in rGO wt% in HC the tensile strength and Young's modulus increases and maximum strength and modulus are obtained at 0.5 wt% rGO. With further increase in rGO wt% the tensile strength and Young's modulus of HC decreases which may be due to aggregations of rGO sheets [22]. At higher wt% of rGO, agglomerations of rGO are may be take place which results in improper dispersion of rGO sheets within the polymer matrix which results in decreasing mechanical properties of HC.

4. Conclusion

In this study solution mixing method is adopted to prepare 2D rGO-1D MWCNTs/PMMA hybrid polymer composites. Due to formation of network structure of MWCNTs and rGO in PMMA, it gives more contact area for stress transfer from polymer matrix to the hybrid nanofiller. As a result of which enhancement in the tensile strength and Young's modulus as compared to MWCNTs/PMMA and rGO/PMMA composites have been observed.

References

[1] Zhu Y, Murali S, Cai W, Li X, Suk J W, Potts J R, Ruoff R S 2010 *Adv. Mater* **22** 3906.
[2] Meyyappan M, Carbon Nanotubes: Science and Applications, CRC Press 2005.
[3] Ajayan P M, Tour J M 2007 *Nature* **447** 1066.
[4] Stankovich S, Dikin D A, Dommett G H B, Kohlhaas K M, Zimney E J, Stach E A, Piner R D, Nguyen S T, Ruoff R S 2006 *Nature* **442** 282.
[5] Yang S, Lin W, Huang Y, Tien H, Wang J, Ma C M, Li S, Wang Y 2011 *Carbon* **49** 793.
[6] Ma P, Siddiqui N A, Marom G, Kim J 2010 *Composites: Part A* **41** 1345.
[7] Kim S W, Kim T, Kim Y S, Choi H S, Lim H J, Yang S J, Park C R 2012 *Carbon* **50** 3.
[8] Kim H, Abdala A A, Macosko C W 2010 *Macromolecules* **43** 6515.
[9] Chatterjee S, Nafezarefi F, Schlagenhauf L, Nu¨esch F A, Chu B T T 2012 *Carbon* **50** 5380.
[10] Li Y, Yang T, Yu T, Lianxi Zheng L, Liao K 2011 *J. Mater. Chem* **21** 10844.
[11] Kim K, Choi K, Park S 2009 *Carbon Letters* **10** 335.
[12] Kumar S, Sun L L, Caceres S, Li B, Wood W, Perugini A, Maguire R G, Zhong W H, 2010 *Nanotechnology* **21** 105702.
[13] Zhang C, Huang S, Tjiu W W, Fan W, Liu T 2012 *J. Mater. Chem* **22** 2427.
[14] Patole A S, Patole S P, Jung S Y, Yoo J B, An J H, Kim T H 2012 *Eur. Polym. J* **48** 252.
[15] Hummers W S, Offeman R E 1958 *J. Am. Chem. Soc* **80** 1339.
[16] Stankovich S, Dikin D A, Piner D A, Kohlhaas K A, Kleinhammes A, Jia Y, Wu Y, Nguyen S T, Ruoff R S 2007 *Carbon* **45** 1558.
[17] Jin Z, Pramoda K P, Xu G, Goh S H 2001 *Chem. Phys. Lett* **337** 43.
[18] Haggenmueller R, Gommans H H, Rinzler A G, Fischer J E, Winey K I 2000 *Chem. Phys. Lett* **330** 219.
[19] Jia Z, Wang Z, Xu C, Liang J, Wei B, Wu D, Zhu S 1999 *Mater. Sci. Eng A* **271** 395.
[20] Chapelle M L, Stéphan C, Nguyen T P, Lefrant S, Journet C, Bernier P, Munoz E, Benito A, Maser W K, Martinez M T, de la Fuente G F, Guillard T, Flamant G, Alvarez L, Laplaze D 1999 *Synth. Met* **103** 2510.
[21] Wan Y, Tang L, Gong L, Yan D, Li Y, Wu L, Jiang J, Lai G 2014 *Carbon* **69** 467.
[22] Amr I T, Al-Amer A, Thomas S, Al-Harthi M, Girei S A, Sougrat R, Atieh M A 2011 *Composites: Part B* **42** 1554.

IOP Conf. Series: Materials Science and Engineering **338** (2018) 012056 doi:10.1088/1757-899X/338/1/012056

Experimental investigation of various surface integrity aspects in hard turning of AISI 4340 alloy steel with coated and uncoated cermet

Anshuman Das*, S. K. Patel, Ch. Sateesh Kumar and B. B. Biswal

[a] Research Scholar, Department of Industrial Design, NIT Rourkela, Odisha-769008, India

[b] Associate Professor, Department of Mechanical Engineering, NIT Rourkela, Odisha-769008, India

[c] Research Scholar, Department of Mechanical Engineering, NIT Rourkela, Odisha-769008, India

[d] Professor, Department of Industrial Design, NIT Rourkela, Odisha-769008, India

* Email address of the Corresponding author: anshuman.das2009@gmail.com;

Contact no: +91-9437104639

Abstract

The newer technological developments are exerting immense pressure on domain of production. These fabrication industries are busy finding solutions to reduce the costs of cutting materials, enhance the machined parts quality and testing different materials, which can be made versatile for cutting materials, which are difficult for machining. High-speed machining has been the domain of paramount importance for mechanical engineering. In this study, the variation of surface integrity parameters of hardened AISI 4340 alloy steel was analyzed. The surface integrity parameters like surface roughness, micro hardness, machined surface morphology and white layer of hardened AISI 4340 alloy steel were compared using coated and uncoated cermet inserts under dry cutting condition. From the results, it was deduced that coated insert outperformed uncoated one in terms of different surface integrity characteristics.

Keywords: **-** Hard turning; Surface integrity; Cermet; Micro hardness; White layer

1. Introduction

The constantly changing and competitive era of science and technology forcing manufacturing sector to search innovative solutions to reduce the cutting costs, increase the quality of the machined parts. Machining accomplishes desired size, shape, surface finish, dimensional accuracy and other functional features by gradual material removal as chips. The system of machining comprises of cutting tool, workpiece and machine tool. Hence, it is significant to evaluate workpiece-tool interaction that directly affects overall performance with the help of cutting parameters such as cutting speed (V_c), feed (f) and depth of cut (d) which largely depends on the materialistic properties of both cutting tool and workpiece. Use of conventional cutting fluids is harmful for environment. Therefore, now researches are focussed widely on the advanced strategies for machining of different types of material alloys so that better surface quality can be achieved without having its adverse effects on the environment while maintaining high productivity.

Pereira et al. [1] compared the performances of positive and negative CBN inserts in dry and cryogenic cooling environment during hard turning of ASP23 steel from which it was deduced that roughness was substantially diminished in cryogenic cooling compared to dry condition. Kalam et al. [2] studied different aspects of machinability while hard turning of AISI D3 steel with CO_2 gas as coolant. The results showed that surface roughness and surface temperature were reduced with the application of CO_2 gas compared to liquid cooling and the higher material removal rate was also achieved with gas coolant application. Grzesik et al. [3] studied the effects of cryogenic cooling on surface layer characteristics produced by hard turning and found that cryogenic cooling substantially affected white layer and micro hardness compared to dry turning. However, dry hard turning produced lower surface roughness compared to cryogenic cooling and fine grain structure was produced in dry turning, whereas coarser grain structure was obtained in cryogenic cooling.

Therefore, the objectives of the present experimental analysis are decided as follows. To study the various surface integrity aspects and assess the results comparatively in hard turning of 4340 alloy steel using both coated cermets and uncoated ones. The surface integrity aspects has micro hardness, surface roughness and formation of white layer to be studied in details.

2. Experimental Details (Materials and Methods)

In this section, experimental methodology, the details of equipment used such as machine tools, cutting tools, workpiece material and machining parameters with the experimental set-up have been described.

2.1 Workpiece

AISI 4340 is a low alloy steel having medium level carbon popular for its application where high strength and toughness is required. This type of alloy steel can be machined in all conventional machining processes. This material has wide industrial applications in automotive and aircraft domains.

2.2 Tool

Cermet is a composite material made out of ceramic and metallic binders. It is preferably intended to have the ideal properties of both ceramic particles and binder materials. For the present experimental investigation, square shaped cermet inserts were used which were of two types, i.e. uncoated and coated. Nomenclature of both coated and uncoated cermets are tabulated below.

2.3 Experimental procedure

The workpiece was heated to heat treament process. After this whole process of heat treatment, there was a significant change in hardness from 18 HRC to 48 HRC of the sample.

- *First stage of experiment*

First dry turning was performed on 4340 alloy steel having diameter of 50 mm and length of 700 mm with uncoated insert as depicted in Fig. 1. Experimental condition is provided in Table 1.

IOP Conf. Series: Materials Science and Engineering **338** (2018) 012056 doi:10.1088/1757-899X/338/1/012056

Fig. 1. Experimental Setup.

Table 1. Experimental condition during the first stage of the experiment.

Experimental condition	Details
Cutting tools	Uncoated Cermet tools
Tool designation	SNMG 120408
Tool holder designation	PSBNR 2525K12
Cutting Speed (V_c) (in m/min)	80,100,120
Feed (f) (in mm/rev)	0.05,0.1,0.15
Depth of Cut (d) (in mm)	0.1,0.2,0.3
Environment	Dry
Machining-length (in mm)	120

- *Second stage of experiment*

  Dry turning was performed on the same material i.e. 4340 alloy steel with coated cermet inserts. And the procedure adopted in first stage repeated here. The experimental condition for second stage of experiment was identical to first one. Combinations of machining parameters for each run using coated and uncoated cermets were given in Table 2. The combinations of machining parameter was set in alignment with Taguchi's L_9 orthogonal array. Every variation has three variations, low, medium and high.

Table 2. Combinations of machining parameters for various runs.

Run No.	Speed (m/min)	Feed (min/rev)	Depth of Cut (mm)
1	80	0.05	0.1
2	80	0.1	0.2
3	80	0.15	0.3
4	100	0.05	0.2
5	100	0.1	0.3
6	100	0.15	0.1
7	120	0.05	0.3
8	120	0.1	0.1
9	120	0.15	0.2

3. Results and discussions

3.1 Surface Roughness

Surface roughness is one of the remarkable parameter to analyze surface integrity aspects. The surface roughness variation at different cutting speeds, depth of cuts and feeds with both coated cermet and uncoated ones is exhibited by Fig.2. It was likely evident that coated inserts enhanced the surface quality than uncoated inserts regarding surface roughness. At higher speed of 120 m/min, higher degree of surface roughness was observed while machining with uncoated inserts.

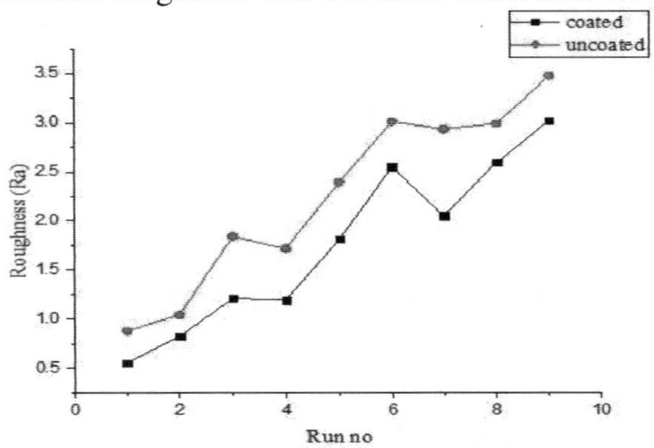

Fig. 2. Variation of surface roughness (Ra) with run numbers.

3.2 Micro hardness

The degree at which machined surface work hardening takes place, highly depends on various machining parameters. Fig.3. is the graph where abscissa depicts distance towards center from edges of the machined specimen and ordinate depicted microhardness. From the graphs, it was observed that there was enormous difference between hardness of subsurface regions and machined surface to that of bulk material. The graph also proved that microhardness caused more by uncoated cermet than with coated ones.

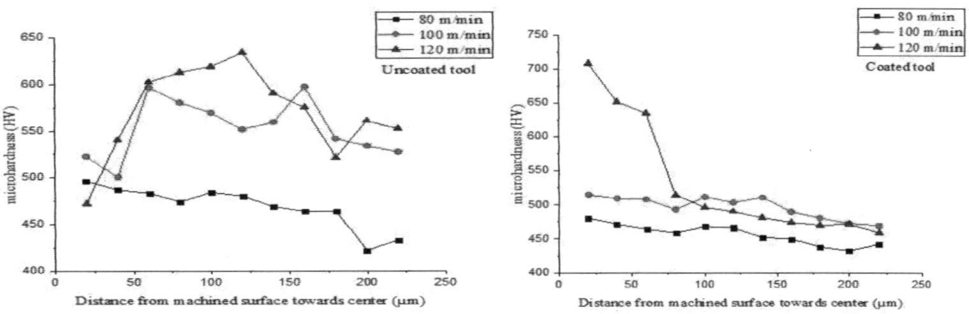

(a) **(b)**

Fig. 3. Variation of Vickers microhardness with reference to the distance from the machined surface towards the center (μm) for **(a)** uncoated insert **(b)** coated insert at variable speeds of 80,100 and 120 m/min.

3.3 Surface and sub-surface analysis

White layer, the unwanted formation on machined surface is observed due to many reasons. Plastic deformation during machining, rapid heating due to steep inclination in cutting speed, quenching and surface tension are prominent phenomena that causes white layer formation. Where higher degree of stress and temperature is required, white layer is highly undesirable. At various cutting speeds, surface analysis and subsurface regions formed during the machining operation are substantiated by Fig. 4. Using both coated and uncoated inserts. White layer is more visible at comparatively higher magnitude of cutting speed, say 120 m/min. With uncoated cermet, whiter layer was shown at relatively lower cutting speed than coated inserts. Coated cermet has anti-friction property that protects the tool to deform by thermal and mechanical properties of work. No white layer was seen when turning was carried out with coated cermet at lower cutting speed. It seemed visible for speed nearly 120 m/min and higher.

V_c (in m/min)	80	100	120
Uncoated cermet insert	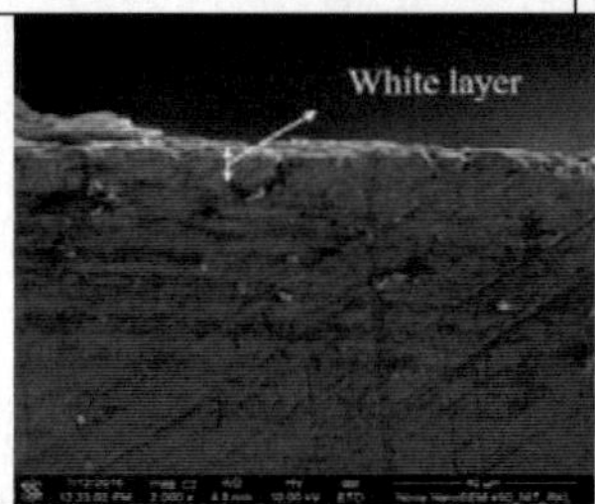	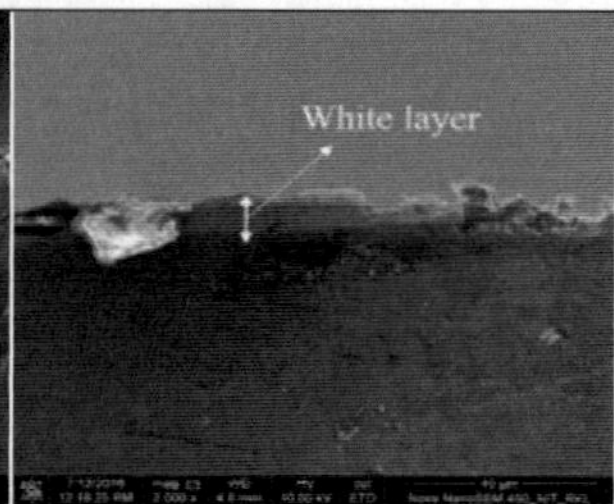	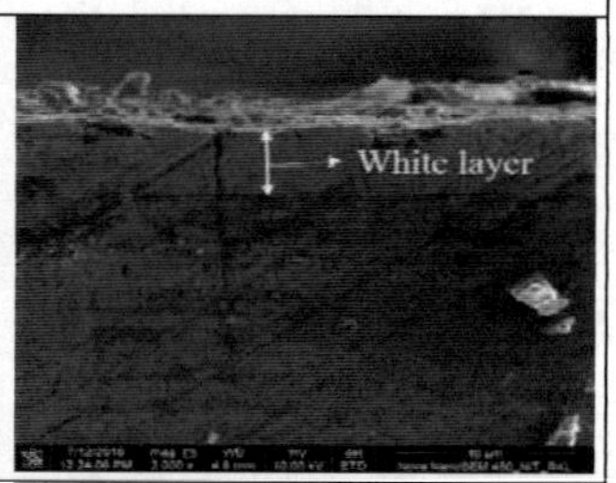
Coated cermet insert	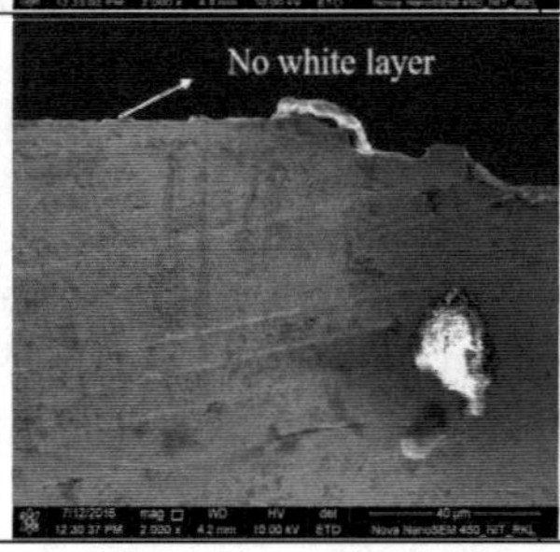		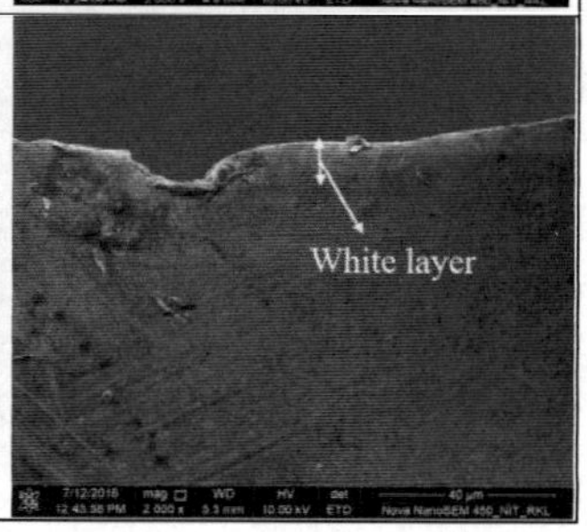

Fig. 4. Images of white layer formed during machining at various speeds with coated and uncoated inserts

3.4 Macromorphology of machined surface

Machined surface morphology manifests the most precise analysis of surface quality. Fig. 5 shows the forms and structures of machined surface examined under Scanning Electron Microscope (SEM). The work used in the experiment was AISI 4340 low carbon alloy steel using both coated and uncoated cermet as cutting inserts. Somewhere material were redeposited. Some debris and feed marks were also found. Marks of material being smeared has been observed at lower cutting speed, nearly 80m/min. At higher cutting velocity, temperature get enhanced and plastic deformation caused feed marks visible on machined surface.

IOP Conf. Series: Materials Science and Engineering **338** (2018) 012056 doi:10.1088/1757-899X/338/1/012056

V_c (in m/min)	80	100	120
Uncoated cermet insert			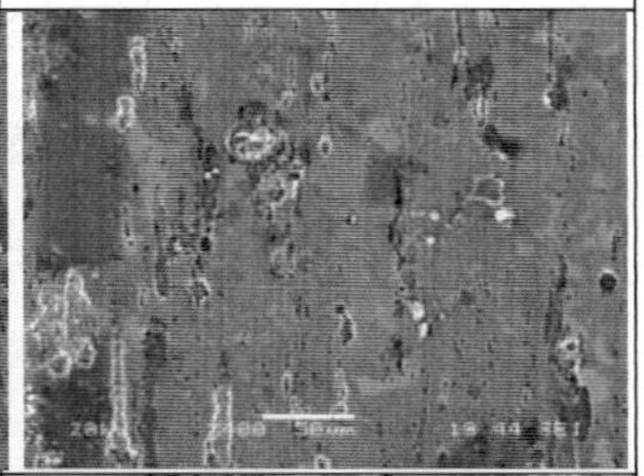
Coated cermet insert		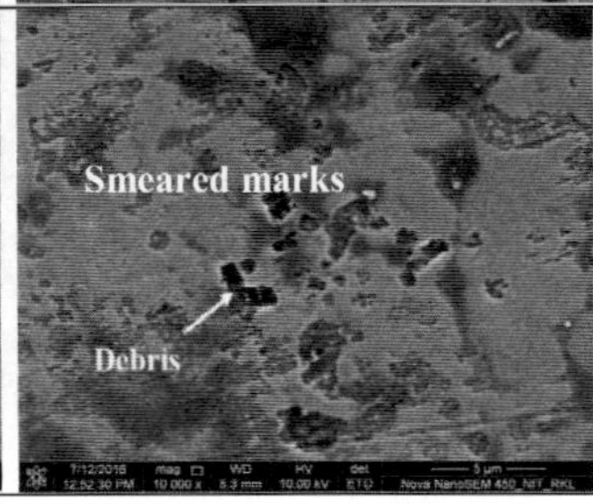	

Fig. 5. SEM images of machined surface morphology at different speed levels using coated and uncoated inserts.

4. Conclusions

Every experimental investigation ends with some rational conclusions and here they are. Machining with coated tool deliveres better surface finish in comparison to uncoated tools. Micro-hardness increases with cutting speed for both the tools. The machined surface hardness is found greater for uncoated cermet than coated one for the given identical cutting speeds. The magnitude of white layer thickness on the machined surface is found lower with coated insert than with uncoated cermet.

References

1. O. Pereira et al. 2015, Cryogenic Hard Turning of ASP23 steel using Carbon Dioxide. *Procedia Engineering*, Volume **132**, Pp. 486 – 491.
2. S. A. Kalam et al. 2015, Elimination of White Layer formation during Hard Turning of AISI D3 Steel to improve Fatigue life. *IOSR Journal of Mechanical and Civil Engineering (IOSR-JMCE)*, Volume **12**, No (3), Pp. 07-14.
3. W. Grzesik et al. 2012, Effects of cryogenic cooling on surface layer characteristics produced by hard turning. *Achieves of Materials science and Engineering*, Volume **54**, No 1, Pp.5-12.

IOP Conf. Series: Materials Science and Engineering **338** (2018) 012057 doi:10.1088/1757-899X/338/1/012057

Effect of severely thermal shocked MWCNT enhanced glass fiber reinforced polymer composite: An emphasis on tensile and thermal responses

K. K Mahato*, A.O. Fulmali, R Kattaguri, K. Dutta, R K Prusty and B C Ray

Composite Materials Group, Department of Metallurgical and Materials Engineering
National Institute of Technology, Rourkela-769008, India

*Email: kishorepce@gmail.com

Abstract. Fiber reinforced polymeric (FRP) composite materials are exposed to diverse changing environmental temperatures during their in-service period. Current investigation is aimed to investigate the influence of thermal-shock exposure on the mechanical behavior of multiwalled carbon nanotube (MWCNT) enhanced glass fiber reinforced polymeric (GFRP) composites. The samples were exposed to +70°C for 36 hrs followed by further exposure to – 60°C for the similar interval of time. Tensile tests were conducted in order to evaluate the results of thermal-shock on the mechanical behavior of the neat and conditioned samples at 1 mm/min loading rate. The polymer phase i.e. epoxy was modified with various MWCNT content. The ultimate tensile strength (UTS) was raised by 15.11 % with increase in the 0.1 % MWCNT content GFRP as related to the thermal-shocked neat GFRP conditioned samples. The possible reason may be attributed to the variation in the coefficients of thermal expansion at the time of conditioning. Also, upto some extent the pre-existing residual stresses allows uniform distribution of stress and hence the reason in enhanced mechanical properties of GFRP and MWCNT filled composites. In order to access the modifications in the glass transition temperature (T_g) due to the addition of MWCNT in GFRP composite and also due to the thermal shock temperature modulated differential scanning calorimeter (TMDSC) measurements are carried out. Scanning electron microscopy(SEM) was carried out to identify different modes of failures and strengthening morphology in the composites.

1. Introduction

Fiber reinforced polymeric (FRP) composites finds a widespread usage in the structural as well as in different filed of applications. The very extensive fields where these are used are aerospace, automobile, structural, sporting goods, oil and gas pipelines, cryogenic containers. In aircrafts and space vehicles, the structural components are come in direct contact with the environment that varies with different temperatures. Hence, investigation of mechanical properties and various failure micro mechanisms of these composites are highly essential with respect to temperature variations. Nowadays different nanofillers are used in polymeric matrix to enhance the mechanical, thermal and electrical properties. Carbon nanotubes are used as nanofiller material as its possesses high modulus (around 1 TPa [1]) and strength (approximately 22 GPa [2]). Various material investigators have performed exploration with the mechanical behavior of GFRP composites with inclusion of different types of nano-fillers [3–5] at numerous environmental parameters and different nature of loadings [6–10]. The present experimental investigation describes the inclusion of various contents of MWCNTs in polymer matrix that resulted in enhancement in tensile behavior over neat GFRP composite. The polymer matrix was reformed with different contents of

MWCNTs i.e. 0.1 wt. %, 0.3 wt. % and 0.5 wt. % and in accordance to that different laminated composites were prepared.

Properties	Epoxy	Glass Fibers (Woven Fabric)
Density (g/cm)	1.162	2.58
Tensile Strength (GPa)	0.11	3.4
Tensile modulus (GPa)	4.1	72.3
Strain at failure (%)	4.6	4.8
Areal weight of fabric (g/m^2)	-	360

Table 1. Various mechanical properties of epoxy and glass fibers.

2. Materials and Methods

2.1 Testing Materials
The present study involves preparation of neat GFRP and MWCNT filled GFRP composites. The resin used as matrix material established with the trade name as Diglycidyl Ether of Bisphenol A (DGEBA) and hardener having the trade name as triethylene tetra amine (TETA) was used. The orientation of E-glass fiber (woven fabric) utilized in the present work was 0/90° and it was procured from Owens Corning Industries. Hardener and resin (epoxy) were acquired from Atul Industries Ltd, Gujarat having the trade name of K-6 and Lapox L-12 respectively. At the time of preparation of laminates, the hardener to resin weight ratio was maintained by 1:10.

2.2 Fabrication of neat GFRP and MWCNT filled GFRP composites
The laminated composites were prepared with 9 stacks of glass fibers having 50:50 weight fractions of epoxy and glass fibers respectively. The different mechanical properties of epoxy resin and glass fibers are illustrated in table 1. Hand layup process was carried out to prepare the GFRP and MWCNT filled composite laminate. To achieve reasonable mechanical properties of the composites, it is vital to ensure proper distribution of MWCNTs in the polymer matrix. Therefore, the different weight fractions of MWCNTs were processed thoroughly in magnetic stirrer and ultra-sonicator. Hot compression press at 60°C was used for the curing of laminated composites at a pressure of 490 kPa for 20 minutes. According to ASTM D3039 the composites were cut with the help of diamond tipped cutter. Further, post curing of composites specimens was carried out in an hot chamber for 6 hours at 140°C [11].

2.3 Testing Process of laminated GFRP and MWCNT filled composite
Universal testing machine having the model number Instron 8862 was accomplished for conducting the tensile tests. The tensile tests are carried out in the room temperature for neat GFRP and MWCNTs filled GFRP composites at 1 mm/min loading rates. Glass transition behavior of the composites are examined by means of TMDSC. SEM was used for determining the several failure and strengthening mechanisms occurred in laminated composite surfaces of the tested fractured samples.

Results and discussion

3.1 *Effect of MWCNT filled GFRP composites on the tensile properties at thermal shock environmental conditions*

It is evident from figure 1 that the stress-strain behavior for neat GFRP, 0.1 wt.% MWCNT, 0.3 wt.% MWCNT and 0.5 wt.% MWCNT at 1 mm/min loading speeds. It can be seen that upto 0.1 wt.% MWCNT content tensile strength was increased (figure1) in comparison to the neat GFRP composite. Further, 0.3 and 0.5 wt. % MWCNT filled composites shows decrease in the tensile strength and the possible reason of decrement in strength may be agglomeration of MWCNTs in the matrix phase. The agglomeration of MWCNTs contributes to improper transmission of stress at the interfacial region.

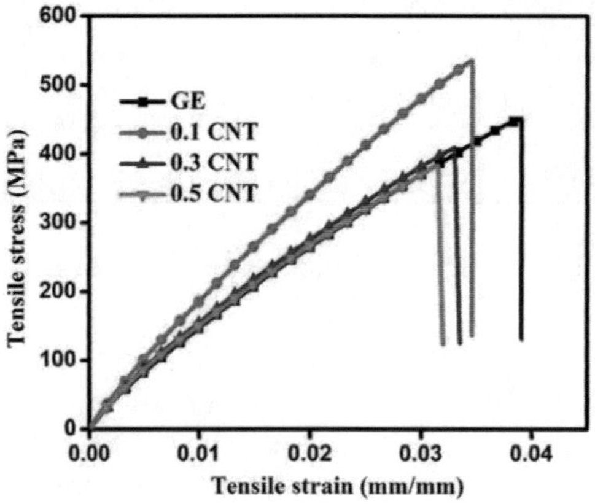

Figure 1. Tensile Stress–strain plot of neat GFRP and MWCNT filled composites at thermal shocked conditioning temperature at 1 mm/min loading speed

The different tensile behavior i.e. strength and modulus acquired from figure 1 are presented along with various MWCNT content for neat GFRP and MWCNTs filled GFRP composite in figure 2. The tensile properties are strongly depending upon the MWCNT content for GFRP composites as shown in figure 2. Incorporation of 0.1 wt % MWCNT/GFRP thermal shocked composite resulted in improvement in tensile strength in comparison to neat GFRP composite as shown in figure 2(a).

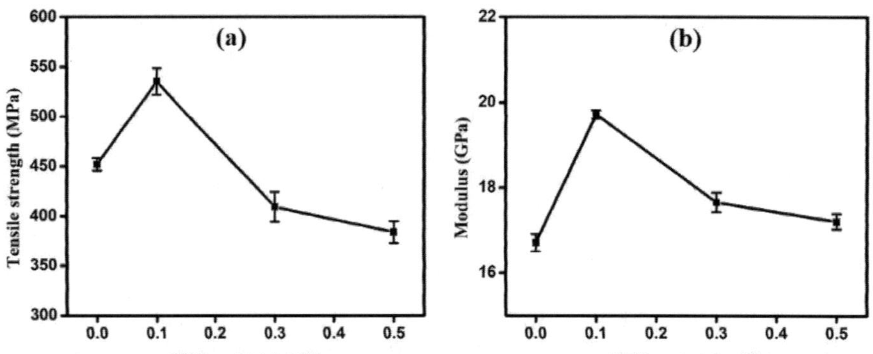

Figure 2. Variation in tensile properties (a) Strength and (b) Modulus with different content of MWCNT in the composites.

An improvement of 15.11 % in tensile strength was observed in case of 0.1 wt. % MWCNT filled GFRP composite in comparison to neat GFRP. The improvement in strength may be owed to high specific surface zone of MWCNTs in the polymer matrix. This imparts higher interfacial space in the ultimate composites. Therefore, transfer of stress is easily facilitated across the interfacial region. This contributes to enhancement in strength of MWCNT filled GFRP composites in comparison to neat GFRP composites even after thermal shocked environment. Further addition of MWCNTs content leads to decrease in strength and modulus values. Agglomeration of the MWCNT nanoparticles in the polymer matrix was the prime cause decrease in strength value. At the fiber/matrix interface due to agglomeration of MWCNTs causes poor wettability. This leads to interfacial debonding between the MWCNT filled fiber/matrix composites. Figure 2 (a) show that the tensile strength MWCNT/GFRP composite is increasing upto 0.1 wt % MWCNT content. The modulus vs MWCNT content value was illustrated in figure 2 (b). This shows decrease in the modulus value for more than 0.1 MWCNT-GFRP composites. Improper adhesion of matrix/MWCNT/fiber in the composite and inadequate transfer of stress from polymer matrix to the fiber. the bulk images of the composites with various failure patterns tested at thermal shock environment at 1mm/min loading rate are illustrated in figure 3. The opening of cracks mainly develops at the middle portion of the samples.

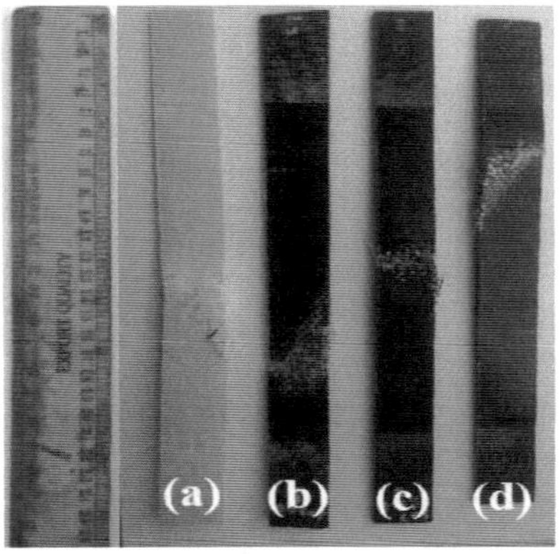

Figure 3. Bulk images of failed tensile specimens i.e. GFRP and MWCNT filled GFRP composites (a) GFRP (b) 0.1 wt. % MWCNT/GFRP (c) 0.3 wt. % MWCNT/GFRP and (d) 0.5 wt. % MWCNT/GFRP

3.2 Fractography analysis

After the tensile tests, different modes of failures and strengthening morphology in the composites were analyzed with the help of nano nova FESEM.

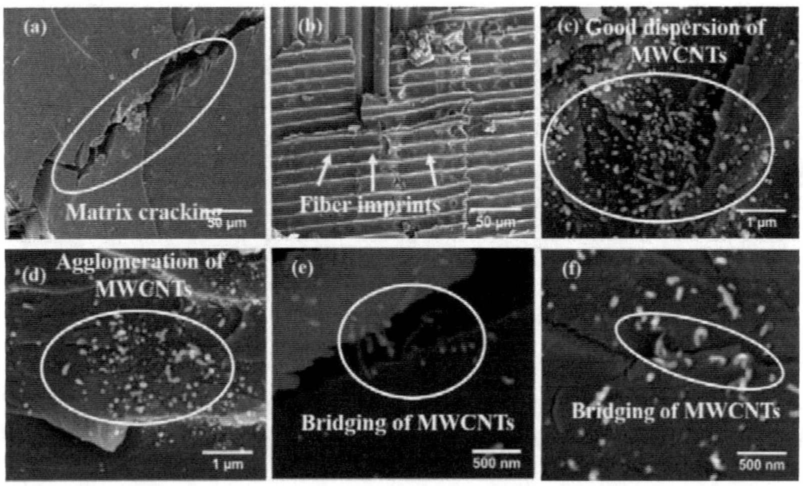

Figure 4. SEM micrographs of (a) Neat GFRP (b) Neat GFRP(c) 0.1 wt. % MWCNT/GFRP (d) 0.5 wt. % MWCNT/GFRP (e) 0.1 wt. % MWCNT/GFRP and (f) 0.1 wt. % MWCNT/GFRP

Figure 4(a) and 4(b) reveals SEM micrographs of neat GFRP composites showing matrix cracking and fiber pullout. From figure 4(c) 0.1 wt. % MWCNT/GFRP composite reveals better dispersion as compared to 0.5 wt.% MWCNT/GFRP composites [figure 4(d)] in polymer matrix. Better distribution of MWCNTs in the polymer matrix delivers proper transfer of stress from the polymer matrix to fiber. Agglomeration of MWCNTs at a specific area leads to improper transfer of stress across the matrix resulting in decrease in strength and modulus value. Bridging of MWCNTs in the polymer matrix certainly enhances the tensile strength value in case of 0.1 wt % MWCNT enhanced GFRP composites [figure 4(e)-4(f)].

3.3 Thermal analysis

TMDSC measurements were carried out in order to know the T_g of the polymer part of the composites material.

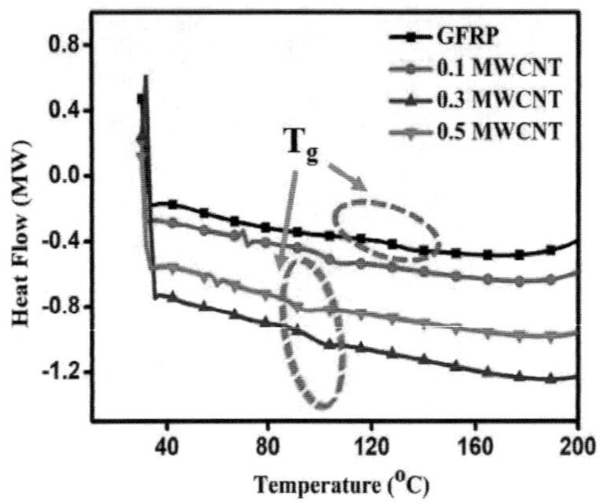

Figure 5. TMDSC graph of neat and MWCNT filled GFRP composites

Figure 5 illustrates the TMDSC curves for neat GFRP and MWCNTs enhanced GFRP composite at 0.1wt. % MWCNT, 0.3 wt. % MWCNT and 0.5 wt. % MWCNT. The neat GFRP composites shows T_g around 119.42 °C, while the MWCNT/GFRP with 0.1 wt. %, 0.3 wt. % and 0.5 wt. % of MWCNT reveals 100.9 °C, 94.59 °C and 88.22 °C respectively. The T_g of 0.5 wt. % MWCNT enhanced GFRP composites is found to be decreasing from 119.32 °C to 88.22 °C as compared neat GFRP composites. This decrease in T_g value may be attributed to increase of MWCNT content in the GFRP composites. The hindrance of crosslink formations due to the entrapment of MWCNTs particles between the polymeric chains of the polymeric network resulting lowering in the T_g value.

4. Conclusions

The current study involves on the tensile behavior of different contents of MWCNTs in the glass reinforced composites at thermal shocked environment and comprises the following conclusions

- MWCNT content upto 0.1% enhanced GFRP composite exhibits improvement in the tensile strength by 15.11 % as compared to the neat GFRP composite.

- Tensile modulus was found to be maximum at 0.1 wt.% MWCNT and with increased weight percentage of MWCNT in the polymer matrix resulted in decrease in modulus value attributed to accumulation of the MWCNT at a particular region in the composites.
- The T_g of 0.5 wt. % MWCNT enhanced GFRP composites was found to be decreasing from 119.42 °C to 88.22 °C in comparison to neat GFRP composites.

Acknowledgement

The authors state their thankfulness to NIT Rourkela for facilitating the essential requirements during the investigation. A great thanks to Council of Scientific and Industrial India (22(0735)/17/EMR-II) for providing financial funding for carrying out the present research work.

References

[1] Salvetat J-P, Briggs G A D, Bonard J-M, Bacsa R R, Kulik A J, Stöckli T, Burnham N A and Forró L 1999 Elastic and Shear Moduli of Single-Walled Carbon Nanotube Ropes *Phys. Rev. Lett.* **82** 944–7
[2] Li F, Cheng H M, Bai S, Su G and Dresselhaus M S 2000 Tensile strength of single-walled carbon nanotubes directly measured from their macroscopic ropes *Appl. Phys. Lett.* **77** 3161–3
[3] Nayak R K, Mahato K K and Ray B C 2016 Water absorption behavior, mechanical and thermal properties of nano TiO2 enhanced glass fiber reinforced polymer composites *Compos. Part Appl. Sci. Manuf.* **90** 736–47
[4] Mahato K K, Rathore D K, Prusty R K, Dutta K and Ray B C 2017 Tensile behavior of MWCNT enhanced glass fiber reinforced polymeric composites at various crosshead speeds *IOP Conf. Ser. Mater. Sci. Eng.* **115** 12017
[5] Kathi J, Rhee K-Y and Lee J H 2009 Effect of chemical functionalization of multi-walled carbon nanotubes with 3-aminopropyltriethoxysilane on mechanical and morphological properties of epoxy nanocomposites *Compos. Part Appl. Sci. Manuf.* **40** 800–9
[6] Mahato K K, Biswal M, Rathore D K, Prusty R K, Dutta K and Ray B C 2016 Effect of loading rate on tensile properties and failure behavior of glass fiber/epoxy composite *IOP Conf. Ser. Mater. Sci. Eng.* **115** 12017
[7] Mahato K K, Rathore D K, Dutta K and Ray B C 2017 Effect of loading rates of severely thermal-shocked glass fiber/epoxy composites *Compos. Commun.* **3** 7–10
[8] Mahato K K, Dutta K and Ray B C 2018 Loading rate sensitivity of liquid nitrogen conditioned glass fiber reinforced polymeric composites: An emphasis on tensile and thermal responses . *Appl. Polym. Sci.* **133**
[9] Ray B C 2006 Loading Rate Sensitivity of Glass Fiber–Epoxy Composite at Ambient and Sub-ambient Temperatures *J. Reinf. Plast. Compos.* **25** 329–33
[10] Ray B C 2004 Effects of crosshead velocity and sub-zero temperature on mechanical behaviour of hygrothermally conditioned glass fiber reinforced epoxy composites *Mater. Sci. Eng. A* **379** 39–44
[11] Kumar D S, Shukla M J, Mahato K K, Rathore D K, Prusty R K and Ray B C 2015 Effect of post-curing on thermal and mechanical behavior of GFRP composites *IOP Conf. Ser. Mater. Sci. Eng.* **75** 12012

IOP Conf. Series: Materials Science and Engineering **338** (2018) 012058 doi:10.1088/1757-899X/338/1/012058

Optical & current-voltage characteristics of dip coated TiO_2 thin film

S.Roy[1], N.Tripathy[2], P.K. Sahu[1], J.P.Kar[2]*

[1]Department of Electrical Engineering, National Institute of Technology, Rourkela, India 769008
[2]Department of Physics and Astronomy, National Institute of Technology, Rourkela, India 769008

E-mail: karjp@nitrkl.ac.in, Phone : +91-9438532157

Abstract:

TiO_2 thin films have been synthesized by dip coating process on the p-type silicon wafer for different withdrawal speeds varied from 1 cm/min to 4 cm/min. The films were annealed at 500°C for 1 hour in air atmosphere. The film thickness was found to be increased with the increase in withdrawal speed. The evolution of the anatase phase of TiO_2 has been confirmed by structural studies. Optical characteristics of the dip coated TiO_2 films were investigated by Raman spectroscopy. For electrical measurement Al/TiO_2/Si MOS structures were fabricated. The current-voltage characteristics have shown the rapid rise in leakage current for initial bias voltage and afterwards, it became quasi saturated. The film, deposited at the withdrawal speed of 1 cm/min, has shown higher leakage current.

Key words: Sol gel process, annealing, Raman, MOS, electrical characterisation

1. Introduction

Electronic materials in form of thin films are now one of the emerging research field [1]. It has vast range of applications such as sensor and actuators [2], memory devices [3], component of microelectronics [4]. Titanium dioxide (TiO_2) being a multifunctional material attracting a lot of interest among researchers due to its promising properties. Due to its chemical stability, it find its application in optical and electronic industries [5,6]. It possess a bandgap of 3.2 eV in anatase form [7]. The optical and electrical properties are two important parameters in order to evaluate the quality of electronic materials. In the context, an attempt has been undertaken to observe the modulation of optical and electrical properties of the TiO_2 layer with growth parameter. The fabrication processes of TiO_2 films strongly affect the structural, optical and electrical properties of the materials. There are several reports of deposition of TiO_2 thin films on silicon substrates by sputtering [8,9], CVD [10], thermal evaporation techniques [11]. However, the vacuum based techniques are expensive and suitable for small area substrates. On the other hand, dip coating has advantages such as extremely easy method, economical apparatus and coating on large substrate, ease of doping etc. In this report, TiO_2 thin films were fabricated by adopting dip coating technique with various withdrawal speed of the substrates. Afterwards, the optical and electrical properties of the films were carried out.

IOP Conf. Series: Materials Science and Engineering **338** (2018) 012058 doi:10.1088/1757-899X/338/1/012058

2. Materials and methods

Suitable amount of titanium (IV) n-butoxide (Ti(O-nBu)$_4$, 97%), n-butanol, acetylacetine (AcAc), glacial acetic acid was used to make a base solution. After 1 hour of continuous stirring, a uniform light yellow solution was obtained. TiO_2 thin films were deposited by dip coating process on p-type silicon [(100), 1-10 Ω.cm] substrate. Silicon substrates were cleaned by standard RCA (Radio Corporation of America) method. The silicon substrates were dipping on the precursor solution by varying withdrawal speed from 1 to 4 cm/min. After completion of one coat, the thin film was annealed for 10 min with a temperature 100°C. The samples were annealed at 500°C for 1 hour in the air ambient after completion of the whole dip coating process. The optical spectroscopy of the as prepared TiO_2 thin film was investigated from micro-Raman spectrometer (Lab- Ram HR Evolution, Horiba) with 532 nm Nd-YAG laser. For electrical characterizations top and bottom aluminum contacts were made by thermally evaporation of aluminum in vacuum ambient. The electrical measurement of the (Al/TiO_2/p-Si) MOS structures were determined by the Keithley 6487 Picoammeter voltage source.

3. Results and discussions

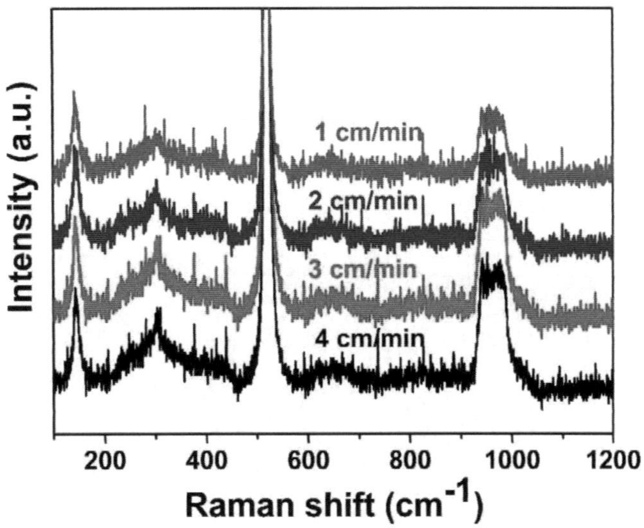

Figure.1 Raman spectra of the TiO_2 thin films.

All the films have shown the evolution of anatase phase irrespective of their withdrawal speed. The films coated at 4 cm/min withdrawal speed has shown the X-ray diffraction peak with highest intensity (not shown here), which may be due to the faster gelation rate during the growth. Raman spectra of the TiO_2 thin films for different withdrawal speeds are shown in figure 1. The Raman bands observed at 305 cm^{-1}, 520 cm^{-1} and 970 cm^{-1} correspond to the silicon wafer [12]. In addition, two Eg Raman modes of TiO_2 at 144 cm^{-1} and around 641 cm^{-1} are clearly appeared. These phonon bands correspond to the anatase phase without the signal of brookite or rutile phase [13]. The peak at 144 cm^{-1} is found to be narrow and intense for the films deposited at a withdrawal speed of 4 cm/min, which indicates the structural improvement of the TiO_2 films. The films coated with higher withdrawal speed have faster gelation rate, where the solvent get evaporated quickly by producing dense and compact films without much defects.

IOP Conf. Series: Materials Science and Engineering **338** (2018) 012058 doi:10.1088/1757-899X/338/1/012058

Figure 2 depicts that leakage current density of the TiO_2 thin film deposited by sol-gel process with various withdrawal speed varied from 1 to 4 cm/min. The leakage current was initially increased rapidly up to 2 volt and afterwards it became quasi-saturated. The leakage current density is found to be decreased with the increase in the withdrawal speed. The leakage current is maximum for 1 cm/min and minimum for 4 cm/min.

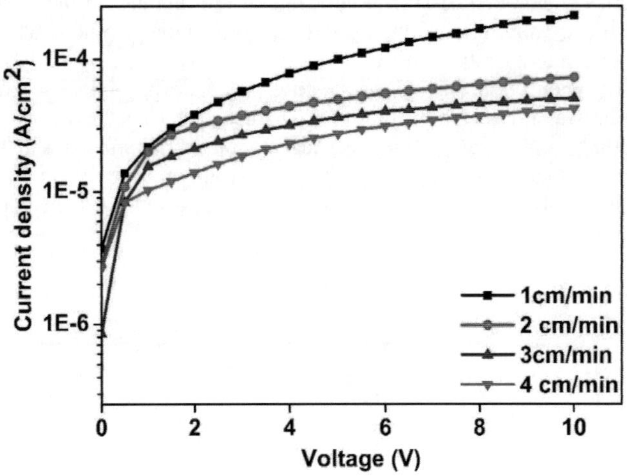

Figure.2 Current - voltage characteristics of TiO_2 thin films.

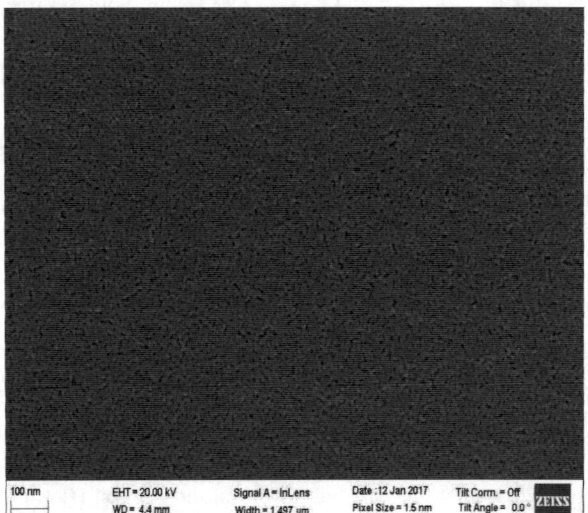

Figure.3 FESEM images of the dip-coated TiO_2 thin films with the withdrawal speed of 4cm/min.

The reduction in the leakage current for 4 cm/min withdrawal speed may be due to the improvement crystallinity. In addition, the lower leakage current may be due to the presence of less number of interface and bulk trap density in the films [14,15]. Figure 3 shows the FESEM micrograph of the TiO_2 films coated at the withdrawal speed of 4 cm/min. The surface microstructures are uniform without presence of any cracks.

4. Conclusions

TiO_2 thin films were coated with various withdrawal speed on the p-type silicon substrate. Post-deposition annealing of all the films were carried out at 500°C for 1 hour in the air ambient. The Raman spectroscopy and the current voltage characteristics of the dip coated films were investigated with the various withdrawal speed. The Raman spectra depicted the anatase phase of TiO_2 thin films. The leakage current density was greatly influenced by the variation of withdrawal speed during the dip coating. The leakage current density is found to be minimum for the films coated at 4 cm/min. The film coated at the withdrawal speed of 4 cm/min has shown uniform surface morphology.

References

[1] Nathan A, Ahnood A, Cole M T, Lee S, Suzuki Y, Hiralal P, Bonaccorso F, Hasan T, Garcia-Gancedo L, Dyadyusha A, Haque S, Andrew P, Hofmann S, Moultrie J, Chu D, Flewitt A J, Ferrari A C, Kelly M J, Robertson J, Amaratunga G A J and Milne W I 2012 Flexible electronics: The next ubiquitous platform *Proc. IEEE* **100** 1486–517

[2] Eslamian M 2017 Inorganic and Organic Solution-Processed Thin Film Devices *Nano-Micro Lett.* **9** 1–23

[3] Strukov D B and Kohlstedt H 2012 Resistive switching phenomena in thin films: Materials, devices, and applications *MRS Bull.* **37** 108–14

[4] Read D T and A V A 2007 Thin Films for Microelectronics and Photonics: Physics, Mechanics, Characterization, and Reliability *Micro- Opto-Electronic Mater. Struct. Physics, Mech. Des. Reliab. Packag.* A135–80

[5] Okada M, Yamada Y, Jin P, Tazawa M and Yoshimura K 2003 Fabrication of multifunctional coating which combines low-e property and visible-light-responsive photocatalytic activity *Thin Solid Films* **442** 217–21

[6] Ennaoui A, Sankapal B R, Skryshevsky V and Lux-Steiner M C 2006 TiO2 and TiO2-SiO2 thin films and powders by one-step soft-solution method: Synthesis and characterizations *Sol. Energy Mater. Sol. Cells* **90** 1533–41

[7] Dette C, Pérez-Osorio M A, Kley C S, Punke P, Patrick C E, Jacobson P, Giustino F, Jung S J and Kern K 2014 TiO $_2$ Anatase with a Bandgap in the Visible Region *Nano Lett.* **14** 6533–8

[8] Kubart T, Jensen J, Nyberg T, Liljeholm L, Depla D and Berg S 2009 Influence of the target composition on reactively sputtered titanium oxide films *Vacuum* **83** 1295–8

[9] Williams L M and Hess D W 1983 Structural properties of titanium dioxide films deposited in an rf glow discharge *J. Vac. Sci. Technol. A Vacuum, Surfaces, Film.* **1** 1810–9

[10] Kamada T, Kitagawa M, Shibuya M and Hirao T 1991 Structure and properties of silicon titanium oxide films prepared by plasma-enhanced chemical vapor deposition method *Jpn. J. Appl. Phys.* **30** 3594–6

[11] Bedikyan L, Zakhariev S and Zakharieva M 2013 Titanium dioxide thin films: Preparation and optical properties *J. Chem. Technol. Metall.* **48** 555–8

[12] Pandya S G and Kordesch M E 2015 Characterization of InSb Nanoparticles Synthesized

Using Inert Gas Condensation *Nanoscale Res. Lett.* **10** 258

[13] Xu J, Wang Z, Li W, Zhang X, He D and Xiao X 2017 Ag Nanoparticles Located on Three-Dimensional Pine Tree-Like Hierarchical TiO2 Nanotube Array Films as High-Efficiency Plasmonic Photocatalysts *Nanoscale Res. Lett.* **12** 54

[14] Seo Y, Lee S, An I, Song C and Jeong H 2009 Conduction mechanism of leakage current due to the traps in ZrO2 thin film *Semicond. Sci. Technol.* **24** 115016

[15] Kumar A, Mondal S and Koteswara Rao K S R 2016 Critical investigation of high performance spin-coated high-κ titania thin films based MOS capacitor *J. Mater. Sci. Mater. Electron.* **27** 5264–70

IOP Conf. Series: Materials Science and Engineering **338** (2018) 012059 doi:10.1088/1757-899X/338/1/012059

Erosion wear response of epoxy composites filled with steel industry slag and sludge particles: A comparative study

Abhilash Purohit, Alok Satapathy

Department of Mechanical Engineering,
National Institute of Technology, Rourkela, Odisha, India

Corresponding author Email : mr.abhilash.purohit@gmail.com

Abstract: In the field of composite research, use of industrial wastes such as slag and sludge particles as filler in wear resistant polymer composites has not been very common. Owing to the very high cost of conventional filler materials in polymer composites, exploring the possibility of using low cost minerals and industrial wastes for this purpose has become the need of the hour. In this context this work explores the possibility of such polymer composites filled with low cost industrial wastes and presents a comparison of mechanical characteristics among three types of epoxy based composites filled with Linz - Donawitz sludge (LD sludge), blast furnace slag (BF slag) and Linz - Donawitz slag (LD slag) respectively. A comparative study in regard to their solid particle erosion wear characteristics under similar test conditions is also included. Composites with different weight proportions (0, 5, 10, 15 and 20 wt. %) of LD sludge are fabricated by solution casting technique. Mechanical properties such as micro-hardness, tensile strength and flexural strength of three types of composites have been evaluated as per ASTM test standards and solid particle erosion wear test is performed following a design of experiment approach based on Taguchi's orthogonal array. Five control factors (impact velocity, erodent size, filler content, impingement angle and erodent temperature) each at five levels are considered to conduct erosion wear tests. The test results for epoxy-LD sludge composites are compared with those of epoxy-BF slag and epoxy-LD slag composites reported by previous investigators. The comparison reveals that epoxy filled with LD sludge exhibits superior mechanical and erosion wear characteristics among the three types of composites considered in this study. This work also opens up a new avenue for value added utilization of an abundant industrial waste in the making of epoxy based functional composites.

1. Introduction

The iron and steel industry is one of the most populated industries in India due to availability of raw materials in huge quantities. Every day enormous quantities of waste materials in the form of sludge and slag are produced from these industries. During the production of unit tonne of iron and steel, two to four tones of sludge and slag are being produced. The solid wastes that are generated during the production of iron and steel are Linz-Donawitz sludge (LDS), Linz-Donawitz slag (LD slag) and blast furnace slag (BF slag). The chemical compositions of these wastes differ widely but they have various recyclable elements such as iron, phosphorous, carbon, calcium, etc. BF slag is produced during the production of pig iron from iron ore in blast furnace. The main chemical components present in BF slag are SiO_2, CaO, Fe_2O_3, MgO, Al_2O_3 etc. Similarly, LD slag is generated from LD converter during the production of steel from pig iron and mainly contains silicon oxide, calcium oxide and iron oxide. LD sludge is generated after processing of the flue dust emerging from basic oxygen furnace. It mainly contains magnetite, hematite, wüstite or FeO, calcite and silica.

Nowadays researchers are giving more attention towards the polymer matrix composites due to their superior mechanical and physical characteristics. They can also be used as wear resistance material in several low load applications. The wear resistance and mechanical properties of neat polymer can be improved by adding a reinforcing phase into the polymer matrix [1–5]. Many reports are available on utilization of polymers and their composites in erosive wear conditions [6-7]. Although a large number of filler materials are used as reinforcing phase in polymer composites for wear resistance application [8], employment of steel industry solid wastes in this context is very rare. The slag and sludge generated from iron and steel industry can be utilized as reinforcing phase in matrix to improve the mechanical and wear characteristics of the composite samples due to presence of hard phases like magnetite, hematite, wüstite, calcite and silica [9-11]. In view of this, this work reports on the comparative study of erosion wear response of steel industry wastes filled epoxy composites.

2. Materials and methods

2.1. Fabrication of composite

The fabrications of composite samples are done by solution casting technique. The hardener HY-951 and the epoxy resin LY-556 are mixed in a ratio of 1:10 by weight. Then the LD sludge particles are added in different proportions and properly mixed with the epoxy-hardener combination. The dough is then discharged into moulds of dissimilar shapes. The moulds are kept to cure for few hours at room temperature, after which the samples are collected by breaking the moulds. Finally, the composites are cut to the required dimensions as per the prescribed standards for the evaluation of mechanical and tribological properties.

2.2. Characterization

The mechanical characterization such as micro-hardness tests of the fabricated composite samples are conducted with the help of a Vaiseshika micro-hardness tester. For each composition of composite samples five experiments are carried out and the mean value is noted as the micro-hardness value for that particular composite. Three point bend and tensile tests of the composite samples are conducted in the UTM Instron 1195 as per ASTM D-2344 and D-638 respectively. Tensile tests are carried out on dog bone shaped specimen whereas; flexural tests are conducted on flat specimen with dimensions 150 mm × 20 mm × 5 mm. The flexural and tensile tests are repeated three times for each sample and the mean value is noted as the flexural and tensile strength of that specimen respectively.

2.3. Erosion wear test

The erosion tests are conducted using an erosion wear equipment shown in Figure 1 as per ASTM G76. It is composed of an erodent flow control knob, an air compressor, an erodent feeder, a nozzle, an erodent collector and a specimen holder. Angle of impingement can be changed by fixing the specimen holder at a prescribed angle. Erodent particles used in the erosion tests are dry silica sand (SiO_2) with average particle size ranging from 50 to 250 μm and are dried in an oven just before the tests.

The specimens of the composites for erosion tests are prepared with the thickness 5 mm and the wear faces being 20 mm × 20 mm. To conduct erosion tests, five major factors have been considered, they are filler content, impingement angle, impact velocity, erodent temperature and erodent size each at five levels as shown in Table 1. Taguchi's L_{25} experimental design is chosen to conduct the erosion wear tests. An electronic balance is used to measure the weight loss from the composite specimen after each test. Each test is repeated for three times and mean value is reported as the erosion rate of that specimen.

Table 1. Process parameters and levels for solid particle erosion tests

Process parameters	Levels					Units
	I	II	III	IV	V	
Impact velocity	32	40	48	56	64	m/s
Impingement angle	30	45	60	75	90	degree
Filler content	0	5	10	15	20	wt. %
Erodent size	50	100	150	200	250	μm

Erodent temperature	30	40	50	60	70	°C

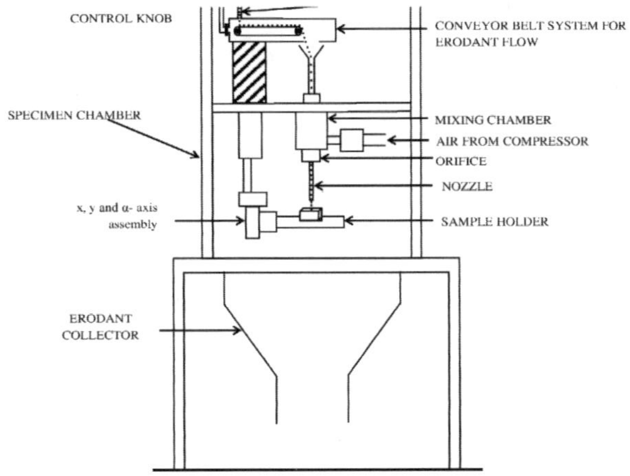

Figure 1. Schematic diagram of the erosion test apparatus

3. Results and Discussion

3.1. Micro-Hardness, Tensile and Flexural Test Results

In the present work, the micro-hardness, tensile and flexural strength values of BF slag, LD sludge (LDS) and LD slag filled epoxy composites have been obtained and are given in Table 2. It is observed that the micro-hardness of the composites increases with increase in filler loading and this improvement is attributed to the hard phases present in the reinforcing fillers. The micro-hardness value of neat epoxy increases to 68.86% with addition of 20 wt. % LD sludge. Further, for same filler loading LD sludge filled epoxy composites show higher micro-hardness values than BF slag and LD slag filled epoxy composites. A marginal drop in tensile and flexural strength values is observed with increase in filler loading for all the three classes of composite samples reinforced with BF slag, LD sludge and LD slag. It is also examined that tensile and flexural strength of the LDS filled epoxy composites exhibit superior outcome than LD slag and BF slag reinforced composites.

Table 2. Comparison of Micro-hardness, flexural and tensile strength of three classes of composites

Filler	Micro-hardness (GPa)			Tensile Strength (MPa)			Flexural Strength (MPa)		
Content (wt. %)	EP-LD Slag	EP-BF Slag	EP-LD Sludge	EP-LD Slag	EP-BF Slag	EP-LD Sludge	EP-LD Slag	EP-BF Slag	EP-LD Sludge
0	0.085	0.085	0.085	48	48	48	22.5	22.5	22.5
5	0.096	0.098	0.12	43.81	45.24	45.71	19.12	20.75	20.9
10	0.108	0.112	0.181	41.23	43.98	44.37	16.38	19.81	20.1
15	0.162	0.174	0.232	39.17	42.47	42.84	14.72	18.54	19.3
20	0.182	0.196	0.273	38.51	41.05	41.62	13.23	17.53	18.8

3.2. Erosion Wear Test Results

Table 3 shows the erosion rates for three different industrial wastes filled epoxy composites. Each row of the table corresponds to a experiment. Total 25 experiments are carried out as per Taguchi's L_{25} orthogonal array. From the erosion rate data given in Table 3 it has been observed that wear resistance of all the three classes of composites increases with increase in filler loading. Further, LD sludge filled composite shows superior wear resistance property than that of other two classes of composites. Figure 2 shows variation in erosion rate for all the three classes of composites.

Table 3. The erosion wear test results for three classes of composites

Test Run	Impact Velocity (m/s)	Impingement Angle (degree)	Filler Content (wt. %)	Erodent Size (μm)	Erodent Temperature (°C)	Erosion Rate (mg/kg)		
						EP-LD sludge	EP-LD slag	EP-BF slag
1	32	30	0	50	30	58.594	65.012	69.808
2	32	45	5	100	40	55.294	63.295	65.877
3	32	60	10	150	50	51.794	57.352	61.707
4	32	75	15	200	60	48.495	58.635	56.776
5	32	90	20	250	70	42.195	51.538	49.271
6	40	30	5	150	60	56.794	63.985	67.664
7	40	45	10	200	70	54.194	59.056	64.567
8	40	60	15	250	30	51.894	58.465	61.827
9	40	75	20	50	40	46.395	51.269	55.275
10	40	90	0	100	50	61.793	69.617	73.621
11	48	30	10	250	40	56.094	65.196	63.834
12	48	45	15	50	50	53.694	60.493	63.971
13	48	60	20	100	60	49.295	55.536	58.729
14	48	75	0	150	70	66.193	74.574	78.862
15	48	90	5	200	30	60.293	65.927	71.835
16	56	30	15	100	70	54.694	63.619	61.162
17	56	45	20	150	30	51.194	56.676	60.993
18	56	60	0	200	40	69.393	80.179	78.674
19	56	75	5	250	50	64.893	75.109	77.313
20	56	90	10	50	60	59.694	67.252	68.119
21	64	30	20	200	50	52.394	59.028	62.422
22	64	45	0	250	60	71.492	78.544	85.176
23	64	60	5	50	70	69.793	78.629	83.151
24	64	75	10	100	30	64.393	72.546	76.718
25	64	90	15	150	40	57.194	64.435	62.142

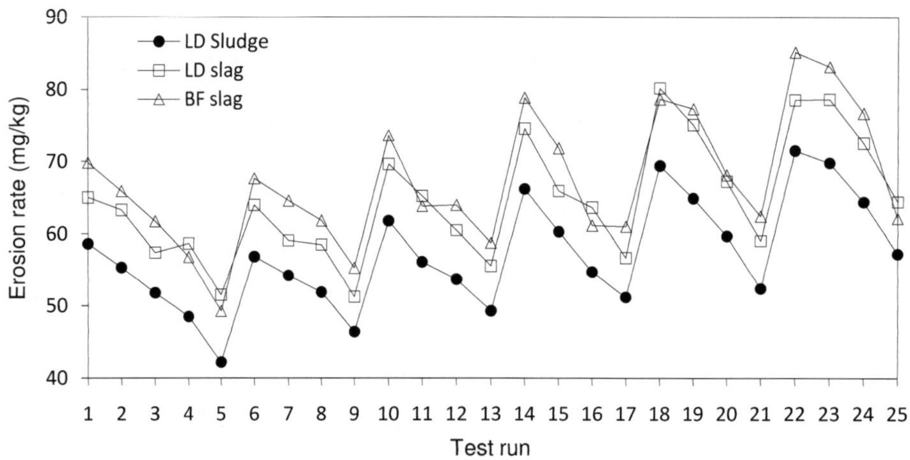

Figure 2. Comparative plot for erosion rates of three classes composites for 25 experiments

4. Conclusions

i. Fabrication of epoxy based composites reinforced with LD sludge is feasible through solution casting route.
ii. Superior mechanical properties are observed in case of LD sludge filler epoxy composites as compared to epoxy-LD slag and epoxy-BF slag composites.
iii. Solid particle erosion wear analysis reveals that wear resistance of neat epoxy improves with the addition of filler and epoxy-LD sludge composite confirmed better erosive wear characteristic than other two classes of composites.

References

[1] Khruschov M M 1974 *Wear* **28** 69.
[2] Lee G Y, Dharan C K H and Ritchie R O 2002 *Wear* **252** 322.
[3] Harsha A P Tewari U S and Venkatraman B 2003 *Wear* **254** 693.
[4] Ahmed S and Jones F R 1990 *J. Mater. Sci.* **25** 4933.
[5] Basavarajappa S and Ellangovan S 2012 *wear* **296** 491.
[6] Purohit A and Satapathy A 2017 *Poly. Compos.*doi: 10.1002/pc.24610.
[7] Bagci M 2016 *Tribol. Int.* **94** 336.
[8] Katz H S and Mileski J V 1987 "Handbook of fillers for plastics", A Von Nostrand Reinhold Book International Journal of Plastics Technology.
[9] Pati P R and Satapathy A 2013 *International Journal on Mechanical Engineering and Robotics* **1** 39.
[10] Padhi P K and Satapathy A 2013 *Tribology Transactions* **56** 789.
[11] Purohit A and Satapathy A 2016 *J. Compos. Mater* **51** 899.

IOP Conf. Series: Materials Science and Engineering **338** (2018) 012060 doi:10.1088/1757-899X/338/1/012060

Mechanical properties enhancement and microstructure study of Al-Si-TiB_2 in situ composites

S K Sahoo*, J Majhi, A B Pattnaik, J K Sahoo and Swagat Das

Department of Metallurgical and Materials Engineering

Indira Gandhi Institute of Technology, Sarang, Odisha, INDIA-759146

*Email: sandeep.talcher@igitsarang.ac.in

Abstract. Al–Si alloy-based composite is one of the most promising MMC materials owing to its outstanding mechanical properties, wear and corrosion resistance, low cost and ability to be synthesized via conventional casting routes. Challenges in achieving clean interface between reinforced particles and matrix alloy have been overcome by means of in-situ techniques of fabrication. Present investigation is concerned with synthesizing Al-Si-TiB_2 in-situ composites through stir casting route using K_2TiF_6 and KBF_4 halide salts for exothermic salt metal reaction. X-Ray diffraction analysis revealed the existence of TiB_2 in the prepared samples. Effect of TiB_2in-situ particles in the Al-Si base alloy has been investigated from the results obtained from optical microscopy as well as SEM study and wear analysis with a pin on disc wear testing apparatus. Improved hardness and wear properties were observed with addition of TiB_2.

1. Introduction

Al-Si Hypereutectic alloys have gained the curiosity of researchers for their capability to substitute CI counter parts in different fields of application. Al–Si hypereutectic alloys are broadly expended in aerospace industries and automotive fields in a variety of product applications demanding high strength to weight ratio, better wear and corrosion resistance, good machinability, superior castability and thermal expansion coefficient [1]. Achievement of improved mechanical properties of these alloys is feasible by concurrent grain refinement and modifying the primary as well as eutectic silicon with the reinforcement of TiB_2. Incorporation of hard TiB_2, SiO_2, SiC or Al_2O_3 particle reinforcement in the base alloy matrix seldom accounts significant ascent the density but rather guarantees appreciable enhancement in the strength to weight ratio, modulus and tribological properties [2-5].

Al-Si-TiB_2 composites reveal beneficial and exceptional features in regard to its properties and synthesis route. Being a ceramic particle TiB_2 reveals some peculiar properties like very high melting point (more than 2750°C), excellent hardness value (above 950 HV) and superior modulus (about 5.3 x 10^5GPa). Strain fields are generally built up at the interface due to the mismatch in the coefficient of thermal expansion between Al and TiB_2 as a result of which rise in dislocation density occurs and serve as preferable sites for nucleation. Moreover, the exothermic reaction of halide salts in the Al melt accounts for a beneficial contribution for formation and reinforcement of in-situ TiB_2 particles in Al based alloys with the use of salt metal reaction. Unlike ex-situ method, incorporation of the in-situ technique for synthesis of the composite eliminates inhomogeneous distribution, poor wettability of the reinforcing phase and provides a clean interface in absence of undesirable reaction products. TiB_2 particles are accountable for grain refinement of the Al-Si alloy, pining of dislocation line, and formation of Orowan dislocation loops and also grain boundary strengthening in Al-Si-TiB_2 metal matrix composites [6-8].

In this work an attempt has been made to explore the synthesis and characterization of Al-Si eutectic and hyper-eutectic alloys with 5wt. % in-situ TiB_2 through salt metal reaction with the help of stir casting method.

2. Materials and methods

Composites of Al-Si alloy with 12.6, 15 and 18 wt.% Si reinforced with 5wt.%TiB_2 have been prepared with the use of halide salts. Physical, mechanical and tribological characterisations have been conducted to evaluate the properties of the samples.

2.1. Composite fabrication

Cp grade aluminium and master alloy of Al-50 wt% Si supplied by Minex Metallurgical Co. Ltd., have been taken in a clay-graphite crucible for melting in a pit type f/c at about 800°C to synthesise Al-x Si (x = 12.6, 15, 18) alloy. The Al-xSi-5TiB_2 composites were manufactured by adding together K_2TiF_6 and KBF_4 halide salts to molten Al-Si alloy and allowed for 30 minutes of reaction time. Titanium Diboride (TiB_2) particles are formed as a result of chemical reactions of the salts in the molten base alloy. Periodic stirring was carried out to ensure the reaction of salts to be complete and dispersion of TiB_2 particles to be uniform within molten aluminium alloy. Before casting, all the lighter dross was discharged to ensure sound casting. The desired composites specimens were casted by pouring into cast iron mould preheated at 450°C [6].

2.2. Characterizations

2.2.1. Microstructural characterization

Samples for microstructural observations were prepared following standard procedures and etched with Keller's reagent. The micrographs of the composites have been studied with the use of optical microscope (CARL ZEISS) at varying magnifications and with Scanning Electron Microscope (SEM) as well.

2.2.2. XRD analysis

XRD analysis of the prepared samples was carried out with Cu-Kα target to recognize presence of various constituents of in the samples by matching the obtained peaks with JCPDS data files.

2.2.3. Hardness and Density Test

Hardness values of the specimens were measured in Microvicker's hardness tester using a square base diamond pyramid as indenter with 1kgf applied load and 15 seconds of dwelling time Samples were tested for their density in Mettler-Toledo digital balance with density kit.

2.2.4. Wear test

A pin-on-disc wear testing machine have been in use to evaluate the wear behaviour of the cylindrical pins of the composite sample with diameter of 10mm and 30mm length. The operating test parameters comprised of varying load (10, 20, 30N) with 40mm track radius for 300 seconds at room temperature without use of any lubricating material. After every test the mass loss due to wear of each specimen was determined. The role of applied load and amount of silicon on wear characteristics of the prepared composites was studied.

3. Results and discussions

3.1. Microstructural characterization

Figure 1 and figure 2 reveals that the composite with eutectic composition of the alloy exhibits needle form of silicon uniformly scattered all over the Al-matrix with little amount of primary silicon due to non equilibrium cooling. But in the Al-Si alloy with hyper eutectic compositions more primary Si particles are evident in micrograph. In presence of TiB_2 particles, those act as the preferred sites for nucleation during solidification,refine the structure and almost rounded shaped primary Si of smaller size were formed unlike cast base alloys.

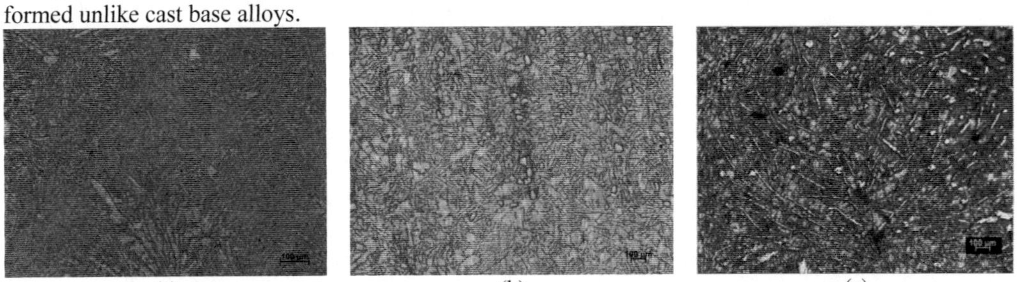

(a) (b) (c)

Figure 1.Optical Micrographs of (a) Al-12.6Si-5TiB_2 (b) Al-15Si- 5TiB_2 (c) Al-18Si-5TiB_2.

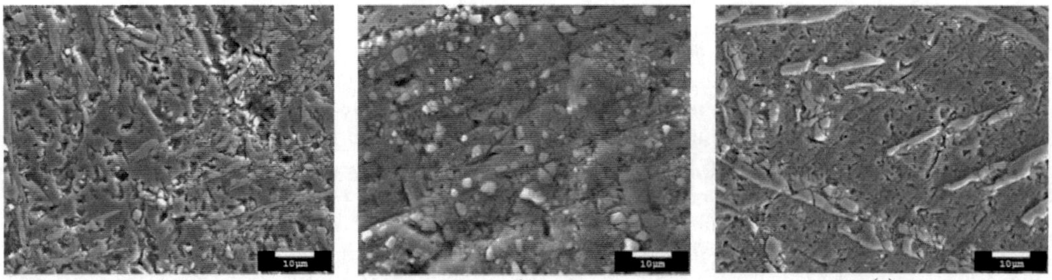

(a) (b) (c)

Figure 2.SEM Micrographs of (a) Al-12.6Si-5TiB_2 (b) Al-15Si- 5TiB_2 (c) Al-18Si-5TiB_2.

3.2. XRD analysis

The largest peaks in the obtained XRD results describe the presence of aluminium. Minor peaks signify the existence of silicon and TiB_2 particles. It is also noticed in figure 3 that the intensity of the peaks of silicon increases with its increase in the weight percentage in the base alloy. Tiny peaks of TiB_2 in the XRD pattern corroborate the occurrence of TiB_2 in the selected alloy matrix.

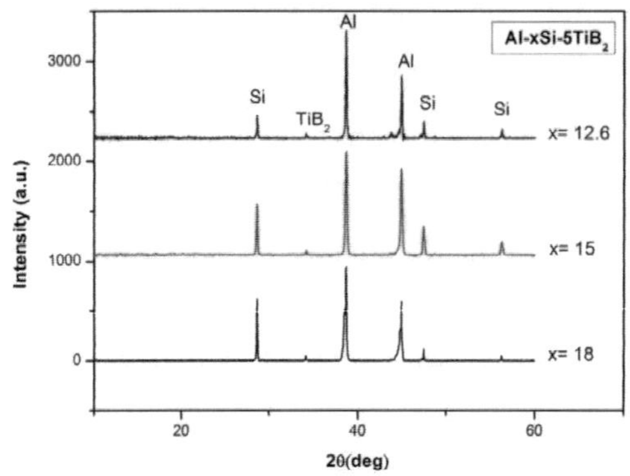

Figure 3.XRD analysis of composites

3.3. Hardness and density

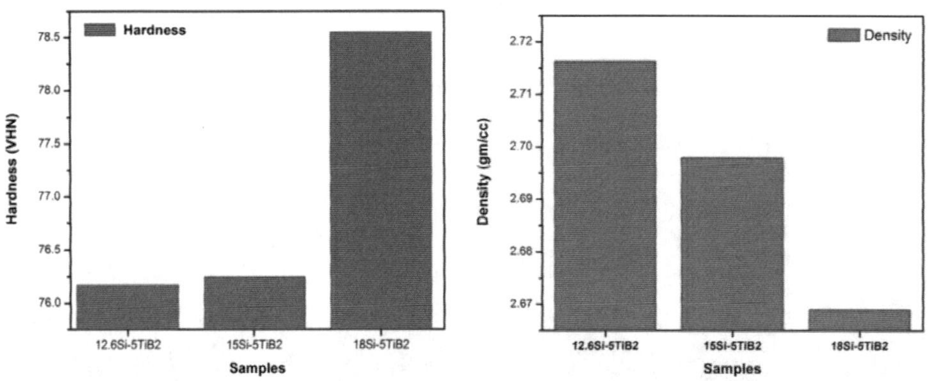

Figure 4 (a) Hardness as a function of amount of silicon (b) Density as a function of amount of silicon of the composites

The VHN and density histograms of figure 4 illustrate that AMC with 18 % Si possesses utmost hardness with minimum density value. Rise in the quantity of the hard primary silicon as well as eutectic Al-Si phases as a result of increasing Si content in the base alloy accounts for enhanced resistance to indentation and provides high hardness values for the composites.

3.4. Wear test

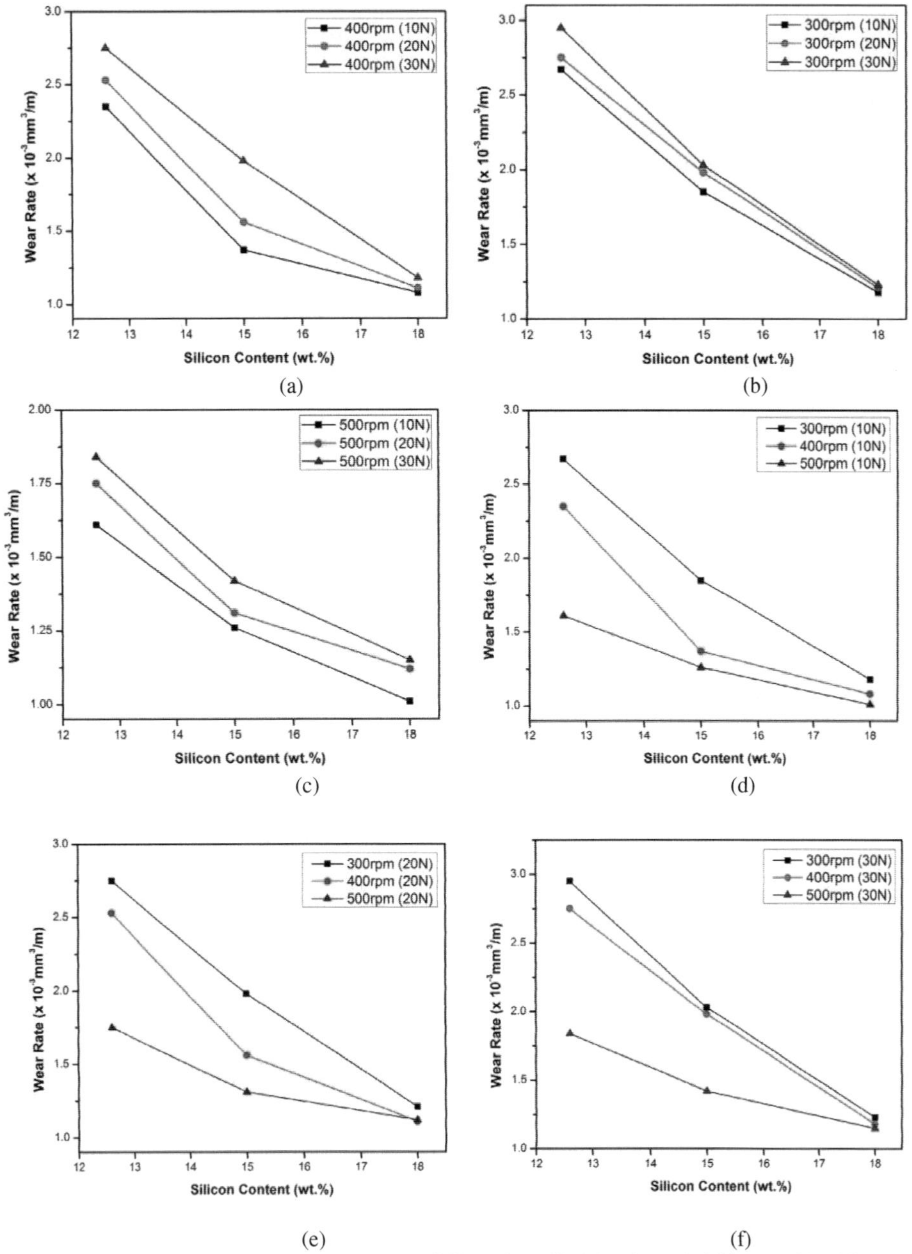

Figure 5.Wear rate with respect to amount of Si and applied load at (a) 300 rpm,(b) 400rpm (c) 500rpm and Wear rate with respect to amount of Si and rpm at (d) 10N,(e) 20N (f) 30N.

IOP Conf. Series: Materials Science and Engineering **338** (2018) 012060 doi:10.1088/1757-899X/338/1/012060

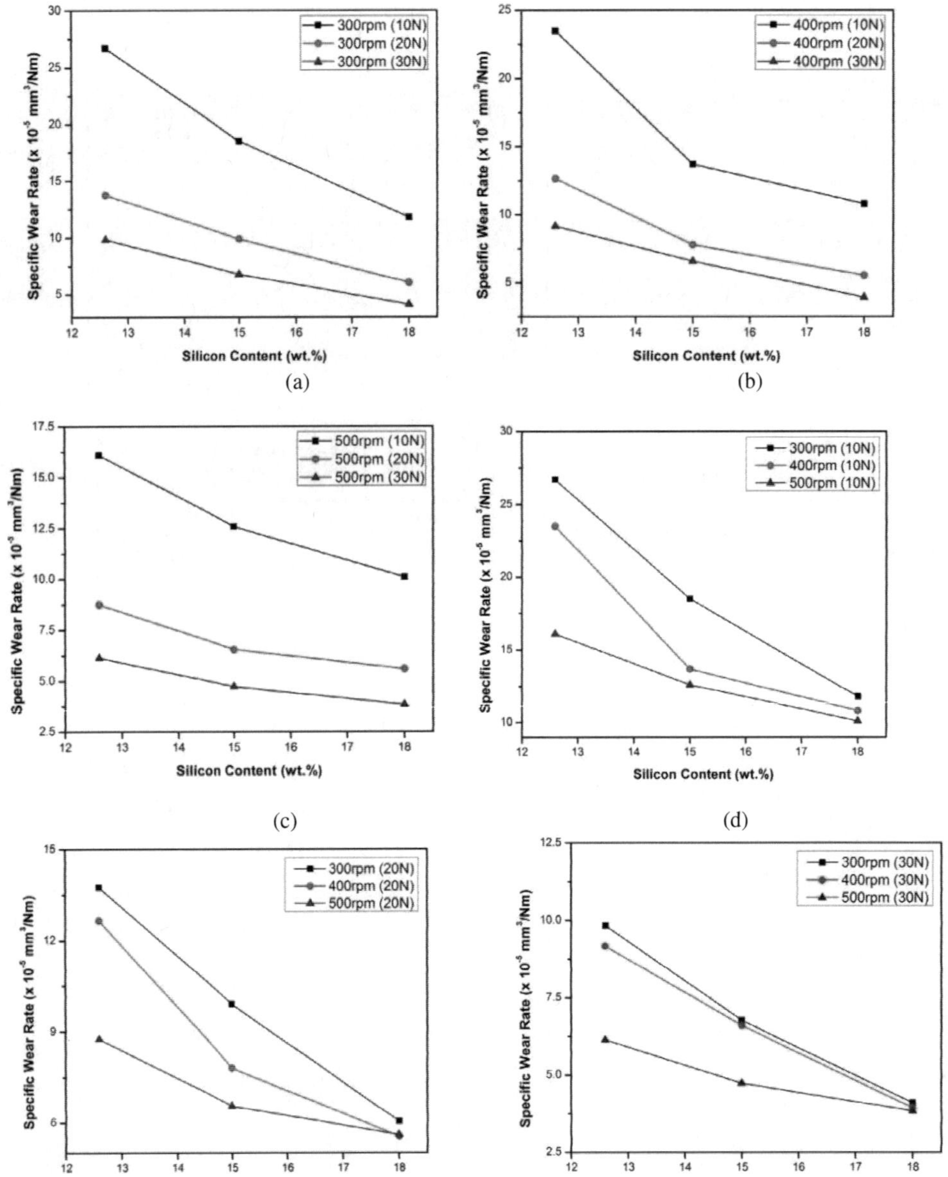

Figure 6 Sp. wear rate w.r.t. amount of Si and load at (a) 300 rpm,(b) 400rpm (c) 500rpm and Sp. wear rate w.r.t. amount of Si and rpm at (d) 10N,(e) 20N (f) 30N.

Figure 5 shows wear rate of prepared samples with three varying wt% of Si, at different applied load and rpm. The wear rate found to decrease with increasing amount of Si because of enhancement in hardness of the composite, refining of primary as well as eutectic silicon and their modification but the wear rate was brought up with higher applied load and found to be maximum at a load of 30N. Figure 6 shows specific wear rate of the composite and a similar trend was observed for varying amount of Si and minimum specific wear rate was observedin composites with increased amount of silicon in all conditions. Uniform dispersion of the TiB_2 particles a result of intermittent stirring and achievement of size reduction of blocky primary silicon as a result of presence of TiB_2 particles contributes to lower sp. wear rate in Al-18Si-5TiB_2 composites.

Generally, with increasing applied load sp. wear rate decreases for a fixed sliding distance due to the work hardening of the sliding surface. Modification in morphology from coarse eutectic Si plates to fine needles and refinement of the primary Si leads to high wear resistance of Al–Si alloys. Wear rate as well as sp. wear rate found to be inversely proportional to sliding speed due to more resident time for intimate contact of the mating surfaces and more material flow required maintain the relative motion. This corroborates the findings of S.A. Kori, T.M. Chandrashekharaiah [9].

4. Conclusions

Present analysis leads to the conclusions as follows:

1) In-situ Composites of Al-Si alloys reinforced with 5wt%TiB_2 have been successfully manufactured through exothermic reaction of halide salts via stir casting route.
2) Occurrence of TiB_2 in the prepared composites was ascertained by XRD analysis.
3) Higher hardness and lower density values were perceived in composites with higher amount of Silicon.
4) Wear rate as well as specific wear rate found to decrease at all the working conditions with higher silicon content.
5) Microstructure study disclosed the role of TiB_2 in alteration in morphology and distribution of primary and eutectic silicon in Al-Si base alloy.

References

[1] Gupta M and Lavernia E J 2001 *Journal of materials processing technology* **54** 1-4 261-270.
[2] Mandal A, Maiti R, Chakraborty M and Murty B S 2004 *Materials Science and Engineering* A **386** 1-2 296-300.
[3] Sahoo S K, Majhi J, Patnaik S C, Behera A, Sahoo J K and Sahoo B P 2017 *Elixir Materials Science* **113** 49066-69.
[4] Mandal A, Murty B S and M. Chakraborty 2009 *Materials Science and Engineering* A **506** 1-2 27-33.
[5] Mandal A and Makhlouf M M 2009 *Transactions of the Indian Institute of Metals* **62 4**-5 357-360.
[6] Mallikarjuna C, Shashidhara S M, Mallik U S and Parashivamurthy K I 2011 *Materials & Design* **326** 3554-59.
[7] Rajan H B M, Ramabalan S, Dinaharan I and Vijay S J 2014 *Archives of civil and mechanical engineering* **14** 1 72-79.
[8] Sahoo J K, Sahoo S K, Sutar H and Sarangi B 2017 *IOP Conference Series: Materials Science and Engineering* **178** 1 (IOP Publishing, 2017) pp1-10.
[9] Kori S A and Chandrashekharaiah T M 2007 *Wear* **263** 1-6 745-755.

Influence of Temperature on Pyrolysis of Biomass

S Kumar[1*], D Pattnaik[1], S K Bhuyan[1], S C Mishra[1]

[1]Metallurgical and materials engineering department,
National Institute of Technology , Rourkela ,India

E-mail*:sudhanshunitrkl@gmail.com

Abstract: The Antimicrobial and anti-inflammatory nature of the castor oil is used for the treatment of constipation and also used to clean the intestine before a surgical treatment. The growing demand castor oil due to its cheaper price and have much demand to grow hairs faster which give proteins to hair and increases the life span of hair and increases the hair capacity and provide hair smoothness because it has high ricinoleic ratio which is beneficial for skin and hair. The main objective of this paper is by observing its usefulness and the current aim for the research work is to produce bio-char/carbon compounds and extraction of higher carbon content. The castor leaves are collected and dried in the sun light to get it moisture free and proximate analysis is studied that reveals 14% of fixed amount of carbon. The dried leaves of castor plant are pyrolized at different temperature ranges 250°C, 350 °C, 450°C, 600°C and 900°C. The pyrolised specimens were studied by Fourier-transform infrared spectroscopy analysis which shows the transformation and formation of several bonds including carbon bonds which are affected during carbonization. Scanning electron microscope and energy dispersive x-ray spectroscopy have been studied on pyrolised specimen to get microstructures elements in the bio-char. The parameter called carbon/oxygen ration gives the advantage of char in the agricultural land for fertility motive and discrepancy of carbon and oxygen contents have been analysed. It has been found that carbon/oxygen ration increased along increment in Pyrolysis tempt. Up to 500°C and therefore along moreover increment in tempt. Beyond 500°C there is a decrement trend has been found.

Keywords: temperature, pyrolysis, FTIR, bio-char.

1. Introduction

Biomass is stated as a industrial word from which energy is generated by burning the wood and organic wastes which releases carbon emission but it has been classified as a renewable source of energy. In the current scenario the bio wastes are substitute as a auxiliary source of energy and several works have completed previously due to degradation of the fossil fuels and increasing demands for the energy. A huge amount of bio wastes and biomass residues generated from various economic action and human activities [1-2]. By burning biomass carbon emission is released. The large amount of biomass and bio-waste are generated from different economic activity. Pyrolysis is stated as the thermal decomposition of substance at elevated temperature in presence of inert atmosphere as a vacuum or nitrogen gas. It turns organic substance in to carbon containing residue where several reactions occur at the same time which effectively applied to the castor leaves resulting in to some different product such as biomass. The products of Pyrolysis might be such as bio-char, fuel gas, oil and tar [3]. Pyrolysis is done to get pyrogenic carbon [4] which include black carbon, soot, bio-char. Bio-char is stated as a carbonaceous product produced by Pyrolysis or gasification in which bio mass is heated in the absence of oxygen or under reduction. Bio-char is used additionally as a soil enhancer which increases the water holding capacity of the agricultural land and also increases the fertility of the soil. Castor leaves provide high char on the basis of its high lignin content and also has metallic quantity high. Polysaccharide gives less char as compared to lignin which has been analyzed in the study [5].

2. Materials and methods

The leaves of castor plant have been gathered from the college compound. These leaves were dried under the sunray for two months to get rid of the moisture and then it was made in fine powder form by the help of grinder machine by running at the speed of 250 rev/min for forty minutes subsequently grinder is stopped and powdered leaves is collected from the machine and then it is stored in a zip packet to avoid from contaminated air and vapors.

2.1 Proximate analysis

Castor leaves powders are subjected for proximate analysis. 2 gram of sample were putted in to uncovered crucible and stored in an oven for 1 hour in a maintained temperature of 105°C to get the moisture content. Again, above amount of powdered leaves were kept into a covered silica cup and after that is placed inside the pit furnace for 7 minutes in a maintained temperature of 910°C to know volatile quantity of matter. Afterwards, same quantity of powdered leaves putted into the uncovered silica cup and placed inside the pit furnace for 30 minutes in a maintained temperature of 770°C. Then after it crucible is seized and amount of ash content was studied and thereafter fixed amount of carbon is studied on the basis of proximate to determine the features of raw specimen which is to be pyrolised.

2.2 Pyrolysis

The char is being obtained after paralyzing the organic substance and stated as a disintegration of biological substance at high temperatures thermal decomposition take place in presence of inert atmosphere or in presence of nitrogen. Bio char is carbon dense solid. This enhances a solid mixture which known as bio char and some liquid amount of substance is called as bio-oil and gaseous amount of substance produced is called syngas. Temperature plays an important role in in yielding the Pyrolysis process. At a low temperature range between 400-500°C incurs more bio char while at higher temperature generally greater than 700 the liquid and gaseous products are obtained.

2.3 Processing and characterization of raw material

The composition and microstructure of Specimen have been examined by applying a Nova Nano SEM 450 branded by FEI fesem machine with an INCA analyzer system before analyzing the sample it has being coated with gold coating in a stutter coating for 300 seconds after that sample is brought up and positioned inside a case to occupy the sample and to read the sample in the machine and capturing space of the specimen is 6mm. after that picture of char is recorded at several magnifications. hydraulic press machine is used to make the pellet which contain powdered samples and potassium bromide which contain the ration of the ration of one: nine and 8-9-ton load was applied in the hydraulic press to make the pellet and this pellet is subjected to IR Radiation and from which chemical functional group is identified by using FTIR Perkin Elmer Spectrum.

3. Result and Discussion

3.1. Proximate analysis

Castor leaves powders are subjected for proximate analysis. 2 gram of sample were putted in to uncovered crucible and stored in an oven for 1 hour in a maintained temperature of 105 to know the content of the moisture present in the specimen. Again, above amount of powdered leaves were kept in a covered silica cup and after that is placed inside the pit furnace for 7 minutes in a maintained temperature of 910°C to know the content of volatile matter. Afterwards, above quantity of powdered leaves kept in the uncovered silica cup and placed inside the pit furnace for 30 minutes in a maintained temperature of 770°C. Then after it crucible is seized and amount of ash content was studied and thereafter fixed amount of carbon is studied on the basis of proximate to determine the features of raw specimen which is to be pyrolised.

The amount of organic substances such as moisture content, ash content, fixed carbon and volatile matter can be determined by applying proximate analysis and gas been tabulated as below: -

Proximate analysis of castor leaves	***Wt.%***
Moisture content	*11*
Ash content	*10*
Fixed carbon	*14*
Volatile matter	*65*

Table 1. Proximate analysis

3.2. Characterization and analysis of bio chars:

3.2.1 Microstructure- analysis

Powdered castor leaves which is carbonized at different temperature is subjected to fesem for the micro structural analysis which reveals the feedstock along with bio char which is described below. It reveals irregular, packed morphology with cross linkage structure shown in given below figures. The cross linkage shows the presence of network consist of lignin, pectin, hemicelluloses and cellulose. In few pores of char lamellae is also observed.

Fig: 1. FESEM microstructure of Bio char carbonized at 250°C

Fig: 2. FESEM microstructure of Bio char carbonized at 350°C

IOP Conf. Series: Materials Science and Engineering **338** (2018) 012061 doi:10.1088/1757-899X/338/1/012061

Fig: 3. FESEM microstructure of Bio char carbonized at 600°C

Fig: 4. FESEM microstructure of Bio char carbonized at 900°C

3.2.2 Analysis of the Element

EDS is employed for the analysis of the element of the sample. It is the analytical method. The element which is analyzed is given below in the Table 2. And plotted in the fig. 5. We observed that in carbonized sample the percentage of atomic weight of oxygen, silicon, carbon, calcium and potassium varies with carbonized temperature. Above all we found that percentage of atomic weight of oxygen and carbon is high on various temperatures.

Temperature	oxygen	silicon	carbon	calcium	potassium
0	34.39	1.01	59.93	0.23	0.88
250	27.52	0.67	55.34	3.23	4.10
350	32.87	1.08	56.52	2.41	4.46
450	30.01	0.45	61.19	2.35	3.23
600	55.01	2.02	12.99	9.98	9.21
900	51.99	3.01	11.01	0.00	0.03

Table 2. Atomic weight percentage of various element with carbonizing temperature.

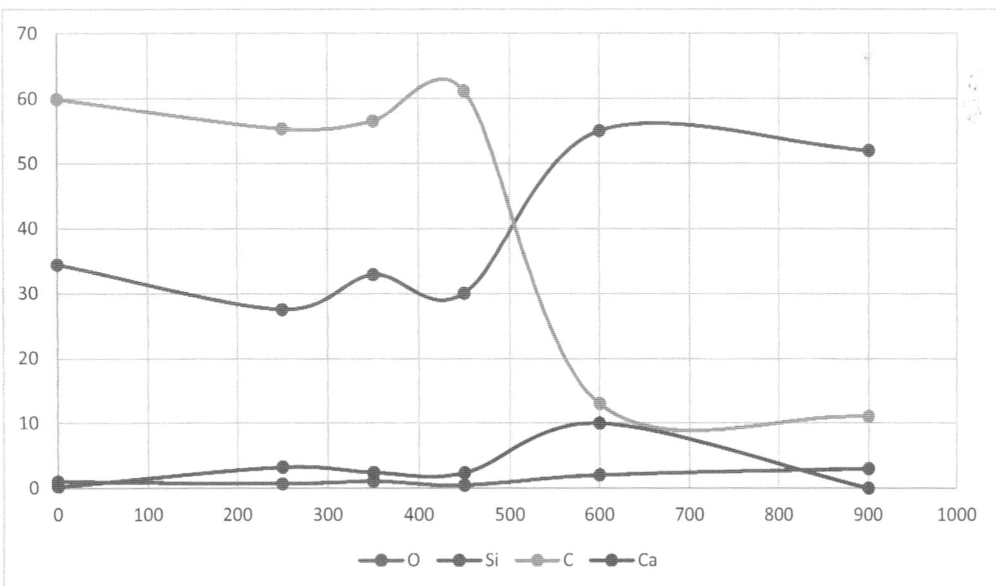

Fig 5. Atomic weight % percentage VS Tempt. Graph

3.3 FTIR Analysis

Fourier transform infrared spectroscopy technique were performed to determine the distribution of functional group present in the pyrolised product (bio char) which is either solid, liquid or gas as infrared spectrum of adsorption or emission. IR radiation is passed through the sample to detect is functional group or various organic system. In given below figure 6 show different functional group present in the castor leaves.

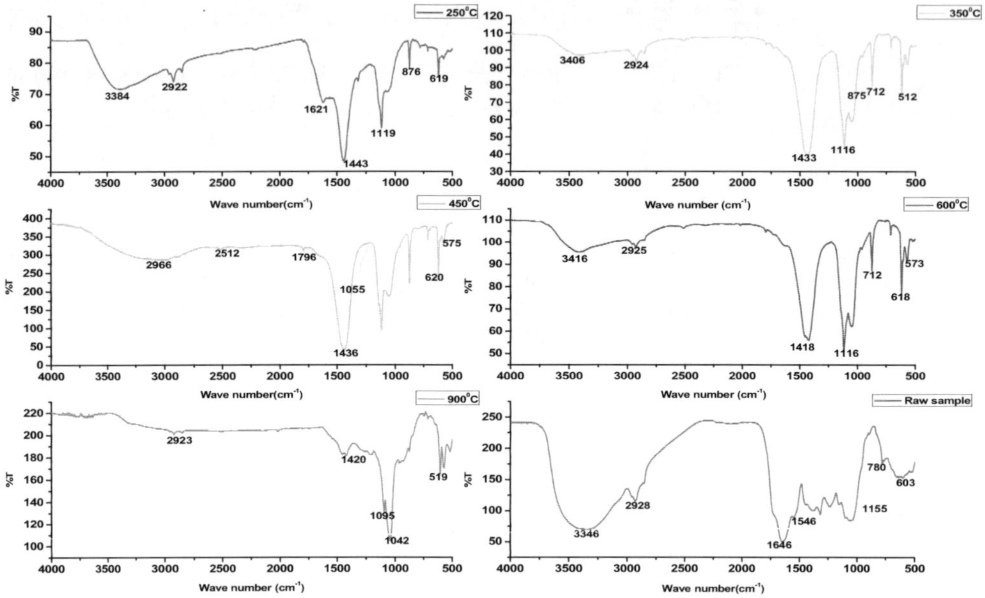

Fig 6. Graph of FTIR of the samples

The absorption peak originating from the molecules as vibrational transitions shows the characteristics of different organic group. Occurrence of absorption band shows the characteristics of class compound. (e.g. Aromatic, alcohols, acid, ethers, amides etc.). The Powdered castor leaves are subjected to ftir spectroscopy which determines the functional group of organic matter shown in above figure 6 which reveals the presence of the alkenyl group (CH_2=CH-), phenols, aliphatic cyanide and nitride. The presence of O-H stretching vibration band between 3250 and 3500 cm^{-1} shows the presence of carboxylic and alcohol as well as water. The C=O stretching vibration band shows the presence of ketones, carboxylic acid and aldehyde between 1870 and 1600 cm^{-1}. The C-O stretching vibrations Band between 920 and 650 cm^{-1}shows the appearance of alcohol and esters. The bio chars spectrum shows the presence of aliphatic character indicated by the absorption band present between 2990 and 2780 and slight peaks have been also observed between 1300 and 1440 cm-1. Little peaks have been also observed between 515 and 700 which reveal the presence of aromatic stretching vibration. These all characteristics have been analyzed in pyrolised specimen at several temperatures.

4. Conclusion

The main purpose of this research work is to use the bio waste product as a useful resources and to make high value product and to conserve it for future generation as well as alternative source of exhausting fossil fuel and to save environment from pollution as well as to generate employees in rural areas and skilled person. The efficient way to utilize the bio waste product thermo chemically for various applications. Bio char is obtained after the Pyrolysis of castor leaves process. The fixed carbon is studied 14%. These bio char is used to enhance the fertility of the soil and used for water holding capacity in the agriculture land, and also there is variation of such element oxygen, nitrogen, silicon, calcium with the variation of temperature is observed and amount of carbon content is also affected with temperature variation is observed. and metallic element etc. There is variation in bonding during temperature treatment is observed. Ftir studied proves the presence of carboxylic, aldehyde, ketones etc. The recent study of this research will also determine bio char properties which would be beneficial for the future work.

References

[1] Hume E P 1949 Economic Botany **3** 42-45.
[2] Tejano E A 1985 Philippine Journal of Coconut studies **1** 1-7.
[3] Festin T F and Jose W I 1979 Conservation & Recycling **3** 383-388.
[4] Shenbagavalli S and Mahimairaja S 2012 International Journal of Plant, Animal and Environmental Sciences **2** 197-201.
[5] Mukherjee A and Lal R 2013 Agronomy **3** 313-339.
[6] Ben A and Kraiem T 2013 Science Direct **12** 1.
[7] Mishra S C and Supriya A 2017 LAP Lambert Publication.

IOP Conf. Series: Materials Science and Engineering **338** (2018) 012062 doi:10.1088/1757-899X/338/1/012062

Synthesis of quaternary chalcogenide CZTS nanoparticles by a hydrothermal route

S Das, K Sa, P C Mahakul, J Raiguru, I Alam, BVRS Subramanyam, P Mahanandia*
Department of Physics & Astronomy, National Institute of Technology, Rourkela-769008, Odisha
*E-mail: pitam@nitrkl.ac.in

Abstract: Cu_2ZnSnS_4 (CZTS) has emerged as a potential absorber towards inorganic photovoltaic device application for its outstanding properties like non toxicity, earth abundancy nature, optimal band gap matched with solar spectrum (1.45- 1.65eV), high absorption coefficient (10^4cm^{-1}). Here, a low cost, environment friendly facile hydrothermal route to synthesize phase pure CZTS nanoparticles using Cu (II), Zn (II), Sn (II) inorganic metal salts and thiourea as Sulphur source in distilled water solution as precursor is reported. The as synthesized samples characterized by X-Ray diffraction (XRD) and RAMAN confirmed structure and phase of CZTS nanocrystals. The morphology of the prepared CZTS have been characterized by scanning electron microscopy (SEM). The particle size is found in the range 4-5 nm with crystalline nature have been characterized by transmission electron microscope (TEM). The optical band gap of the as prepared samples is calculated to be 1.65eV from UV-Visible analysis which proves it can be used towards photovoltaic applications.

1. Introduction

Increasing in worldwide clean energy crisis calls for the development of harvesting the renewable solar energy through photovoltaic cell (PVC). Thin film solar cells (TFSCs) belong to second generation PV, in which active absorber layer attracts the attention of research community due to large versatility in deposition techniques, device fabrication process and tuning the photo conversion efficiency (PCE). Mostly, TFSCs include the high efficient absorber materials like cadmium telluride (CdTe), copper indium gallium selenium (CIGS), copper indium selenium (CIS) [1, 2]. The highly toxicity of elements like indium, gallium, selenium urgently calls for the non- toxic, earth abundant, low cost absorber materials towards PVC application [3]. Being a P- type semiconductor, Cu_2ZnSnS_4 referred as CZTS has emerged as a potential absorber to overcome the existing glitches along with its outstanding specifications like high absorption coefficient (>10^4cm^{-1}) and the tunable optical band gap of 1.45-1.65eV which has favorably match with the solar spectrum [4]. According to physical limitation of Shockley Queisser (SQ) photon balance the PCE for a single – junction CZTS solar cells should be 32.2% [5]. On the other hand, in practical the efficiency has reached to the level of 12.6% by Todorov et al. [6] which is noticeably less in comparison to the predicted theoretical value. To enhance the PCE of CZTS based solar cells there are continuous research effort to optimize the material properties and device architecture. There has been lot of research focus on the methods of preparation of CZTS which controls the phase and sizes. Since the Vacuum as well as non- vacuum methods are utilized to deposit and fabricate CZTS based TFSCs. Vacuum routes consists of atomic beam sputtering, RF magnetron sputtering, thermal evaporation, and pulse laser deposition (PLD), chemical vapor deposition (CVD) etc. Vacuum based deposition methods suffer certain drawbacks like high input capita (like high vacuum, high temperature, expensive chemicals etc. sophisticated experimental set up), low output and difficulties during deposition into large area based devices. These shortcomings are the limitations for achieving the ultimate goal of maximal PCE of low cost, environmentally sustainable TFSCs. Thus non vacuum deposition technique such as electrochemical deposition, spray pyrolysis, SILAR method, sol gel method, direct liquid coating are promising methods for synthesis of CZTS. The high performance direct liquid coating deposition routes for deposition of CZTS either utilize binary metal sulphides solution based inks or sulfurization of metal precursor [7]. The maximal efficiency of 12.6% was obtained for CZT(S,Se) nanocrystals based solar cells fabricated by Todorov et al. [6] through a hydrazine hydrate-based solution approach. The hydrazine-based process suffers from major drawback i.e hydrazine is massively explosive, hepatotoxic, and carcinogenic. Zinc sulphides typically have negligible solubility in hydrazine hydrate solvent under ambient conditions. One way to prefabricate the photovoltaic absorber material by using nanocrystal ink and these inks could be ideally deposited under mild conditions by screen paint, knife coating, roll to roll, dip coat and spin coating etc. for large scale production. Guo et al. [8] reported that the PCE of CZTS based photovoltaic device up to 7.2% by simply coating colloidal CZTS nanoparticle ink with subsequent annealing and post selenization steps. As material synthesis procedure greatly affects the device performance, it should be cost effective, environment friendly, easily hand able. On the other hand, nanocrystals seek tremendous attention over bulk materials as they are solution processible, tunable band gap with respect to nanocrystal size which realizes multi junction with multiple carrier generation. Various methods are employed to synthesize CZTS nanocrystals such as hot injection [9], sol-gel [10], microwave assisted method [11], solvothermal [12] etc. Among them one pot facile solvothermal wet chemical approach provides the uniform precursor formation with controlled phase and growth nanocrystal size. Mostly CZTS nanoparticles were prepared via amine group solvents as surfactant, capping agent and stabilizer which are quite expensive. To overcome this, synthesis of size controlled CZTS nanoparticles via a simple, environment friendly, low cost method should employed.

Chunrui Wang et al. [13] synthesized CZTS nanocrystals by hydrothermal route using distill water as solvent and they have found~7nm range particles confirmed through transmission electron microscopy. Here we report, a low cost green hydrothermal synthesis approach of CZTS nanocrystals using $CuCl_2.2H_2O$, $ZnCl_2$, $SnCl_2$ $.2H_2O$ metal salts and thiourea as reacting agent. As for our concern there is only one paper on this context is reported. Further improvement in physical properties and sized of the prepared CZTS could be optimized correlating with the preparation conditions.

2. Materials and methods

In an environment friendly synthesis process, Copper chloride dihydrate ($CuCl_2{\cdot}2H_2O$), Zinc chloride ($ZnCl_2$) and Stannous chloride dihydrate ($SnCl_2{\cdot}2H_2O$) and thiourea (CH_4N_2S) were all of analytical grade and used as received. All distilled water used throughout synthesis was procured from Milli-Q purification system. Copper chloride dihydrate, Zinc chloride, stannous chloride dihydrate and thiourea as Sulphur source (with molar ratio 2:1:1:4) are dissolved in adequate amount of distilled water to make precursor solution. The precursor has been stirred for 45 mins till the color of the solution turns clear, slightly yellowish. Then the precursor solution was transferred into a 100ml stainless steel with Teflon liner autoclave kept inside in a muffle furnace for 20 hours at 200°C temperature. After the hydrothermal reaction the autoclave is allowed to cool to ambient conditions naturally. The obtained precipitate was washed 4 times with ethyl alcohol and dried at 80°C under vacuum condition. Then the obtained black powder was collected for further characterizations. The schematic diagram of the synthesis procedure is given in Figure 1.

The synthesized CZTS powder has been characterized through various characterization techniques. The crucial factors like crystallinity and the phase of the synthesized sample was confirmed by XRD, RAMAN (Renishaw Raman microscope)and the particle size and the information regarding the nano crystal distribution was revealed by the microscopy study Technai G220 (transmission electron microscopy (TEM), Nano nova scanning electron microscopy (SEM)). Structural characteristics of powder were determined by XRD analysis carried out using a diffractometer (Philips X'pert) with a scan rate of 5° per minute with step size 0.05 in the 2θ range from 10° to 80°. The analytical diffractometer (Rigaku Japan, Ultima-IV) operating in Bragg configuration is equipped with graphite-monochromitized with Cu Kα radiation (λ = 1.54184 A°). The morphology, size and shape of the derived particles were examined by TEM and from its selected area electron diffraction (SAED) and high resolution TEM (HRTEM) results the crystal structure of the samples can be further identified. Morphological view made by TEM displays of nano-sized crystals well dispersed (c= 0.03mg/mL) in N, N dimethylformamide (DMF, C_3H_7NO) as a dispersant medium. Specimens for transmission electron microscopy were obtained by taking CZTS sample-DMF dispersion onto a 300-mesh Cu grid, coated with a lacey carbon film and subsequent evaporation of the solvent. For optical analysis c= 0.01mg/ml was prepared in 15ml valve. The band gap was extracted from the data obtained by the analysis of UV-Visible spectroscopy.

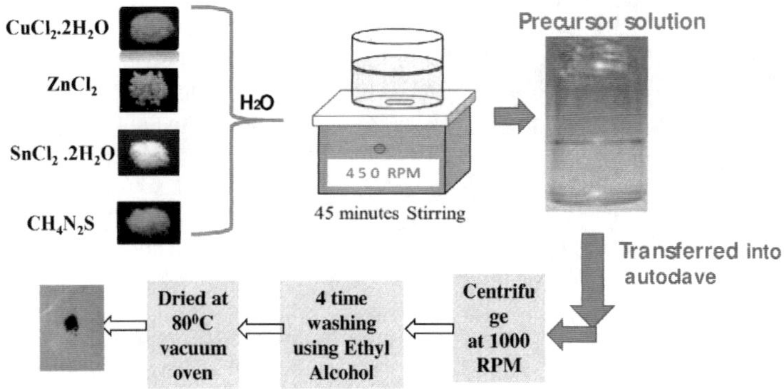

Figure 1. Schematic diagram of CZTS nanocrystal synthesis

3. Result and Discussion

Figure 2(a) shows the diffraction pattern of CZTS nanocrystals prepared at low temperature 200°C for 24 hours. It exhibits prominent peaks at the positions 2θ = 27.98°, 31.75°, 38.23°, 45.65°, 54.73°, 74.13° assigned to the orientation of (112), (200), (211), (220), (312) and (332) diffraction planes corresponds to the kesterite phase of CZTS (JCPDS card no-26 0575) respectively. From the Debye-Scherrer formula:

$$d = 0.89\lambda/\beta \, \text{Cos}\theta$$

(Where, 0.89 is a constant called shape factor, λ is X- ray wavelength, θ is Bragg angel, β is full width half maximum (FWHM) of significant diffraction peak and d is the crystallite size) the nano crystallite size of as synthesized CZTS sample is calculated to be 4.95 nm from the strongest. The lattice constant of quaternary kesterite CZTS is quite close to binary phase β ZnS and Cu_3SnS_4 structure [14, 15]. The tetragonal CZTS unit cell inside which each sulfur anion is bonded to four cations and each cation is bonded to four sulfur anions has lattice parameters a=b= 5.4270A°, c = 10.8480A° . The kesterite CZTS has a tetragonal unit cell with S atoms located at face - centered – cubic sub lattice and Cu, Zn, Sn atoms occupy the half the tetrahedral interstitial site within S sub-lattice. The lattice parameters for tetragonal unit cell were calculated by using the expression (h, k, and l are the miller indices of the plane):

$$\frac{4Sin^2\theta_{hkl}}{\lambda^2} = \frac{h^2+k^2}{a^2} + \frac{l^2}{c^2}$$

Raman scattering gives a more definitive structure analysis. The information regarding the electronic state excitations of lattice in a material providing the information of detecting secondary phases can be obtained by extremely useful tool Raman Spectroscopy (RS) analysis. The decrease in the particle dimensions can cause a wavenumber shift and Raman peak broadening as a result of phonon confinement observed in nanometer scale. Raman spectra of the nanocrystals shown in Figure 2(b) have a peak around 334 cm^{-1} which correspond to kesterite phase of CZTS. Figure 3 (a) shows the SEM photograph of as prepared CZTS nanocrystals of 5μm scale. The uniform distribution of CZTS nanocrystals is confirmed from SEM photograph. The CZTS agglomeration consists of several small nanocrystals is shown in Figure.

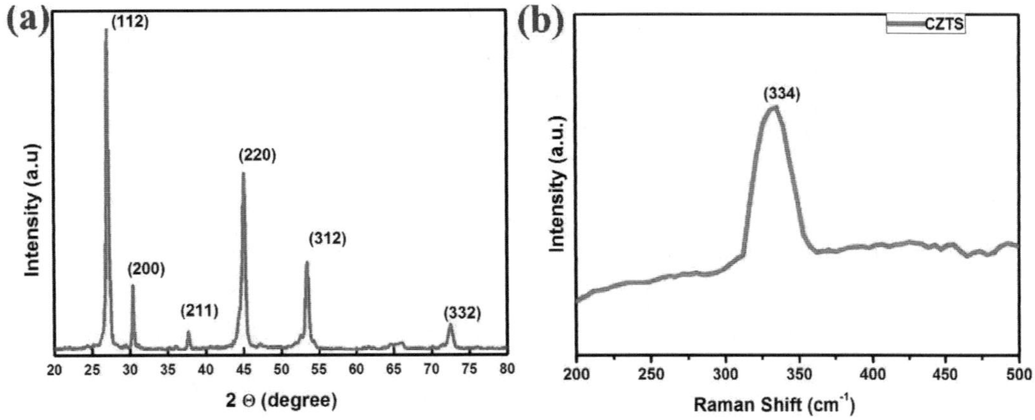

Figure 2. (a) X-Ray diffraction pattern and **(b)** Raman spectra of CZTS nanocrystals synthesized at 200°C for 20h

The low resolution TEM (LRTEM), shown in Figure 3 (b) gives a clear cut picture of the distribution of CZTS nanoparticles. The HRTEM (high resolution TEM) figure given in Figure 3 (c) gives the information regarding crystallinity, size and shape of the synthesized CZTS nanocrystals. From TEM photograph it is clear that the synthesized CZTS nanocrystals are spherical in shape with average nanocrystal size ~4.23nm, which is in agreement with obtained value from XRD data. Figure 3 (d) is the selected area electron diffraction (SAED) pattern of CZTS. The SAED pattern reveals the poly-crystallinity nature of sample. The HRTEM images show inter planar spacing of 3.13A° of kesterite CZTS.

The optical properties of colloidal nanoparticles have quite vital significance for their further use in optoelectronics. Figure 4(a) shows the absorption spectra of as synthesized CZTS QDs as a function of nanocrystal size which features a broad shoulder with a trail towards long wavelength direction. The band gap of the CZTS nano particles stably dispersed in DMF was obtained by plotting $(\alpha h\nu)^2$ as a function of hν also depicted in the Figure 4(b). The band gap of the synthesized sample is estimated from the absorbance plot. The absorbance graph shows a strong absorbance in visible range and a long tail extending to longer wavelengths. From the long wavelength extrapolation of the band edge of the absorption spectra, the optical band gap of the synthesized sample is calculated to be 1.65eV Figure 4 (b). The good absorption in visible range makes the material suitable for the absorber layer towards photovoltaic device application.

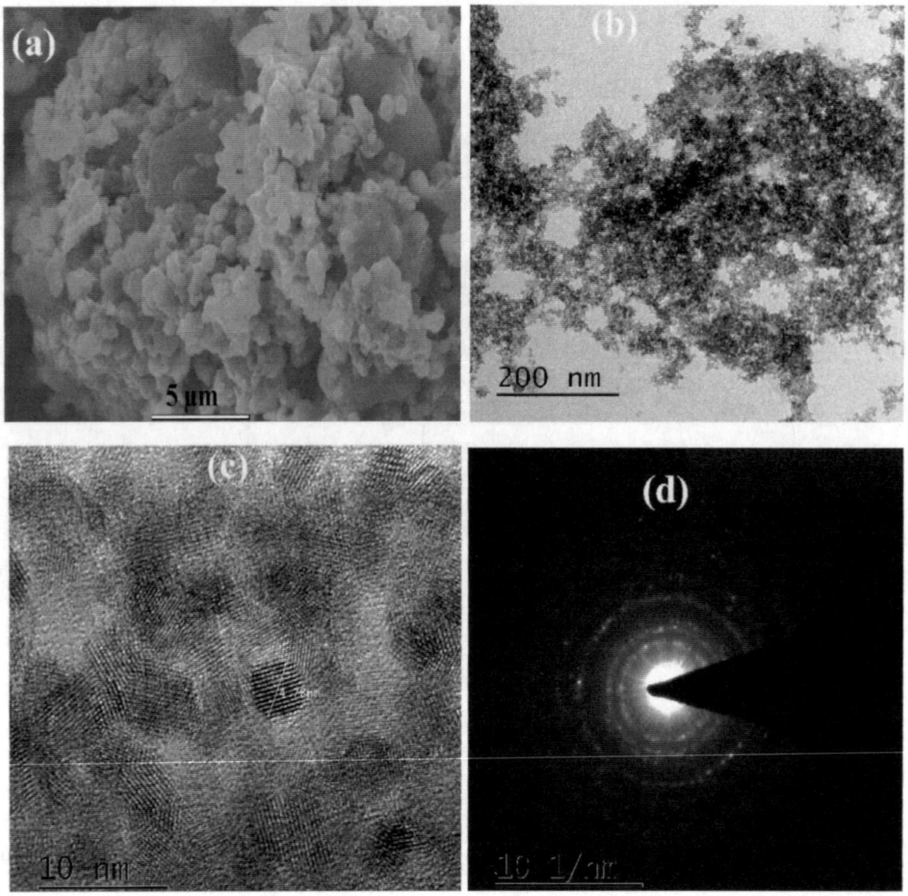

Figure 3. (a) SEM (b) LRTEM (c) HRTEM (d) SAED pattern of the synthesized CZTS nanocrystals

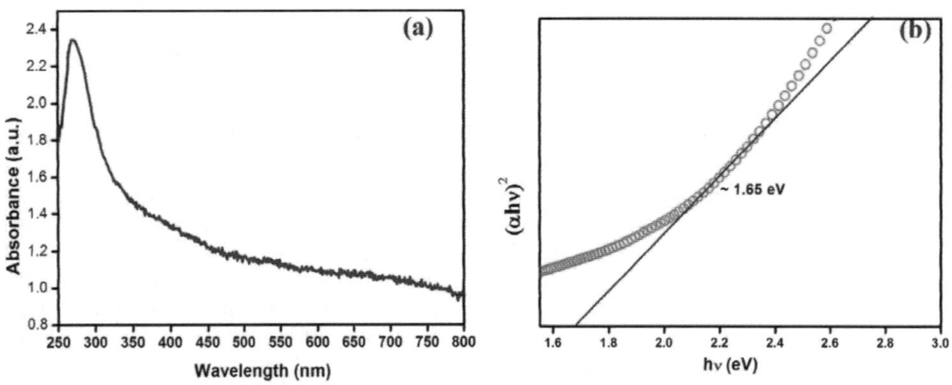

Figure 4. (a) UV-Visible absorbance spectra **(b)** band gap of CZTS nanocrystals

IOP Conf. Series: Materials Science and Engineering **338** (2018) 012062 doi:10.1088/1757-899X/338/1/012062

4. Conclusion

In conclusion, kesterite CZTS nanocrystals are synthesized by a simple hydrothermal environmental friendly approach using thiourea as Sulphur source. XRD, SEM, LRTEM, HRTEM characterizations have confirmed the structure, phase, morphology of synthesized CZTS nanocrystals. X-ray diffraction study confirmations about the crystalline phase of synthesized CZTS. From the HRTEM analysis the average size of CZTS nanocrystal is found to be around 4nm. The band gap of synthesized sample is calculated to be 1.65eV which indicates that the synthesized material has good applicability in visible range. The synthesized CZTS nanocrystals could be used as absorber layer material in photovoltaic device for various applications.

References

[1] Tanaka K, Oonuki M, Moritake N, Uchiki H 2009 *Sol. Energy Mater. Sol. Cells* **93** 583–587.

[2] Ming Wei, Qingyang Du, Dacheng Wang, Weifeng Liu, Guoshun Jiang, Changfei Zhu 2012 *Mater. Lett.* **79** 177–179.

[3] Emsley J 1998 *the Elements, 3rd ed., Oxford Univ. Press* p. 289.

[4] Matsushita H, Maeda T, A. Katsui, and T. Takizawa 2000 *Journal of Crystal Growth* **208** 416–422.

[5] Shockley W, Queisser H. J 1961 *J. Appl. Phys.* **32** 510–519.

[6] Wang W, Winkler M.T, Gunaman 2013 *Advanced Energy Materials.*

[7] Song X, Ji Xu, Li M, Lin W, Luo Xi, Zhang H 2014 *Article ID* **613173** 11.

[8] Guo Q, Ford G. M, Yang W. C, Walker B.C, Stach E.A, Hillhouse H.W, Agrawal R 2010 *J. Am. Chem. Soc.* **132** 17384–17386.

[9] Guo Q, Hillhouse H.W, Agrawal R 2009 *J. Am. Chem. Soc.* **131** 11672–11673.

[10] Mitzi D.B, Gunawan O, Todorov T.K, Wang K, Guha S 2011 *Sol. Energy Mater. Sol. Cells* **95** 1421–1436.

[11] R. Sarvana Kumar, B.D. Ryu, S. Chandramohan, J.K. Seol, S.K. Lee, C-.H. Hong 2012 *Mater. Lett.* **86** 174–177.

[12] T.K. Todorov, K.B. Reuter, D.B. Mitzi 2010 *Adv. Mater.* **22** E156–E159.

[13] Wang C, Cheng C, Cao Y, Fang W, Lijuan Zhao L, Xiaofeng Xu. 2011 *Japanese Journal of Applied Physics* **50** 065003.

[14] Fernandes P A, Salome P M, Da Cunha A F 2009 *Thin Solid Films* **517** 2519.

[15] Fontane X, Calvo-Barrio L, Izquierdo-Roca V, Saucedo E, Perez-Rodriguez A, Morante J R, Berg D M, Dale P J Siebentritt S 2011 *Appl. Phys. Lett.* **98** 181905.

IOP Conf. Series: Materials Science and Engineering **338** (2018) 012063 doi:10.1088/1757-899X/338/1/012063

Preparation of few layer graphene sheets (FLGS) prepared by an electrochemical method

I Alam, K Sa, S das, J Raiguru, BVRSSubramanyam, PC Mahakul and P Mahanandia*

Department of Physics & Astronomy, National Institute of Technology Rourkela, Orissa-769008, India

*Email id-pmahanandia@gmail.com

Abstract: Graphene with two dimensional structures is a very exceptional material, since it has advantage of being excellent conducting material. Few layer of graphene sheets (FLGS) have been successfully prepared by an electrochemical method directly from graphite using oxalic acid as electrolyte. The prepared few layer FLGS have been characterized by X-ray diffraction(XRD), UV-Vis spectroscopy, field emission scanning electron microscope(FESEM), transmission electron microscope(TEM). The characterization reveals that the prepared FLGS is very good in quality. The measured current (I) vs voltage (v) shows that FLGS are highly conducting in nature. The hydrophobicity/Hydrophilicity of the prepared FLGS have been investigated for various applications.

Introduction: Since the discovery of one-atom-thick single layer of SP^2 bonded carbon atomsgraphenesheet in 2004 [1], closely packed honeycomb crystal lattice has paying attention from fundamental research and its application [2].The unique various properties such as electrical, physical, optical etc. has great application in Field-effect transistor [3,4], ultrasensitive sensors [5], electrochemical devices [6], transparent electrodes [7], different nanocomposite [8], batteries [9], supercapacitors [10] and hydrogen storage [10]. Main research focus is for the preparation of graphene in large scale with low cost due to the exotic physical properties and the predicted applications in science and technology. The methods toprepare a graphene sheets are mechanical exfoliation [1], Epitaxial growth [11], and chemical vapour deposition [12].However, the above methods are not effectively influenced of the mass production due to their low yield and high cost. There must be required another technique to produce the mass production of graphene with low cost, scalability, reproducibility, process ability and quality for the commercial application of the industrial purposes.Reduced graphene oxide(RGO) is the same like as graphene which is obtained from graphene oxide (GO) using another type of reducing agents [13,14].There are to be found heavily oxygenated by way of hydroxyl and epoxide functional groups on the surface and at the edges [15] of the obtained graphene sheet in which the electrical property is degraded.There are another method to prepare graphene straightforwardly from graphite exclusive of functional groups. This method is a one-step amalgamation of ionic-liquid-functionalized graphene sheet, also called electrochemical method [16]. In this method awful impurities for example nitrogen and water arises on the surface of graphene sheet. Highly eminence graphene was prepared by Su et al. by electrochemical exfoliation with strong sulfuric acid like as the electrolyte [17] which can smash up the honeycomb lattice throughout preparation.For the high quality few layer of graphene sheet, an electrochemical method from graphite via oxalic acid as electrolyte has been reported to avoid impurities and functional groups [18]. We have investigated the hydrophobic and hydrophilic nature of the graphene prepared by electrochemical method as reported earlier. The graphene sheets obtained in these methods do not contain any functional groups of carbon with maintaining the hexagonal structure with better electrical property.

Materials and Methods:

High quality pure graphite rod and Oxalic acid (Sigma Aldrich) and a DC voltage source.

FLGS were prepared by electrochemical method using two high quality graphite rods like electrodes and oxalic acid like electrolytes. A dc voltage of 60 V was appliedto the electrodes which was immersed in the solution of oxalic acid (0.3M) maintained at 0° C about 20 hours. At first the electrodes were wettedand then the interaction of $C_2O_4H^-$with the grain boundary of graphite were happened. After a few times, the graphite rods rust and dissociates into small pieces in the solution. The upper portion of the solution was filtered after sonication and then neutralized to recover the graphene flakes. The collected samples were dried in an oven at 130^{0}C about 10 hours.

Result and Discussion:

The FESEM image of graphene sheets prepared by electrochemical is shown in Figure 1(a). From the FESEM images of graphene sheets, it is mainly observed that exfoliated crumpled graphene sheets are randomly oriented.The TEM image of graphene sheets are shown in Figure 1(b).A few layers of sheets through a sheet distance about 0.335nm is confirmed by the TEM images. The HRTEM image (Figure 1(c))reveals that the prepared materials are few layers graphene sheets(FLGS). The many spots in diffraction pattern confirms the crystalline nature of FLGS (figure 1(d)).

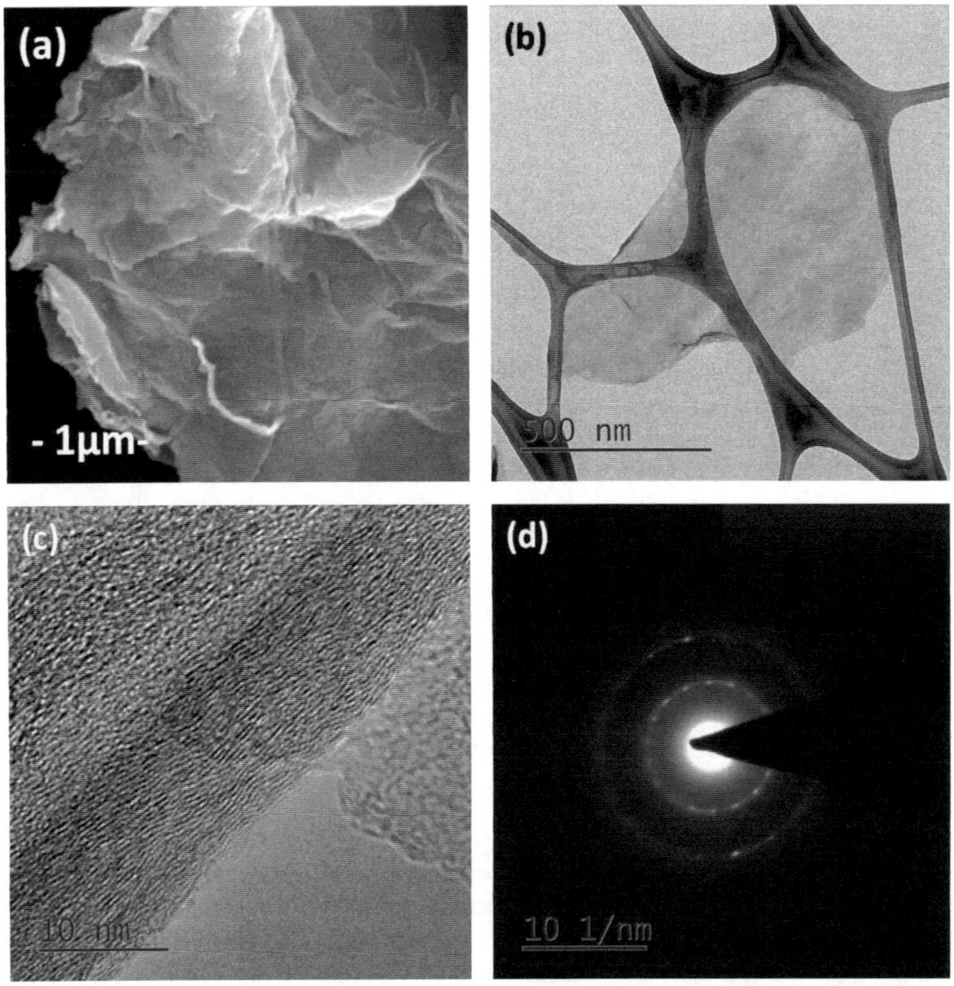

Figure1: (a) FESEM, (b)TEM, (c)High resolution TEM images, (d) SAED pattern of FLGS

The XRD pattern of FLGS shown in figure 2(a) assured the crystalline structure which shows a sharp diffraction peak at 26.148°.UV-Vis absorption spectrum of few layer of graphene sheets shown in figure 2(b) shows the absorption peak at 273 nm in relation to C=C bonds admit that there is no other functional group.Current (I) – Voltage (V) characteristics of few layer of graphene sheets is shown in the figure 2(c). The obtained higher current in FLGS is due to sp^2 conjugation in FLGS without any functional group.

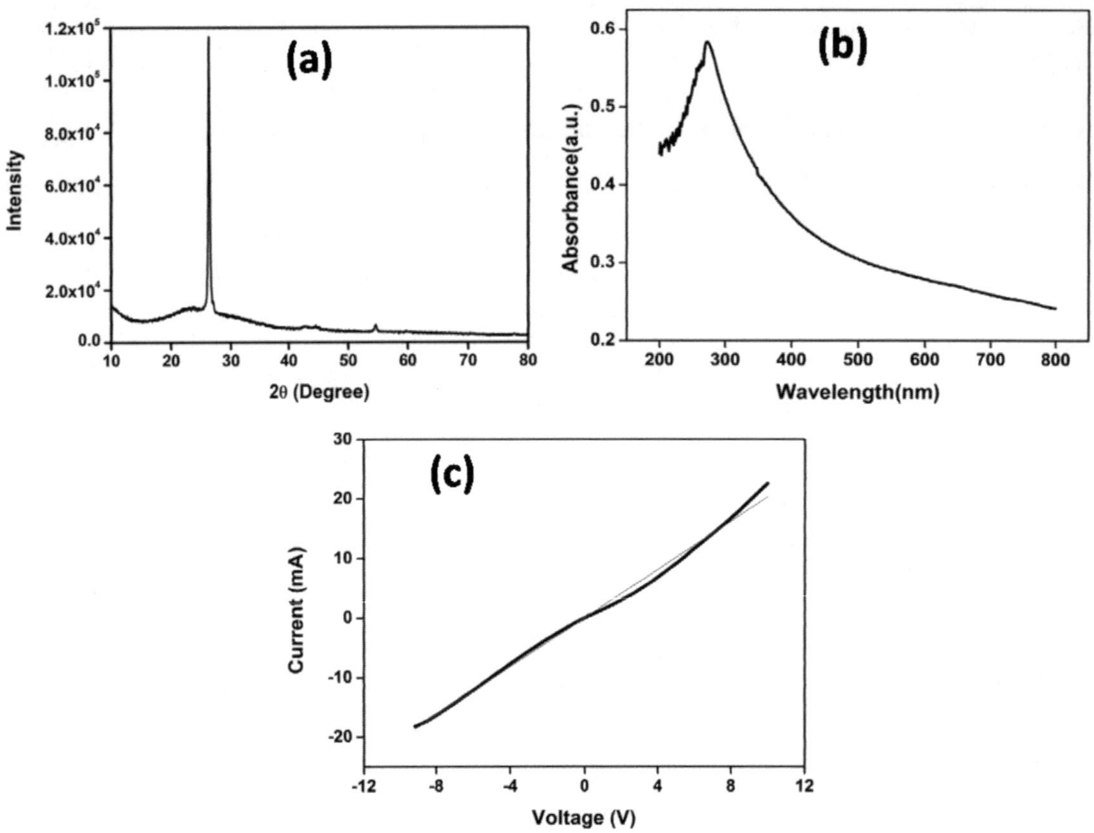

Figure 2:(a)XRD pattern, (b) UV-Vis, (c) I-V characteristics of FLGS.

The reversible wettability transition fromhydrophobic to high hydrophilic nature of FLGSwas accomplished by Ultraviolet (UV) irradiation and stored in vacuum [19]. The irradiation duration–dependent contact angle (CA) of water on FLGS shown in figure 3 in ambient temperature was measured by an optical Contact angle meter. According to this figure CA slowly decreases from 87.16^0 to 52.93^0 for FLGS showing to UV irradiation from 0 to 120 minutesin step of 30 minutes. Therefore the hydrophobic state can be transferred to the highly hydrophilic state under UV irradiation within 2 hour.

Then the samples were heated at 80^0 C for 5 hours in vacuum and measured the contact angle. Figure 4 shows the transition from hydrophilic to hydrophobic of the obtained graphene sheet. From the figure, it can be observed that the contact angle increases towards hydrophobic using vacuum storage after 5 days of the hydrophilic graphene sheet which was obtained by the UV irradiation. From the above experiment it is clearly seen that the transition from hydrophobic to hydrophilic state under UV irradiation (2 hours) is faster than the vacuum storage (5 days). From this result, It can be suggested that after applying irradiation the transition has been occurred by the chemical adsorption before physical adsorption on graphene, as the physically adsorbed molecules desorbed within a short time. ApplyingUV irradiation, the chemisorptions of O_2 molecules is induced on the surface of graphene sheet. When the UV irradiation is not applying and the graphene sheet was placed in vacuum then the O2 molecules were gradually replaced by the physicallyadsorbed water molecules on the surface.But there is possible transition from hydrophobic to hydrophilic and vice versa of graphene using Density Functional theory (DFT) which was already reported [20].

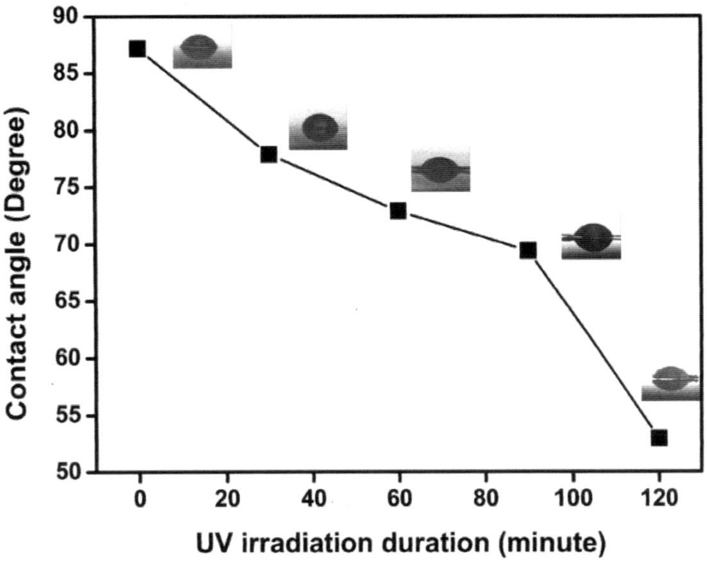

Figure 3: The contact angle of graphene with different duration of UV irradiation.

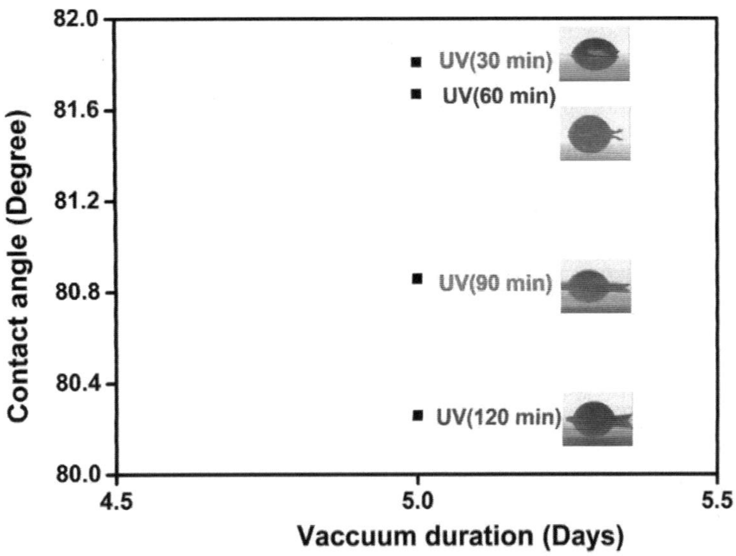

Figure 4: The contact angle of graphene sheet in the vacuum storage.

IOP Conf. Series: Materials Science and Engineering **338** (2018) 012063 doi:10.1088/1757-899X/338/1/012063

Conclusion:

Here a significant synthesis of FLGS was prepared directly from graphite by electrochemical method. From the above characterization, it is clear that the obtained FLGS are better in quality than rGO [13,14].The transitions of graphene sheets under UV irradiation and vacuum has been observed experimentally. From hydrophobic to hydrophilic transition,it can be found that UV irradiation workedlike a switch by means of thedissociative adsorption of H_2O on the graphene surface and also concluded that it can be controlled.The obtained result gives a new insights into the elemental principles of water splitting through graphene-based materials.

References:

1. NovoselovK. S., GeimA. K., MorozovS. V., JiangD., ZhangY., DubonosS. V., Grigorieva I. V. and FirsovA. A., 2004*Science***306**666–669.
2. Katsnelson M. I. 2007*MaterToday***10** 20–27.
3. ZhangY., TanY. W., StormerH. L. and KimP., 2005*Nature***438**201–204.
4. GiljeS., SongH., WangM., Wang K. L. and KanerR. B., 2007*Nano Lett.* **7** 3394–3398.
5. SchedinF., GeimA. K., MorozovS. V., HillE. W., BlakeP., Katsnelson M. I. and NovoselovK. S., 2007*Nat. Mater.* **6**652–655.
6. BunchJ. S., ZandeA. M., VerbridgeS. S., FrankI. W., TanenbaumD. M., ParpiaJ. M., Craighed H. G. and McEuenP. L.,2007*Science***315** 490–493.
7. WangX., ZhiL. and MullenK., 2008*Nano Lett.* **8**323–327.
8. StankovichS., DikinD. A., DommettG. H. B., KohlhaasK. M., ZimneyE. J., StachE. A., PinerR. D., Nguyen S. T. and RuoffR. S.,2006*Nature***442** 282–286.
9. Takamura T., Endo K., Fu L., Wu Y. P., Lee K. J. and Matsumoto T.,2007*Eletrochim.Acta***53**1055–1061.
10. Novoselov K. S., Jiang D., Schedin F., Booth T. J., Khotkevich V. V.,Morozov S. V. and Geim A. K. 2005*Proc. Natl. Acad. Sci. U.S.A.* **102** 10451–10453.
11. BergerC., SongZ., LiT., LiX., OgbazghiA. Y., FengR., DaiZ., MarchenkovA. N., ConradE. H., First P. N. and de HeerW. A.,2004*J. Phys. Chem. B***108** 19912–19916.
12. KimK. S., ZhaoY., JangH., LeeS. Y., KimJ. M., KimK., AhnJ. H., KimP., ChoiJ. Y. and HongB. H., 2009*Nature***457** 706–710.
13. StankovichS., DikinD. A., PinerR. D., KohlhaasK. A.,KleinhammesA., JiaY., WuY., NguyenS. T. and RuoffR. S., 2007*Carbon***45** 1558–1565.
14. BarmanK., MahanandiaP. and NandaK. K., 2013*RSC Adv.***3**12621–12624.
15. LiuP. G., GongK. C., XiaoP. and XiaoM., 2000*J. Mater. Chem.***10**933–935.
16. LiuN., LuoF., XuH., LiuY., ZhangC. and ChenJ., 2008*Adv. Funct. Mater.***18** 1518-1525.
17. SuC., LuA., XuY., ChenF., Khlobystov A. N. and LiL., 2011*ACS Nano***5** 2332–2339.
18. Mahanandia P., Simon F.,Heinrich G. and Nanda K.K. 2014 *RSC Chem. Commun.***50**4613.
19. Zhang X., Wan S., Pu J., Wang L. and Liu X.,2011*J. Mater. Chem.***21** 12251–12258.
20. Jiang Q. G., Ao Z. M., Chu D. W. and Jiang Q., 2012*J. Phys.Chem. C***116** 19321–19326.